QS 4 COL

anatomy and

p

for

he

DATE DUE

19/10/6	
16/2/17	
GAYLORD	PRINTED IN U.S.A.

anatomy and physiology

for nursing and healthcare professionals

second edition

Bruce Colbert
Director of Allied Health, University of Pittsburgh and Johnstown

Jeff Ankney
Director of Clinical Education, University of Pittsburgh and Johnstown

Karen T. Lee
Associate Professor of Biology, University of Pittsburgh and Johnstown

Adapted by

Martin Steggall
Associate Dean, Director of Undergraduate Studies at School of Health Sciences, City University, London

Maria Dingle
Senior Lecturer in Applied Biological Sciences, School of Health Sciences, City University, London

Harlow, England • London • New York • Boston • San Francisco • Toronto • Sydney • Auckland • Singapore • Hong Kong
Tokyo • Seoul • Taipei • New Delhi • Cape Town • São Paulo • Mexico City • Madrid • Amsterdam • Munich • Paris • Milan

Pearson Education Limited
Edinburgh Gate
Harlow
Essex CM20 2JE
England

and Associated Companies throughout the world

Visit us on the World Wide Web at:
www.pearson.com/uk

Authorised adaption from the United States edition, entitled *Anatomy and Physiology for Health Professionals: An Interactive Journey*, 1st edition, ISBN: 0131512684 by Colbert, Bruce, J., Ankney, Jeff, Lee, Karen, published by Pearson Education, Inc., publishing as Prentice Hall, copyright © 2007.

First published 2009
Second edition published 2012

© Pearson Education Limited 2012

ISBN: 978-0-273-75803-7

British Library Cataloguing-in-Publication Data
A catalogue record for this book is available from the British Library

Library of Congress Cataloging-in-Publication Data
Colbert, Bruce J.
  Anatomy and physiology for nursing and healthcare professionals / Bruce Colbert, Jeff Ankney, Karen T. Lee ; adapted by Martin Steggall, Maria Dingle. – 2nd ed.
     p. ; cm.
  "Authorised adaption from the United States edition, entitled Anatomy and Physiology for Health Professionals: An Interactive Journey, 1st edition, ISBN: 0131512684 by Colbert, Bruce, J., Ankney, Jeff, Lee, Karen, published by Pearson Education, Inc., publishing as Prentice Hall, copyright © 2007." – Galley.
  Includes index.
  ISBN 978-0-273-75803-7 (pbk.)
  I. Ankney, Jeff. II. Lee, Karen T. III. Steggall, Martin. IV. Dingle, Maria.
V. Colbert, Bruce J. Anatomy and Physiology for Health Professionals. VI. Title.
  [DNLM: 1. Anatomy. 2. Physiological Phenomena. QS 4]

  612–dc23
                                                          2012004164

10 9 8 7 6 5 4 3 2 1
16 15 14 13 12

Typeset in 11.5/13.5pt Goudy by 35
Printed and bound by Rotolito Lombarda, Italy

Brief contents

About the authors		xiv
Preface		xv
Guided tour		xvi
Mastering walkthrough		xviii
Publisher's acknowledgements		xx

Chapter 1	Anatomy and physiology: learning the language	1
Chapter 2	The human body	22
Chapter 3	Biochemistry	40
Chapter 4	The cells	54
Chapter 5	Tissues and systems	82
Chapter 6	The skeletal system	110
Chapter 7	The muscular system	138
Chapter 8	The integumentary system	162
Chapter 9	The nervous system: part I	184
Chapter 10	The nervous system: part II	212
Chapter 11	The special senses	238
Chapter 12	The endocrine system	256
Chapter 13	The cardiovascular system	282
Chapter 14	The lymphatic and immune systems	318
Chapter 15	The respiratory system	352
Chapter 16	The gastrointestinal system	388
Chapter 17	The urinary system	422
Chapter 18	The reproductive system	448
Chapter 19	Ageing	482

Appendices:		501
	A Answers to test your knowledge	502
	B Medical terminology, word parts, and singular and plural endings	508
	C Clinical abbreviations	516
	D Laboratory reference values	523
Glossary		526
Index		549

Contents

About the authors xiv
Preface xv
Guided tour xvi
Mastering walkthrough xviii
Publisher's acknowledgements xx

Chapter 1   Anatomy and physiology: learning the language   1

Learning objectives 1
What is anatomy and physiology? 2
The language 4
The language of disease 11
Anatomy and physiology concepts 12

Summary 18
Case study 18
Review questions 19
Suggested activities 20
Answers 21

Chapter 2   The human body   22

Learning objectives 23
The map of the human body 24

Summary 37
Case study 37
Review questions 38
Suggested activities 39
Answers 39

Chapter 3   Biochemistry   40

Learning objectives 41
Introduction 42
Composition of matter 42
Chemical bonds and reactions 44
Types of chemical bonds 45
Electrolytes 45

Acids and bases   46
Organic compounds   48

Summary   51
Review questions   51
Answers   52

Chapter 4   The cells   **54**

Learning objectives   55
Overview of cells   56
Cell structure   57
Mitosis   69
The cell cycle   70
Micro-organisms   72

Summary   77
Case study   78
Review questions   79
Suggested activities   80
Answers   81

Chapter 5   Tissues and systems   **82**

Learning objectives   83
Overview of tissues   84
Organs   90
Systems   94

Summary   106
Case study   107
Review questions   107
Suggested activities   108
Answers   109

Chapter 6   The skeletal system   **110**

Learning objectives   111
System overview: more than the 'bare bones' about bones   112
The skeleton   123
Common disorders of the skeletal system   130

Summary   133
Case study   134
Review questions   135
Suggested activities   136
Answers   136

Chapter 7 The muscular system **138**

Learning objectives 139
Overview of the muscular system 140
Skeletal muscle movement 147
Muscular movement at the cellular level 151
Visceral or smooth muscle 153
Cardiac muscle 154
Muscular fuel 156
Common disorders of the muscular system 156

Summary 158
Case study 158
Review questions 159
Suggested activities 160
Answers 161

Chapter 8 The integumentary system **162**

Learning objectives 163
System overview 164
How skin heals 168
Nails 172
Hair 173
Temperature regulation 175
Common disorders of the integumentary system 176

Summary 180
Case study 180
Review questions 181
Suggested activities 182
Answers 183

Chapter 9 The nervous system: part I **184**

Learning objectives 185
Organisation 186
Nervous tissue 188
How neurons work 193
Spinal cord and spinal nerves 200
Common disorders of the nervous system 206

Summary 208
Case study 209
Review questions 209
Suggested activities 211
Answers 211

Chapter 10   The nervous system: part II   **212**

Learning objectives   213
The brain and cranial nerves   214
Overall organisation   214
Cranial nerves   221
The big picture: integration of brain, spinal cord and PNS   224
The big picture: the motor system   226
The autonomic nervous system   229
Other systems   232
Common disorders of the nervous system   232

Summary   234
Case study   234
Review questions   235
Suggested activity   236
Answers   236

Chapter 11   The special senses   **238**

Learning objectives   239
The different senses   240
Sense of sight   241
The sense of hearing   245
Other senses   249
Common disorders of the eye and ear   252

Summary   253
Case study   254
Review questions   254
Suggested activity   255
Answers   255

Chapter 12   The endocrine system   **256**

Learning objectives   257
Organisation of the endocrine system   258
Control of endocrine activity   261
The major endocrine organs   266
Common disorders of the endocrine system   275

Summary   277
Case study   278
Review questions   278
Suggested activities   279
Answers   280

Chapter 13   The cardiovascular system   **282**

Learning objectives   283
System overview   284
Incredible pumps: the heart   285
Blood   294
Blood vessels: the vascular system   300
Common disorders of the cardiovascular system   308
The lymphatic connection   312

Summary   313
Case study   314
Review questions   314
Suggested activities   315
Answers   316

Chapter 14   The lymphatic and immune systems   **318**

Learning objectives   319
The defence zone   320
The lymphatic system   320
The immune system   325
How the immune system works   333
The big picture   343
Common disorders of the immune system   344

Summary   347
Case study   348
Review questions   349
Suggested activities   350
Answers   350

Chapter 15   The respiratory system   **352**

Learning objectives   353
System overview   354
The respiratory system   358
The lower respiratory tract   367
Common disorders of the respiratory system   380

Summary   383
Case study   384
Review questions   384
Suggested activities   385
Answers   386

Chapter 16   The gastrointestinal system   **388**

Learning objectives   389
System overview   390

The mouth and oral cavity   391
Pharynx   397
Oesophagus   397
The walls of the alimentary canal   398
Stomach   399
Small intestine   403
Nutrition   407
Large intestine   409
Accessory organs   412
Common disorders of the digestive system   415

Summary   418
Case study   419
Review questions   420
Suggested activities   421
Answers   421

Chapter 17   The urinary system   422

Learning objectives   423
System overview   424
The anatomy of the kidney   425
Urine formation   431
Common disorders of the urinary system   443

Summary   444
Case study   445
Review questions   446
Suggested activities   447
Answers   447

Chapter 18   The reproductive system   448

Learning objectives   449
Tissue growth and replacement   450
Sexual reproduction   451
The human life cycle   453
The human reproductive system   453
Reproductive physiology: female   459
Male anatomy   465
Reproductive physiology: male   468
Pregnancy   471
Common disorders of the reproductive system   473

Summary   477
Case study   479
Review questions   479
Suggested activities   480
Answers   481

Chapter 19   **Ageing**   **482**

   Learning objectives   483
   Older adults   484
   Well-being   487
   Cancer prevention and treatment   495
   More amazing body facts   497

   Summary   499
   Review questions   499
   Suggested activities   500
   Answers   500

Appendices   **501**

   A   Answers to test your knowledge   502
   B   Medical terminology, word parts, and singular and plural endings   508
   C   Clinical abbreviations   516
   D   Laboratory reference values   523

   Glossary   526
   Index   549

About the authors

Bruce Colbert is the Director of the Allied Health Department at the University of Pittsburgh at Johnstown. He holds a Master's in Health Education and Administration, has authored four books, written several articles, and given over 175 invited lectures and workshops at both the regional and national level.

Many of his workshops provide teacher training involving techniques to make the health sciences engaging and relevant to today's students. In addition, he does workshops on developing effective critical and creative thinking, stress and time management, and study skills.

Jeff Ankney is the Director of Clinical Education for the University of Pittsburgh at Johnstown Respiratory Care Program where he is responsible for the development and evaluation of hospital clinical sites. In the past, Jeff has served as a public school teacher, Assistant Director of Cardiopulmonary Services, Program Coordinator of Pulmonary Rehabilitation, and a member of hospital utilisation review and hospital policy committees. He is a consultant on hospital management and pulmonary rehabilitation concerns. Jeff was also the recipient of the American Cancer Society Public Education Award.

Karen Lee is an Associate Professor in the Biology Department at the University of Pittsburgh at Johnstown, where she teaches all the anatomy courses, including Anatomy and Physiology for Nursing and Allied Health students. She presents regularly at scientific conferences and has published several articles on crustacean physiology and behaviour. An active member of the Council on Undergraduate Research, she is also the chair of the University of Pittsburgh at Johnstown's annual undergraduate symposium.

Martin Steggall is Associate Dean, Director of Undergraduate Studies at City University, London. He continues in clinical practice and specialises in the clinical management of erectile dysfunction and premature ejaculation.

Maria Dingle is currently a Senior Lecturer in Biological Sciences in the School of Health Sciences at City University, London. Maria obtained a BA in Education from The Victoria University of Manchester and then went on to complete a MSc in Physiology from University College London, focusing on the physiological response to orthostatic hypotension and the role of dangling in nursing. Maria has been teaching biological sciences to nurses (student and qualified) and biomedical engineers at City University for 16 years. Maria has a particular interest in cardio-respiratory physiology and pathophysiology, and the biological basis of tissue viability. She is also a Governor in two local schools. She enjoys gardening, watercolour painting and jewellery making.

Preface

By 2013, all nurse education will be at BSc or post-graduate diploma level. We acknowledge that science may be daunting for some and this text is aimed at prospective nurses who have either not studied anatomy and physiology before or who have not studied it for some time. As such, we have tried to get across the essential features of the body systems and have avoided complex detail. In contemporary nursing, there is a need for a sound foundation in anatomy and physiology, and we hope that this book will help with *introducing* you to this complex subject. Acknowledging that science changes, and the application of science changes with it, we have provided a series of mini-lectures that provide both an overview of each system, but some more indepth information for those wishing to develop their understanding of the human sciences and how this knowledge can be applied to clinical practice.

Special features

This text contains terminology guides, clinical applications, amazing facts and learning hints, that we hope you find engaging and supportive as you learn what is, in effect, a whole new language. There are case studies and self-assessment questions that are linked to UK-based practice and there is a MasteringA&P website, which has pronunciation guides, PowerPoint presentations and some interactive material. The key to studying anatomy and physiology is to adopt a combination approach: a little reading; some practical exercises; some reflection and further reading; and some review of learning by utilising the self-assessment tests in the book and companion website.

We hope that you enjoy reading the book!

Acknowledgements

To our students who challenge, surprise and enthuse us with their thirst for knowledge. To our families, whose support and tolerance have been invaluable. To David Harrison at Pearson Education for his guidance and support. To Ms J. Watkinson, for invaluable help and guidance with phonics, and finally, a special thanks to Iñaki, Aiden, Emily, William and Harriet for their encouragement.

Guided tour

This textbook has been designed to be fun, interesting and rich in features to help your understanding of this challenging topic.

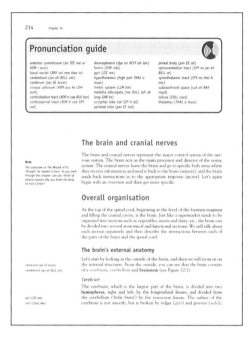

Pronunciation guide
Understanding and pronouncing medical words is critical to your success. This box shows you the correct pronunciation for the key terms that you will come across in each chapter.

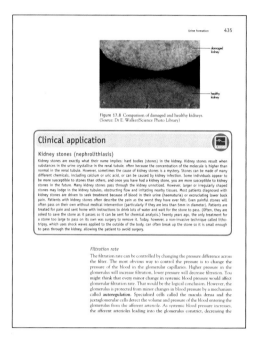

Clinical application
This shows the relevance of what you are learning to your clinical practice.

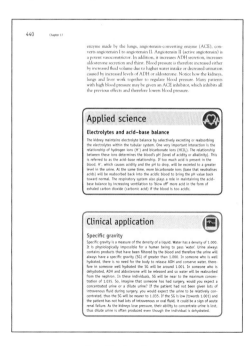

Applied science
This shows how the sciences are integrated into the study of anatomy and physiology.

Amazing body facts
These are the 'that's amazing!' facts to give you an appreciation of just how wonderful the human design is.

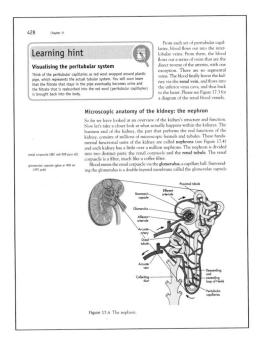

Learning hint

These are useful hints to help you to learn difficult concepts.

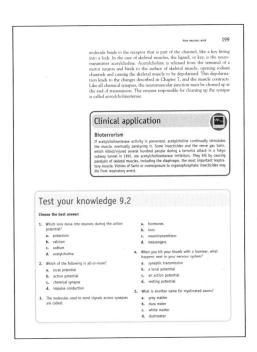

Test your knowledge

These sections throughout every chapter enable you to check that you have understood what you have just read, before moving on to the next section.

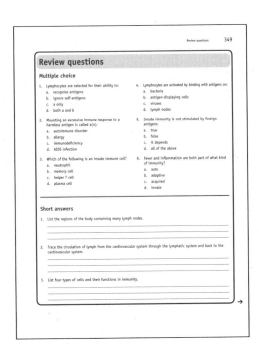

Review questions

The end-of-chapter Review Questions test your understanding of the subject. Answers and guided solutions are provided at the end of the chapter.

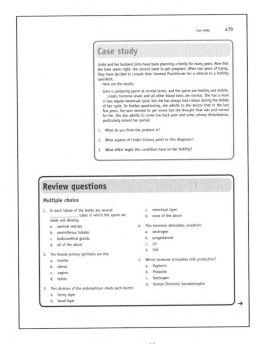

Case studies

Understanding and pronouncing medical words is critical to your success. This box shows you the correct pronunciation for the key terms that you will come across in each chapter.

Mastering walkthrough

Everyone's doing it.
Your friends are doing it.
More than 8 million students worldwide are doing it.

The MasteringA&P homework and tutorial system will help you reach that crucial place of understanding, whenever and wherever you are. Register using the access card that comes with this textbook.

"It got me to think more about A&P rather than just listening," – Jade Hill, Student, University of East Anglia, UK

Chapter Guides

Listen to an expert explain tough topics, so you can make sure that you understand the concepts at your own pace.

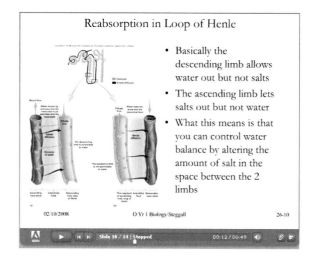

Interactive Case Studies

Understand the link between anatomy and physiology and clinical practice by working through interactive case studies. They will also improve your clinical decision making skills!

Animation and Simulation

See anatomy and physiology come to life through A&P Flix animations and Interactive Physiology simulations. MasteringA&P gives you hints and feedback just like your tutor would, so that you learn from your mistakes.

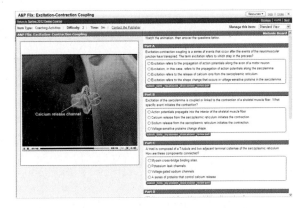

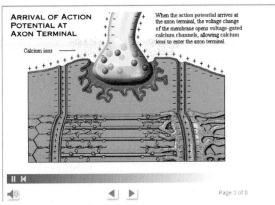

Diagram Labelling

Master the names of structures with fun labelling exercises that coach you to the correct answer.

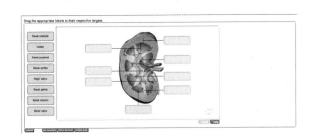

eText

Access your textbook on the go with the eText. You can quickly search for specific content, and add notes and highlights.

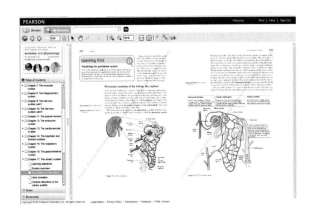

"**More informative than lectures. The ability to structure my learning as I want it is has been invaluable in my studies.**" –
Michael Rivett, Student, Anglia Ruskin University

Publisher's acknowledgements

The publisher would like to thank the following for their kind permission to reproduce their photographs:

(Key: b-bottom; c-centre; l-left; r-right; t-top)

5 Stockbyte. 24 Pearson Education, Inc: Mike Gallitelli. **74 Getty Images:** Dr. Arthur Siegelman. **75 Science Photo Library Ltd:** Thomas Deerinck, NCMIR. **76 Alamy Images:** © Medical-on-Line / Alamy. **91 Science Photo Library Ltd. 128 Science Photo Library Ltd:** DR P. MARAZZI. **131-a Science Photo Library Ltd:** Doug Sizemore, Visuals Unlimited / Science Photo Library. **131-b Science Photo Library Ltd:** HR Bramaz ISM / Science Photo Library. **131-c Alamy Images:** © Medical-on-Line / Alamy. **131-d Science Photo Library Ltd:** DR M.A. ANSARY / SCIENCE PHOTO LIBRARY. **131-e Science Photo Library Ltd:** LIVING ART ENTERPRISES, LLC / SCIENCE PHOTO LIBRARY. **131-f Science Photo Library Ltd:** MATT MEADOWS / PETER ARNOLD INC. / SCIENCE PHOTO LIBRARY. **150-a Pearson Education, Inc:** Mike Gallitelli. **150-b Pearson Education, Inc:** Michal Heron. **178 Science Photo Library Ltd:** Scott Camazine / PHOTOTAKE. **178-b Science Photo Library Ltd:** JAMES STEVENSON / SCIENCE PHOTO LIBRARY. **178-c Alamy Images:** © Bubbles Photolibrary / Alamy. **178-d Alamy Images:** © Medical-on-Line / Alamy. **178-e Alamy Images:** © Doug Diamond / Alamy. **178-f Alamy Images:** © Hercules Robinson / Alamy. **178-g Science Photo Library Ltd:** DR P. MARAZZI / SCIENCE PHOTO LIBRARY. **255-a Alamy Images:** © Medical-on-Line / Alamy. **255-b Alamy Images:** SHOUT / Alamy. **280-b Science Photo Library Ltd:** BETTINA CIRONE / SCIENCE PHOTO LIBRARY. **280 Science Photo Library Ltd:** JOHN PAUL KAY / PETER ARNOLD INC. / SCIENCE PHOTO LIBRARY. **420 Science Photo Library Ltd:** CNRI / SCIENCE PHOTO LIBRARY. **441 Science Photo Library Ltd:** DR. E. WALKER / SCIENCE PHOTO LIBRARY. **460 Alamy Images:** © PHOTOTAKE Inc. / Alamy. **128b Science Photo Library Ltd:** Medical Photo NHS Lothian

Cover images: *Front:* **Getty Images**

Anatomy and physiology

Learning the language

Imagine getting ready to travel to a foreign country where we do not speak the language. To maximise success, one of the most important preparatory steps is to develop a basic understanding of the native language. The key language upon which health professions and the study of anatomy and physiology are based is medical terminology. Therefore, this chapter lays the foundation of learning the native language (medical terminology) of medicine. Future chapters build on this foundation so that, by the end of this book, we will not only understand anatomy and physiology, but will be fluent in medical terminology.

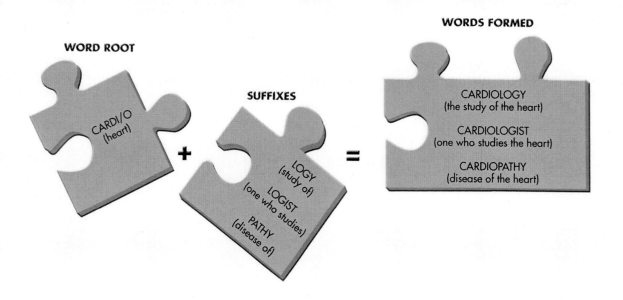

WORDS FORMED

WORD ROOT

CARDI/O
(heart)

SUFFIXES

LOGY
(study of)
LOGIST
(one who studies)
PATHY
(disease of)

CARDIOLOGY
(the study of the heart)

CARDIOLOGIST
(one who studies the heart)

CARDIOPATHY
(disease of the heart)

Learning objectives

At the end of this chapter, you will be able to:

◆ Understand the terms *anatomy* and *physiology* and their various related topics
◆ Understand the structural hierarchy of the human body
◆ Relate the importance and purpose of medical terminology to anatomy and physiology
◆ Construct and define medical terms using word roots, prefixes and suffixes
◆ Identify the survival needs of the human body
◆ Explain the concept and importance of homeostasis
◆ Contrast the metabolic processes of anabolism and catabolism

Pronunciation guide

aetiology (*ee tee OLL oh jee*)
anabolism (*an AH bol izm*)
anatomy (*ah NAH toe me*)
catabolism (*ka TAH bol izm*)
diagnosis (*dye ag NO siss*)

homeostasis (*hoe mee yoh STAY siss*)
macroscopic anatomy (*MAK row SCOP ik ah NAH toe me*)
metabolism (*meh TAH bol izm*)
microscopic anatomy (*MY krow SCOP ik ah NAH toe me*)

pathology (*pa THOL oh jee*)
physiology (*fizz ee OLL oh jee*)
prognosis (*prog NO siss*)
syndrome (*SIN drome*)

What is anatomy and physiology?

You're probably so accustomed to hearing the words anatomy and physiology used together that you may not have given much thought to what each one means and how they differ. They each have unique meanings. Let's take a closer look.

Anatomy

anatomy (*ah NAH toe me*)

Anatomy is the study of the internal and external *structures* of plants, animals or, for our focus, the human body. The human body is an amazing and complex structure that can perform an almost limitless number of tasks. To truly understand how something works, it is important to know how it is put together. Leonardo da Vinci, in the 1400s, correctly drew the human skeleton and could be considered one of the earliest anatomists (one who studies anatomy). The word anatomy is from the Greek language and literally means 'to cut apart', which is exactly what you must do to see how something is put together. For example, the study of the arrangement of the bones that comprise the human skeleton, which is the anatomical framework for our bodies, is considered anatomy.

microscopic anatomy (*MY krow SCOP ik ah NAH toe me*)
  micro = *small*
  scope = *instrument to examine*
macroscopic anatomy (*MAK row SCOP ik ah NAH toe me*)
  macro or **gross** = *large*
  cyto = *cells*
  histo = *tissues*
  logy = *the study of*

Just as we can subdivide biology into more specific areas, such as cell biology, plant biology and animal biology, we can also broadly divide anatomy into microscopic anatomy and macroscopic anatomy (also known as **gross anatomy**). Microscopic anatomy is the study of structures that can be seen and examined only with magnification aids such as a microscope. The study of cellular structure (**cytology**) and tissue samples (**histology**) are examples of microscopic anatomy.

Gross anatomy represents the study of the structures visible to the unaided or naked eye. For example, the study of the various bones that make up the human body is gross anatomy. Viewing an X-ray of the arm to determine the type and location of a broken bone is considered an examination of gross anatomy.

Physiology

physiology (*fizz ee OLL oh jee*)
  physio = *relationship to nature*
  logy = *study of*

Physiology focuses on the *function* and vital processes of the various structures making up the human body. These physiologic processes include muscle contraction, our sense of smell and sight, how we breathe, and so on. We will focus on each of these processes in their respective chapters. Physiology is closely related to anatomy because it is the study of how an anatomical structure such as a cell or bone actually functions. Physiology deals with all the vital processes of life and is more complex and, therefore, has many sub-specialties. Human physiology, animal physiology, cellular physiology and neurophysiology are just some of the specific branches of physiology.

Putting it all together

In summary, anatomy focuses on *structure* and how something is put together, whereas physiology is the study of how those different structures work together to make the body function as a whole. For example, anatomy would be the study of the structure of the red blood cells (RBCs), and physiology would be the study of how the RBCs carry vital oxygen throughout our body. Figure 1.1 shows deformed RBCs (sickle shaped) that are present in the disease sickle cell anaemia. Because of the anatomical deformity, the physiological process of effectively carrying oxygen is adversely affected.

You will notice that the design of a structure is often related to its function. For example, the type of joint located between bones is dictated by the functions of those bones: hinge joints are located at the knees where back and forth bending movement is required, while a ball and socket joint at the hip provides for a greater range of motion.

Learning hint

Using the margins of this book

Notice that the margin notes present a breakdown of the medical terms discussed in the text. Sometimes you may already know the term and may not need to refer to the margin note, but it is always there to help reinforce the word. On occasion, you may even see a short story on the word origin where it is of interest or helps to further explain the term.

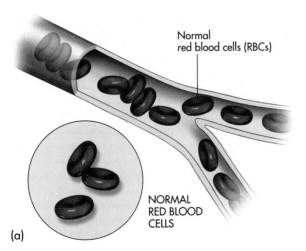

Normal red blood cells (RBCs)

NORMAL RED BLOOD CELLS

(a)

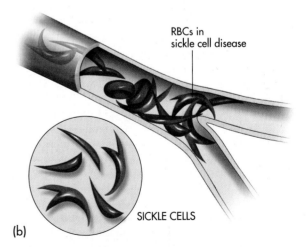

RBCs in sickle cell disease

SICKLE CELLS

(b)

Figure 1.1 (a) Normal red blood cells (RBCs) are flexible and doughnut-shaped and move with ease through blood vessels. (b) The anatomical distortion of the *structure* of RBCs in sickle cell anaemia affects their normal *function* to carry oxygen. In addition, the sickle cells lose their ability to bend and pass through the small blood vessels, thereby causing blockages to blood flow.

Test your knowledge 1.1

Indicate whether the following examples are gross anatomy or microscopic anatomy by putting a G or M in the space provided.

1. _____ viewing an X-ray to determine the type of bone fracture

2. _____ classifying a tumour to be cancerous by cell type

3. _____ viewing bacteria to determine what disease is present

4. _____ examining the chest for any obvious deformities

5. _____ a histologist and cytologist primarily study this type of anatomy

---

pathology (pa THOL oh jee)
patho = disease; disease literally means not (dis) at ease

Therefore, it makes sense to combine these two sciences into anatomy and physiology (A&P). Human anatomy and physiology forms the foundation for all healthcare-related practice. Anything that upsets the normal structure or functioning can be called disease, and the study of disease is pathology. The study of abnormal function is **pathophysiology**.

Structural hierarchy of the body

The simplest level of the hierarchy of the body is the **chemical** level. At this level we see atoms combine to form molecules (water, proteins etc).

The **cellular** level consists of all of the cells of the body which are made up of molecules. There are many different types of cells found in the body (muscle cells, skin cells) and they have different functions based on their structure and location in the body.

The **tissue** level – similar cells group together to form tissues – there are four types found in the body – epithelial, muscle, connective and nervous.

The **organ** level – organs in the body are formed from two or more different types of tissues – this enables quite complex functions to be performed. The walls of a blood vessel are made up of connective, epithelial and muscle tissue.

The **organ system** level – a group of organs work together to accomplish a particular function – the heart and the blood vessels work together to ensure that blood is delivered around the body.

The **organismal** level – there are 11 organ systems that contribute to and make up the human body. The interaction between these body systems is vital to ensure that the human body works effectively. Each of these organ systems will be discussed in more detail throughout this book (Figure 1.2).

The language

Anatomy and physiology also has its own unique language that you must learn before you can converse comfortably. Some words, like *heart*, *lungs* and *blood pressure*, are already familiar to you. Others will seem strange and foreign. Let's take a closer look.

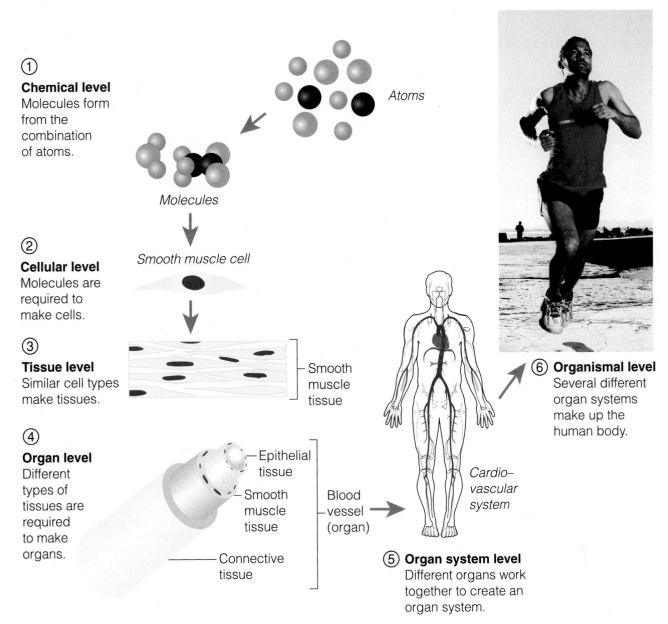

① Chemical level
Molecules form from the combination of atoms.

Atoms

Molecules

② Cellular level
Molecules are required to make cells.

Smooth muscle cell

③ Tissue level
Similar cell types make tissues.

Smooth muscle tissue

④ Organ level
Different types of tissues are required to make organs.

Epithelial tissue

Smooth muscle tissue

Connective tissue

Blood vessel (organ)

Cardio– vascular system

⑥ Organismal level
Several different organ systems make up the human body.

⑤ Organ system level
Different organs work together to create an organ system.

Figure 1.2 The structural hierarchy of the human body.

Medical terminology

As stated earlier, the language of anatomy and physiology is primarily based on medical terminology. Understanding medical terminology may seem like an overwhelming task because, on the surface, there appears to be SO many terms. In reality, there are only a relatively few root terms, prefixes and suffixes, but they can be put together in a host of ways to form numerous terms.

Each medical term has a basic structure upon which to build, and this is called the word root. For example, *cardi* is the word root for terms pertaining to the heart. Rarely is the word root used alone. Instead, it is combined with prefixes and suffixes that can change its meaning. Prefixes come before the word

cardi = *heart*

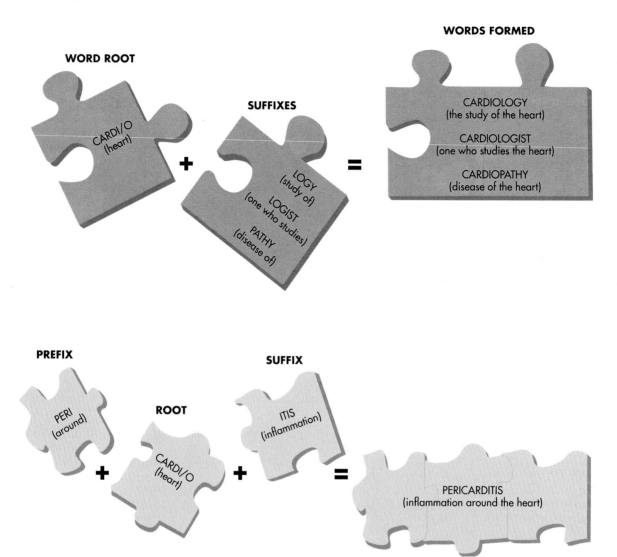

Figure 1.3 How prefixes and suffixes can be combined with a word root to form many medical terms.

logy = *study of*

tachy = *fast*
logist = *one who studies*

root, while suffixes come after the word root. The suffix **logy** means 'study of', and therefore, we can combine *cardi* and *ology* to form **cardiology**, which is the study of the heart. The prefix *tachy* means 'fast' and can be placed in front of the word root to form **tachycardia**, which means a fast heart rate. Figure 1.3 shows the components of a medical term.

Often you will be given a combining form, which is the word root and a connecting vowel (usually *o*), to make it easier to pronounce and combine with possible suffixes. For example, the combining form for heart is *cardi/o*. Listed in Table 1.1 are some common combining forms to get you started.

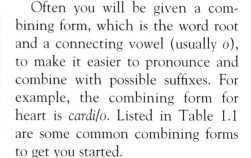

Learning hint

Combining and forming medical terms

If a suffix begins with a vowel, drop the vowel in the combining form. For example, the combining form for stomach is *gastr/o*, and if we add the suffix for inflammation, *itis*, the medical term becomes *gastritis*.

Learning hint

General hints on forming medical terms

While you can learn the various word roots, prefixes and suffixes, it gets confusing trying to put them together correctly. In most instances, the medical definition indicates the last part of the term first, especially when suffixes are used. For example, an inflammation of the stomach is gastritis, not itisgastro, and one who studies the stomach is a gastrologist, not an ologistgastro. When using prefixes, you usually put the parts together in the order you say the definition. For example, slow heart rate is bradycardia, not cardiabrady. As with general rules, there are exceptions, but with practice it will become familiar to you.

Table 1.1 Common combining terms

Word root/combining form	Meaning
abdomin/o	abdomen
angi/o	vessel
arthr/o	joint
cardi/o	heart
cyan/o	blue
cyt/o	cell
derm/o	skin
erythr/o	red
gastr/o	stomach
glyc/o	sugar
hepat/o	liver
hist/o	tissue
leuc/o	white
mamm/o	breast
nephr/o	kidney
neur/o	nerve
oste/o	bone
path/o	disease
phag/o	to swallow
rhin/o	nose

Table 1.2 Common prefixes

Prefix	Meaning
a/an	without
acro	extremities
brady	slow
dia	through
dys	difficult
electro	electric
endo	within
epi	upon or over
hyper	above normal
hypo	below normal
macro	large
micro	small
peri	around
tachy	fast

Now let's add some common prefixes that can be placed before the word roots to alter their meaning (see Table 1.2).

Finally, let's add some common suffixes (Table 1.3) and then see what kinds of words we can form with just these few parts.

Using Tables 1.1 to 1.3, look at all the terms you can make from just the one word root, *cardi/o*. Cardiology is the study of the heart, and a cardiologist is one who studies the heart. Bradycardia is a slow heart rate, tachycardia is a fast heart rate, and an electrocardiogram is an electrical recording of the heart. If your heart were enlarged due to inflammation (carditis), you would have cardiomegaly, which would mean you have heart disease (cardiopathy). The Tin Man from *The Wizard of Oz* thought he had no heart (acardia), but realised that he had a heart all the time.

Abbreviations

Abbreviations are used extensively in the medical profession. They are useful in simplifying long, complicated terms for disease, diagnostic procedures and therapies that require extensive documentation. For now, review Table 1.4 for some common abbreviations you may have heard in a healthcare setting or on television.

Of course you will learn many more terms and abbreviations as we explore the following chapters and become fluent in conversational medical language. This will help you to avoid using lay terms (common, everyday terms) to

Table 1.3 Common suffixes

Suffix	Meaning
algia	pain
cyte	cell
ectomy	surgical removal of
gram	the actual recorded record
graphy	the process of recording
ist	one who specialises
itis	inflammation of
logist	one who studies
logy	study of
otomy	cutting into
ostomy	surgically forming an opening
megaly	enlargement of
pathy	disease
phobia	fear of
plasty	surgical repair
penia	decrease or lack of
scope	instrument to view or examine
sis	disease or condition of

describe medical and anatomical concepts. Now you know that the correct term for 'getting a nose job' is *rhinoplasty*.

The metric system

Whereas medical terminology represents the written and spoken language for understanding anatomy and physiology, the metric system is the 'mathematical language' of anatomy and physiology. For example, blood pressure is measured in millimetres of mercury (mmHg), and organ size is usually measured in centimetres (cm). Medication doses are prescribed in milligrams (mg) or micrograms (μg) and fluids are given in millilitres (ml). Weight is usually measured in kilograms (kg). What exactly does it mean when you are taught that normal cardiac output is 6 litres per minute? You can now see why you must be familiar with the metric system in order to truly understand anatomy and physiology and related healthcare. While the metric system may seem complicated if you are not familiar with it, it really isn't if you have a basic understanding of numeracy.

Test your knowledge 1.2

Define the medical terms:

1. gastritis _____

2. rhinoplasty _____

3. bradycardia _____

4. mammogram _____

5. cytomegaly _____

Give the correct medical term:

6. inflammation of the kidneys _____

7. removal of the stomach _____

8. enlarged heart _____

9. disease of the bones _____

10. one who studies the nerves _____

Table 1.4 Common medical abbreviations

Abbreviations	Meaning
A&E	accident and emergency department
A&P	anatomy and physiology
BD	Latin *bis die*, which means 'twice daily'
BP	blood pressure
CPR	cardiopulmonary resuscitation
GI	gastrointestinal
ICU	intensive care unit
NPO/NBM	Latin *nil per os*, which means 'nothing by mouth'
OD	Latin *omni die*, which means 'every day'
O/E	on examination
PRN	Latin *pro re nata*, which means 'when required'
QDS	Latin *quarter die sumendum*, which means 'to be taken four times daily'
SOB	shortness of breath
SOBOE	shortness of breath on exertion
STAT	Latin *statim*, which means 'immediately'
TDS	Latin *ter die sumendum*, which means 'to be taken three times daily'

In the UK the International System of Units (SI) is commonly used in healthcare. The SI system is also known as the international or **metric system** and is based on the power of 10. The metric system is also the system used by drug manufacturers, and is widely used in healthcare around the world.

The language of disease

This chapter is about learning the language. Ideally, the body works to make things function smoothly and in balance. Sometimes things happen to alter those functions and problems occur. Eating habits, smoking, inherited traits, trauma, environmental factors and even ageing can alter the body's balance and lead to **disease**. Disease, simply put, is a condition in which the body fails to function normally.

While this is an anatomy and physiology course that focuses on *normal* function and structure, it is often helpful to reinforce the concepts with some elaboration of what can go wrong. Therefore, at the end of each system chapter, a *brief* discussion on some of the major diseases associated with that system is provided. An even further in-depth discussion is contained in the companion website. For now, a brief discussion on some of the unique language of disease is needed to lay the foundation for future discussions.

Signs and symptoms of disease

Think back to a time when you were ill. You may have had a fever, cough, nausea, dizziness, joint aches or a generalised weakness. These are examples of what we call **signs** and **symptoms** of disease. While the terms 'signs' and 'symptoms' are often used interchangeably, each has its own specific definition. Signs are more definitive, objective, obvious indicators of an illness. Fever or monitoring the change in the size or colour of a mole are good examples of signs. **Vital signs** are common, measurable indicators that help us to assess the health of our patients. Vital signs are the signs vital to life and include pulse (heart rate), blood pressure, body temperature and respiratory rate. The standard values of vital signs can change according to the patient's age and sex.

Clinical application

The vital sign of pulse

The pulse is commonly taken by applying slight finger pressure over the radial artery located in each wrist (on the thumb side) and counting the number of beats in a 60-second period. The normal heart rate for an adult is 60–80 beats per minute, a child's rate is approximately 70–120, and a newborn's rate is 90–170 beats per minute. If an adult has a heart rate of 165 beats per minute, what medical term would you use to describe that condition?

Test your knowledge 1.3

Answer the following questions:

1. Check which of the following are vital signs:

   a. _____ pulse

   b. _____ pain

   c. _____ blood pressure

   d. _____ age

   e. _____ indigestion

   f. _____ respiratory rate

   g. _____ body temperature

2. Which of the following is the medical term for the cause of a disease?

   a. prognosis

   b. diagnosis

   c. aetiology

   d. syndrome

3. Which of the following is the medical term for the outcome of a disease?

   a. prognosis

   b. diagnosis

   c. aetiology

   d. syndrome

Symptoms, on the other hand, are more subjective and more difficult to measure consistently. A perfect example of a symptom is pain. The response to pain varies amongst individuals, so an equal amount of pain (as in a headache) applied to a number of people could be perceived as a light, moderate or an intense level of pain depending on each individual's perception. In spite of the fact that symptoms are hard to measure, they are still very important in the diagnosis of disease. Sometimes a disease exhibits a set group of signs and symptoms that may occur at about the same time. This specific grouping of signs and symptoms is known as a **syndrome**. Signs, symptoms and syndromes are further explained throughout the rest of our textbook as they relate to the anatomy and physiology of the various body systems.

Discovering as many signs and symptoms as possible can help to **diagnose** a disease. A diagnosis is an identification of a disease determined by studying the patient's signs, symptoms, history and results of diagnostic tests. Getting the medical history can help in determining the **aetiology**, or cause, of the disease. The **prognosis** is the prediction of the outcome of a disease. Hopefully, your *prognosis* for doing well in this anatomy and physiology course is excellent.

Anatomy and physiology concepts

In this section, we take a closer look at some additional concepts related to the study of anatomy and physiology that you will learn more about as you work through this book.

Metabolism

If you travel to other countries, you will see many different cultures and customs. Even though each culture is unique, they all share certain similarities. The same can be said in anatomy and physiology. We all share certain functions that are vital to survival. All humans, for example, need food in order to produce complex chemical reactions necessary for growth, reproduction, movement, and so on. **Metabolism** refers to all of the chemical operations going on within our bodies. Metabolism requires various nutrients or fuel to function and produces waste products much like a car consumes petrol for power and produces waste, or exhaust. Metabolism, for now, can be thought of as 'all the life-sustaining reactions within the body'.

Metabolism is further subdivided into two opposite processes. **Anabolism** is the process by which simpler compounds are *built up* and used to manufacture materials for growth, repair and reproduction, such as the assembly of amino acids to form proteins. This is the *building* phase of metabolism. **Catabolism** is the process by which complex substances are *broken down* into simpler substances. For example, the breakdown of food into simpler chemical building blocks for energy use is a catabolic process. An abnormal and extreme example of catabolism is someone with anorexia nervosa (someone who is starving themselves) whose body 'feeds upon itself', actually consuming the body's own tissues.

Homeostasis

For the body to remain alive, it must constantly monitor both its internal and external environment and make the appropriate adjustments. In order for cells to thrive, they must be maintained in an environment that provides a proper temperature range, balanced oxygen levels and adequate nutrients. Heart rate and blood pressure must also be monitored and maintained within a certain range or set point for optimal functioning depending upon the body activity. **Homeostasis** is the physiologic processes that monitor and maintain a stable internal environment or equilibrium. Survival depends upon the body's ability to maintain homeostasis. The body has a number of survival needs – things that must be available in order for the body to survive:

Oxygen – we cannot survive at all without this – cell death occurs very quickly if we are deprived of oxygen. There is 20.9% oxygen in the air that we breathe, and it is the job of the cardiovascular and respiratory systems to ensure that all the cells of the body receive the oxygen they need.

Water – there is approximately 40 litres of water in the average adult body. It provides a base for body fluids (eg: digestive juices) and excretions (urine) and secretions (sweat). We obtain our water requirements from food and drink.

Nutrients – these are obtained from the food we eat and they provide the body with the necessary substances for all of the processes that go on in the body. Carbohydrates provide energy, proteins are for repair and growth, and fats are for building cells and providing some food reserves.

Stable body temperature – body temperature in humans is 37°C – this must be maintained in order to allow chemical processes to take place effectively and efficiently. As body temperature increases, metabolic processes increase and as body temperature decreases so metabolic activity slows down. Body temperature changes can have a significant effect on the way in which enzymes in the body are able to work. Enzymes are protein-based catalysts that enable a wide variety of chemical reactions to take place throughout the body. As body temperature goes up so enzyme activity increases, however if body temperature goes too high then enzymes become denatured (irreversibly damaged). As body temperature declines so too does enzyme activity. Both these changes in body temperature will therefore affect a number of chemical reactions in the body and can therefore have far-reaching effects on how the body functions overall.

Exposure to appropriate atmospheric pressure – breathing and the way in which gas exchange takes place in the lungs is dependent on appropriate atmospheric pressure. At sea level it is 760 mmHg. At high altitudes where atmospheric pressure may be lower, gas exchange may be inadequate which may lead to breathing difficulties in some people. Our respiratory system is designed to facilitate changes in pressure within the lungs so that gas exchange can take place effectively and efficiently.

All of these survival needs are vital to the body and it's internal environment and homeostatic regulation refers to the adjustments made in the human organism to maintain this stable internal environment.

Irrespective of what needs to be regulated, all homeostatic control systems in the body have three key components:

1. **Receptor** – this is a sensor that is responsible for monitoring changes and responding to them. They may be receptors in the skin that respond to external temperature changes or they may be receptors that respond to changing levels of substances in the blood. Those changes are referred to as *stimuli*.

2. **Control centre** – once a receptor has responded to a stimuli, that information needs to be conveyed to a control centre. The information is sent via a pathway known as an *afferent* pathway – these may be nerves or the blood – to the control centre. Which control centre is involved will depend on what change has taken place. When blood sugar levels have changed then the control centre is the pancreas. If body temperature has changed then the control centre is the hypothalamus in the brain. Once the control centre has decided what to do then information needs to be sent out to effect a change. This information goes out via an *efferent* pathway to a relevant effector.

3. **Effector** – this puts into effect the response that has been decided by the control centre. In the case of blood sugar changes then the effector will be the response of cells in the pancreas to secrete hormones to change the blood sugar level back to what it should be. In changes to body temperature then the effector may be the shivering of muscles to warm us up if we are cold, or sweating to cool us down (see Figure 1.4 for more detail on thermoregulation). In both examples the mechanisms that have been initiated are designed to ensure that blood sugar and body temperature are bought back to normal levels.

This is known as *negative feedback*, and most homeostatic control systems operate in this way. What this ensures is that the original stimulus is stopped or the intensity of its effect is reduced.

The thermostat in your house functions like a homeostatic mechanism. A temperature is set and then maintained by a sensor that monitors the internal environmental temperature and either heats the house if the sensor registers too cold or cools the house if the sensor registers too hot. There is a continuous feedback loop from the sensor to the thermostat to determine what action

Clinical application

Metabolic syndrome or syndrome X

There is an emerging and controversial (in terms of agreement) syndrome affecting individuals in the Western world called the *metabolic syndrome*, or *syndrome X*. A patient with this syndrome exhibits three of the following five common conditions: high blood sugar levels (hyperglycaemia); high blood pressure (hypertension); abdominal obesity; high triglycerides (a lipid substance in the blood); and low blood levels of HDL (which is the 'good' form of blood chole sterol). Individuals who exhibit this syndrome are at an increased risk from a form of diabetes, and from heart attacks and/or strokes. This is essentially a syndrome that has been created as a result of poor diet and lack of exercise.

is needed. Because the feedback loop opposes the stimulus (cools down if too hot, heats up if too cold), it is referred to as a **negative feedback loop**.

The body relies on negative feedback to continually sense the internal and external environment and make adjustments to maintain homeostasis.

Thermoregulation as a negative feedback mechanism

Amazing body facts

Bizarre signs and symptoms!

Here are some strange signs and symptoms that have been indications of diseases. Note that there are other signs, symptoms and tests to determine specific diseases. So do not use this list of oddities as a sole diagnostic tool!

1. Generalised itching skin can be an indication of Hodgkin's disease.
2. Sweating at night may indicate tuberculosis.
3. A desire to eat clay or starchy paste may indicate an iron deficiency in the body.
4. Breath that smells like pear drops or a fruit-flavoured chewing gum may be an indication of diabetes.
5. A magenta-coloured tongue is indicative of a riboflavin deficiency.
6. A patient with profound kidney disease often doesn't have moons (cuticles) on his or her fingernails.
7. A hairy tongue may mean that a patient's normal mouth flora has died from improper use of antibiotics.
8. Spoon-shaped fingernails may point to an iron deficiency in the body.
9. Brown linear streaks on the fingernails of fair-skinned people may indicate melanoma (skin cancer).

The hypothalamus in the brain represents the body's thermostatic control. If the hypothalamus senses a very cold environment, it opposes this cold stimulus (negative feedback loop) and performs physiologic processes to gain heat within the body to maintain an internal temperature near 37.0°C. The body begins to shiver, and this increased muscular activity generates heat. In addition, since most heat loss is through peripheral areas (head, arms and legs), the body decreases the size of the peripheral blood vessels (vasoconstriction), causing the blood to be deeper from the skin surface where the heat would be lost to the cold environment. This keeps the blood closer to the core of the body where it is warmer. Of course, we can assist the body by wearing a heavy coat and hat, which would remove much of the stress of the cold environment, or simply get out of the cold to a warmer environment.

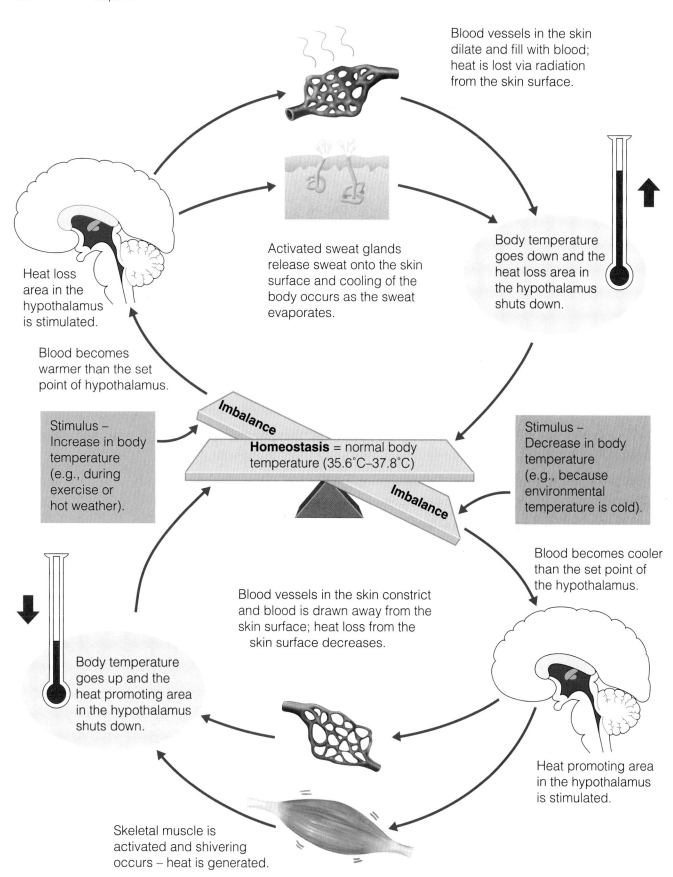

Blood vessels in the skin dilate and fill with blood; heat is lost via radiation from the skin surface.

Heat loss area in the hypothalamus is stimulated.

Blood becomes warmer than the set point of hypothalamus.

Stimulus – Increase in body temperature (e.g., during exercise or hot weather).

Activated sweat glands release sweat onto the skin surface and cooling of the body occurs as the sweat evaporates.

Body temperature goes down and the heat loss area in the hypothalamus shuts down.

Imbalance

Homeostasis = normal body temperature (35.6˚C–37.8˚C)

Imbalance

Stimulus – Decrease in body temperature (e.g., because environmental temperature is cold).

Blood becomes cooler than the set point of the hypothalamus.

Body temperature goes up and the heat promoting area in the hypothalamus shuts down.

Blood vessels in the skin constrict and blood is drawn away from the skin surface; heat loss from the skin surface decreases.

Heat promoting area in the hypothalamus is stimulated.

Skeletal muscle is activated and shivering occurs – heat is generated.

Figure 1.4 Mechanisms of body temperature regulation.

Conversely, if you are in the desert and the temperature is 50°C, the body senses this as too hot and stimulates physiologic processes to cool you down. These processes include sweating (evaporation is a cooling process – heat is lost as the sweat becomes a gas) and enlarging the peripheral vessels (peripheral vasodilatation) in order to radiate the body heat into the external environment. In healthcare practice, if a patient presents with a very high temperature, then measures can be employed to assist in reducing the high temperature. This may include the administration of an antipyretic medication or the provision of a rotating fan in order to make the individual feel more comfortable. What is important is that the cause of the high temperature is determined. Much of healthcare practice is just that – assisting the body in returning it to homeostasis.

Your body is also capable of **positive feedback**, which increases the magnitude of a change. Positive feedback is not a way to regulate your body, because it increases a change away from the ideal set point. Positive feedback can be harmful if the cycle cannot be broken, but sometimes positive feedback is necessary for a process to run to completion.

A good example of necessary positive feedback is the continued contraction of the uterus during childbirth. When a baby is ready to be born, the hypothalamus releases the hormone oxytocin from the posterior pituitary (neurohypophysis) into the blood. Oxytocin is circulated to the uterus and increases the intensity of uterine contractions. As the uterus contracts, the pressure inside the uterus caused by the baby moving down the birth canal increases the signal to the hypothalamus. More oxytocin is released, and the uterus contracts harder. Pressure gets higher inside the uterus, the hypothalamus is signalled to release more oxytocin and the uterus contracts yet harder. This cycle of ever-increasing uterine contractions due to an ever-increasing release of oxytocin from the hypothalamus continues until the pressure inside the uterus decreases – that is, until the baby is born. Once that happens, the levels of oxytocin reduce.

Clinical application

'Breaking' a fever

It is believed that most fevers are the body's way of making an inhospitable environment for a pathogen to survive. Why is it when someone begins sweating after a prolonged fever (increase in body temperature), the fever is said to be 'breaking'? A fever sets the hypothalamus to a higher set point temperature. The body then increases its metabolism to generate more heat to reach this new higher temperature. Once whatever is causing the fever is gone, the hypothalamus set temperature is turned back down to normal. The body must now rapidly get rid of the excess heat by the cooling process of evaporation through sweating.

Summary

- Anatomy is the study of the actual internal and external structures of the body, and physiology is the study of how these structures normally function. Pathology is the study of the disease processes by which abnormal structures and abnormal body functions can occur.

- The human body exhibits a structural hierarchy that goes from a simple level to a complex level – chemical → cellular → tissue → organ → organ system → organismal.

- Medical terminology is the language of medicine and combines word roots, prefixes and suffixes to construct numerous medical terms to describe conditions, locations, diagnostic tools, and so on.

- The metric system is the mathematical language of healthcare based on powers of 10.

- Metabolism refers to all of the chemical operations going on within the body and can be broken down into two opposite processes. The building phase of metabolism is anabolism, in which simpler compounds are *built up* and used to manufacture materials for growth, reproduction and repairs. The tearing down phase is catabolism, in which complex substances are *broken down* into simpler substances, such as food broken down for energy use.

- The body has some key survival needs – oxygen, water, nutrients, maintenance of ideal body temperature and exposure to appropriate atmospheric pressure.

- The body tries to maintain a balanced or stable environment called homeostasis. It must constantly monitor the environment and make changes to maintain this balance. It often accomplishes homeostasis through negative feedback.

Case study

A 66-year-old Asian male involved in a car accident is taken to the ICU with SOB and abdominal pain. He has acrocyanosis, tachycardia and a past medical history of cardiopathy. He weighs 68 kg and is 170 cm tall. His chest X-ray shows an enlarged heart. His facial injuries will require future rhinoplastic surgery. An electrocardiogram and abdominal X-ray are ordered.

1. Where exactly in the hospital was the patient taken?

   _____

2. Describe the patient's colour, heart rate and breathing.

   _____

   _____

3. What is the medical term for what the X-ray showed?

   _____

4. What future facial surgery will he need?

   _____

   _____

Review questions

Multiple choice

1. Which of the following is an example of microscopic anatomy?
   a. viewing an X-ray
   b. examining the shape of an organ during an autopsy
   c. classifying a type of bacterial cell
   d. watching how the pupils in the eyes react to light

2. Acromegaly means which of the following?
   a. a large stomach
   b. enlarged extremities
   c. an inflamed stomach lining
   d. a large acrobat

3. The breakdown of sugar in the body for energy is called
   a. anabolism
   b. catabolism
   c. hypobolism
   d. hyperbolism

4. Which of the following is a measurement system based on the power of 10?
   a. English system
   b. British Imperial system
   c. metric system
   d. weights and measures system

5. The cause of a disease is referred to as the
   a. prognosis
   b. diagnosis
   c. pathology
   d. aetiology

Fill in the blanks

1. Ted's knee injury occurred at last night's football game. Today his doctor wants to make a small incision and use a device to 'look around the joint' to assess the damage. What is the term for this device? _____

2. _____ is the study of the structures of the body, and _____ is the study of the functions of these structures.

3. Pulse and temperature represent two _____ signs of the body.

4. Raheem had blood tests carried out that showed a normal number of white blood cells (WBCs) and red blood cells (RBCs). What are the respective medical terms for these cell types? _____

5. _____ is the term used to describe the study of disease.

Short answers

1. Explain the difference between diagnosis and prognosis.
   _____
   _____
   _____
   _____

2. Knowing that difficulty swallowing is called dysphagia, what do you think the function of a phagocyte is?

_____
_____
_____
_____

3. Contrast negative and positive feedback loops.

_____
_____
_____
_____

4. Describe one example of homeostasis in your body.

_____
_____
_____
_____

Suggested activities

1. Using a medical dictionary, find five new medical terms and give their definition.

_____   _____
_____   _____
_____   _____
_____   _____

2. Make up 3 × 10-cm note cards with five word roots discussed in this chapter and see how many medical words you can make using either prefixes or suffixes in the tables. For example, the word root *arthr/o* can be used to make the following: arthritis, arthralgia, arthroscope and arthroplasty. Confirm that you made a real word by looking it up in a medical dictionary.

Answers

Answers to case study

1. The patient was taken to the Intensive Care Unit.
2. The patient has blue extremities, a fast heart rate and he is short of breath.
3. His X-ray shows cardiomegaly
4. He will require surgery to repair his broken nose.

Answers to multiple choice questions

1. c
2. b
3. b
4. c
5. d

Answers to fill in the blanks

1. Arthroscope
2. Anatomy, physiology
3. Vital signs
4. Leucocytes, erythrocytes
5. Pathology

Answers to short answer questions

1. Diagnosis – identification of a disease by looking at the patient's signs, symptoms, history and test results.
   Prognosis – the outcome of a disease.
2. A phagocyte engulfs (swallows) foreign substances such as bacteria.
3. Positive feedback – increases the magnitude of a change.
   Negative feedback – opposition to a stimulus in order to return things to normal.
4. Example of an homeostatic mechanism includes temperature control. If the temperature rises, the blood vessels dilate, facilitating heat loss to the surrounding areas. Blood then cools, returning the body temperature to 'normal'.

The human body

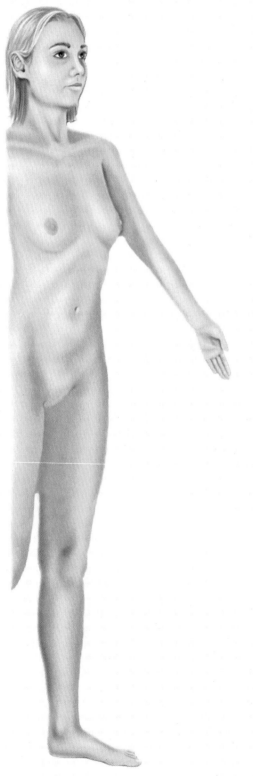

We now have a basic understanding of medical language and some basic anatomy and physiology concepts. This chapter provides the major external map of the human body that will serve as a guide for future chapters. Each chapter will map in detail the internal regions. The directional terms and body locations are a foundation upon which to build as we investigate further this wondrous creation called the human body. Isn't it ironic that if there is one thing we should know better than anything else, it should be our own bodies? To borrow from an old saying, by the end of this textbook, you will know your entire body like the back of your hand.

Learning objectives

At the end of this chapter, you will be able to:

◆ List and describe the various body positions
◆ Define the body planes and associated directional terms
◆ Locate and describe the body cavities and their respective organs
◆ List and describe the anatomical divisions of the abdominal region
◆ Identify and locate the various body regions

Pronunciation guide

abdominopelvic cavity (*abDOM in oh PELL vik CAH vih tee*)
antecubital (*an tee KEW bit all*)
buccal (*BUCK all*)
caudal (*CORD all*)
cephalic (*keff AL ik*)
coronal plane (*kor ROW nall*)

cranial (*KRAY nee yall*)
crural (*CRUR all*)
distal (*DISS tall*)
dorsal (*DOR sall*)
gluteal (*GLOO tee yall*)
mediastinum (*meh dee yuh STY num*)

midsagittal plane (*mid SAJ it all*)
pleural cavities (*PLERR all*)
superficial (*super FISH all*)
thoracic cavity (*thor AH sick CAH vih tee*)
transverse (*tranz VERS*)

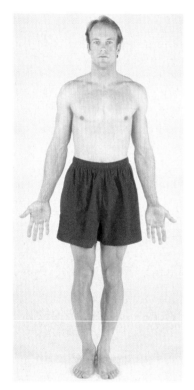

Figure 2.1 The anatomical position. (*Source*: Mike Gallitelli/Pearson Education Inc.)

anatomical position (*ah nah TOM ik all*)

supine position (*soo PINE*)

The map of the human body

When reading a map, you need certain universal directional terms, such as north, east, south and west, to help you understand and use the map. A map is often made to represent a specific region so that more details can be included about that particular region, making it easier to explore. Similarly, scientists have created standardised body directional terms and split the body into distinct regions, sections and cavities so that we can more clearly and rapidly locate and discuss anatomical features. Having certain anatomical landmarks on the body also provides needed points of reference. For example, the spinal cord is a major anatomical landmark for many structures in the centre of our bodies.

If a patient states, 'I have pain in my stomach', does that really tell you a lot of information? Location of pain can help in determining what is wrong with a patient. It is helpful to know the type of pain (dull, sharp or stabbing) and *exactly* where in that region the pain is located to help determine its cause. For example, pain in the general stomach area can indicate a variety of problems, including an ulcer, heart attack, appendicitis, indigestion or liver problems. Knowing the exact region can help a clinician better determine the exact problem.

Body positions

The body can assume many positions and therefore have different orientations. To standardise the orientation for the study of anatomy, scientists developed the **anatomical position**. The anatomical position, as shown in Figure 2.1, is a human standing erect, face forward, with feet parallel and arms hanging at the side, and with palms facing forward.

Other body positions that are important to discuss because of clinical assessments and treatments in healthcare are the **prone**, **supine** and **Fowler's** positions. The supine position is laying face *upward*, or on your back. The prone position is laying face *downward*, or on your stomach. The Fowler's position is sitting in bed with the head of the bed elevated 45 to 60 degrees. This position is often used in the hospital to facilitate breathing and for

Test your knowledge 2.1

Answer the following questions:

1. Try standing in the actual anatomical position.

2. Give the best body position (prone, supine or Fowler's) for the following circumstances:

   a. having a back massage _____

   b. eating in a hospital bed _____

   c. watching television in bed _____

   d. watching the stars at night _____

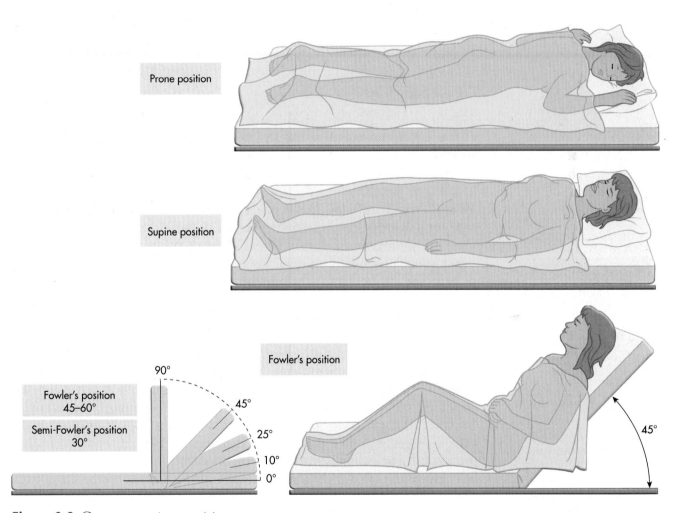

Prone position

Supine position

Fowler's position

Fowler's position 45–60°

Semi-Fowler's position 30°

90°

45°

25°

10°

0°

45°

Figure 2.2 Common patient positions.

comfort of the bed-bound patient while eating or talking. See Figure 2.2 for these body positions.

Body planes and directional terms

Sometimes it is necessary to divide the body or even an organ or tissue sample into specific sections to further examine it. A plane is an imaginary line drawn through the body or organ to separate it into specific sections. For example, in

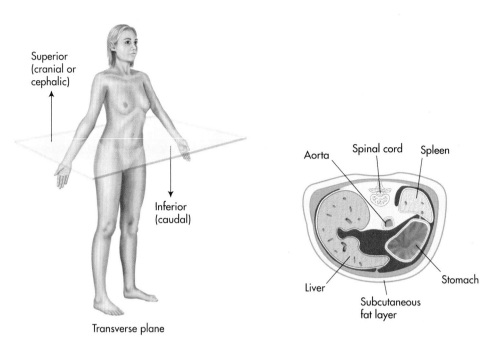

Figure 2.3 Transverse plane and a cross-sectional view of the upper abdominal region.

transverse (*Tranz VERS*)

Figure 2.3, we see the **transverse plane**, or **horizontal plane**, dividing the body into top (**superior**) and bottom (**inferior**) sections. This can also be called **cross-sectioning** the body. Cross-sectioning is often done with tissue and organ samples to further examine internal structures.

Notice in Figure 2.3 that certain directional terms can be used to describe areas divided by the transverse plane. One more analogy that relates to a map is the concept of a reference point. If you were travelling from London to Dover, you would have to travel in a south-easterly direction. London is your starting point and serves as your reference point. However, if you were travelling from Dover to London, you would travel in a north-westerly direction because Dover is now your point of reference. In Figure 2.3, you can see that superior (**cranial** or **cephalic**) means toward the head or upper body and inferior (**caudal**) means away from the head or toward the lower part of the body. Any body part can be either superior or inferior depending upon your reference point. For example, the knee is superior to the ankle if the ankle is the reference point. Turning this around, the ankle is inferior to the knee if the knee is the reference point. Two other terms from this illustration are *cranial*, which refers to the skull, and *caudal*, which refers to body parts near the tail (coccyx).

cranial (*KRAY nee yall*)
   cranio = *skull*

cephalic (*keff AL ik*)
   cephalo = *toward the head*

caudal (*CORD all*)
   cauda = *tail*

-*ic* and -*al* are adjective endings that mean 'pertaining to'

The **median plane**, or **midsagittal plane**, divides the body into right and left halves. Figure 2.4 shows this plane and the directional terms associated with it. **Medial** refers to body parts located near the middle or midline of the body. **Lateral** refers to body parts located away from the midline (or on the side). If a pathologist were to section and examine an organ, he or she might make a midsagittal cut (cut the organ into equal right and left halves) in order to examine the internal parts of the organ or might simply make several sagittal (vertical or lengthwise) cuts to slice the organ into smaller sections for closer examination.

midsagittal (*mid SAJ it all*)

medial (*MEE dee yall*)
lateral (*LAT er all*)

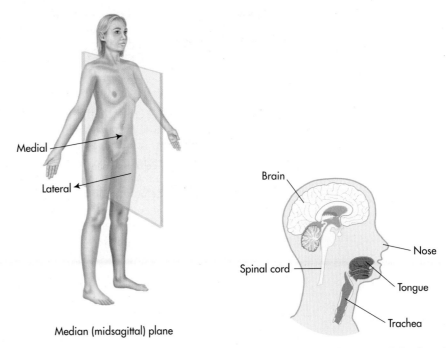

Median (midsagittal) plane

Figure 2.4 Midsagittal or median plane along with a sagittal view of the head.

coronal (*kor ROW nall*)

The **frontal plane**, or **coronal** **plane**, divides the body into front and back sections. **Anterior** and **ventral** refer to body parts toward or on the front of the body, and **posterior** and **dorsal** refer to body parts toward or on the back of the body. Figure 2.5 demonstrates the coronal plane and associated directional terms.

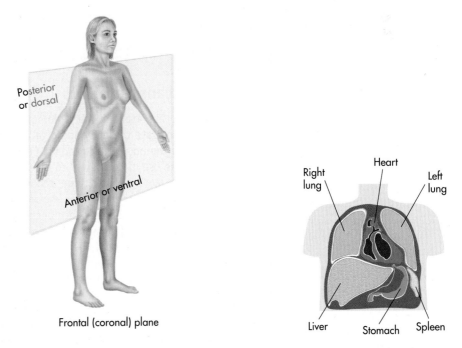

Frontal (coronal) plane

Figure 2.5 Frontal or coronal plane along with a coronal view of the chest and stomach.

Clinical application

Do you know your left from your right?

By now, it should be clear that a precise, standardised language with directional terms is needed to study anatomy and physiology and apply it in a healthcare setting. Something as simple as left and right can become critical. For example, suppose you are a surgeon and need to put a label around a patient's right leg to designate it as the leg to be amputated in forthcoming surgery. If you approach the patient from the bottom of the bed and place the tag on the leg on YOUR right side, you have erroneously placed it on the patient's left leg, and this could have disastrous results. The message to take home is that left and right *always* refer to the patient's left and right, *not yours*.

Additional directional terms

proximal (*PROK sim all*)
distal (*DISS tall*)

There are some additional directional terms that are important in healthcare. Proximal refers to body parts close to a point of reference of the body. This is contrasted by distal, which refers to body parts away from a point of reference. For example, using your shoulder as a reference point, your elbow is proximal to your shoulder, while your fingers are distal to your shoulder. **External** means on the outside and **internal** refers to structures on the inside. Did you know that your external skin is actually the body's largest organ? Most other organs are located internally within body cavities.

Superficial means towards or at the body surface. When a clinician draws blood from you, he or she looks for superficial veins that are easy to see and easy to access with the needle. **Deep** means away from the body surface. The

Applied science

X-rays, CT scans and MRIs

Positions are also important in the radiologic sciences. X-rays are high-energy radiation that penetrate the body and give a two-dimensional view of the bones, air and tissues in the body. The standard X-ray (much like a photograph) can be enhanced with the use of computers to give much greater detail and contrast and to allow for a more realistic three-dimensional view. For example, if a golfball-sized tumour in the lung was shown on a standard chest X-ray, you would have no idea of its actual depth because it would look flat, like a coin. Computed tomography (CT) scanning uses a narrowly focused X-ray beam that circles rapidly around the body. The computer constructs thin-slice images and combines them to give much greater detail and allow for a more three-dimensional view; much like a loaf of sliced bread gives a better idea of the total shape of the loaf than does a single slice. The CT scan reveals the true depth of the coin-shaped tumour shown on the normal X-ray. A magnetic resonance imager (MRI) produces even greater detail of tissue structures, even down to individual nerve bundles. Another possible advantage of the MRI is a decrease in radiation exposure.

large veins in your legs are deep veins and are more protected than superficial veins because injury to them can be more critical to survival than can injury to a smaller, superficial blood vessel. **Central** refers to locations around the centre of the body (torso and head), and **peripheral** refers to the extremities (arms and legs) or surrounding or outer regions (see Table 2.1).

Table 2.1 Directional terms

Directional term	Meaning	Use in a sentence
proximal	near point of reference	The wrist is *proximal* to the fingers.
distal	away from point of reference	The shoulder is *distal* to the fingers.
external	on the outside	An *external* defibrillator is used on the outside of the chest.
internal	on the inside	He received *internal* injuries from the accident.
superficial	at the body surface	The cut was only *superficial*.
deep	under the body surface	The patient had *deep* wounds from the knife blade.
central	locations around the centre of the body	The patient had *central* chest pain.
peripheral	surrounding or outer regions	The patient had *peripheral* swelling of the feet.

Test your knowledge 2.2

Answer the following questions:

1. Give the opposite directional term:
   a. superior _____
   b. posterior _____
   c. caudal _____
   d. ventral _____
   e. distal _____
   f. external _____
   g. superficial _____
   h. peripheral _____
   i. medial _____

2. A spinal tap is performed on the (_____) portion of the body.

3. The plane that divides the body into upper and lower regions is called the _____ plane.

4. Cutting an organ into two equal halves (right and left) requires a _____ incision.

5. A scratch on the surface of the skin is called a _____ wound.

6. The wrist is _____ to the hand and _____ to the elbow.

7. The nose is _____ to the mouth.

8. A pain in your side can also be referred to as _____ pain.

9. If your hands and feet are swollen with fluid (oedema), you are said to have _____ oedema.

Body cavities

The body has two large spaces or cavities that house and protect organs. Located in the back of the body is the dorsal cavity and in the front, the ventral cavity. Figure 2.6 illustrates these cavities. The larger anterior cavity is subdivided into two main cavities called the **thoracic cavity** and the **abdominopelvic cavity**. These cavities are physically separated by the large, dome-shaped muscle called the diaphragm that is used for breathing. The thoracic cavity contains the heart, lungs and large blood vessels. The heart has its own small cavity called the pericardial cavity. The abdominopelvic cavity contains the digestive organs – stomach, intestines, liver, gallbladder, pancreas and spleen – in the upper or abdominal portion. The lower portion, called the pelvic cavity, contains the urinary and reproductive organs and the last part of the large intestine. A posterior or dorsal cavity is located in the back of the body and consists of the **cranial cavity**, which houses the brain, and the **spinal cavity**, which contains the spinal cord.

There are also smaller body cavities that designate specific areas, and these are further explored in later chapters. For example, the nasal cavity is the space behind the nose, the oral or buccal cavity is the space within the mouth, and the orbital cavity houses the eyes.

thoracic cavity (*thor AH sick CAH vih tee*)

abdominopelvic cavity (*abDOM in oh PELL vik CAH vih tee*)

cranial cavity (*KRAY nee yall CAH vih tee*)

spinal cavity (*SPY nall CAH vih tee*)

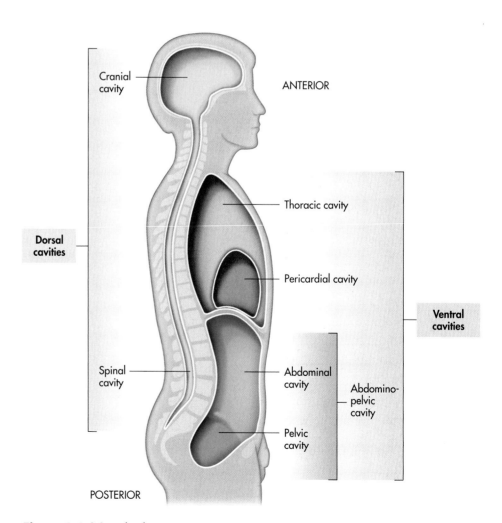

Figure 2.6 Main body cavities.

Test your knowledge 2.3

Identify the major body cavity where the following organs are located.

a. heart _____

b. spinal cord _____

c. stomach _____

d. lungs _____

e. reproductive organs _____

f. brain _____

Clinical application

The central landmark – the spinal column

The spinal or vertebral column is a major, centrally located anatomical landmark and has five sets of vertebrae (spinal bones) labelled for the region of body location (see Figure 2.7). The seven cervical (C) vertebrae are located in the neck; the twelve thoracic (T) vertebrae are located in the chest; the five lumbar (L) vertebrae are located in the lower back; and the five fused sacral (S) vertebrae (sacrum) are located near the final coccyx vertebra (tailbone). For example, the T5 vertebra is used to help locate on a chest X-ray the area where the right and left lung begin to branch. You'll learn even more about the spinal column and cord in Chapter 6, 'The skeletal system', and Chapters 9 and 10, 'The nervous system' (parts I and II).

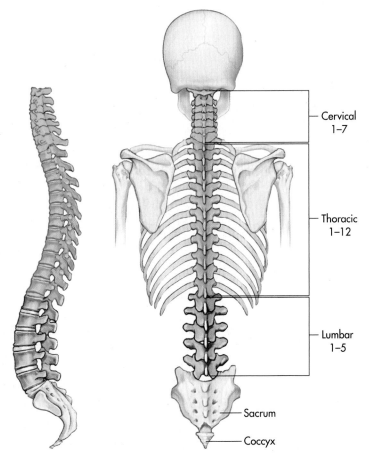

Cervical
1–7

Thoracic
1–12

Lumbar
1–5

Sacrum

Coccyx

Figure 2.7 The spinal column.

Body regions

epi = *above*
gastric = *stomach*
hypo = *below*
chondriac = *refers to ribs*
umbilical = *belly button*

lumbar = *lower back*
inguinal = *referring to groin*

The abdominal region houses a number of organs. Anatomists have divided this region according to Figure 2.8. Notice that understanding directional terms assists in locating the regions. For example, the **epigastric** region (**epi** – above; **gastric** – stomach) is located superior to the umbilical region. The right and left **hypochondriac** regions are located on either side of the epigastric region and contain the lower ribs. The centrally located **umbilical** region houses the naval or belly button. You may not remember your umbilical cord being cut as a newborn, but your belly button is a reminder that it occurred. Lateral to this region are the right and left lumbar regions at the level of the **lumbar** vertebrae. The **hypogastric** region lies inferior to the umbilical region and is flanked by the right and left iliac or **inguinal** regions. The **iliac** region is where the thigh meets the body trunk and is also called the groin region.

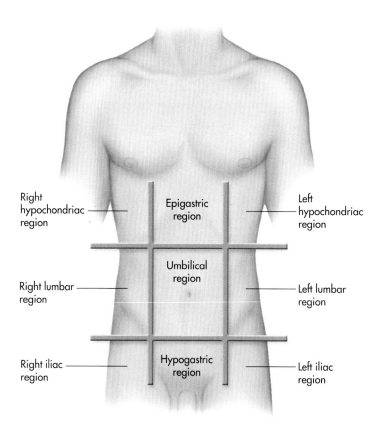

Figure 2.8 The nine divisions of the abdominal region.

Clinical application

Hernias

You may have heard of an umbilical (belly button) bulge, or inguinal hernia, and now you know exactly where such hernias are located. Just what is a hernia? A hernia is a tear in the muscle wall that allows a structure (usually an organ) to protrude through it. Sometimes this can be a minor nuisance, but a hernia can also be very dangerous if the blood flow is restricted to the portion of the organ that is protruding. Restricted blood flow can lead to death of the tissue and to serious consequences. Death of a tissue is called **necrosis**. Figure 2.9 shows an inguinal and umbilical hernia with a protrusion of the intestines.

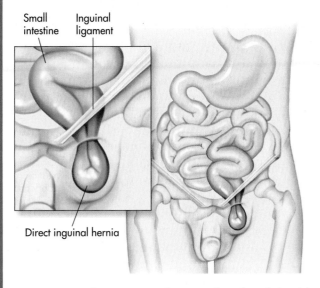

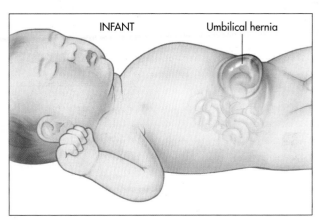

Figure 2.9 Illustrations of inguinal and umbilical hernias.

Amazing body facts

Psoas test

This test – with its strange name – is one way to help determine if a patient has appendicitis. The patient is placed in a supine position and instructed to raise his or her right leg while the practitioner places a hand on the patient's right thigh and gives a slight opposing downward force. If it is appendicitis, the patient will usually experience pain in the right lower quadrant. The term psoas refers to the name of the muscle that is located in this region.

A more practical way for health professionals to compartmentalise the abdominal region is to separate it into anatomical quadrants. Figure 2.10 illustrates these quadrants, which are helpful in describing the location of abdominal pain. Knowing the organs located in the quadrant where the pain occurs can provide a clue to what type of problem the patient has. For example, tenderness in the right lower quadrant (RLQ) can be a symptom of appendicitis, because that is where the appendix is located. RUQ (right upper quadrant) pain may mean a liver or gallbladder problem.

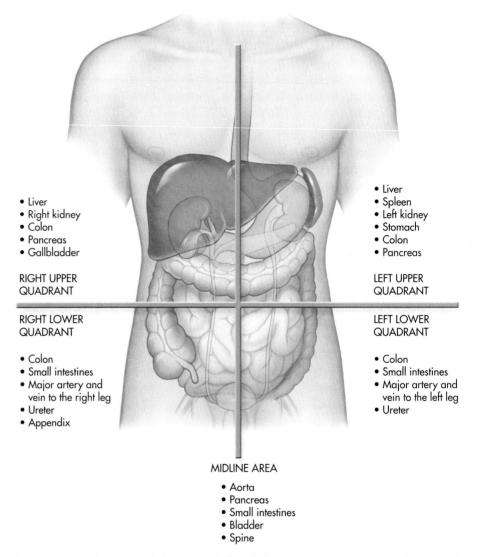

- Liver
- Right kidney
- Colon
- Pancreas
- Gallbladder

RIGHT UPPER QUADRANT

RIGHT LOWER QUADRANT

- Colon
- Small intestines
- Major artery and vein to the right leg
- Ureter
- Appendix

- Liver
- Spleen
- Left kidney
- Stomach
- Colon
- Pancreas

LEFT UPPER QUADRANT

LEFT LOWER QUADRANT

- Colon
- Small intestines
- Major artery and vein to the left leg
- Ureter

MIDLINE AREA

- Aorta
- Pancreas
- Small intestines
- Bladder
- Spine

Figure 2.10 The clinical division of the abdominal region into quadrants with related organs and structures.

There are additional body regions that further aid in locating areas and structures. For example, what if you were asked to obtain an axillary temperature on an infant? Just where is the brachial or femoral pulse? What part of the body does carpal tunnel syndrome affect? See Figure 2.11 for other common body regions and parts that are discussed in later chapters. In addition, review Table 2.2 for further practical examples of the importance of the various body regions.

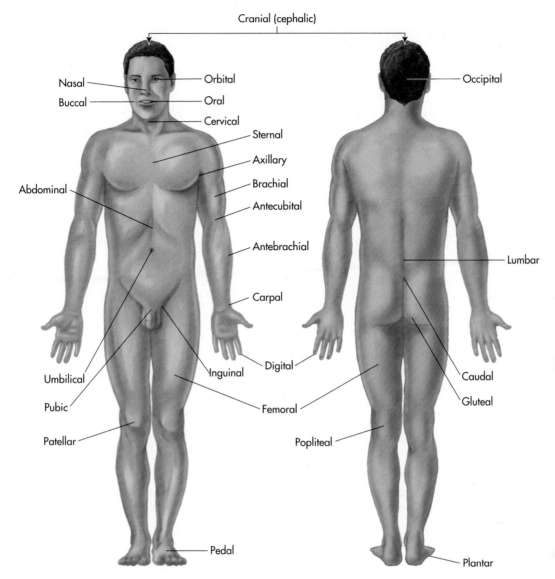

Figure 2.11 Anterior and posterior body regions.

Test your knowledge 2.4

Fill in the blanks with the appropriate medical term for the body region. These can have more than one answer.

1. People who chew paan, betel nut, smokeless tobacco or snuff are more susceptible to _____ cancer.

2. Antiperspirant sprays are usually used in the _____ region.

3. Belly button rings are usually found in the _____ region.

4. If you sit too long at your desk, you can develop _____ pain.

5. During a health check, your reflexes are tested with a little rubber hammer that taps your _____ region.

Table 2.2 Examples of body regions and their location

Body region	Location	Health-related example
antebrachial	forearm	between the wrist and elbow
antecubital	depressed area in front of elbow	area used to take blood or start an intravenous infusion (IV)
axillary	armpit	can be used to measure temperature
brachial	upper arm	take blood pressure
buccal	cheek	check buccal region for central cyanosis
carpal	wrist	carpal tunnel syndrome
cervical	neck	cervical collar needed for neck injuries
digital	fingers	place digital oxygen sensors
femoral	upper inner thigh	check femoral pulse for effective CPR
gluteal	buttocks	the buttock is an injection site
lumbar	lower back	lumbar pain often occurs on long car trips
nasal	nose	medications can be given by nasal spray
oral	mouth	oral route is the most common route for medications
orbital	eye area	orbital injury can cause damage to sight
patellar	knee	patellar injuries are very common in sports
pedal	foot	people with heart problems may have pedal oedema (swelling)
plantar	sole of foot	plantar warts can be painful
pubic	genital region	the pubic region is often checked for body lice
sternal	breastbone area	the sternal area is used for CPR
thoracic	chest	the thoracic area is used to listen to heart and lung sounds

Summary

- The body can assume many different positions and, to standardise the study of anatomy, scientists often refer to the anatomical position. In the anatomical position, the person stands with face and toes forward, hands at sides, and palms facing forward. Other positions, such as the prone, supine and Fowler's positions are used in healthcare for assessment and treatments.

- The body can be divided by the use of planes into different sections. For example, the transverse, or horizontal, plane divides the body into superior and inferior sections. The median, or midsagittal, plane divides the body into equal right and left halves, and the frontal, or coronal, plane divides the body into anterior and posterior sections.

- Directional terms, such as internal and external, proximal and distal, superficial and deep, central and peripheral, help us to navigate the body.

- It is important to always remember that directions such as right and left are referenced from the *patient's* perspective and NOT yours.

- The body has several cavities that house anatomical structures (mainly organs). For example, the cranial cavity houses the brain, the thoracic cavity houses the heart and lungs, the abdominopelvic cavity houses the digestive and reproductive organs, and the spinal cavity houses (guess what) the spinal cord.

- The body has many specific regions. For example, the umbilical region is found around your naval or belly button, and the femoral region is located in the upper inner thigh area.

- The directional terms, anatomical landmarks, body regions and body cavities are all important to know so that healthcare professionals can communicate in specific terms that leave no room for confusion.

Case study

A 50-year-old female patient presents with sternal pain radiating to the left brachial area. Peripheral cyanosis is noted in the digital areas, and she exhibits pedal oedema. No epigastric pain is noted. She reports that she became dizzy and fell, bruising the right orbital region, and she received superficial cuts to the right patellar region. The doctor prescribes IV fluids to be commenced via the left antecubital space (fossa) after inserting an IV cannula. Please answer the following questions in common lay terms.

1. Where would you suggest placing a dressing?
2. Where did her pain begin?
3. Where does the pain move to?
4. Does she have stomach pain?
5. Where will the IV be sited?
6. What part of her body is swollen?
7. Which part of her body has become bluish in colour?
8. Where has she sustained bruising?

Review questions

Multiple choice

1. A masseur would ask you to assume which position for a back massage?
   a. prone
   b. supine
   c. Fowler
   d. lotus

2. Which of the following is *not* in the abdominopelvic cavity?
   a. stomach
   b. liver
   c. reproductive organs
   d. heart

3. Carpal tunnel syndrome occurs in what region of the body?
   a. head
   b. cheek
   c. armpit
   d. wrist

4. The midsagittal plane divides the body into
   a. top and bottom
   b. front and back
   c. upper and lower
   d. left and right

5. An organ contained in the right lower quadrant (RLQ) would be:
   a. appendix
   b. heart
   c. lungs
   d. brain

Fill in the blanks

1. A standard position in which a human stands erect, face forward, with feet parallel, arms at sides and palms forward, is called the _____ position.

2. The _____ position is laying face upward and on your back.

3. The mouth is located _____ to the nose, whereas the nose is located _____ to the mouth.

4. The organ found in the cranial cavity is the _____.

5. _____ indicates blueness of the extremities and therefore affects the peripheral areas of the body.

Short answers

1. List the organs found in the abdominal cavity.

   _____

   _____

2. Contrast the differences between the prone, supine and Fowler positions.

   _____

   _____

3. List and describe two specific body regions that are found on the legs.

_____

_____

4. List and describe the location of the nine abdominal regions using directional terms.

_____

_____

Suggested activities

1. Using a white t-shirt, draw and label the abdominal quadrants and the related organs.

2. Play pin the tail on the donkey by guiding the blindfolded person using only directional terminology.

Answers

Answers to case study

1. Place a dressing on her right knee for the cuts she sustained due to the fall.
2. The pain started in her chest region.
3. The pain has moved to her left upper arm.
4. There is no stomach pain.
5. The IV will be commenced in the front of the elbow in the left arm.
6. Her feet are swollen.
7. Her fingers have become blue.
8. She has a bruise around her right eye.

Answers to multiple choice questions

1. a
2. d
3. d
4. d
5. a

Answers to fill in the blanks

1. anatomical
2. supine
3. inferior, superior
4. brain
5. peripheral cyanosis

Answers to short answer questions

1. Liver; kidneys; stomach; spleen; large intestine; pancreas; small intestine; ureters; appendix
2. Prone – lying on your stomach; supine – lying on your back; Fowler's position – knees bent and torso at a 45° angle
3. Any from:
   Femoral – upper inner thigh
   Patellar – knee
   Popliteal – behind the knee
   Plantar – sole of the foot
4. Right hypochondriac region; epigastric region; left hypochondriac region; right lumbar region; umbilical region; left lumber region; right iliac region; hypogastric region; left iliac region.

Biochemistry

It may seem strange to have a chapter with a discussion of what cannot be seen with the naked eye, but as all living things are made up of chemicals, a brief introduction to the types of chemicals that you are likely to encounter would seem to be a good place to start. In this chapter, we will introduce you to some of the chemicals that you will see later in the book, and explore how these chemicals join together and perform particular activities in the body. We shall also look at how foods release the vital building blocks that allow us to continue to survive in a hostile environment.

Learning objectives

At the end of this chapter, you will be able to:

◆ List the main elements that form body matter
◆ Identify and describe some common chemical reactions
◆ Define what a molecule is, and how it relates to compounds
◆ List the common chemical bonds
◆ Explain the difference between an acid and an alkali
◆ Describe the pH scale
◆ Define what an enzyme is
◆ Describe the base pairs found in DNA and RNA

Pronunciation guide

anion (*ANN eye on*)
atom (*AT om*)
cation (*CAT eye on*)
cholesterol (*coe LESS ter oll*)
compound (*COMM pound*)
covalent (*coe VAY lent*)
deoxyribonucleic acid (*dee OX ee RYE bow new KLEE ik*)
disaccharides (*dye SACK are ides*)
electrolytes (*eh LEK troh lights*)
elements (*ELL eh ments*)

enzymes (*EN simes*)
extracellular fluid (*EX trah SELL yoo lar*)
hydrogen (*HIGH droh jen*)
hydrolysis (*high DROL ih siss*)
interstitial fluid (*IN ter STI shall*)
intracellular fluid (*IN trah SELL yoo lar*)
isotopes (*EYE soh topes*)
molecule (*MOLL eck yule*)
monosaccharides (*moh no SACK are ides*)

nucleotides (*NEW klee oh tides*)
phospholipids (*FOSS foe lip ids*)
polysaccharides (*POLL ee SACK are ides*)
ribonucleic acid (*RYE bow new KLEE ik*)
solute (*SOLL yoot*)
solvent (*SOLL vent*)
steroids (*STARE royds*)
triglycerides (*try GLISS er ides*)

Introduction

Chemistry is the study of matter, matter being the basic building blocks of life, and biochemistry is the study of how these building blocks make up living material. Matter is simply anything that occupies space and can be made up of liquids, solids or gases. Liquid matter is, for example, blood, whereas an example of solid matter would be bone. Can you guess an example of gaseous matter? The air we breathe is gaseous matter – it surrounds us but we cannot see it.

Composition of matter

Matter is made up of a limited number of substances, which are called **elements**. Elements cannot be broken down to smaller substances by conventional chemical means. There are 112 known elements and common examples of these include oxygen, carbon, nitrogen, iron and hydrogen. Elements that are joined together are called **compounds**. You probably know the chemical formula for water, H_2O, which means there are two hydrogen elements and one oxygen element that combine together and form water (see Table 3.1).

Elements are built from the smallest particles that retain special properties, which are called **atoms**. *Atom* comes from the Greek word meaning 'incapable of being divided'. Atoms are unique and are identified by a single capital letter, for example hydrogen (H), or by two letters, for example magnesium (Mg). Atoms can be made up of three things, protons, neutrons and electrons. Each atom has a central nucleus where the protons and neutrons are found, and surrounding the nucleus are electrons. (see Figure 3.1)

Hydrogen (H) is made up of one proton and one electron. Protons carry a positive electrical charge and electrons carry a negative charge. Each atom has to be balanced, which means that there needs to be an equal

Amazing body facts

Body composition

Body weight is made up of combinations of four elements – carbon, oxygen, hydrogen and nitrogen.

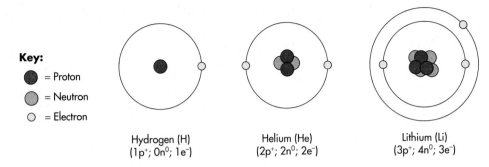

Key:
- ● = Proton
- ● = Neutron
- ○ = Electron

Hydrogen (H)
($1p^+$; $0n^0$; $1e^-$)

Helium (He)
($2p^+$; $2n^0$; $2e^-$)

Lithium (Li)
($3p^+$; $4n^0$; $3e^-$)

Figure 3.1 Structure of the smallest atoms.

number of positive and negative charges, which are balanced out. The electrons orbit the nucleus, rather like the way the moon orbits the Earth, (however in biochemistry there can be several moons), but the positive and negative charges need to balance out. For example, if we take hydrogen as our first example, we find one proton (positively charged) and one electron (negatively charged), so the positive and negative balance. If we take another element, say helium, we find two protons, two neutrons (which have no charge) and two electrons – therefore, the atom remains balanced.

So, what makes atoms different? Each atom is made up of different numbers of protons, neutrons and electrons. If we take lithium as an example: lithium (Li) has three electrons orbiting the nucleus – two in an inner shell and one in an outer shell (see Figure 3.1).

Each element has a different number of protons, and the number of these protons gives the element its atomic number. For example, hydrogen has one proton so it has an atomic number of 1; lithium has three protons, so it has an atomic number of 3. Returning to the common elements, oxygen (O) has an atomic number of 8, carbon (C) has an atomic number of 6, and nitrogen (N) has an atomic number of 7. Atoms can join together, and when this occurs they form a **molecule**, so if two hydrogen atoms join together, forming H_2, it

Table 3.1 Examples of common elements found in the body

Element	Atomic symbol	Percentage of body mass	Comment
oxygen	O	65	found in water and air, needed for respiration
carbon	C	18	found in organic molecules
hydrogen	H	9.5	found in water and organic molecules
nitrogen	N	3.2	found in proteins, DNA and RNA
calcium	Ca	1.9	found in bones and teeth
phosphorus	P	1.0	found in ATP (energy-containing chemical in cells) and in bones and teeth
potassium	K	0.4	found inside cells
sodium	Na	0.2	found in blood
chlorine	Cl	0.2	found in blood
magnesium	Mg	0.1	found in enzymes
iron	Fe	0.1	found in haemoglobin in red blood cells

is called a **molecule**. Atoms with the same number of protons but a different number of neutrons are called **isotopes**.

Returning now to water, the chemical formula is H_2O, which tells us that there are two hydrogen atoms combining with one oxygen atom. But how do these combine together? These atoms are able to combine by sharing electrons. The electron orbiting each hydrogen atom joins with the six electrons orbiting the oxygen atom. The result of the chemical combination of atoms is called the formation of **compounds**.

Chemical bonds and reactions

Reactions occur when atoms combine or dissolve. This is achieved by transferring or sharing the outer shell electrons. There is a maximum of electrons that each shell can contain. There are only two spaces in the inner shell, so the maximum number of electrons found in the inner shell is two. The next shell can contain up to eight electrons, and the shell after that can contain 18 electrons.

It is only the outer shell that is important in reactions, because it is here that electrons can be shared and molecules or compounds formed. If we take salt as an example, the chemical formula is NaCl, which tells us that salt is formed from Na and Cl as shown in Figure 3.2. As we can see, there is only a

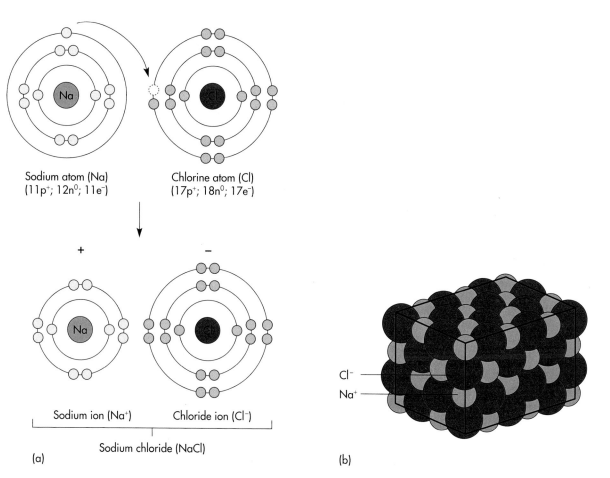

Sodium atom (Na)
($11p^+$; $12n^0$; $11e^-$)

Chlorine atom (Cl)
($17p^+$; $18n^0$; $17e^-$)

Sodium ion (Na$^+$)   Chloride ion (Cl$^-$)

Sodium chloride (NaCl)

(a)

Cl$^-$
Na$^+$

(b)

Figure 3.2 Ionic bond formation. (a) Sodium loses 1 electron and chlorine gains 1 electron. This stabilises both atoms. (b) Salt (NaCl) crystals are formed as substantial numbers of Na$^+$ and Cl$^-$ associate.

single electron in the outer shell of the Na atom but seven electrons in the chlorine atom. The sodium electron is attracted to the chlorine atom, and forms NaCl, or salt. This is an example of an **ionic** bond. The outer electron from the Na is *transferred* to complete the outer Cl shell.

Types of chemical bonds

Learning hint

Bonds in salt

Take some table salt, chemical formula NaCl. Look at the crystals formed – they are white and grainy. Now, try and crush the salt – it should be possible to make the compound smaller, but if you taste it, it retains its characteristic taste. Now, take another sample of NaCl and drop it into a glass of clear water. Wait a few minutes and sip the water – it should taste salty. What these experiments demonstrate is the durability of ionic compounds.

There are three types of chemical bond: ionic, covalent and hydrogen.

Ionic bonds form whenever one atom, or group of atoms, *transfers* electrons to another atom or group of atoms. The resulting compound has some specific properties. Ionic compounds have high melting and boiling points, are soluble in water, conduct electricity when in solution, and can shatter easily. Sodium transferring one electron to chloride to form sodium chloride is an example of an ionic bond.

The next type of bond is a **covalent** bond. These atoms *share* electrons from their outer shells. There are some specific properties of covalent bonds too; they have a low melting and boiling point, they don't mix well with water molecules, and they do not conduct electricity. When two hydrogen atoms share an electron this forms a molecule of hydrogen gas.

Finally we have **hydrogen** bonds. These bonds are not very strong, so they break apart quite easily. A hydrogen atom forms a bridge between atoms that can easily fill their electron shells. A good example of hydrogen bonding is water, although proteins and DNA are held together by hydrogen bonding too.

Throughout this book, you will find lots of examples of chemical bonding and reactions. The most common reactions that you will see are in the ionic compounds and, as we have seen, ionic compounds can conduct electricity, and so we also call these compounds **electrolytes**.

Electrolytes

Electrolytes are found in body fluids, which is a water-based solution. Water is an example of a **solvent**, meaning that it can hold another substance in a solution form. This 'other' substance is called a **solute**, meaning a substance that is dissolved or suspended in a solution. Approximately 60–70 per cent of our body weight is made up of water, which is divided into water found inside cells, called **intracellular fluid** or **ICF**, and water found outside the cells, which is called **extracellular fluid** or **ECF**. Extracellular fluid is further subdivided into fluid in the blood (called **plasma**) and fluid that bathes or surrounds cells (called **interstitial fluid**).

Table 3.2 Common cations and anions

Cations	Normal values in the blood and locations
sodium (Na^+)	135–145 mmol/l main extracellular fluid cation
potassium (K^+)	3.5–5.0 mmol/l main intracellular fluid cation
calcium (Ca^{2+})	2.2–2.6 mmol/l (in the blood) found in both fluid compartments
magnesium (Mg^{2+})	0.7–1.0 mmol/l (in the blood) found in both fluid compartments
hydrogen (H^+)	found in both fluid compartments (measured using the pH scale)
Anions	
chloride (Cl^-)	100–108 mmol/l main anion in extracellular fluid
phosphorus (P^-)	2.5–4.5 mmol/l main anion in intracellular fluid
bicarbonate (HCO_3^-)	22–28 mmol/l found in both fluid compartments

When the electrolytes are dissolved in water, ions are formed. Positively charged ions are termed **cations**, whereas negatively charged ions are called **anions**. Examples of the common cations and anions are shown in Table 3.2.

Acids and bases

Acids and bases are also electrolytes because they can conduct electricity and break down (dissociate) in water. You are probably familiar with the term *acid* and know that it is quite harmful to other substances. Acids can dissolve metals and literally burn a hole through material (*don't* try this at home!). The definition of an acid is something that can *release* hydrogen ions. Acids taste sour. Acids dissolved in water release hydrogen ions that can easily react with other atoms. For example, if we take hydrochloric acid, which is found in the stomach, the chemical formula is HCl, meaning one hydrogen atom and one chlorine atom. If we break the compound HCl apart, the released hydrogen atom can react with other atoms and potentially harm the body.

Bases, which have a bitter taste, can *accept* hydrogen ions; for example, bicarbonate (HCO_3^-) can accept hydrogen ions, forming carbonic acid (H_2CO_3), which is a weak acid. Hydroxides are common bases, which accept spare hydrogen ions.

The concentration or amount of hydrogen and hydroxides are measured using the pH scale. The pH scale is from 0 to 14, where a value between 0 and 6.9 means there are more hydrogen ions compared to hydroxide or hydroxyl ions, and so it is said to be acidic. Neutral pH, where there are the same number of hydrogen and hydroxyl ions, is 7. A pH greater than (>) 7 indicates that there are more hydroxyl ions, and therefore it is said to be alkali. To give some examples, the pH of blood is between 7.35 and 7.45; bleach has a pH of 12 and vinegar has a pH of 3. This means that our bodies are exposed to potentially harmful substances, and therefore we must have systems to protect the body as much as possible from changes in pH (see Figure 3.3).

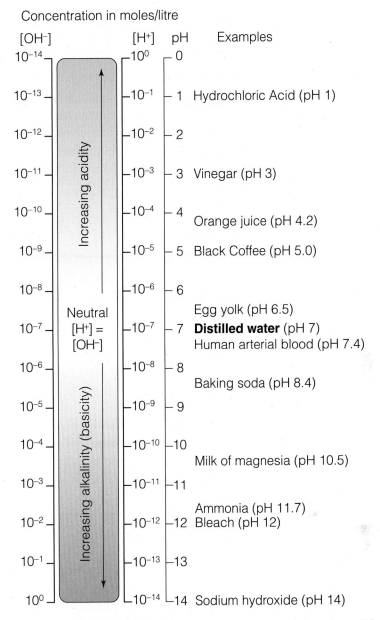

Figure 3.3 The pH scale and pH values of representative substances. The pH scale is based on the number of hydrogen ions in solution. The actual concentration (expressed in moles per litre) of hydrogen ions [H⁺] and the corresponding hydroxyl ion concentration [OH⁻] are indicated for each pH value noted. At a pH of 7, [H⁺] = [OH⁻], and the solution is neutral. A solution with a pH below 7 is acidic; above 7, basic or alkaline.

The systems that help regulate the acid/base balance include the respiratory and renal systems. The role of the respiratory system (discussed in more detail later) is to take in air and get rid of carbon dioxide (CO_2). Carbon dioxide is a weak acid. If we stop breathing, there is a build-up of CO_2 in the body and this build-up of acid is detected in the brain (the cardio-respiratory centre), which gives you an overwhelming desire to breathe. This is the simplest way of getting rid of excess acid, but if there is a problem with the respiratory system, the kidneys come to the rescue and enable excess acid to be lost in the urine.

Organic compounds

We have discussed how atoms form together to make compounds which allow certain processes in the body to work, but we now need to consider where the body gets its energy from. When we eat food a whole series of chemical steps are needed to break down the food into smaller and smaller parts, but we also need to break down the chemical bonds that make up the food too. We will talk in more detail about the gastrointestinal system later in the book, but the food we eat is in different forms; for example, carbohydrates, lipids (fats) and proteins.

Carbohydrates

These are sugars and starches that contain carbon, hydrogen and oxygen. There are twice as many hydrogen atoms as carbon and oxygen. Examples of carbohydrates are glucose (formula $C_6H_{12}O_6$) and ribose (formula $C_5H_{10}O_5$). Carbohydrates are classified according to size and so can be monosaccharides, disaccharides or polysaccharides.

Monosaccharide means one (*mono* meaning one) sugar (*saccharide*) and so monosaccharides are also called simple sugars. The most important simple sugar is glucose, which is carried in the blood and utilised by cells (in the presence of insulin).

Disaccharides are double sugars (*di* meaning two), which are joined together by a mechanism called dehydration synthesis, which basically means that a water molecule is lost. Essential disaccharides are sucrose, lactose and maltose. Sucrose is a combination of glucose and fructose and is found in cane sugar; lactose is a combination of glucose and galactose and is found in milk; maltose is a combination of glucose and glucose, and is found in malt sugar. These disaccharides are too big to get into cells, unlike glucose, and so they must be broken apart. This is achieved by a process called **hydrolysis**, the addition of a water molecule to the bond.

Polysaccharides, meaning many sugars, are long groups of simple sugars. These are large molecules that are insoluble. The two main polysaccharides are starch and glycogen. We get starch from food such as potatoes and root vegetables, and glycogen from animal tissues.

Lipids

Lipids are found in fat in various forms. The main types of lipids are **triglycerides, phospholipids** and **steroids**. Triglycerides are also called neutral fats, and contain fatty acids and glycerol. The function of triglycerides in the body is insulation and protection, cushioning organs. They also are a major source of stored energy and are strongly water-phobic. Phospholipids are similar to triglycerides; their function in the body is formation of the cell membrane, which serves to separate the two body fluid compartments. Steroids, our final group of lipids, are structurally different from other lipids. An example of a steroid is **cholesterol**. Cholesterol helps to stabilise the cell membrane, but is also used: in the production of vitamin D, which is produced in the skin following exposure to sunlight; in oestrogen, progesterone and testosterone formation;

and bile salts, which are made from the breakdown of cholesterol, and help the gut to digest and absorb fat.

Proteins

Almost half of the organic compounds in the body are proteins. Proteins have many functions in the body, and are formed by amino acids. Amino acids are joined together to make larger molecules, each having a specific role; for example, haemoglobin, the molecule that helps to carry oxygen and carbon dioxide, is a protein molecule. There are two main types of proteins found in the body: **structural** (for example connective tissue fibres found in the skin) and **functional** (for example, haemoglobin). Other proteins that are formed can speed up chemical reactions. These proteins are called **enzymes**, and an example of an enzyme is angiotensin-converting enzyme (abbreviated to ACE) which helps to narrow arteries and therefore has a role in the control of blood pressure.

Nucleic acids

These compounds are made up of carbon, oxygen, hydrogen, nitrogen and phosphorus atoms, and are formed from **nucleotides**. A nucleotide has three parts: a nitrogen-containing base, a pentose sugar, and a phosphate group (see Figure 3.4).

There are five different types of base: adenine (A), guanine (G), cytosine (C), thymine (T) and uracil (U). The two main types of nucleic acids are **deoxyribonucleic acid (DNA)** and **ribonucleic acid (RNA)**. DNA is the genetic material contained in the nucleus of your cells, and RNA is found outside the nucleus. Your DNA is replicated each time the cell divides and is the control centre, issuing instructions on how to build proteins. RNA builds the proteins. DNA is formed from the bases A, G, T and C, whereas RNA is formed from G, C, U and A. The bases that form DNA and RNA are linked in a specific way. Basically, in the case of DNA, G only ever combines with C, or C with G, whereas T only combines with A. For RNA, C combines with G as usual, but RNA doesn't have thymine but has uracil, so adenine combines with uracil (the easy way to remember this is as 'uncle and auntie' for the base pairs uracil and adenine in RNA).

There are three types of RNA – messenger, ribosomal and transfer. Messenger RNA carries information for building proteins from the instructions given by the DNA to ribosomes. Transfer RNA takes amino acids to the ribosomes, and ribosomal RNA actually forms part of the ribosomes which is where proteins are made inside cells.

Adenosine triphosphate (ATP)

We've said already that glucose is a vital source of energy to the body and that insulin is needed to get the glucose out of the body and into the cells. But then what? Glucose cannot be used by cells in its present state, and needs to be broken down and stored in a different way. This is where ATP comes in. ATP is a modified nucleotide that has some high energy phosphate bonds. As these bonds are broken, the energy released can be used by the cell to do a particular activity (see Applied Science box on p. 68).

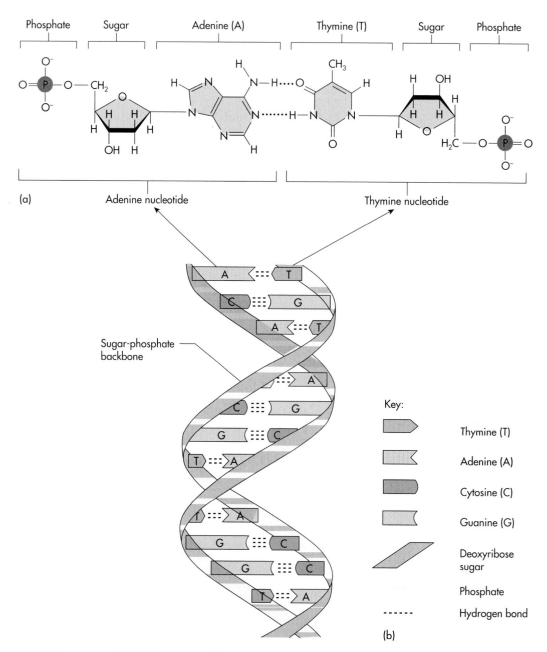

Figure 3.4 DNA Structure. (a) Adenine and thymine are 2 of the 4 nucleotides that make up DNA. Each is made up of a base attached to a deoxyribose sugar molecule which is linked to a phosphate group. The two nucleotides are joined together (via their bases) by hydrogen bonds. (b) DNA is a coiled structure made up of repeating nucleotides which form rungs suspended from the backbone of DNA. The backbone is made up of alternate phosphate and sugar molecules. The nucleotides are bound together by hydrogen bonds – adenine (A) with thymine (T) and guanine (G) with cytosine (C).

Summary

- Biochemistry is the study of how chemicals are used in the body.
- Atoms have protons, neutrons and electrons and can share or transfer electrons from their outer shells to form compounds.
- There are three types of chemical bond: ionic, covalent and hydrogen.
- Electrolytes, positively or negatively charged atoms, are found in the body and help in conducting electricity, as well as maintaining the separation of the fluid compartments in the body.
- pH is the scale by which the number of hydrogen ions and hydroxyl ions are measured. A neutral pH is 7.
- Organic compounds provide the body with the vital nutrients. These include carbohydrates, lipids and proteins.
- Proteins are built from amino acids, and types of proteins include enzymes.
- Nucleic acids are composed of oxygen, carbon, hydrogen, nitrogen and phosphorus, and are formed from nucleotides. The five bases in nucleotides are adenine, guanine, cytosine, thymine and uracil.

Review questions

Multiple choice

1. What is the chemical symbol for sodium?
   a. S
   b. Na
   c. So
   d. Cl

2. Which definition best describes a covalent bond?
   a. the bonds are weak and have high melting points
   b. they mix well with water
   c. they share electrons and have low melting points
   d. they form a bridge between atoms

3. What are the main types of lipid?
   a. Monosaccharides and polysaccharides
   b. Disaccharides and polysaccharides
   c. Triglycerides and steroids
   d. Carbohydrates and glucose

4. Which of the following is a structural protein?
   a. collagen
   b. enzymes
   c. antibodies
   d. steroids

5. What is the pH of the blood?
   a. 7.2–7.3
   b. 8.35–8.45
   c. 7.35–7.45
   d. 8.2–8.3

6. Which of the following combination of monosaccharides is sucrose?
   a. glucose and glucose
   b. glucose and galactose
   c. glucose and fructose
   d. galactose and fructose

Answers

Answers to multiple choice questions

1. b
2. c
3. c

4. a
5. c
6. c

The cells

Just as a town is a complex combination of structures and systems, and a brick or cement block is a basic structure with which these buildings are constructed, so the human body is a complex combination of structures and systems, and a cell is the basic building block with which the body is built. Just as there are different types, shapes and sizes of building blocks, there are different types, shapes and sizes of cells. Blood cells, skin cells, nerve cells, and so on, are all different from each other. We will learn more about each kind of cell in later chapters.

Although a cement block is a basic building block, it could not exist without sand, lime and water, the components necessary to make cement. As this chapter explains, cells also consist of component parts, tiny cell structures called organelles that are needed to perform specific functions to keep the cell alive.

On a more involved level, cells of a similar type form tissue that functions together in an organ, and organs perform specific functions to create a system. For example, cardiac cells form heart tissues, which form the heart organ, which is part of the circulatory system. Finally, systems work together to form a functioning human body! When you think about it, we are very much like a town. Towns need transportation systems, control systems, systems to import food and water and export waste, and heating and cooling systems.

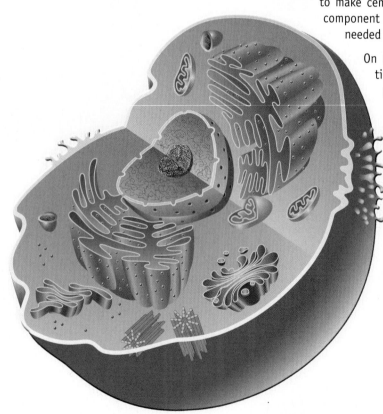

Learning objectives

At the end of this chapter, you will be able to:

◆ List and describe the various parts of a cell
◆ Explain the function of each organelle found within the cell
◆ Explain the process of cellular mitosis
◆ Describe the types of active and passive transport within cells
◆ Describe the structures required for cell motility
◆ Differentiate between bacteria, viruses, fungi and protozoa

Pronunciation guide

benign (*ben INE*)
capsid (*CAP sid*)
centrioles (*SEN tree yollz*)
centrosomes (*SEN troh soams*)
chromatin (*CROW mah tin*)
cilia (*SILL ee yah*)
cytoplasm (*SIGH toe plazm*)
deoxyribonucleic acid (*dee OX ee RYE bow new KLEE ik*)
endocytosis (*EN doe sigh TOE siss*)
endoplasmic reticulum (*EN doh PLAZ*

mik ree TICK yoo lum)
exocytosis (*EX oh sigh TOE siss*)
flagella (*flah JELL ah*)
fungi (*FUN ghee*)
Golgi apparatus (*GOAL ghee app ah RAH tuss*)
lysosomes (*LIE soh soams*)
malignant (*mah LIG nant*)
metastasis (*meh TASS tah siss*)
mitochondria (*my toe CON dree yah*)
mycelia (*my SEE lee yah*)

organelles (*ore gah NELLZ*)
osmosis (*oz MOW siss*)
phagocytosis (*FAG oh sigh TOE siss*)
pinocytosis (*PIN oh sigh TOE siss*)
protozoa (*pro toe ZOE wah*)
ribonucleic acid (*RYE bow new KLEE ik*)
ribosomes (*RYE bow soams*)
vesicle (*VEE sickle*)

Overview of cells

Cells are units formed from chemicals and structures and are found in all living things. Some organisms are composed of only a single cell. Practically all of the cells in our body are microscopic in size, ranging from about one third to one thirteenth of the size of the dot in this full stop. What is amazing is that certain cells, for example nerve cells, can be half a metre in length or longer. When we refer to cells as the 'building blocks' of our bodies, we immediately think of brick-shaped objects. However, cells can be flat, round, thread-like or irregularly shaped. While the approximately *7.5 trillion* cells found in the human body vary in size, shape and purpose, they normally work together to allow for proper functioning of processes necessary for life, such as digestion, respiration, reproduction, movement and production of heat and energy. Figure 4.1 represents a typical cell with its major components. We now discuss the individual components.

Applied science

Atoms and molecules

Although cells are composed of small structures called organelles, it is important to note that these organelles are composed of even smaller substances. Atoms, which are the tiny building blocks of all matter, combine to form molecules such as water, sugar and proteins, which are then used to build cellular structures and facilitate cellular functions.

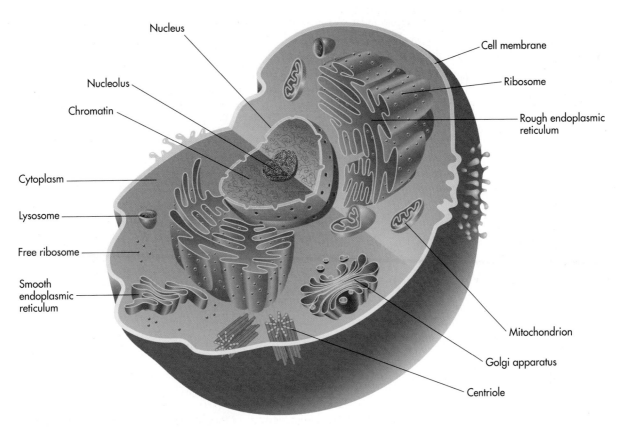

Figure 4.1 Cellular components.

Cell structure

Even though cells in our bodies can vary greatly in size, shape and function, they share certain common traits. As previously stated, we consider cells the basic building blocks of the human body. However, to better understand them, let's look at cells as miniature towns with a variety of systems, structures and organisations that are necessary for proper function. Almost all human cells possess a nucleus (except mature red blood cells), organelles, cytoplasm and a cell membrane. Each component of a cell has a special purpose.

Cell membrane

We can think of the **cell membrane** as the town boundary. This is a defined boundary that possesses a definite shape and actually holds the cell contents together. The cell membrane acts as a protective covering. For a town to thrive, people and materials must be able to travel in and out of the town. A cell membrane is responsible for allowing materials in and out of the cell. What is interesting is that the membrane allows only certain things into or out of the cell. Because it chooses what may pass through, we call it a *selectively permeable* (or *semi-permeable*) membrane.

In addition, the cell membrane has identification markers on it to show that it comes from a certain person, much like GB is an 'identification marker'

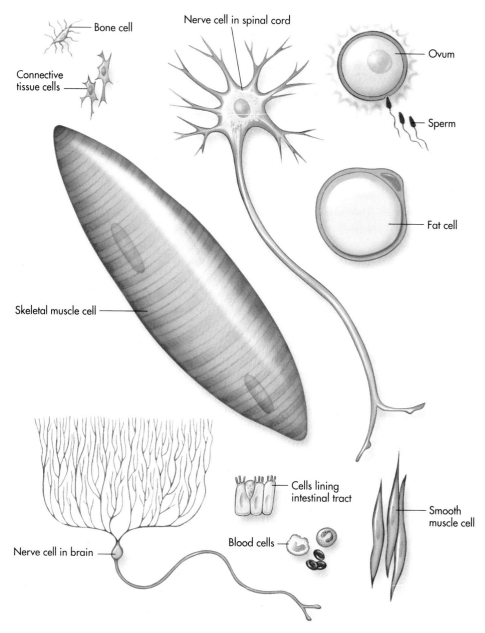

Figure 4.2 Various types of cells within the human body.

on a car that tells us that the car is from Great Britain. If a foreign cell shows up in an individual (such as in a transplanted organ), the body signals an attack on that cell or group of cells. For all that it is responsible for, the cell membrane is only 5×10^{-7} millimetres thick. Each cell, regardless of its shape or function, must have a cell membrane in order to maintain its integrity and survive. See Figure 4.2, which shows examples of various cell types found within the human body.

Transport methods

If we think of the cell membrane as the town boundary, we need to discuss how things get across this boundary. It can be done in two broad ways. **Passive**

Learning hint

Concentration gradient

If a substance moves from higher concentration to lower concentration, it is said to be moving *with* the concentration gradient (difference). For example, if you are in a long queue in the supermarket, the queue is 'concentrated' with people, then three checkouts open so the people will quickly move from the highly concentrated line to the lower concentrated areas (the new open checkouts). Soon, you will notice, all the lines will equalise, demonstrating that even people move with a concentration gradient and reach equilibrium.

transport and **active transport** describe the movement of materials across the cell membrane. *Passive transport*, as the words suggest, requires no extra form of energy to complete. It is similar to having a yo-yo in your hand and simply letting go. The yo-yo simply falls as it unwinds from the string. No input of energy is needed. *Active transport* requires some addition of energy to make it happen. Throwing your yo-yo requires the addition of energy from your arm to make it happen.

Passive transport

Passive transport can be further divided into four types:

- diffusion
- facilitated diffusion
- osmosis
- filtration

Diffusion Diffusion is the most common means of passive transport by which a substance of higher concentration travels to an area of lesser concentration. The difference between these two concentrations is called the concentration gradient. This is like putting a packet of powdered drink mix into a jug of water. The water gradually assumes the colour and flavour of the powder until the entire contents of the container are the same colour and taste (nature likes a nice, equal balance!). Another example may be one of your classmates overusing perfume or cologne. Once in the classroom, the smell diffuses from the high concentration on the individual to the low concentration throughout the classroom.

Diffusion is necessary in the transportation of oxygen from the lungs and into the blood. It is also needed to transport the waste (carbon dioxide) from the blood to the lungs and eventually out into the air. This vital process is further discussed in Chapter 15, 'The respiratory system'. See Figure 4.3 for examples of diffusion.

Facilitated diffusion Facilitated diffusion is a variation of diffusion in which a substance is helped in moving across the membrane, in a similar way to an usher helping you to your seat at a cinema. Glucose is the substance that is often helped in the body. You may wonder if it is helped, how can it be considered *passive* transport? It can best be thought of as a situation in which the glucose was already moving in an attempt to cross the membrane and conveniently encounters an already revolving door. Once it steps into the door, it is quickly 'pushed' along until it comes out on the other side of the membrane. See Figure 4.4, which illustrates facilitated diffusion.

Figure 4.3 Two examples of diffusion.

Figure 4.4 Facilitated diffusion.

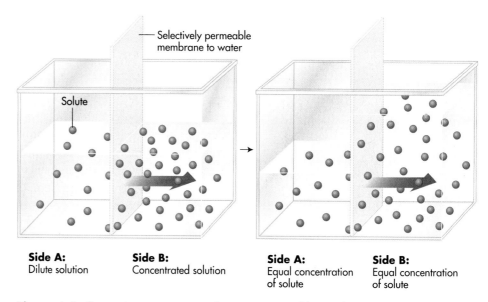

Figure 4.5 Osmosis: water moves from an area of low solute concentration to an area of higher solute concentration.

Osmosis Osmosis is another form of passive transport in which water travels through a selectively permeable membrane to equalise concentrations of a substance. The substance that is dissolved in the water is called the *solute*. Remember that nature likes things balanced and equal. Water tends to travel across a membrane from areas that have a low concentration of a solute to areas that have a higher concentration of the solute until the concentration is the same on both sides of the membrane. Keep in mind that the water is moving with *its* concentration gradient. See Figure 4.5 for a visual description. This ability of a substance to 'pull' water toward an area of higher concentration of the solute is called **osmotic pressure**. The greater the concentration of the solute, the greater the osmotic pressure it exerts to bring in water.

Filtration Filtration is the final member of the passive transport group. This method differs from osmosis in that pressure is applied to force water and its

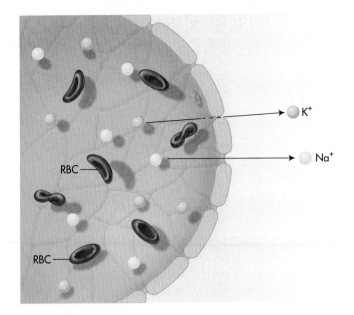

Figure 4.6 The process of filtration in the kidneys, where smaller solutes such as the electrolytes sodium and potassium pass through the membrane, while the larger red blood cells normally do not.

dissolved materials across a membrane. Filtration caused by pressure is similar to a rush of people being pushed through the turnstiles during rush hour or the effect you get when you squeeze the trigger on a water pistol. The major supplier of force in the body is the pumping of the heart, which forces blood flow into the kidneys where filtration takes place. This concept is expanded in Chapter 17, 'The urinary system'. For now, see Figure 4.6, which illustrates the process of filtration.

Test your knowledge 4.1

Fill in the blanks:

1. This form of passive transportation is like combining drink mix and a jug of water: _____.

2. When considering osmosis, water travels across a semi-permeable membrane from an area of _____ concentration of solute to an area of _____ concentration of solute.

3. _____ allows only certain sizes of particles to pass through.

4. The act of removing carbon dioxide from the blood to the lungs is achieved through _____.

5. Glucose is often transported by _____.

Active transport

Active transport can be broken down into three types:

- active transport pumps
- endocytosis
- exocytosis

Active transport pumps Active transport pumps require the addition of energy in the form of an energy molecule called adenosine triphosphate, or ATP (which we will soon discuss) to move a substance. Energy is needed because the cell is trying to move a substance into an area that already has a high concentration of that substance. It's like trying to put 600 grams of sugar into a 500-gram bag! It can be done, but you have to apply a lot of pressure (energy). A common example in our cells is the need to transport potassium (K). Cells contain a good amount of potassium. The only way to get more into a cell is to apply energy to 'push' it in.

endocytosis (*EN doe sigh TOE siss*)
   endo = *within*
   cyt = *cell*
   osis = *condition*
   phago = *eating*
   pino = *to drink*

Endocytosis Endocytosis is utilised by the cells for the *intake* of liquid and food when the substance is too large to diffuse across the cell membrane. The cell membrane actually surrounds the substance with a small portion of its membrane, forming a chamber, or **vesicle**, which then separates from the membrane and moves into the cell. If it is a solid particle being transported, we call it **phagocytosis**. That is also the term that describes what white blood cells do to bacteria to prevent infections in our bodies. If the intake involves water, it is called **pinocytosis**.

exocytosis (*EX oh sigh TOE siss*)
   exo = *outside*

Exocytosis In some situations, the cell needs to transport substances *out* of itself. This is called **exocytosis**. Some cells may produce a substance needed outside the cell. Once this substance is made, it is surrounded by a membrane forming a vesicle (bladder or sac) and moves to the cell membrane. This vesicle becomes a part of the cell membrane and expels its load out of the cell. For further explanation of active transport, see Figure 4.8.

Please see Table 4.1, which puts all of the methods of transport together for your viewing pleasure.

In order for a group of cells to work together effectively as a tissue, the cells need to be anchored to each other. This is achieved by the presence of membrane junctions. There are three types (see Figure 4.7):

- Tight junctions – impermeable and bind cells together to prevent substances from passing through the spaces between the cells. Found in cells in the small intestine to prevent enzymes leaking into the blood.
- Desmosomes – anchoring junctions. These prevent cells from being pulled apart from each other. This is important in cells that are exposed to mechanical stress, for example skin cells.
- Gap junctions – found in the heart. These allow chemicals to pass from one cell to another, particularly nutrients and ions.

Test your knowledge 4.2

Answer the following:

1. Differentiate between phagocytosis and pinocytosis.

2. Identify whether the following processes are active or passive:

   a. endocytosis _____

   b. facilitated diffusion _____

   c. osmosis _____

   d. phagocytosis _____

   e. filtration _____

Figure 4.7 Intercellular connections.

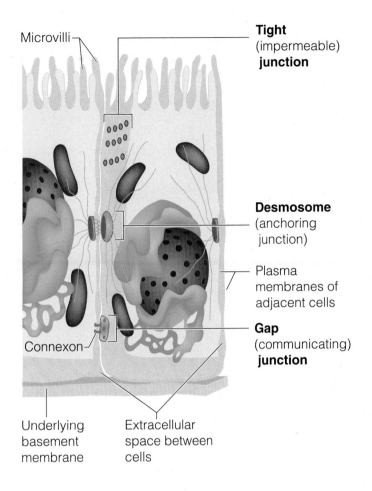

Microvilli

Tight (impermeable) **junction**

Desmosome (anchoring junction)

Plasma membranes of adjacent cells

Gap (communicating) **junction**

Connexon

Underlying basement membrane

Extracellular space between cells

Table 4.1 Methods of cellular transportation

Cellular transportation methods	Description
Passive transport (no energy required)	
diffusion	moving a substance from an area of high concentration to an area of low concentration
facilitated diffusion	substance is assisted via 'revolving door' in a direction it was already travelling from an area of high concentration to an area of low concentration
osmosis	water travels across a membrane from areas that have a low concentration of a solute to areas that have a higher concentration of solute until the concentration is the same on both sides of the membrane
filtration	pressure is applied to force water and dissolved materials across a membrane
Active transport (energy required)	
active transport pumps	require additional energy in the form of ATP to move substances against the concentration gradient (from low concentration to high concentration)
endocytosis	ingesting substances that are too large to diffuse across the cell membrane
phagocytosis	form of endocytosis in which *solid* particles are being brought into the cell via vesicles
pinocytosis	form of endocytosis in which *liquid* is being brought into the cell via vesicles
exocytosis	transportation of material outside of the cell

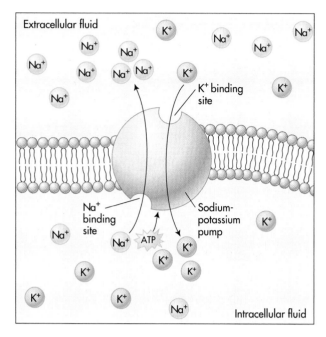

ACTIVE TRANSPORT PUMP

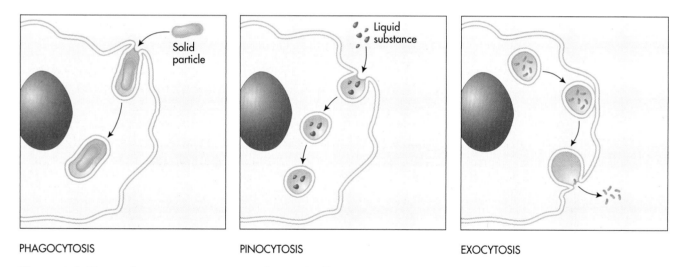

PHAGOCYTOSIS   PINOCYTOSIS   EXOCYTOSIS

Figure 4.8 Types of active transport in and out of cells.

Cytoplasm

Living organisms require balanced environments in which to thrive. Humans require the right mixture of oxygen and nitrogen; sea creatures require the right balance of salt and water; a chick embryo requires albumen, or 'egg white', in which to develop. Likewise, the internal parts of a cell require a special environment, called **cytoplasm**, in order to survive.

Nucleus and nucleolus

The nucleus has been described as the 'brain of the cell'. In our case, we will consider it to be the town hall: a control centre. A town hall dictates the

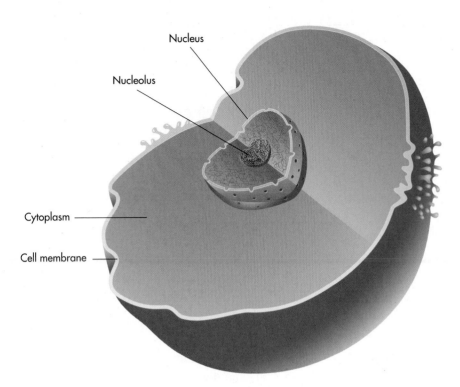

Figure 4.9 The cell membrane, cytoplasm, nucleus and nucleolus.

activity of the town departments much as the nucleus dictates the activities of the organelles in the cell. Since the town hall is crucial to the town's function, security is important to protect it from attack or damage. That is why metal detectors are installed at the doors. The nucleus of a cell is similar. It is surrounded by a double-walled nuclear membrane. Even though this membrane is composed of two layers, it has large pores that allow certain materials to pass in and out of the nucleus.

Somewhere in the town hall are blueprints of the town showing the buildings, streets, water and gas pipes, and so forth. The nucleus also contains 'blueprints' – and 'building codes' too. **Chromatin** is the material found in the nucleus that contains DNA (**deoxyribonucleic acid**). DNA contains the specifications (blueprints) for the creation of new cells. Chromatin eventually forms chromosomes, which contain **genes**. Genes determine our inherited characteristics (do you remember your grandparents saying how much you look like your mother or father?).

A spherical body made up of dense fibres called the **nucleolus** is found within the cell nucleus. Its major function is to synthesise the **ribonucleic acid (RNA)** that forms ribosomes. Now we have the blueprints, but what about the materials that we need? Who's in charge of getting these materials together? This is where **ribosomes** and **centrosomes** come into play. Figure 4.9 shows the cell membrane, cytoplasm, nucleus and nucleolus. We will continue to rebuild the cell's infrastructure as we discuss their specific functions.

Ribosomes

Ribosomes are organelles that are found on the endoplasmic reticulum or floating around in the cytoplasm. Ribosomes are made of RNA and assist in the production of enzymes and other proteins that are needed for cell repair and reproduction. Following our analogy, ribosomes can be considered to be building material suppliers for remodelling and repair.

Centrosomes

In a town, completely new structures are often needed to replace old ones. Therefore, the cell needs a building contractor to build new structures. The **centrosomes** fill this need. Centrosomes contain centrioles that are involved in the division of the cell. Cellular division, or reproduction, will be discussed once we cover all the cell parts. Centrioles are tubular-shaped and usually found in pairs. See Figure 4.10, which now adds the ribosomes, centrosomes and mitochondria to the cell.

Mitochondria

Could you possibly imagine what it would be like if we had no electricity in our town? Things would come to a standstill. The mitochondria, tiny bean-shaped organelles, act as power plants to provide up to 95 per cent of the body's energy needs for cellular repair, movement and reproduction. As a town's need for power increases, more power plants are built. Similarly, if a given cell type is very active and needs more power, there are a larger number of mitochondria in that cell. Liver cells, which are quite active, can have up to 2000 mitochondria in each cell. Sperm cells 'swim' with a tail (flagellum), and have mitochondria coiled around the tail for energy. Special enzymes in the mitochondria help to take in oxygen and use it to produce energy. While

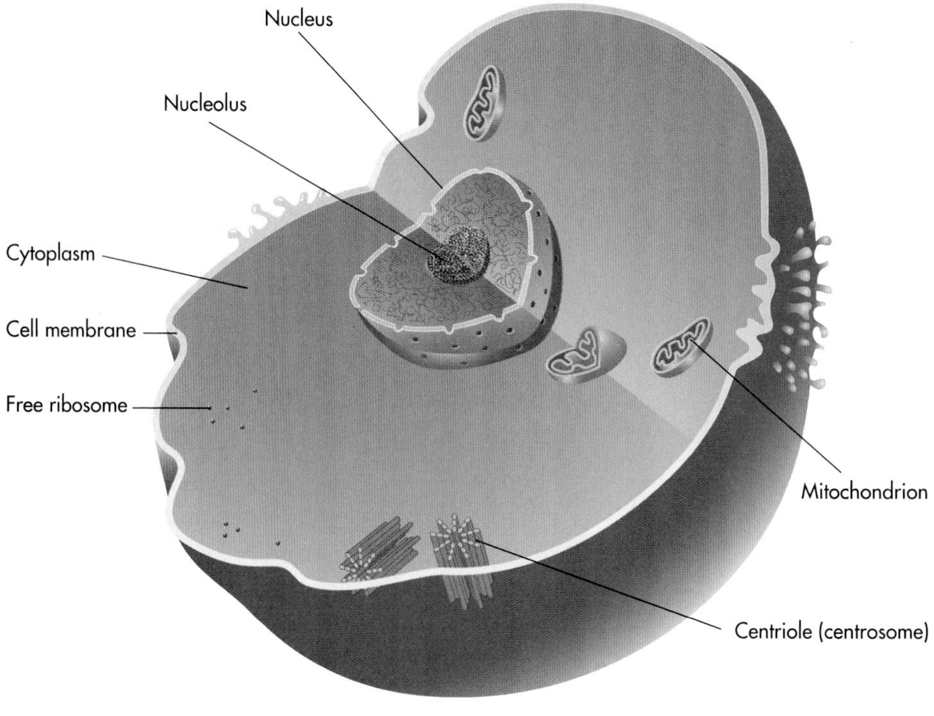

centrosomes (*SEN troh soams*)
centrioles (*SEN tree yollz*)

mitochondria (*my toe CON dree yah*)

Figure 4.10 The cell membrane, cytoplasm, nucleus, nucleolus, ribosomes, centrosomes and mitochondria.

our town uses electricity for most of its power, the cell uses adenosine triphosphate (ATP), which is created by the mitochondria (see Figure 4.10).

Endoplasmic reticulum

endoplasmic reticulum (*EN doh PLAZ mik ree TICK yoo lum*)

Although there are paths, walkways and pavements in our town, the main structure for travel is the road system. The endoplasmic reticulum is a series of channels set up in the cytoplasm that are formed from folded membranes. The endoplasmic reticulum has two distinct forms. One has a sandpaper-like surface, the result of ribosomes on its surface, which we call the *rough* endoplasmic reticulum, and it is responsible for the synthesis of protein. Once the protein is synthesised, it is sent to the Golgi apparatus for processing. The second form has no ribosomes on its surface, making it appear smooth, so we call it the *smooth* endoplasmic reticulum (complex stuff!). The smooth endoplasmic reticulum synthesises lipids (fats) and steroids. Think of this as a series of gravel and paved roads with butcher shops and food processing plants along the way (see Figure 4.11).

Golgi apparatus

Golgi apparatus (*GOAL ghee app ah RAH tuss*)

Towns have factories with their own fleets of lorries. The Golgi apparatus is very similar to these factories. This organelle looks like a bunch of flattened, membranous sacs. Once the Golgi apparatus receives protein from the endoplasmic reticulum, it further processes and stores it as a shippable product, much like a packaging plant does. Not only does it prepare the protein for shipping, a part of the Golgi apparatus surrounds the protein that then

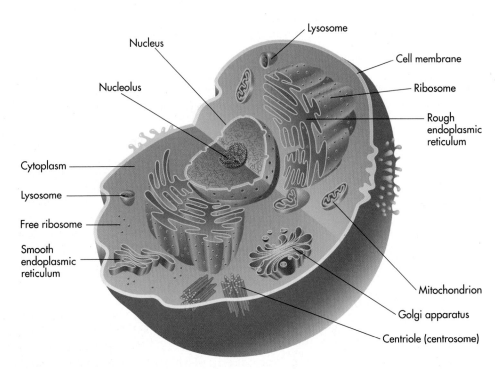

Figure 4.11 The cell membrane, cytoplasm, nucleus, nucleolus, ribosomes, centrosomes, mitochondria, endoplasmic reticulum, Golgi apparatus and lysosomes.

separates itself from the main body of this organelle. That portion, with its load, then travels to the cell membrane where it releases (secretes) the protein! This is an actual example of exocytosis. Cells that constitute organs with a high level of secretion or storage (like the digestive system) contain higher numbers of the Golgi apparatus. Salivary glands and pancreatic glands, for example, are made of cells containing many Golgi apparati.

Lysosomes

lysis = *to break down*

All towns create waste that must be removed. Cells are no different. **Lysosomes** are vesicles containing powerful enzymes that take care of cleaning up intracellular debris and other waste. Lysosomes are multi-talented. They also aid in maintaining health by destroying unwanted bacteria through the process of phagocytosis. See Figure 4.11, which contains all the organelles.

Other interesting parts

cytoskeleton (*SIGH toe SKELL et on*)

Similar in some ways to the skeleton, the cytoskeleton is a network of microtubules and interconnected filaments that provide shape to the cell and allow the cell and its contents to be mobile.

vesicle (*VEE sickle*)

Vesicles, often created by the Golgi apparatus, can be thought of as small lorries. They can be loaded up with substances, travel to another site in the cell or cell membrane, and then drop off their loads.

flagella (*flah JELL ah*)

Suppose we could build towns that float on large bodies of water. How would we propel them to new locations? Certain cells can solve this problem through the use of flagella. Flagella are whip-shaped tails that move some cells in a fashion similar to that of a tadpole.

The future may offer new ways of transporting people and materials. For example, perhaps towns will install moving conveyor belts, similar to those in

---

Applied science

Cell energy and ATP – the energy molecule

We all know that we need to eat to obtain energy, but how does energy get from food to cells? In simple terms, the body takes in food and breaks it down (digestion). During this process, energy is released from the food. Now, the problem is that cells can't use this energy directly. Only food converted to glucose (simple sugar) can be used to make energy. Glucose can be used by your cells during a series of chemical reactions called *cellular respiration*. During cellular respiration, glucose is combined with oxygen and is transformed in your mitochondria into the high-energy molecule called adenosine triphosphate (ATP). ATP is made up of a base, a sugar and three (hence, *tri*phosphate) phosphate groups. The phosphate groups are held together by high-energy bonds. When a bond is broken, a high level of energy is released. Energy in this form can be used by the cells. When a bond is used, ATP becomes ADP (adenosine diphosphate), which has only two phosphate groups. ADP now is able to go and pick up another phosphate and form a high-energy bond so energy is stored and the process can begin again. Life is good.

cilia = (*SILL ee yah*)

some airports. Certain cells have solved such transport problems already through the use of **cilia**. Cilia are short, microscopic, hair-like projections located on the outer surface of some cells. They begin a wave-like motion that carries particles in a given direction. This is one way our lungs stay clean from the dust particles and germs that we inhale every day. Cilia lining our bronchioles trap dust and other inhaled particles to prevent them travelling further into the lungs where they may be problematic.

Test your knowledge 4.3

Choose the best answer:

1. What organelle most closely follows the given analogy?
   a. _____ the town hall
   b. _____ road system
   c. _____ power plant
   d. _____ packing and shipping plant
   e. _____ the 'cell propeller'

Mitosis

Cellular reproduction is the process of making a new cell. Cellular reproduction is also known as **cell division**, because one cell divides into two cells when it reproduces. Cells can only come from other cells. When cells make *identical* copies of themselves *without the involvement* of another cell it is called **asexual reproduction**. Most cells are able to reproduce themselves asexually whether they are animal cells, plant cells or bacteria.

a = *without*

The cells that make up the human body are a type of cell known as a **eukaryotic cell**. Eukaryotic cells have a nucleus, cellular organelles and, usually, several chromosomes in the nucleus. (Reminder: the genetic material of the cell, DNA, is bundled into 'packages' of chromatin known as chromosomes.) Since chromosomes carry all the instructions for the cells, all cells must have a complete set after reproduction. These instructions include how the cell is to function within the body and blueprints for reproduction. No matter whether a cell has one chromosome, like bacteria, or 46 chromosomes, like humans, all the chromosomes must be copied before the cell can divide.

eukaryotic cell (*YOO care ee YOH tik*)

Let's start off with the simple cellular division of a bacterium. Bacterial cells, which do not have a nucleus or organelles, reproduce very easily, through a process known as *binary fission*. Bacterial cells simply copy their DNA, divide up the cytoplasm and split in half.

Now let's go to the more complex cellular reproduction. Eukaryotic

---

Amazing body facts

Bacterial reproduction

Bacteria can reproduce so rapidly that they can double their population every hour. No wonder a bacterial infection can get out of hand so fast.

cells, like yours, must go through a more complicated set of manoeuvres in order to reproduce. Not only do your cells have to duplicate all 46 of their chromosomes, they have to make sure that each cell gets all of the chromosomes and all of the right organelles. The process of sorting the chromosomes, so that each new cell gets the right number of copies of all of the genetic material, is called **mitosis**. Mitosis is the only way that eukaryotic cells can reproduce asexually.

mitosis (*my TOE siss*)

The cell cycle

The total life of a eukaryotic cell can be divided into two major phases, known collectively as the **cell cycle**. Most of the cell cycle is devoted to a phase known as **interphase**. During interphase, a cell is not dividing, but is performing its normal function along with stockpiling needed materials and preparing for division by also copying DNA and making new organelles. Only a brief portion of the cell cycle, the **mitotic phase**, is devoted to actual cell division. The mitotic phase is divided into two major portions. Mitosis is the division and sorting of the *genetic material*, while **cytokinesis** is the division of the *cytoplasm*.

cytokinesis (*SIGH toe kin EE siss*)

Mitosis, the division of the genetic material, is the most complicated part of cell division for a eukaryotic cell. What further complicates mitosis is that the process is divided into four specific phases. As you'll soon see, the four phases are based on the position of the chromosomes relative to the new cells. They are: **prophase**, **metaphase**, **anaphase** and **telophase**. To put all these processes and phases in perspective, please refer to the flowchart in Figure 4.12.

The phases of mitosis

Let's proceed step by step through the four phases of mitosis using Figure 4.13 as a reference.

1. Prophase (*pro* = before) – the nucleus disappears, the chromosomes become visible, a set of chromosomal anchor lines or guide wires, the spindle, forms.

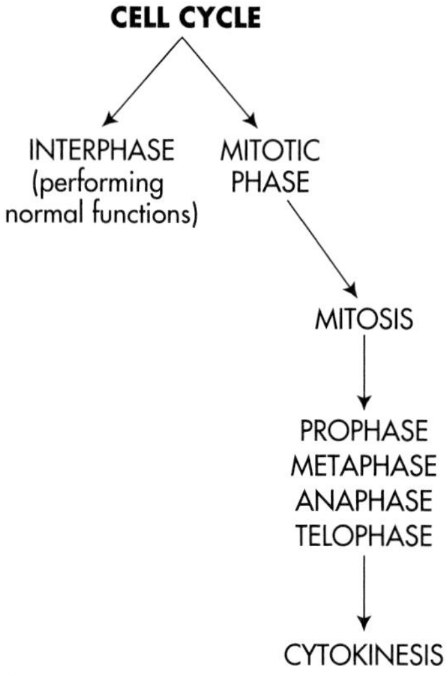

CELL CYCLE

INTERPHASE   MITOTIC
(performing   PHASE
normal functions)

MITOSIS

PROPHASE
METAPHASE
ANAPHASE
TELOPHASE

CYTOKINESIS

Figure 4.12 Flowchart of the cell cycle.

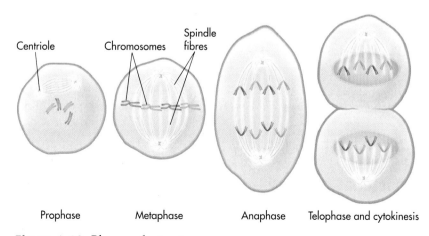

Prophase   Metaphase   Anaphase   Telophase and cytokinesis

Figure 4.13 Phases of mitosis.

2. Metaphase (*meta* = between) – the chromosomes line up in the centre of the cells.

3. Anaphase (*an* = without) – the chromosomes split and the spindles pull them apart.

4. Telophase (*telo* = the end) – the chromosomes go to the far end of the cell, the spindle disappears and the nuclei reappear.

During or directly after telophase, cytokinesis happens and the cell divides in half. The original cell was the mother cell that has now formed into two new identical daughter cells. Thus mitosis, asexual reproduction in eukaryotic cells, results in two new daughter cells identical to the original mother cell.

Mitosis in your body

Mitosis, asexual cellular reproduction, serves many purposes in your body. Any time your cells need to be replaced, mitosis is the method used to replace them. Many of your tissues are replaced on a regular basis. Bone, epithelium, skin and blood cells all replace themselves on a regular basis. Repair and regeneration of damaged tissue is accomplished by mitosis as well. If you cut your hand, the skin is replaced, first by collagen, but eventually by the original tissue. Mitosis increases in cells near the injury so that the damaged or destroyed cells can be replaced. A broken bone is replaced in much the same way.

Test your knowledge 4.4

Choose the best answer:

1. Cells are reproducing themselves during this portion of the cell cycle

   a. metaphase
   b. mitotic phase
   c. meiotic phase
   d. manic phase

2. During this phase of mitosis the nucleus disappears and the spindle appears

   a. prophase
   b. metaphase
   c. anaphase
   d. telophase

3. During this phase of mitosis the chromosomes begin to pull apart

   a. prophase
   b. metaphase
   c. anaphase
   d. telophase

4. Which of the following is not a function of asexual reproduction?

   a. tissue repair
   b. replacement of cells
   c. tissue growth
   d. cloning humans

Growth is also accomplished by mitosis. Lengthening of bones as you grow, increases in muscle mass due to exercise, indeed, most ways that tissue gets bigger, are due to mitosis of cells in the tissues or organs. Without mitosis, your body would not be able to grow or replace old or damaged cells.

Learning hint

Mitosis versus meiosis

These words sound and look alike and are therefore often confused. Remember, *me*iosis produces ga*me*tes or sexual cells which contain half of the chromosomes because the sexual union of male and female will contribute the other half. M*i*tosis ('I reproduce myself') is asexual and produces exact copies of the cell and the full complement of chromosomes since no union is needed.

Sexual reproduction is different from mitosis. **Meiosis** is the term for the division of cells in order for reproduction to occur. This process will be discussed in Chapter 18, 'The reproductive system'.

Clinical application

Mitosis run amok

When the body is healthy, cells grow in an orderly fashion: they grow at the appropriate rate in the appropriate number, shape and alignment. Sometimes conditions are altered in the body (either internally or externally) that trigger changes in the way the cells grow. Cell growth can become wild and uncontrolled, leading to too many cells being produced and resulting in a lump, or tumour. Generally, tumours are classified as either benign or malignant. **Benign** tumours typically grow slowly, push healthy cells out of the way and are usually not life-threatening. **Malignant**, or cancerous, tumours grow rapidly and invade rather than push aside healthy cell tissue. Cancer actually means 'crab' (remember your Zodiac signs?) and is a good description of cancer cells in that they spread out into healthy tissue like the legs and pincers of a crab. Cancerous tumours also differ from benign tumours in that parts of a cancerous tumour can break off and travel through the blood system or the lymphatic network and start new tumours in other parts of the body. This breaking off and spreading of malignant cells is called **metastasis**. One reason that lung cancer is so deadly (more women die from it than from breast cancer) is that it can metastasise for a fairly long time before it is even diagnosed in lungs. By the time it is discovered, tumours may be growing in the liver and brain and in other parts of the body, making survival very difficult.

benign (*ben INE*)
malignant (*mah LIG nant*)

Micro-organisms

As in any town, you have a large and diverse population. While the vast majority of the population are good citizens who provide positive contributions to the town, there are some who can cause problems. The same can be said for the world of micro-organisms – this means 'little creatures' and they are too small to be seen with the naked eye. The following 'micro-citizens' of our town will now be discussed:

- bacteria (the study of bacteria is called bacteriology)
- viruses (the study of viruses is called virology)
- fungi (the study of fungi is called mycology)
- protozoa (the study of protozoa is called protozoology)

Bacteria

bacteria (*back TEER ree yah*)

pathogen (*PATH oh jenn*)
    path/o = *disease*
    gen = *producing*

When you hear the word bacteria, you probably first think of an organism that produces disease, or what is called a pathogen. You are generally correct in assuming so because bacteria make up the largest group of pathogens. Some bacterial pathogens even release toxic substances in your body. Bacteria grow rapidly and reproduce by splitting in half, sometimes doubling as rapidly as once every 30 minutes!

However, you will learn throughout this book that bacteria are often harmless and, in fact, essential for life. These bacteria live within or on us, and are part of what we call our **normal flora**. They are often referred to as 'commensal' to indicate that providing they remain in their normal environment they will not caused us any problems. For example, certain bacteria in your intestine help to digest food, and some help to synthesise vitamin K, which helps clot blood so we don't bleed to death when we get a cut or scrape. However should any of the commensal bacteria move to a different environment within the human body they may become pathogenic and therefore able to cause disease. For example, *E. coli* usually resides in the large intestine as a commensal – however should it be able to access the urinary tract it is likely to cause a urinary tract infection such as cystitis.

There are three main different types of bacteris – cocci, bacillus and curved rods. Cocci are usually round in shape and include streptococci and staphylococci – these may be implicated in problems such as a sore throat or wound infections. Bacillus are rod shaped and include the bacteria that cause tuberculosis and tetanus. Curved rods includes bacteria that causes cholera and syphilis. See Figure 4.14, which shows examples of various types of bacteria.

Many bacteria can be treated very effectively with antibiotics. However, over the last ten to twenty years there has been an increase in the number and type of bacteria that have become resistant to the effects of these medicines. The most well known is of course Methicillin Resistant *Staphylococcus Aureus* (MRSA) – if an infection is caused by this it becomes extremely difficult to treat. Reasons for bacteria becoming resistant include: the ability of the bacteria to change itself (mutate) so that the next time it encounters an antibiotic it is not so easily affected by it and so can survive; people not taking all of their antibiotic medication – this gives the bacteria an opportunity to develop resistance as it has not been effectively destroyed and it can then be spread to other people; use of antibiotics inappropriately for infections caused by a virus – antibiotics do not treat viral infections; use of the incorrect type of antibiotic for an infection – for this reason it is important that antibiotics are prescribed by a doctor so they can chose the most appropriate one.

One of the reasons that we are susceptible to bacterial infection is that many types of bacteria require a certain temperature in order to survive and grow and that temperature is body temperature! And so we provide an ideal breeding ground for many bacteria. Other factors that they require for growth

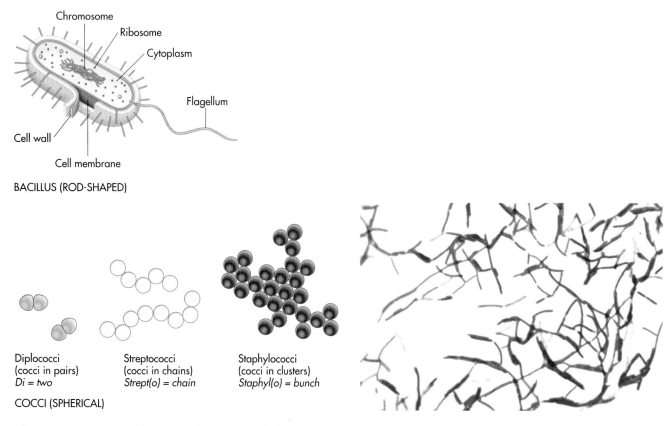

Figure 4.14 Types of bacteria. (*Source*: Rod-shaped Bacteria © Getty Images/Dr. Arthur Siegceman)

include the correct environmental pH, available nutrients (e.g. glucose, proteins, lipids) and access to water.

Viruses

An even more basic pathogen is a **virus** (see Figure 4.15). Viruses (from a Latin term meaning 'poison') are infectious particles that have a core containing genetic material (codes to replicate) surrounded by a protective protein coat called a **capsid**. Some viruses have an additional layer, or membrane, surrounding the capsid. Viruses are interesting because they cannot grow, 'eat' or reproduce by themselves. They must enter another cell (host cell) and hijack that cell's parts, energy supply and materials to do all of the aforementioned activities. In addition, each virus must target specific cells in the body to claim as hosts. For example, the viruses that cause a cold target the cells found in the respiratory system, and the viruses that cause herpes target cells that are found in the tissues of our nervous system. Most of the upper respiratory infections that people get are caused by viruses and, like all viral infections, they do not respond to **antibiotics**. Other diseases caused by a virus include Acquired Immune Deficiency Syndrome (AIDS) caused by the Human Immunodeficiency virus (HIV), mumps and rubella (German measles).

It is interesting to note (or disturbing for those of you who worry about everything) that viruses can stay dormant in the body and become active once again later in life. This is true for all of us who have had chicken pox. These viruses stay in the body and may later become active and cause a potentially

capsid (*CAP sid*)

anti = *against*
bios = *life*

Amazing body facts

Magnetotaxis – some bacteria respond to magnetic fields

Believe it or not, some bacteria can sense and respond to a magnetic field. These types of bacteria are sensitive to the Earth's magnetic field and can orientate themselves to this force. This ability to move in response to magnetic forces is called *magnetotaxis*. This discovery was made by Richard Blakemore as he observed bacteria living in sulphide-rich mud from a lake. As he changed the position of the mud, the bacteria would reorientate themselves to the Earth's magnetic field. Upon further examination, Blakemore determined that these bacteria possessed particles of iron oxide, a magnetic metal compound that is stored in a cell structure called a *magnetosome*.

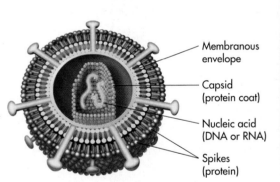

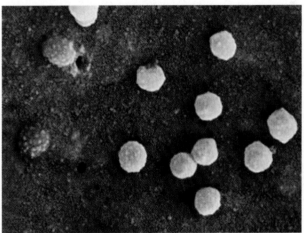

Figure 4.15 A virus. (*Source*: © Thomas Deerinck, NCMIR/Science Photo Library)

painful skin condition called *shingles*, caused by the herpes zoster virus. It is also interesting to note that the actual virus is relatively easy to kill by itself, but once it becomes part of a cell, it is hard to kill without harming the individual cell.

Fungi

fungi (*FUN ghee*)
mycelia (*my SEE lee yah*)

Fungi, the plural form of fungus, can be either a one-celled or multi-celled organism. These plant-like organisms have tiny filaments, called **mycelia**, that travel out from the cell to find and absorb nutrients. Like bacteria, fungi, such as edible mushrooms, can be good, but in certain situations they can also cause problems (see Figure 4.16). Fungi also can spread through the release of **spores**. Normally, we are not affected by fungi, but if the body has a problem with its immune system, then fungi have a chance to cause an infection. If tissue becomes damaged, fungi can more easily create an infection. Spores can be carried by the wind and then inhaled, potentially causing lung infections.

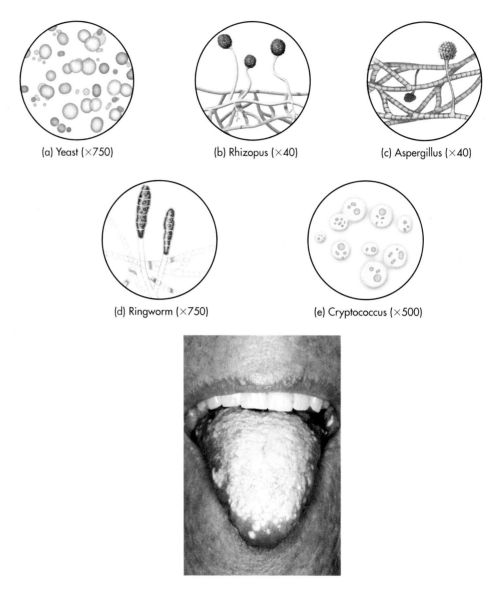

Figure 4.16 Types of fungi and a fungal infection of the tongue.
(*Source*: © Alamy Images/Medical-on-Line)

Interestingly, you can inhale fungi spores and still not develop an infection. Examples of fungal infections are athlete's foot and a mouth fungus called thrush or candidiasis (also known as a yeast infection). Other types include: rhizopus which causes mould on fruit and vegetables; aspergillus which can cause respiratory problems in some individuals, especially if they already have an underlying respiratory condition such as asthma; ringworm which causes athlete's foot; cryptococcus which can cause meningitis.

Protozoa

Protozoa are one-celled animal-like organisms that can be found in water, such as ponds, and in soil. Disease caused by these micro-organisms can result from swallowing them (such as by drinking contaminated water) or from being bitten by insects that carry them in their bodies (such as malaria-carrying

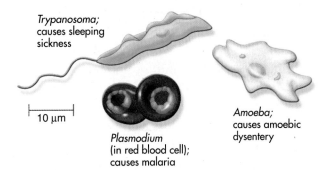

Trypanosoma; causes sleeping sickness

10 μm

Plasmodium (in red blood cell); causes malaria

Amoeba; causes amoebic dysentery

Figure 4.17 Protozoa.

mosquitoes). Other diseases caused by protozoa in humans includes giardiasis (infection of small bowel causing diarrhoea), sleeping sickness transmitted by the tsetse fly, and dysentery which causes severe diarrhoea with abdominal pain and fever.

See Figure 4.17.

Summary

- All living organisms are made of one or more cells. Cells are the fundamental units of living organisms.
- Even though cells are the fundamental units, they are composed of a variety of parts that are necessary for proper cellular function. These small parts are called organelles.
- Substances can cross the cell membrane via passive or active transport.
- Passive transport can occur through diffusion, facilitated diffusion, osmosis or filtration.
- Active transport can occur through active transport pumps, endocytosis or exocytosis.
- One form of cellular movement is through the use of flagella.
- Cilia found in the lungs aid in the removal of foreign particles in the airways through rhythmic movement.
- Cells and tissues grow, and are replaced and repaired by asexual reproduction. Cells make identical copies of themselves. This takes place all over your body whenever tissues grow or are repaired.
- Some tissues, such as epidermis, blood and bone, replace themselves continually, always by asexual reproduction.
- The life of the cell can be depicted as a cell cycle. Only about 10 per cent of the cell cycle is devoted to mitosis. The rest of the time the cell is in interphase, preparing to divide and carrying on day to day cellular activities.
- Asexual reproduction in eukaryotic cells is accomplished by a relatively complex process called mitosis and cytokinesis. Mitosis, the division of the genetic material, takes place in four phases: prophase, metaphase, anaphase and telophase. Cytokinesis is the division of the cytoplasm and organelles. Mitosis produces two daughter cells, identical to each other.

→

- Meiosis is the sexual reproduction of cells. If an organism is going to reproduce sexually it must use specialised cells called gametes with only half the typical number of chromosomes for that organism. The chief difference between mitosis and meiosis is that mitosis is asexual and produces the exact number of chromosomes, whereas meiosis is sexual and combines two cells with half of the needed chromosomes.

- Not all bacteria are bad. In fact, the body needs bacteria to survive.

- Although a virus is not a one-celled organism, it needs another cell to replicate.

- Fungi can be single-celled or multi-celled organisms and can cause infections in the body. Spores can be immune to a harsh environment, thus allowing fungi to spread.

- Protozoa are one-celled and can cause disease through ingestion or through insect bites.

Case study

Given the following mini-scenarios, identify what type of micro-organism may be the causative agent.

1. Two young boys complain of a stomach ache and severe diarrhoea after drinking pond water.

   _____

   _____

2. Barbara is a 13-year-old with a compromised immune system due to an inherited disease. Two days after returning home from a school trip to a bread-making factory, Barbara complains of shortness of breath and is diagnosed with a respiratory infection.

   _____

   _____

3. Stan has had a stubborn cold for three days and is given an antibacterial agent. However, he doesn't respond to the treatment and the cold persists.

   _____

   _____

Review questions

Multiple choice

1. The cell membrane can best be described as:
   a. permeable to all materials
   b. non-permeable
   c. rigid
   d. semi-permeable

2. All of the following are passive forms of transport *except*:
   a. facilitated diffusion
   b. exocytosis
   c. osmosis
   d. filtration

3. With a greater concentration of a solute, what will happen to osmotic pressure?
   a. it will become less
   b. it will become greater
   c. it will remain the same
   d. there is no relation between osmotic pressure and the concentration of solute.

4. Where are the specifications for the creation of new cells found?
   a. DNA
   b. RNA
   c. ATP
   d. ADP

5. Which micro-organisms can cause disease?
   a. bacteria
   b. fungi
   c. virus
   d. all of the above

Short answers

1. List and describe the four methods of passive transport.

   _____
   _____
   _____
   _____

2. Why do viruses need cells?

   _____
   _____
   _____
   _____

3. How does passive transport differ from active transport?

   _____
   _____
   _____
   _____

→

4. Discuss how cells can provide motility for themselves.

    _____

    _____

    _____

    _____

5. Explain what is meant by mitosis.

    _____

    _____

    _____

    _____

Matching

1. Match the following organelles with their function:

    _____ nucleus   a. the 'powerhouse' of the cell

    _____ cell membrane   b. processing, packaging and shipping of materials

    _____ Golgi apparatus   c. the gatekeeper of the cell

    _____ mitochondria   d. the 'brain' of the cell

    _____ cytoplasm   e. the internal environment

    _____ lysosome   f. sanitation engineers

Suggested activities

1. Research the various cell types within the body and see how many you can list.

2. Research five sexually transmitted diseases (STDs) and identify the causative micro-organism for each.

Answers

Answers to case study

1. Protozoa
2. Fungi
3. Virus

Answers to multiple choice questions

1. d
2. b
3. b
4. a
5. d

Answers to short answer questions

1. The four methods are:

   Diffusion – most common – passive transport by which a substance of a higher concentration moves to an area of less concentration.

   Facilitated diffusion – a variation of diffusion where a substance is helped in moving across a membrane.

   Osmosis – water moves from an area of low solute concentration to an area of higher solute concentration through a selectively permeable membrane.

   Filtration – caused by pressure; pressure is applied to force water and its dissolved materials across a membrane.

2. Viruses cannot live outside of a body; they cannot 'eat', grow or reproduce on their own.
3. Active transport requires energy derived from adenosine triphosphate (ATP). Other examples of active transport are endocytosis and exocytosis – endocytosis occurs when the cell surrounds the substance to be absorbed whereas exocytosis transports a substance out of itself.
4. Some cells have flagella or cilia that help with movement.
5. Mitosis is the process of making a new cell. The new cell is an exact copy of the original cell with the full complement of chromosomes (23 pairs).

Answers to matching

a. Mitochondria
b. Golgi apparatus
c. Cell membrane
d. Nucleus
e. Cytoplasm
f. Lysosome

Tissues and systems

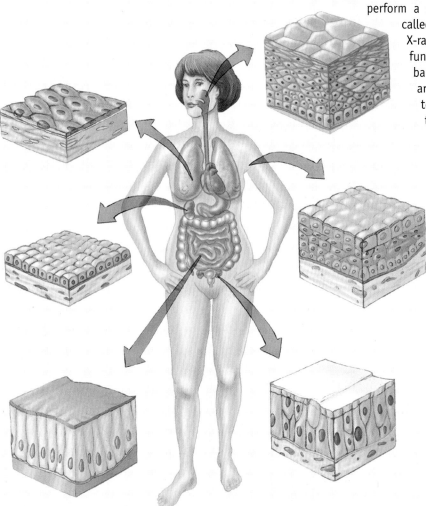

Previously, we discussed cells as the basic building blocks of the body. In this chapter, we explore **tissues**, which are collections of similar cells. We can consider cells to be individual employees. At the next level, we can think of tissue as a group of employees who have the same or similar educational background, such as radiologists at a hospital. A combination of tissues designed to perform a specific function or several functions is called an **organ**. This could be compared to an X-ray department in the hospital, where specific functions such as X-rays, CT scans, MRIs and barium swallows (a procedure, not a bird!) are performed. Organs that work together to perform specific activities, often with the help of accessory structures, form what we call **systems**. We can compare a system to a hospital that provides a service (healthcare) to the citizens of our city through the combination of all of the departments: laboratory, nursing, physiotherapy, dietetics, domestic services/housekeeping, and so on. This chapter provides an overview of tissues, organs and systems, which will be expanded upon in later chapters.

Learning objectives

At the end of this chapter, you will be able to:

◆ Explain the relationship between cells, tissues, organs and systems
◆ List and describe the four main types of tissues
◆ Identify and describe the various body membranes
◆ Differentiate the three main types of muscle tissues
◆ Describe the main components of nerve tissue
◆ List and describe the main functions of the body's systems

Overview of tissues

Just as there are many different types of cells with various functions and responsibilities, tissues come in different shapes and sizes, again with the structure dependent upon the function. Let's begin to explore the different tissue types.

Tissue types

histo = *tissue*

Tissue is a collection of similar cells that act together to perform a function. Imagine individual cells as bricks. Placing these bricks (cells) in a specific pattern creates the functional wall (tissue) of a building. There are many different types of tissues depending upon the required function. The four main types of tissues are:

- epithelial
- connective
- muscle
- nervous

We will now discuss these major types of tissue and their subdivisions.

Epithelial tissue

Similar in purpose to the clingfilm we use to keep food fresh or to cover bowls in the refrigerator, epithelial tissue not only covers and lines much of the body but also covers many of the parts found in the body. The cells in this form of tissue are packed tightly together, forming a sheet that usually has no blood vessels in it (avascular).

squamous (*SKWAY muss*)
cuboidal (*kew BOYD all*)
transitional (*tranz ISH on all*)
stratified (*STRAH tih fide*)

We can further classify epithelial cells by their shape and arrangement. These cells can be flat or scale-like (**squamous**), cube-shaped (**cuboidal**), column-like (**columnar**), or stretchy and variably shaped (**transitional**). If these cells are arranged in a single layer and are all the same type of cell, we classify them as **simple**. If they are arranged in several layers, we say they are **stratified**, and they are named by the type of cell that is on the outer layer (such as stratified columnar). The function required of the cell dictates which type of cell formation is utilised. For example, simple squamous cells are utilised in the lungs because of their flat, thin design, which makes for easy

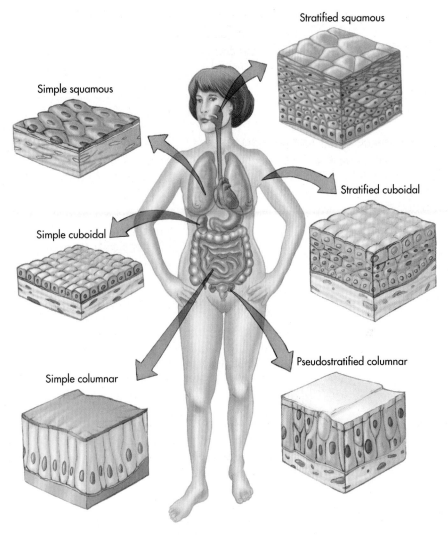

Figure 5.1 Types and locations of epithelial tissues.

transfer of oxygen from the lungs to the blood. Functions of epithelial tissue include protection, absorption, secretion and filtration. Figure 5.1 shows the types and locations of epithelial tissues. These are further discussed in later chapters.

Amazing body facts

Skin and vitamin production

You know that you get vitamins and minerals from the foods you eat, but did you know that your skin produces vitamin D when you are exposed to sunlight? Calcitriol is one of two calcium regulating hormones (the other is parathormone). It is derived from vitamin D, but vitamin D is made in the skin and converted to calcitriol. Individuals who habitually cover themselves, or don't expose themselves to sunlight can develop hypocalcaemia.

Membranes Generally, *membranes* are sheet-like structures found throughout the body that perform special functions. Although membranes can be classified as organs, we discuss them along with tissues for ease of explanation. Membranes classified as *epithelial* membranes possess a layer of epithelial tissue and a bottom layer of a specialised connective tissue. Epithelial membranes are classified into three general categories, as you can see in Table 5.1.

Table 5.1 Types of epithelial membranes

1. **cutaneous**		• functions like a tarpaulin placed over a boat • the main organ of the integumentary system, commonly known as your skin • makes up approximately 16 per cent of the total body weight • skin is largest, visible organ
2. **serous**		• a two-layered membrane with a potential space in between • comprised of the parietal and visceral layers
	parietal	• lines the wall of the cavities in which organs reside • produces serous fluid, which reduces friction between different tissues and organs (without this friction-reducing fluid, each beat of your heart and every breath you take would be uncomfortable. The effect is similar to running water over a sheet of plastic placed on the grass. You can run, jump onto the plastic, and slide. Imagine how that would feel without the water running over the plastic)
	visceral	• wraps around the individual organs • also produces serous fluid, which reduces friction between different tissues and organs
3. **mucous**		• lines openings to the outside world, such as your digestive tract, respiratory system, and urinary and reproductive tracts • called mucous membranes because they contain specialised cells that produce mucus (mucus can act as a lubricant like the oil in a car; mucus also serves several other important purposes, as you will see in later chapters)

serous **membranes** *(SEER uss)*

parietal = *(pah RIE eh tall) wall; therefore, the parietal membrane lines the wall of the cavity*

visceral *(VISS er all)*

viscero = *organs; therefore, visceral membranes enclose organs*

muc/o; myx/o = *combining forms indicating relative to mucus*

Amazing body facts

Blood and lymph as connective tissues

Even though blood and lymph are fluid, they are considered to be connective tissue because they are a liquid mixture comprised of a group of cells that have specialised functions. These fluid connective tissues contain specialised cells and dissolved proteins suspended in a watery substance. We expand upon these two very important tissues in later chapters.

Figure 5.2 shows the location of the serous and mucous membranes of the body.

Connective tissue

Connective tissue is the most common of the tissues and is found throughout the body more than any other form. That is because it is found in organs, bones, nerves, muscles, membranes and skin. The job of connective tissue is to hold things together and provide structure and support. Fine, delicate webs of loosely connected tissue (areolar tissue) hold organs together and help to

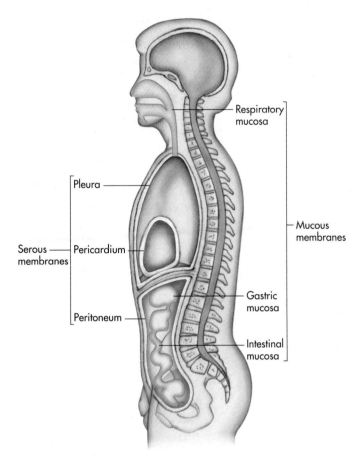

Figure 5.2 Location of serous and mucous membranes.

hold other connective tissues together. Fat is also a connective tissue known as adipose tissue. Although we always seem to want to lose fat, we truly need some fat in our bodies for proper functioning. Connective tissue can also be more densely packed and form strong cord-like structures similar to wire cables on suspension bridges. Tendons and ligaments are composed of dense connective tissue. The skin also needs to be densely packed to form a protective barrier, and it therefore uses dense connective tissue. Functions of connective tissue are protection, support and the binding together of other body structures. Most connective tissue has a very good blood supply (well vascularised), except for tendons and ligaments which have a poor blood supply and cartilage which is avascular. Because connective tissue is so versatile and found throughout the body, we will discuss it in more depth in the relevant chapters. Please see Figure 5.3, which illustrates the various types of connective tissues and shows some of the places where they are found.

    The membrane type associated with connective tissue is the **synovial** membrane. This important membrane type is found in the spaces between bone joints and produces a slippery substance called synovial fluid, which greatly reduces friction when joints move. Imagine runners without synovial membranes: their knees would burst into flames during athletic events. Figure 5.4 shows a synovial joint and membrane.

synovial (*sigh NO vee yall*)

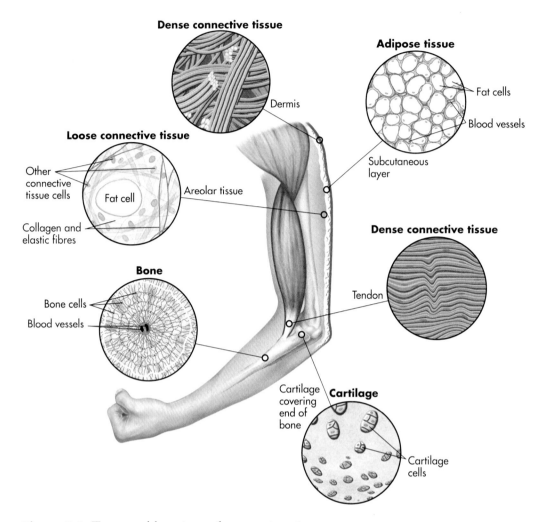

Figure 5.3 Types and locations of connective tissues.

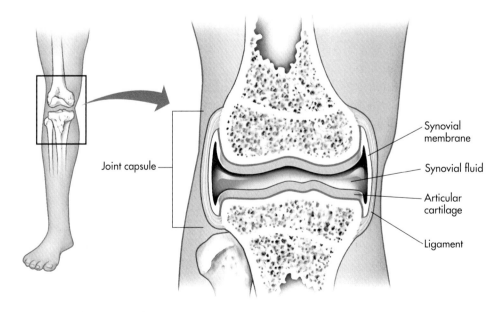

Figure 5.4 The synovial joint.

Learning hint

Mucous isn't always mucus

'A rose by any other name is still a rose' might work well in Shakespeare's plays, but mucous isn't always mucus. Although they sound the same, mucous and mucus are two different things. Mucous is an adjective that describes the type of membrane that produces *mucus*, the actual substance.

Muscle tissue

Muscle tissue provides the means for movement by and in the body. This form of tissue has the ability to shorten itself (contractility). There are three types of muscle tissue: **skeletal**, **cardiac** and **smooth**.

Skeletal muscle Skeletal muscle (often described as **striated** because of its striped appearance) is attached to bones and causes movement by contracting and relaxing. It also surrounds certain openings of the body, such as the mouth, and controls the size of the opening. Unless you are a ventriloquist, it is very hard to speak clearly without moving your lips. The cells that make up this tissue type are long and fibre-like with many nuclei in each cell. Our brain thinks about moving or speaking and causes the correct muscle to contract or relax as necessary. Since this is a conscious effort, we call these muscles *voluntary* muscles.

Cardiac muscle Cardiac muscle is only found in the walls of the heart. Our hearts beat without our conscious thought. Since cardiac muscle contracts and relaxes without conscious thought, this muscle type is considered *involuntary* muscle tissue. Contraction and relaxation of cardiac muscle enables the heart to pump blood around the body and to the lungs. The cells in this tissue type interlock with each other, promoting more efficient contraction, as you will learn in Chapter 13, 'The cardiovascular system'.

Smooth muscle Smooth muscle tissue forms the walls of hollow organs, such as in our digestive system (which is why it is often called *visceral* tissue) and blood vessels. Since we don't have to consciously think about digesting food, we consider this muscle type to also be involuntary muscle tissue. Cells forming this tissue are not as long and fibrous as skeletal muscle and each cell has only one nucleus. Smooth muscle tends to contract more slowly than the other types of muscle. Figure 5.5 shows the various types of muscle tissue.

Nervous tissue

Nerve tissue acts as a rapid messenger service for the body, and its messages can cause actions to occur. There are two types of nerve cells. **Neurons** are the conductors of information, and **glia** (sometimes called **neuroglia**) cells function as support by helping to hold the neurons in place. The branch-like formations that make up part of the neuron, called **dendrites**, receive sensory

---

Note = *smooth muscle is named such as it has no striations*

striated (*stry ATE ed*)

neuro = *nerves*

neurons (*NEW ronz*)

glia = *(GLY ah) glue, hence the name for cells that hold the nerve cells together*

dendr/o = *tree, hence the name of the branching dendrite structure*

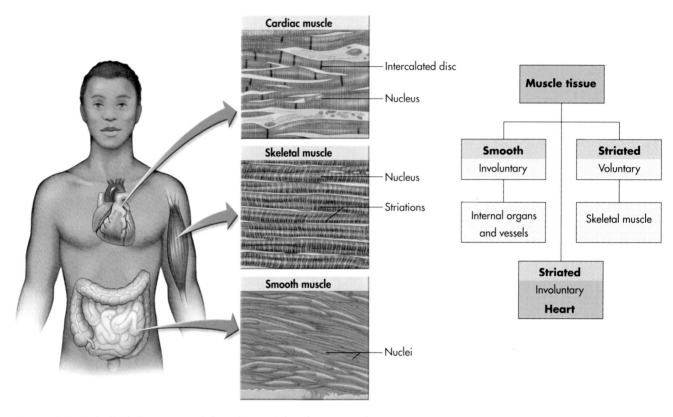

Figure 5.5 Labelled diagram and flowchart of the three muscle tissue types.

meninges (*men IN jeez*)

information. The trunk-shaped structure, called the axon, transports information *away* from the cell body. The membranes associated with covering the brain and spinal cord are called **meninges**. Many nerves have an insulating layer called the myelin sheath, which is further discussed in Chapter 9, 'The nervous system' (part I). Figure 5.6 shows the two types of nerve cells.

Organs

As mentioned earlier, a hospital department is made up of employees who work together to perform specific functions, much like an organ. An organ is the result of two or more types of tissues organised in such a way as to accomplish a task that the tissues cannot do on their own. Some organs occur singly, such as the heart, and some occur in pairs, such as the lungs. It is interesting to note that we can survive quite well with only one healthy organ from a paired group (such as a lung or kidney). Your heart, lungs, stomach, liver and kidneys are all examples of organs found in your body. It is important to understand that there are organs and *vital* organs. Vital organs are the ones that you can't live without. Your heart, brain and lungs are vital organs. Organs that you can live without include your appendix, spleen and gallbladder. Some vital organs come in pairs, so if one is damaged or removed, you can still survive. Lungs and kidneys are a good example. Table 5.2 is a nice, quick reference to the various organs of the body.

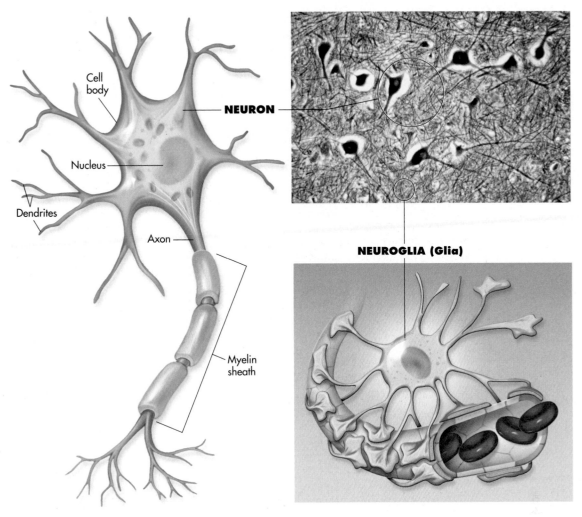

Figure 5.6 The two main types of nerve cells. (*Source*: © Science Photo Library)

Test your knowledge 5.1

Complete the following:

1. List and describe the four types of epithelial cells.

   _____

   _____

2. Lubrication for joints is produced by which type of membrane?

   _____

3. Why is connective tissue so prominent in the body?

   _____

   _____

   _____

4. Explain the difference between the terms mucous and mucus.

   _____

   _____

   _____

   _____

Table 5.2 Systems and organs of the human body

Body system	Organs in the system	Combining form	Medical specialty
integumentary	skin hair nails sweat glands sebaceous glands	dermat/o, cutane/o trich/o ung/o sud/o, hidr/o seb/o	**dermatology** (der ma TOLL oh jee)
musculoskeletal	muscles bones joints	my/o, muscul/o oste/o arthr/o	**orthopaedics** (or thoh PEE diks)
endocrine	thyroid gland pituitary gland testes ovaries adrenal glands pancreas parathyroid glands pineal gland thymus gland	thyr/o pituit/o test/o, orchid/o ovari/o, oophor/o adren/o pancreat/o parathyroid/o pineal/o thym/o	**endocrinology** (en doh krin OLL oh jee) **gynaecology** (guy nee KOLL oh jee)
cardiovascular	heart blood arteries veins	cardi/o haemat/o, haem/o arteri/o phleb/o, ven/o, veni/o	**cardiology** (car dee OLL oh jee) **haematology** (hee ma TOLL oh jee)
lymphatic and immune	spleen lymph thymus gland	splen/o lymph/o thym/o	**immunology** (im yoo NOLL oh jee)
respiratory	nose pharynx larynx trachea lungs bronchial tubes	nas/o, rhin/o pharyng/o laryng/o trache/o pneum/o bronch/o	**otorhinolaryngology** (oh toe rye no larr ing GOLL oh jee) **thoracic** (thor A sick) **surgery** **pulmonary** (pull MOAN a ree)
gastrointestinal	mouth pharynx oesophagus stomach small intestine colon liver gallbladder pancreas	or/o pharyng/o oesophag/o gastr/o enter/o col/o colon/o hepat/o cholecyst/o pancreat/o	**gastroenterology** (gas troh en ter OLL oh jee)

Table 5.2 (*continued*)

urinary	kidneys	nephr/o, ren/o	**nephrology** (*nh FROL oh jee*)
	ureters	ureter/o	
	bladder	cyst/o, vesic/o	**urology** (*you ROLL oh jee*)
	urethra	urethr/o	
reproductive	ovaries	oophor/o	**gynaecology** (*guy nee KOLL oh jee*)
	uterus	uter/o, hyster/o	
	Fallopian tubes	salping/o	**obstetrics** (*ob STET riks*)
	vagina	vagin/o	
	mammary glands	mamm/o	
	testes	orchid/o	
	prostate	prostat/o	
	urethra	urethr/o	
nervous	brain	encephal/o	**neurology** (*new ROL oh jee*)
	spinal cord	myel/o, spin/o	
	nerves	neur/o	**neurosurgery** (*new row SIR jer ee*)
special senses	eye	ocul/o, ophthalm/o	**ophthalmology** (*op thal MOLL oh jee*)
	ear	ot/o	**otolaryngology** (*oh toe larr ing GOLL oh jee*)

uro = *urine*

gynaec/o = *woman*

melano = *black*

oma = *tumour*

Clinical application

Melanoma

Melanoma, a skin cancer, has had a rapid increase in the rate of incidence in the past several decades. Melanoma is significant for two main reasons. First, it has a very high mortality rate and, in fact, is the cancer with the most rapidly increasing mortality rate. In the UK just fewer than 9000 people are diagnosed with melanoma each year. About 3 out of every 100 cancers diagnosed (3%) are melanomas. Second, it is one of our more preventable cancers. The culprit? Excessive sun exposure and tanning! Although the classic patient has fair skin, blue eyes, and blonde or red hair, more darkly pigmented individuals are also at risk, but at a somewhat lower rate. The effects of blistering sunburns have an additive effect. This means that each childhood sunburn you had increases your chance of developing melanoma in adulthood. Protection from excessive sun and early detection are keys to survival.

Systems

A body system is formed by organs that work together to accomplish something more complex than what a single organ can do on its own. Take the heart, for example. Even if your heart is functioning perfectly, you would die without all of the other parts that make up the cardiovascular system. Much like a road system in the city, you need the arteries, veins and blood to get vital oxygen and nutrients to the cells and to remove the waste products produced by those cells.

Although we discuss the body systems separately, it is extremely important that you understand that all of the body systems are interrelated, often depending on each other for proper functioning.

Skeletal system

Most people think that the skeleton's only job is to provide support and structure to the body, much like the framework of a house, but it does much more. The bones of the skeleton protect organs such as the brain. In combination with muscles, it provides movement. It acts as a storage vault for a variety of minerals, such as calcium and phosphorus, *and* it also produces blood cells. That's pretty impressive stuff! The main components of this system are bones, joints, ligaments and cartilage. Figure 5.7 shows the main components of the skeletal system.

Muscular system

'All of this dry reading makes me thirsty. I think I'll go and get something to drink.' The muscular system is responsible for getting you up and over to that refrigerator (see Figure 5.8). This voluntary action is made possible by skeletal muscles that are attached to your bones. Two general classifications of muscle are *voluntary* (of which we just had an example) and *involuntary*. Involuntary muscles perform without consciously being told to do so. The smooth muscle found in the walls of organs, often called *visceral muscle*, and the muscle found in the heart, called cardiac muscle, are examples of involuntary muscles. Smooth muscle is also found in blood vessels and airways, where it helps control the diameter of passageways. So, the main parts of the muscle system are skeletal, smooth and cardiac muscles. We were just joking about the 'dry' reading and know you're anxious to 'move' on to the next system.

Integumentary system

All cities must have services important for Public Health, such as paramedics, police, fire-fighters and safety inspectors, who cover and protect the city and its inhabitants. The body's first line of protection is your skin. Skin is the main part of the integumentary system. Besides protecting your body from invasion, the integumentary system also helps to regulate body temperature through sweating, shivering and changes in the diameter of blood vessels in the skin. Much of the sensory information received from the outside world (heat, cold, pain, pressure, etc.) comes from sensors in the skin. Glands in the skin help to

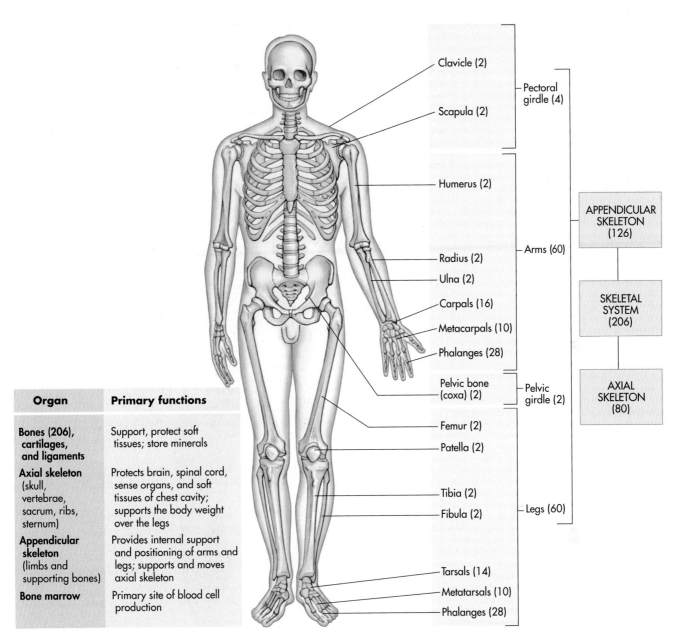

Organ	Primary functions
Bones (206), cartilages, and ligaments	Support, protect soft tissues; store minerals
Axial skeleton (skull, vertebrae, sacrum, ribs, sternum)	Protects brain, spinal cord, sense organs, and soft tissues of chest cavity; supports the body weight over the legs
Appendicular skeleton (limbs and supporting bones)	Provides internal support and positioning of arms and legs; supports and moves axial skeleton
Bone marrow	Primary site of blood cell production

Figure 5.7 The skeletal system.

lubricate and waterproof the skin and also inhibit the growth of unwanted bacteria. The main components of this system include skin, hair, sweat glands, sebaceous glands and nails (see Figure 5.9).

Nervous system

Much like the activities of local government/town hall, the nervous system is the rapid messenger system of the body that both receives and sends messages for activities to occur. The messages conducted by the nervous system are stimulated by the body's internal and external environments. This is important not only so we may experience the world around us, but to also protect

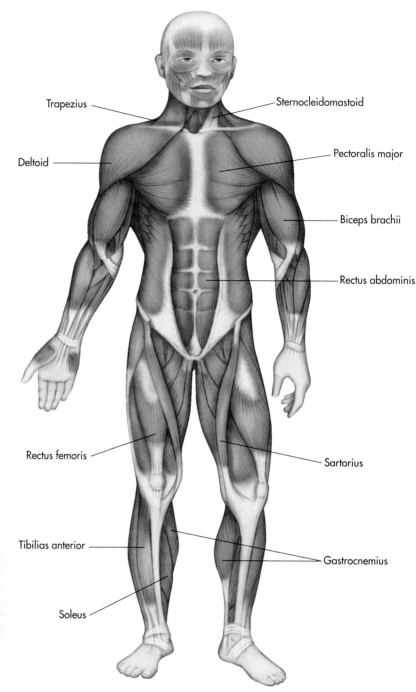

Organ	Primary functions
Skeletal muscles (700)	Provide skeletal movement, control openings of digestive tract, produce heat, support skeletal position, protect soft tissues

Figure 5.8 The muscular system.

us from harm. The nervous system also monitors what is going on inside the body. How do we know when we are hungry or when we have had enough to eat? This information is obtained through **sensations**, which are conscious feelings or an awareness of conditions that occur inside and outside of the body. These sensations are caused by stimulation of our sensory receptors. So, then, the three main functions of the nervous system are sensory (receiving messages), processing and interpreting messages, and motor (acting on those

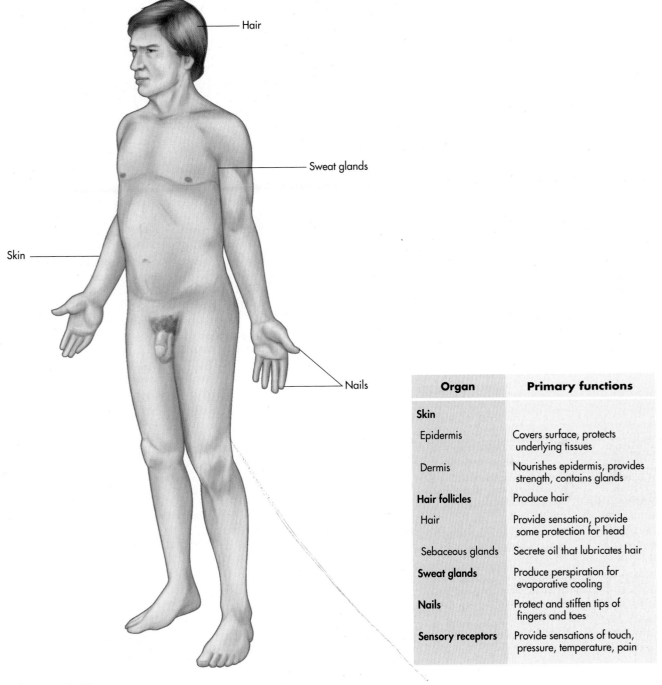

Organ	Primary functions
Skin	
Epidermis	Covers surface, protects underlying tissues
Dermis	Nourishes epidermis, provides strength, contains glands
Hair follicles	Produce hair
Hair	Provide sensation, provide some protection for head
Sebaceous glands	Secrete oil that lubricates hair
Sweat glands	Produce perspiration for evaporative cooling
Nails	Protect and stiffen tips of fingers and toes
Sensory receptors	Provide sensations of touch, pressure, temperature, pain

Figure 5.9 The integumentary system.

messages). The main parts of the nervous system are the nerve cells (glial cells and neurons), the spinal cord with its spinal fluid, peripheral nerves and, of course, the brain. Since we are dealing with sensations, our special sensory organs include the eyes (sight), nose (smell), tongue (taste) and ears (for hearing *and* balance). We look at the special senses in Chapter 11, 'The special senses: the sights and sounds'. Figure 5.10 depicts the nervous system.

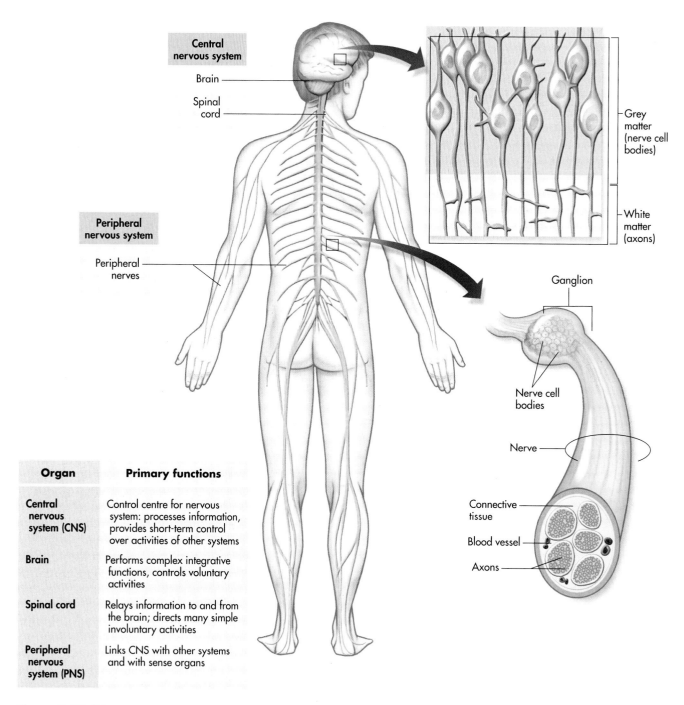

Organ	Primary functions
Central nervous system (CNS)	Control centre for nervous system: processes information, provides short-term control over activities of other systems
Brain	Performs complex integrative functions, controls voluntary activities
Spinal cord	Relays information to and from the brain; directs many simple involuntary activities
Peripheral nervous system (PNS)	Links CNS with other systems and with sense organs

Figure 5.10 The nervous system.

Endocrine system

While not as quick acting as the nervous system, the endocrine system also acts as a control centre for virtually all of the body's organs (see Figure 5.11). This control is accomplished through endocrine glands that release chemical substances called hormones that are circulated via the cardiovascular system. The endocrine system helps to regulate the body's metabolic processes that utilise carbohydrates, fats and proteins, and it plays an important role in the

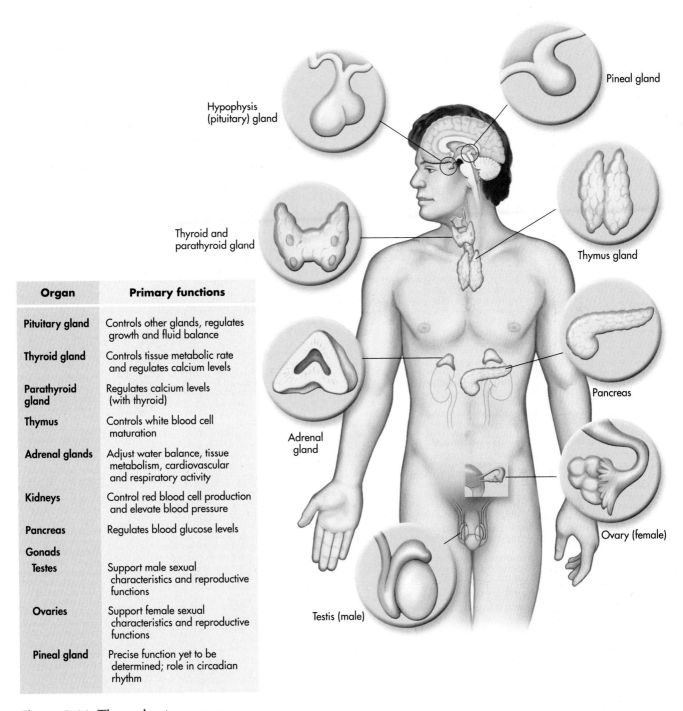

Organ	Primary functions
Pituitary gland	Controls other glands, regulates growth and fluid balance
Thyroid gland	Controls tissue metabolic rate and regulates calcium levels
Parathyroid gland	Regulates calcium levels (with thyroid)
Thymus	Controls white blood cell maturation
Adrenal glands	Adjust water balance, tissue metabolism, cardiovascular and respiratory activity
Kidneys	Control red blood cell production and elevate blood pressure
Pancreas	Regulates blood glucose levels
Gonads	
Testes	Support male sexual characteristics and reproductive functions
Ovaries	Support female sexual characteristics and reproductive functions
Pineal gland	Precise function yet to be determined; role in circadian rhythm

Figure 5.11 The endocrine system.

rate of growth and reproduction. In addition, the endocrine system helps to regulate the fluid and electrolyte balances of the body. If that weren't enough, the hormones produced by the endocrine system help you to deal with general stress and the stresses produced by infection and trauma! The main parts of the endocrine system include the hypothalamus, the pineal, pituitary, thyroid, parathyroid, thymus and adrenal glands, the pancreas and the gonads (testes in males and ovaries in females), plus a variety of hormones.

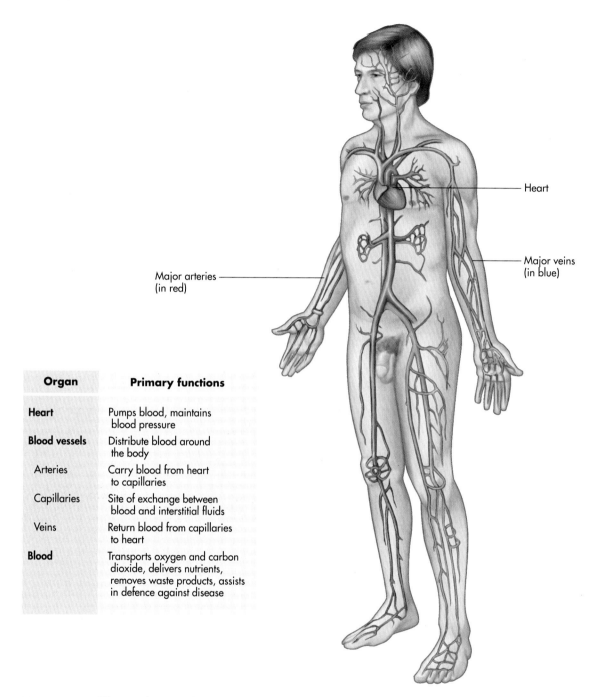

Figure 5.12 The cardiovascular system.

Organ	Primary functions
Heart	Pumps blood, maintains blood pressure
Blood vessels	Distribute blood around the body
Arteries	Carry blood from heart to capillaries
Capillaries	Site of exchange between blood and interstitial fluids
Veins	Return blood from capillaries to heart
Blood	Transports oxygen and carbon dioxide, delivers nutrients, removes waste products, assists in defence against disease

Heart

Major veins (in blue)

Major arteries (in red)

Cardiovascular system

cardio = *heart*
vasculo = *blood vessels*

Often referred to as the circulatory system, the **cardiovascular** system is the main transportation system to each cell of our body, much like the roads, pavements and railways of our city. (see Figure 5.12). Through this system, water, oxygen and a variety of nutrients and other substances necessary for life are transported to the cells, and waste products are transported away from the cells. Also, like our city, these routes can become clogged or blocked, causing major problems. Imagine what happens to a busy three-lane motorway if two of the lanes are closed because of roadworks or an accident. The traffic slows and pressure builds up due to the congestion, much like the blood flow does when the arteries become partially obstructed. The build-up of pressure (hypertension) can be very dangerous. The main components of this system are the heart, arteries, veins, capillaries and, of course, the blood.

Respiratory system

If you think about a time you went swimming and stayed a little too long under water, you will understand how important our respiratory system is. We all know the old concept of being able to live weeks without food and days without water, but think about how long you would last without oxygen. Without conscious effort, your lungs move approximately 9000 litres of air a day. Our respiratory system, much like the ventilation system in an office building, not only supplies us with fresh oxygen but performs several other important functions as well. Our lungs eliminate the carbon dioxide created as a result of cellular metabolism. The respiratory system filters, warms and moistens air as it is inhaled. The mucous lining of the airway – together with the cilia – helps trap foreign particles and germs. This system also helps to maintain the proper acid–base balance of the blood by changing breathing patterns when necessary. The main parts of the respiratory system are the pharynx, larynx, trachea, bronchial tubes and lungs (see Figure 5.13).

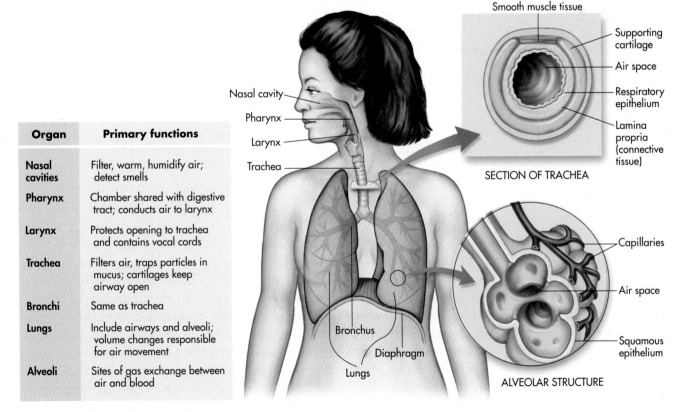

Organ	Primary functions
Nasal cavities	Filter, warm, humidify air; detect smells
Pharynx	Chamber shared with digestive tract; conducts air to larynx
Larynx	Protects opening to trachea and contains vocal cords
Trachea	Filters air, traps particles in mucus; cartilages keep airway open
Bronchi	Same as trachea
Lungs	Include airways and alveoli; volume changes responsible for air movement
Alveoli	Sites of gas exchange between air and blood

Figure 5.13 The respiratory system.

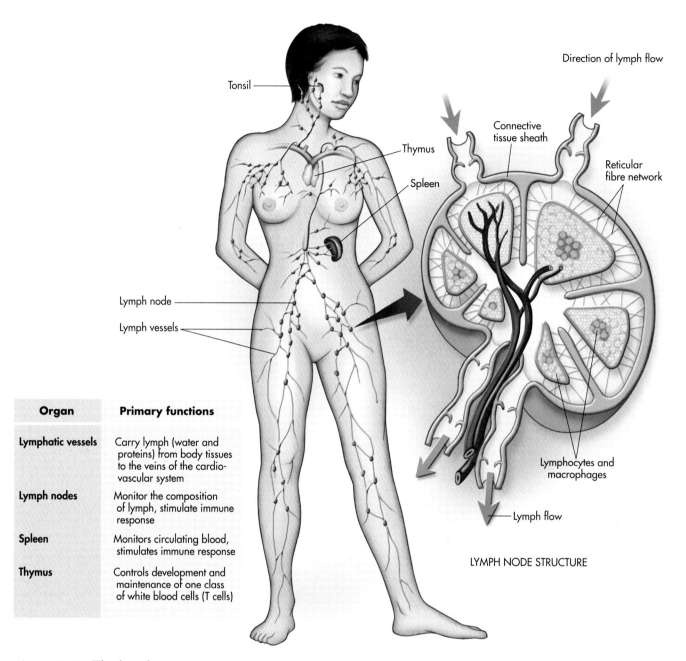

Organ	Primary functions
Lymphatic vessels	Carry lymph (water and proteins) from body tissues to the veins of the cardio-vascular system
Lymph nodes	Monitor the composition of lymph, stimulate immune response
Spleen	Monitors circulating blood, stimulates immune response
Thymus	Controls development and maintenance of one class of white blood cells (T cells)

Figure 5.14 The lymphatic system.

Lymphatic and immune system

Much like the drainage system of our city, this very important, but often forgotten, system is responsible for helping to maintain proper fluid balance in our body and to protect it from infection. Excess fluid that may collect in places it shouldn't in the body is brought back into the lymphatic system, cleaned and processed, and then recirculated. Special structures called lymph nodes act as filters to capture unwanted infective agents. Lymph vessels and ducts, lymph nodes, the thymus gland, tonsils and the spleen are the major parts of the immune system. In addition, the immune portion of the lymphatic system produces specialised infection-fighting white blood cells called lymphocytes. The immune system is the police force of the human body, patrolling for harmful invaders (see Figure 5.14).

Organ	Primary functions
Salivary glands	Provide lubrication, produce buffers and the enzymes that begin digestion
Pharynx	Passageway connected to oesophagus
Oesophagus	Delivers food to stomach
Stomach	Secretes acids and enzymes
Small intestine	Secretes digestive enzymes, absorbs nutrients
Liver	Secretes bile, regulates blood chemistry
Gallbladder	Stores bile for release into small intestine
Pancreas	Secretes digestive enzymes and buffers; contains endocrine cells
Large intestine	Removes water from faecal material, stores waste

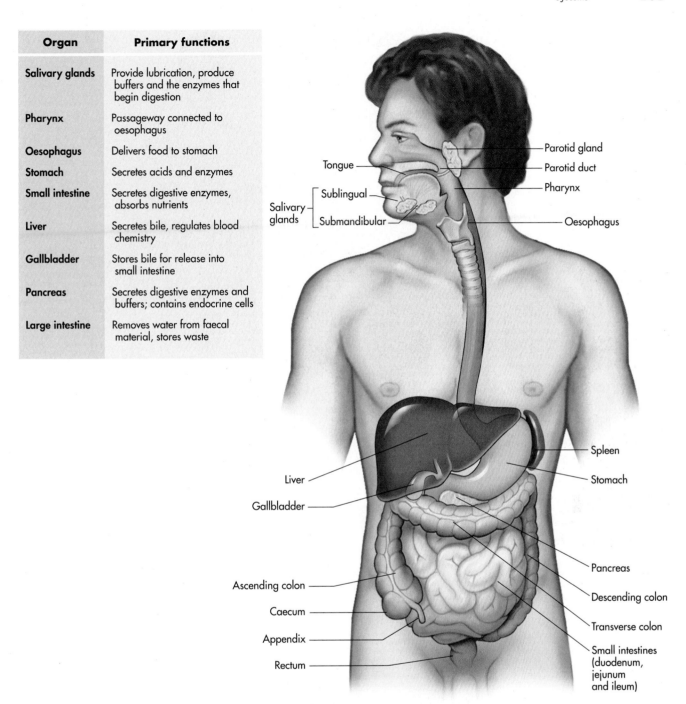

Figure 5.15 The digestive system.

Gastrointestinal, or digestive, system

The digestive system (often called the GI system) breaks down raw materials (food), both mechanically and chemically, into usable substances (see Figure 5.15). Once these usable substances are created, this system absorbs them for transportation via the blood to the cells of the body. Materials that aren't used, as well as cellular waste, are transported out of the body by this system as faeces, much like the waste disposal and sewage system of our city. The main parts of the digestive system are the mouth, pharynx, oesophagus, stomach, intestines, accessory organs and anal canal.

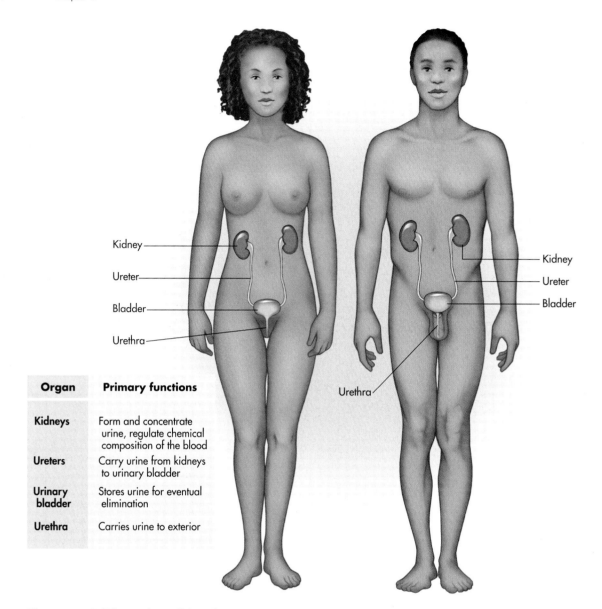

Organ	Primary functions
Kidneys	Form and concentrate urine, regulate chemical composition of the blood
Ureters	Carry urine from kidneys to urinary bladder
Urinary bladder	Stores urine for eventual elimination
Urethra	Carries urine to exterior

Figure 5.16 The male and female urinary systems.

Urinary system

While the digestive system plays a large role in the elimination of certain digested waste, the urinary system plays an important role in the elimination of waste products in the form of nitrogen. In addition, electrolytes, drugs and other toxins, and excessive water are removed. This system is crucial for maintaining the proper balance of water you have in your body and regulating your blood pressure. The urinary system helps regulate the number of red blood cells and the acid–base and electrolyte balance of blood. The main parts of the urinary system are the kidneys, ureters, urinary bladder and urethra (see Figure 5.16).

Reproductive system

We build new cities or new buildings to accommodate growing needs or to replace worn out structures. The reproductive system does the same thing.

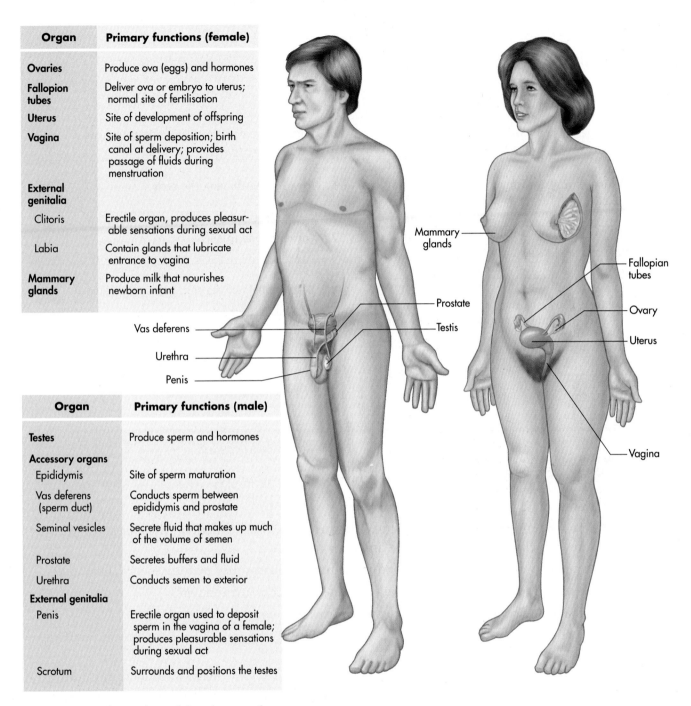

Organ	Primary functions (female)
Ovaries	Produce ova (eggs) and hormones
Fallopion tubes	Deliver ova or embryo to uterus; normal site of fertilisation
Uterus	Site of development of offspring
Vagina	Site of sperm deposition; birth canal at delivery; provides passage of fluids during menstruation
External genitalia	
Clitoris	Erectile organ, produces pleasurable sensations during sexual act
Labia	Contain glands that lubricate entrance to vagina
Mammary glands	Produce milk that nourishes newborn infant

Organ	Primary functions (male)
Testes	Produce sperm and hormones
Accessory organs	
Epididymis	Site of sperm maturation
Vas deferens (sperm duct)	Conducts sperm between epididymis and prostate
Seminal vesicles	Secrete fluid that makes up much of the volume of semen
Prostate	Secretes buffers and fluid
Urethra	Conducts semen to exterior
External genitalia	
Penis	Erectile organ used to deposit sperm in the vagina of a female; produces pleasurable sensations during sexual act
Scrotum	Surrounds and positions the testes

Figure 5.17 The male and female reproductive systems.

genitourinary (*jenn ih toe YOO rin air ee*)

Quite simply, without this system, we would not exist. The reproductive system is often combined with the urinary system to create the **genitourinary**, or **GU** system. Humans require a male and female to produce offspring. The male is needed to provide sperm that contains certain genetic traits of that individual, while the female provides an egg with her traits and a place for the fertilised egg to grow to maturity. The main female parts of this system are the ovaries, eggs, Fallopian tubes, uterus and vagina. For men, the main parts are the testes, sperm and penis (see Figure 5.17).

Test your knowledge 5.2

List the correct system for the following activities:

1. Exchanges carbon dioxide for oxygen

   _____

2. Eliminates nitrogen, drugs and excessive water from the body

   _____

3. Main storage for calcium

   _____

4. Maintains body temperature and provides much of the sensory information from the external world

   _____

5. Protects the body from invading pathogens

   _____

6. Moves blood through the body

   _____

7. Converts food to energy

   _____

8. With the help of the sun, produces vitamin D

   _____

Summary

- Cells are the basic building blocks of the body.
- Tissue is a collection of similar cells that act together to perform a function. The four main types of tissues are epithelial, connective, muscle and nervous.
- Membranes are sheet-like structures found throughout the body; they perform specific functions.
- The four major membrane types are cutaneous, serous, mucous and synovial.
- Tissues that combine to perform a specific function or functions are called an organ.
- Organs that work together, often with the help of accessory structures, to perform specific activities create a system.
- There are 11 major body systems: skeletal, muscular, integumentary, nervous, endocrine, cardiovascular, respiratory, lymphatic/immune, gastrointestinal, urinary and reproductive. Even though these are distinct systems, they are interrelated, and their relationships are highlighted in later chapters.

Case study

A 73-year-old male presents to the Accident and Emergency department of a local hospital. Initial assessment reveals the following:

- afebrile
- mild tachypnoea with mild shortness of breath
- cyanosis of the hands and feet
- mild tachycardia
- history of smoking
- history of diabetes
- moderately overweight

1. Based on the information given, identify which system or systems of the body they would want to further investigate to determine why this individual has come to the Accident and Emergency department. You have had some medical terminology already but may need additional help from the text, a medical dictionary or MasteringA&P.
2. Compile a list of specialists or healthcare professionals to whom you might refer this patient, and explain why.

Review questions

Multiple choice

1. Blood can be classified as which type of tissue?
   a. connective
   b. cardiac
   c. nerve
   d. muscle

2. This membrane lines body cavities and covers the organs found in those cavities.
   a. cutaneous
   b. serous
   c. mucous
   d. mucus

3. This muscle type has interlocking cells for more efficient contraction.
   a. skeletal
   b. neuroglial
   c. cardiac
   d. smooth

4. The acid–base balance found in your body is mainly controlled by the following system:
   a. skeletal
   b. urinary
   c. endocrine
   d. none of the above

5. Which of the following organs belong to the digestive system?
   I. urethra
   II. gallbladder
   III. spleen
   IV. small intestine
   a. I, II
   b. I, III, IV
   c. III, IV
   d. II, IV

6. Which of the following membranes cover organs?
   a. parietal
   b. visceral
   c. smooth
   d. mucous

Fill in the blanks

1. The system that co-ordinates all the body's functions is the _____ system.

2. The _____ system is the 'motion' system.

3. Cartilage is a specific type of _____ tissue.

4. The layer of serous membrane that covers specific organs is called the _____ layer.

5. The skin is composed of _____ _____ tissues.

6. The lymphatic system removes excess _____ and returns it to the circulatory system.

Short answers

1. List in order from simplest to most complex the following: organs, cells, systems, tissues.

   _____

   _____

   _____

2. What is the purpose of dendrites?

   _____

   _____

   _____

3. What is the purpose of synovial fluid?

   _____

   _____

   _____

4. Contrast the three types of muscle tissues and identify where they are found.

   _____

   _____

   _____

Suggested activities

1. Choose one system to research. Identify all the tissue types and membranes found within that system and describe the individual tissue and membrane functions.

2. Create five to ten multiple choice quiz questions related to this chapter and have a quiz show contest.

Answers

Answers to case study

1. He will require investigation of his cardiac and respiratory systems to find out why he is short of breath and has blue hands and feet. His history of smoking makes him at greater risk of respiratory disorders such as chronic bronchitis. As he is also diabetic and overweight he is also at risk of cardiac disorders such as ischaemic heart disease.

   He has no increase in body temperature so it is unlikely he has an infection.

2. He may need to be referred to a cardiologist or a respiratory specialist to confirm the presence or absence of related diseases.

   Referral to a dietician may also be useful to help him lose weight and ensure that his diabetes is under control.

   Referral to a smoking cessation programme may also help him to stop smoking to reduce his risk of respiratory or cardiac disease.

Answers to multiple choice questions

1. a
2. a
3. c
4. b
5. d
6. b

Answers to fill in the blanks

1. Nervous
2. muscular
3. Connective
4. Visceral
5. dense connective
6. Fluid

Answers to short answer questions

1. Cells – tissues – organs – systems.
2. Dendrites are branches at the end of a neuron that carry information to the cell body.
3. Synovial fluid reduces friction between joints when they move.
4. Types are:

   Skeletal muscle – also referred to as striated muscle – attached to bones and causes movement by contracting and relaxing.

   Smooth muscle – forms the walls of hollow organs. It is considered to be an involuntary muscle group.

   Cardiac muscle – only found in the heart, and contracts/relaxes without conscious thought.

The skeletal system

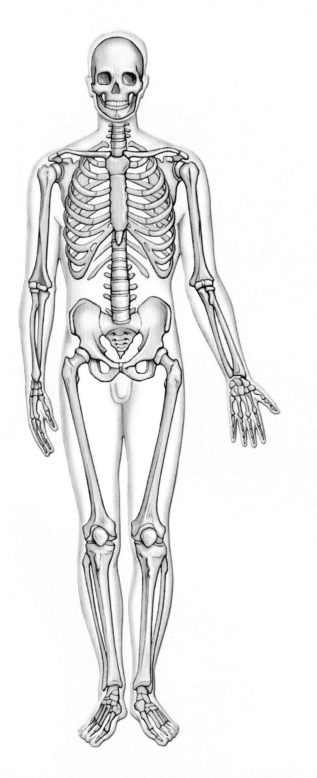

Although we cannot see the framework that holds up a house, without it everything would fall apart. We can think of the human body as a house. The wood framework is the skeleton, composed of bones that provide shape and strength, upon which the muscles and skin are layered, much like stone or wooden cladding.

What's the first thing you think of when someone asks about the function of the skeleton? From our analogy, you would probably say, 'to provide support and allow us to move'. But the skeleton does so much more. As explained in this chapter, the bones that make up your skeleton also protect the soft body parts, produce blood cells, and act as a storage unit for minerals and fat. In this chapter, we discuss the make-up and importance of the 206 bones in the adult skeleton as well as cartilage, ligaments and joints. It's time to 'bone up' on the skeletal system!

Learning objectives

At the end of this chapter, you will be able to:

◆ Describe the functions of the skeletal system
◆ Identify and describe the anatomy and physiology of bone
◆ Locate and describe the various bones within the body
◆ Differentiate between bone, cartilage, ligaments and tendons
◆ Locate and describe the various joints and types of movement of the body
◆ Explain common diseases and disorders of the skeletal system

Pronunciation guide

appendicular skeleton (*app en DICK yoo lar SKELL it on*)
arthritis (*arth RYE tiss*)
articulation (*ar TICK yoo LAY shun*)
axial skeleton (*AK see yall SKELL it on*)
cancellous bone (*CAN sell uss*)
diaphysis (*dye AFF ih siss*)

epiphyseal plate (*eh PIFF ih SEAL*)
epiphysis (*eh PIFF ih siss*)
medullary cavity (*meh DULL air ree*)
osseous tissue (*OSS ee yuss*)
ossification (*OSS iff ih KAY shun*)
osteoarthritis (*OSS tee oh arth RYE tiss*)

osteocyte (*OSS tee oh sight*)
osteons (*OSS tee yonz*)
periosteum (*pair ee OSS tee yum*)
synovial fluid (*sigh NO vee yall*)
trabeculae (*tra BECK yoo lay*)
vertebrae (*VERT eb ray*)

System overview: more than the 'bare bones' about bones

The skeleton has many more uses than to just scare people on Hallowe'en. It is a wondrous structure that serves more functions than simply providing a framework for the human body. It also produces red blood cells, provides protection for organs – the hard skull protects the delicate brain, helps us to breathe via the ribs, acts as a warehouse for mineral storage (e.g. calcium) and, along with the muscular system, allows for movement.

General bone classification

The primary components of the skeleton are bones. Although they may seem lifeless and are composed of non-living minerals such as calcium and phosphorus, they are very much alive, constantly building and repairing themselves. This is kind of ironic because the word *skeleton* is derived from the Greek word meaning 'dried-up body'.

We can classify bone types according to their shape:

- long bones
- short bones
- flat bones
- irregular bones

Long bones are longer than they are wide and are found in your arms and legs. *Short bones* are fairly equal sized in width and length, similar to a cube, and are mostly found in your wrists and ankles. *Flat bones* are thinner bones that can be either flat or curved and are plate-like in nature. Examples of flat bones are the skull, ribs and breastbone (**sternum**). *Irregular bones* are like the parts of a jigsaw puzzle. These are the odd-shaped bones needed to connect to other bones. Some examples of irregular bones are the hip bones and the **vertebrae** that make up your spine. See Figure 6.1 for a view of the various bone shapes.

sternum (*STER num*)

vertebrae (*VERT eb ray*)

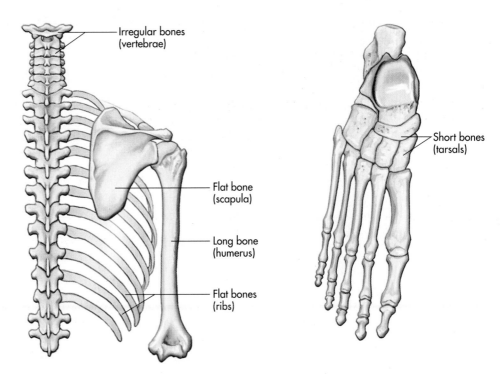

Figure 6.1 Various bone shapes.

Basic bone anatomy

Let's look at the overall construction of a bone by examining the long bone in Figure 6.2. Bone is covered with **periosteum**, which is a tough and fibrous connective tissue. This cover contains blood vessels, which transport blood and nutrients into the bone to nurture the bone cells. It also contains lymph vessels and nerves. In addition, the periosteum acts as anchor points for ligaments and tendons, which we discuss later. Note in Figure 6.2 that both ends of the long bone increase in size. Each bone end is called an **epiphysis**. The region between or 'running through' the two ends is called the **diaphysis**.

You can see that the diaphysis is hollow. This hollow region is called the **medullary cavity** and acts as a storage area for bone marrow; much like a kitchen cabinet that contains food is hollow but well stocked. There are two kinds of bone marrow, yellow and red. Yellow marrow has a high fat content. In emergencies – for instance, in the event of massive blood loss – when you need more red blood cells, some of the yellow marrow can convert to red bone marrow to help in red blood cell production.

Bone tissue

There are two types of bone tissue: compact and spongy. **Compact bone** is a dense, hard tissue that normally composes the shafts of long bones and the outer layer of other bones. Microscopic examination reveals that the material of compact bone is tightly packed. This makes for a dense and strong structure. This material forms microscopic cylindrical-shaped units called **osteons**, or *Haversian systems*. Each unit has mature bone cells (**osteocytes**) forming concentric circles around blood vessels. The area around the osteocytes is

periosteum (*pair ee OSS tee yum*)
  peri = *around*
  osteum = *bone*

epiphysis (*eh PIFF ih siss*)
  epi = *over or upon*
diaphysis (*dye AFF ih siss*)
  dia = *through*
medullary cavity (*meh DULL air ree*)

osteons (*OSS tee yonz*)
osteocytes (*OSS tee oh sight*)

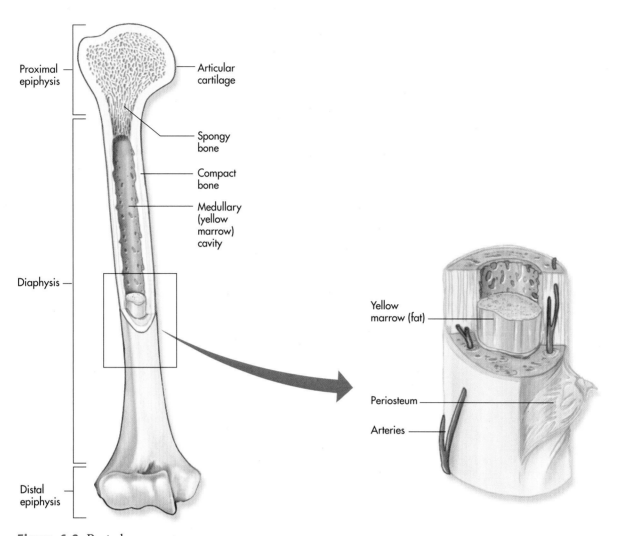

Figure 6.2 Basic bone anatomy.

filled with protein fibres, calcium and other minerals. The osteons run parallel to each other with blood vessels laterally connecting with them to ensure sufficient oxygen and nutrients for the bone cells.

Spongy (or **cancellous**) bone is different from compact bone. Instead of Haversian systems, spongy bone tissue is arranged in bars and plates called **trabeculae**. Irregular holes between the trabeculae give the bone a spongy appearance. Spongy bone is lined with endosteum, a tissue similar to periosteum. This serves two purposes: it helps make the bones lighter in weight and it provides a space for red bone marrow, which produces red blood cells (see Figure 6.3).

Surface structures of bones

Bone is not perfectly smooth. If you examine one closely, you will find a variety of projections, bumps and depressions on its surface. Generally, projecting structures act as points of attachment for muscles, ligaments or tendons, while grooves and depressions act as pathways for nerves and blood vessels. Both projecting structures and depressions can work together as joining or articulation points to form joints such as the ball and socket joint in the hip. Table 6.1 lists many of these bone features.

cancellous **bone** (*CAN sell uss*)
trabeculae (*tra BECK yoo lay*)

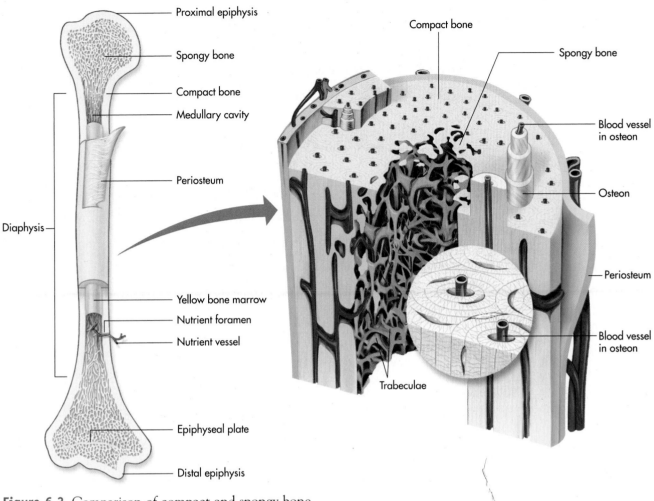

Figure 6.3 Comparison of compact and spongy bone.

Amazing body facts

Red bone marrow and red blood cells

The areas of bone with red bone marrow, which produces red blood cells, are found in the skull, clavicles (collar bones), vertebrae of the spinal column, sternum (breastbone), ribs and pelvis and in the spongy bone that makes up the epiphysis of the long bones. The production of red blood cells, known as **erythropoiesis**, is truly an incredible process! Since red blood cells last only about 120 days, red blood cell production is a constant job in order to maintain the 25,000,000,000,000 (give or take a few) red blood cells contained in the human body. As a result, it has been calculated that approximately 3 million new red blood cells are created every second! What's more, your body can step up production to ten times that rate in cases of severe blood loss. If the red marrow can't maintain the needed production, some of the yellow marrow can be converted to red marrow to assist.

erythropoiesis (*eh RITH roh poy EE siss*)
erythro = *red*
poiesis = *to make*

Table 6.1 Bone features

Bone surface structures	Descriptions
Projecting structures and processes	
condyle	a large, rounded knob, usually articulating with another bone
crest	a narrow ridge
epicondyle	an enlargement near or superior to a condyle
facet	a small, flattened area
head	an articulating end of a bone that is rounded and enlarged
process	a prominent projection
spine	a sharp projection
trochanter	located only on the femur; a larger version of a tubercle
tubercle	a knob-like projection
Depressions and openings	
foramen	a passageway through a bone for blood vessels, nerves and ligaments; a hole
fossa	either a groove or shallow depression
meatus	a tube or tunnel-like passageway through bone
sinus	a hollow area

Test your knowledge 6.1

Complete the following:

1. Label the following on a diagram:
   a. diaphysis
   b. proximal and distal epiphysis
   c. periosteum
   d. spongy bone
   e. compact bone
   f. medullary cavity
   g. epiphyseal plate

2. Where is red bone marrow found in the body and what is its function?

3. Discuss three functions of bone in your body.

4. Mature bone cells in compact bone are called
   a. lymphocytes
   b. osteocytes
   c. leucocytes
   d. monocytes

5. The end of a long bone is called the
   a. epiphysis
   b. periosteum
   c. diaphysis
   d. tubercle

Bone growth and repair

ossification (OSS iff ih KAY shun)

Ossification, or osteogenesis, is the formation of bone in the body. Bones grow *longitudinally* in order to develop height, and they grow *horizontally* (wider and thicker) so they can more efficiently support body weight and any other weight we support when we work or play. There are four types of cells involved in the formation and growth of bone:

- osteoprogenitor cells
- osteoblasts
- osteocytes
- osteoclasts

osteo = *bone*
progeny = *offspring*
blast = *immature stage of cell development*

Osteoprogenitor cells are non-specialised cells found in the periosteum, endosteum and central canal of compact bones. Non-specialised cells can turn into other types of cells as needed. **Osteoblasts** are the cells that actually form bones. They arise from the non-specialised osteoprogenitor cells and are the bone cells that secrete a matrix of calcium with other minerals that give bone its typical characteristics. Osteocytes are considered mature bone cells that were originally osteoblasts. In other words, osteoblasts surround themselves with a matrix of calcium to then become the mature osteocytes. So bone is built up or formed by osteoprogenitor cells becoming osteoblasts, which surround themselves with a mineral matrix to become full-blown osteocytes or bone cells.

clast = *causing breakage into parts*

Not only does the body constantly build bone but it must also constantly be able to tear down old bone. This is the job of the **osteoclast**. It is believed that osteoclasts originate from a type of white blood cell called a monocyte that is found in red bone marrow. Amazingly, the osteoclasts' job is to tear down bone material and help move calcium and phosphate into the blood! You can think of osteoblasts and osteoclasts as employees of a house remodelling company: the osteoblasts are masons laying down brickwork to make new exterior walls, and the osteoclasts are tearing out the inside to remodel! As explained shortly, the job the osteoclasts do is very important for bone growth and repair.

intra = *within*
endo = *within*
chondral = *referring to cartilage*

Bone development and growth begins when you are growing in the womb through **intramembranous** and **endochondral** ossification. *Intramembranous* ossification occurs when bone develops between two sheets composed of fibrous connective tissue, such as in the development of your skull. Cells from connective tissue turn into osteoblasts and form a matrix that is similar to the trabeculae of spongy bone, while other osteoblasts create compact bone over the surface of the spongy bone. As discussed previously, once the matrix surrounds the osteoblasts, they become osteocytes, and this is how bones of the skull develop.

The majority of your skeletal bones are created through endochondral ossification in which shaped cartilage is replaced by bone. Refer to Figure 6.4. In this situation, which begins several months before birth, periosteum surrounds the diaphysis of the 'cartilage' bone as the cartilage itself begins to break down. Osteoblasts come into this region and create spongy bone in an area that is then called the *primary ossification centre*. Meanwhile, other osteoblasts begin to form compact bone under the periosteum. Here is where the osteoclasts come into play. Their job is to break down the spongy bone of the diaphysis to create the medullary cavity. After you are born, the epiphyses on your long bones continue to grow. However, shortly after birth, secondary ossification of this area begins with spongy bone forming and not breaking down. A thin band

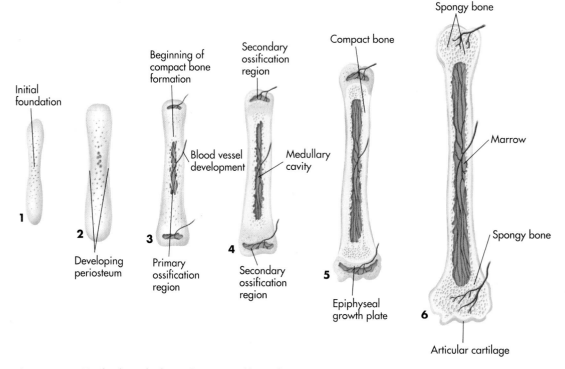

Figure 6.4 Endochondral ossification of long bone.

epiphyseal plate *(eh PIFF ih SEAL)*

of cartilage forms an **epiphyseal plate** (often called the *growth plate*) between the primary and secondary ossification centres. This plate is important because, as long as it exists, the length and width of the bone will increase. As explained in Chapter 12, 'The endocrine system', hormones such as growth hormone (GH) control the growth of bones, which means that eventually the plates become ossified, thereby stopping bone growth.

Amazing body facts

Ageing and bone building

Even in adults, bone continues to be broken down and rebuilt. In fact, about 10 per cent of the body's bone is broken down and rebuilt each year! Your bones continue to increase in mass well into your twenties. The osteoclasts break down and remove worn-out bone cells and deposit calcium into the blood. They last for approximately three weeks. Osteoblasts then pull the calcium out of the blood, form and surround themselves with that mineralised matrix we talked about, and mature into osteocytes. Because of this continual breakdown of old and creation of new bone, adults actually need more calcium in their diets than do children! The process of breaking down and rebuilding bone continues well into you forties so there normally is no net gain or loss of bone mass.

This continuing process allows your body to sculpt bone into shapes that accommodate the body's activity. For example, exercise, such as running or weightlifting, causes calcium to stay in the bone, making it thicker, denser and stronger than those of a sedentary person (couch potato) or an individual who is in outer space. Continuous or repeated actions or postures tend to cause bone to be re-sculpted. For example, due to constant squatting, a certain pattern of bumps forms on the bones of the hips, shins and knees. As a result of this pattern, it was determined that Neanderthal man squatted rather than sat.

Cartilage

cartilage *(KAR tih lij)*

Cartilage is a special form of dense connective tissue that can withstand a fair amount of flexing, tension and pressure. Cartilage has many roles throughout the body. The flexible parts of your nose and ears are cartilage (imagine how many people would have broken off ears and noses if it weren't for cartilage!).

Cartilage also makes a flexible connection between bones. For example, the cartilage between the breastbone and ribs allows your chest to flex and give so you don't break your ribs when you run into things or collide with another player during a rugby game. Something as simple as taking a deep breath, could become a major struggle if it weren't for this flexibility.

This amazing tissue also acts as a cushion between the bones. As you can see in Figure 6.5, *articular cartilage* is located on the ends of bones and acts as a shock absorber, preventing the bone ends from grinding together as they move. In addition at this location, a small sac, called the **bursa**, contains a lubricant called **synovial fluid**. Even with cartilage and synovial fluid protecting the area between bones, joints can wear out and become inflamed, resulting in a condition called **arthritis**, or **osteoarthritis**.

bursa *(BER sah)*

synovial fluid *(sigh NO vee yall)*

arthritis *(arth RYE tiss)*
  arth = *joint*
  itis = *inflammation of*

osteoarthritis *(OSS tee oh arth RYE tiss)*
  osteo = *bone*

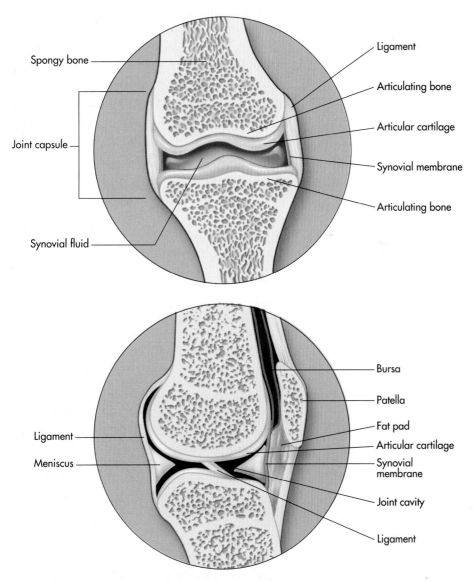

Figure 6.5 Articular cartilage and synovial joint.

Joints and ligaments

articulation (ar TICK yoo LAY shun)

Without **joints**, the body could not move. When two or more bones *join* together, a joint, or an **articulation**, is formed. Freely moving joints have to be held together and yet still be movable. This is accomplished through the use of another specialised connective tissue called a **ligament**. (Again, see Figure 6.5.) Ligaments are very tough, whitish bands that connect from bone to bone and can withstand pretty heavy stress. Do not confuse ligaments with **tendons**. While ligaments hold bone to bone, tendons are cord-like structures that attach muscle to bone. There are several types of joints, and each works in a specific way.

Joints are classified either by function or structure. In terms of function, joints can be immobile, can move a little, or can move freely. For example, skull sutures are immobile, the pubic symphysis between your pelvic bones moves a little, and your elbow moves freely. If we characterise joints by structure, we divide them based on the type of connective tissue that links the bones together. **Fibrous joints** are held together by short connective tissue strands. They are either immobile or slightly movable. The sutures in your skull are fibrous joints. **Cartilaginous joints** are held together by cartilage discs. The pubic symphysis and the joints between your ribs and sternum are cartilaginous joints. Cartilaginous joints are either immobile or slightly movable. Finally, **synovial joints** are joined by a joint cavity lined with a synovial membrane and filled with synovial fluid. All synovial joints are freely moving. Synovial joints are constructed in various ways that determine how they can move.

Test your knowledge 6.2

Choose the best answer:

1. A term that can be used to describe the formation of bone is:
   a. ossification
   b. periosteum
   c. maturation
   d. osteoclasts

2. These cells actually form bones:
   a. osteoclasts
   b. pericytes
   c. generator cells
   d. osteoblasts

3. Another name for the 'growth plate' is:
   a. tectonic plate
   b. epiphyseal plate
   c. upper plate
   d. periosteum plate

4. This special connective tissue composes your ears and nose:
   a. tendons
   b. ligaments
   c. cartilage
   d. adipose

5. This lubricant helps to prevent wear between the joints:
   a. pleural fluid
   b. synovial fluid
   c. lymphatic fluid
   d. interstitial fluid

6. These structures attach bone to bone:
   a. ligaments
   b. tendons
   c. muscles
   d. articulations

- *Pivot joints* (which act like a turnstile) are the type of joint found in your neck and forearm. Pivot joints can only rotate.
- *Ball and socket joints* are located in your hips and shoulders and can perform all types of movement, including rotation.
- *Hinge joints* are found in your knees and elbows. They can either open or close like a door.
- *Gliding joints* are the flat, or slightly curved, plate-like bones found in your wrists and ankles. Gliding joints slide back and forth.
- *Saddle joints* have a bone shaped just like a saddle and another bone similar to a horse's back. This joint type is found in the base of your thumb. Saddle joints rock up and down and side to side.
- *Condyloid joints* occur as a result of an oval-shaped bone end fitting into an elliptical cavity in the other end so there is movement from one plane to another but no rotation. The knuckles of your finger are condyloid joints.
- *Ellipsoidal joints* provide two axes of movement through the same bone, like the joint formed at the wrist with both the radius and ulna.

To better visualise these various joints, look at Figure 6.6.

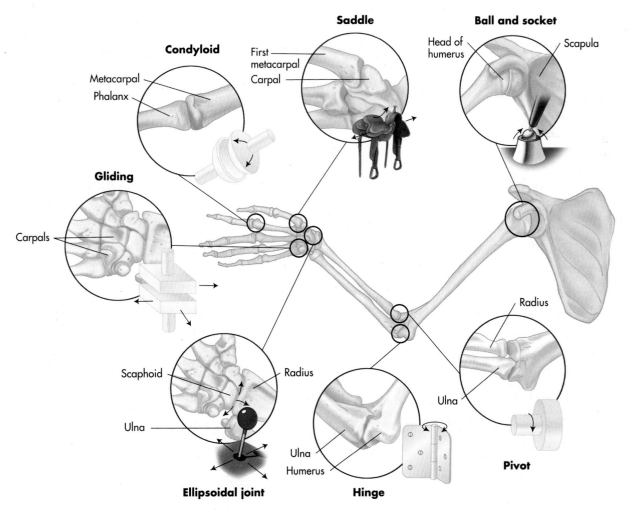

Figure 6.6 Types of joints.

Movement classification

Since joints allow for various types of movement, these individualised movements can also be classified, as you can see in the examples in Figure 6.7. *Flexion* occurs when a joint is bent, decreasing the angle between the involved bones, as when the leg is bent at the knee. *Extension* is a result of straightening a joint so the angle between the involved bones increases, as occurs with a kicking motion. Ballerinas utilise *plantar flexion* when they dance on their toes. *Dorsiflexion* occurs when the foot is bent up toward the leg. If the joint is forced to straighten beyond its normal limits, *hyperextension* occurs.

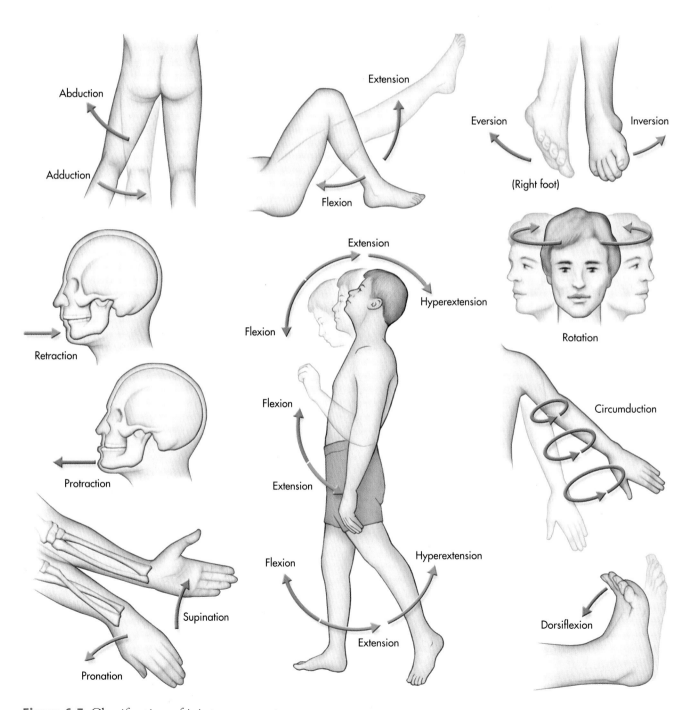

Figure 6.7 Classification of joint movements.

Abduction and *adduction* can be confusing. A**b**duction means to move away from the body's midline (think, '**B**e gone!' as you move your arm up and away to swat a bee). A**dd**uction means to move *toward* the midline of the body. To remember adduction, think of your a**dd**ress, where packages and post come to you. Then when you move your arm back toward yourself after swatting at a bee, you are a**dd**ucting your arm.

Inversion results when the sole of one foot is turned inward so it points to the other foot, while *eversion* is the opposite: the foot is turned outward, pointing away from the opposite foot. *Supination* occurs when the hand is turned to the point where the palm faces upward; *pronation* turns the palm downward. Although you may not have heard of *circumduction*, you have seen this combination of movements in the circular arm movement that a bowler in cricket utilises.

Protraction is the motion of drawing a part forward. *Retraction* is the motion of drawing backward. Figure 6.7 shows the protraction and retraction of the jaw. The movements are analogous to a turtle sticking his head out (protracting) and drawing it back in (retracting) to the shell.

Finally, *rotation* is when a bone 'spins' on its axis. An example is when your head rotates (looking left and right) before you cross the road.

The skeleton

axial skeleton (*AK see yall SKELL it on*)

appendicular skeleton (*app en DICK you lar SKELL it on*)

Anatomically, the skeleton can be divided into two main sections. The **axial skeleton** includes the bones of the bony thorax, spinal column, hyoid bone, bones of the middle ear, and skull. This part of the skeletal system protects the organs of the body and is composed of 80 bones. The **appendicular skeleton** is, as its name implies, the region of your appendages (arms and legs), as well as the connecting bone structures of the hip and shoulder girdles, and contains 126 bones (see Figure 6.8). Interestingly, nearly half of the total number of your bones can be found in your hands and feet!

Special regions of the skeletal system

The skeleton consists of many different special regions. These distinctions make it easier to locate and discuss the hundreds of bones and associated components in this system. In this section, we make a quick tour of each region, beginning at the top.

The human skull

The skull protects and houses the brain and has openings needed for our sensory organs, such as the eyes, nose and ears. It also forms the mouth, which is a common passageway for both the digestive and respiratory systems. The skull contains fibrous connective tissue joints called *suture* lines that hold the bony plates of the skull together. While these joints are not actually movable, they do provide some degree of flexibility, which is important to absorb shock from a blow to the head, thus decreasing the chance of a skull fracture. Figure 6.9 shows the bones of the skull in greater detail.

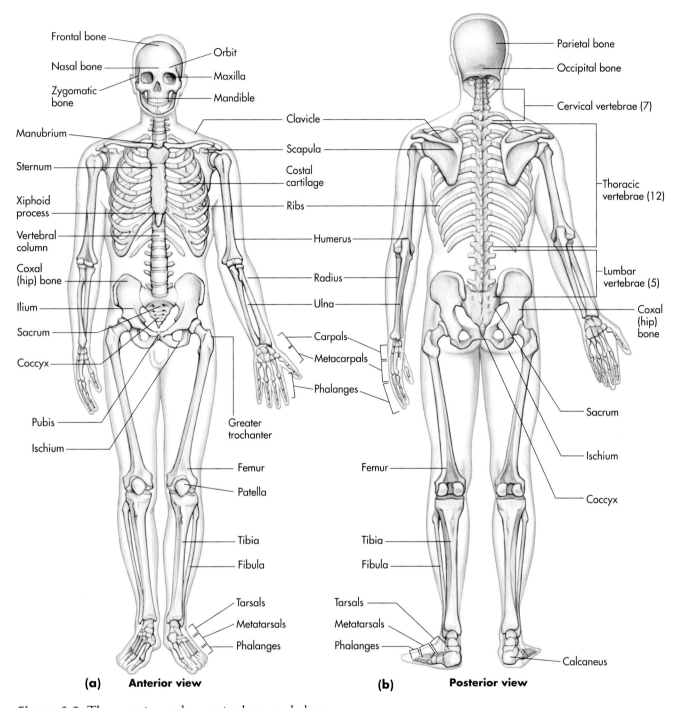

Figure 6.8 The anterior and posterior human skeleton.

The bony thorax

The bones of the chest form a thoracic 'cage' that provides support and protection for the heart, lungs and great blood vessels (see Figure 6.10). This cage is flexible because of cartilaginous connections that allow for movement during the process of breathing. The sternum, or breastbone, is the anatomical location for conducting compressions of the heart during cardiopulmonary resuscitation (CPR). The sternum is composed of three distinct areas. The *manubrium* is the superior portion, and the *body* is the largest, central portion. The *xiphoid* is the final and inferior portion that ossifies (hardens) by age 25

Figure 6.9 Bones of the skull.

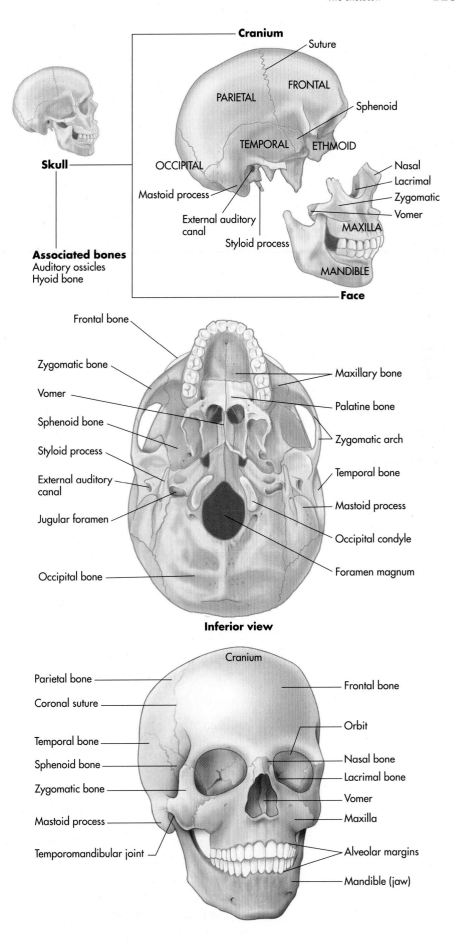

Cranium
Suture
FRONTAL
PARIETAL
Sphenoid
TEMPORAL   ETHMOID
OCCIPITAL
Nasal
Lacrimal
Zygomatic
Vomer
Mastoid process
MAXILLA
External auditory canal
Styloid process
MANDIBLE
Face

Skull

Associated bones
Auditory ossicles
Hyoid bone

Frontal bone
Zygomatic bone
Vomer
Sphenoid bone
Styloid process
External auditory canal
Jugular foramen
Occipital bone

Maxillary bone
Palatine bone
Zygomatic arch
Temporal bone
Mastoid process
Occipital condyle
Foramen magnum

Inferior view

Cranium
Parietal bone
Coronal suture
Temporal bone
Sphenoid bone
Zygomatic bone
Mastoid process
Temporomandibular joint

Frontal bone
Orbit
Nasal bone
Lacrimal bone
Vomer
Maxilla
Alveolar margins
Mandible (jaw)

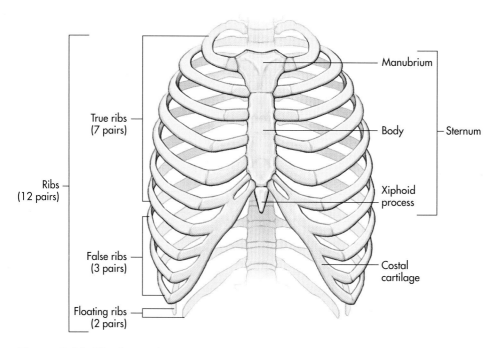

Figure 6.10 The bony thorax.

and can be broken off if CPR is not performed correctly. During cardiac compressions, the heart is compressed anteriorly by the body of the sternum and posteriorly by the bones of the vertebral column.

The thoracic cage consists of 12 pairs of elastic arches of bone called ribs. The ribs are attached by cartilage to allow for their movement when we breathe. The true ribs are pairs 1 to 7 and are called *vertebrosternal* because they connect anteriorly to the sternum and posteriorly to the thoracic vertebrae of the spinal column. Pairs 8 to 10 are called the false ribs, or *vertebrocostal*, because they connect to the costal cartilage of the superior rib and again posteriorly to the thoracic vertebrae. Rib pairs 11 and 12 are called the floating ribs because they have no anterior attachment.

The spinal column

The spinal, or vertebral, column protects the spinal cord which is the superhighway for information coming to and from the central nervous system. The individual bones, or vertebrae, are numbered and classified according to the body region where they are located (see Figure 6.11). For example, there are seven vertebrae found in the cervical or neck region, and they are numbered C–1 to C–7 respectively.

As seen in Figure 6.11, there are seven vertebrae in the neck region, twelve in the upper back, five in the lower back, five fused vertebrae in the mid-buttock region, and three to five small bones at the very end (coccyx). At birth, the vertebral column is concave to the front, like a fetal position (primary curvature) but bends in the opposite direction as the infant starts to raise and holds its head as well as starts to walk. In other words, there will be secondary curvatures by the time a child is two. From two years onward, the vertebral column develops a secondary curvature in the neck, a primary curvature in the upper back, a secondary curvature in the lower back, and a primary curvature in the mid-buttocks and coccyx regions. If the body is not in balance, whether

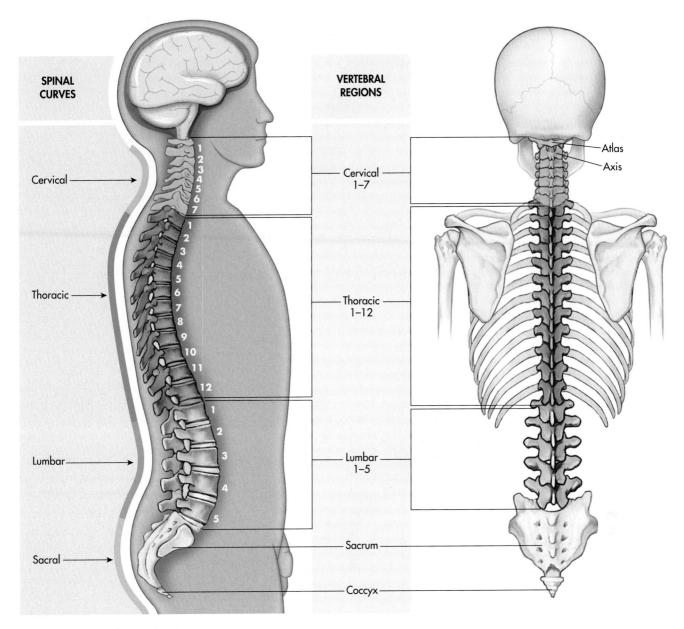

Figure 6.11 The spinal column.

congenital = *born together (the child and the condition)*

due to **congenital** deformity, trauma, poor posture or disease, these curvatures may be exaggerated, leading to kyphosis (humpback, usually in the thorax) or lordosis (usually in the lumbar region). Scoliosis is when there is a sideways bend and sway in the spinal column. Figure 6.12 illustrates these conditions.

Learning hint

Number of vertebrae

To remember the number of vertebrae in each region, think of breakfast at 7 (for cervical), lunch at 12 (for thoracic) and tea at 5 (for the lumbar region).

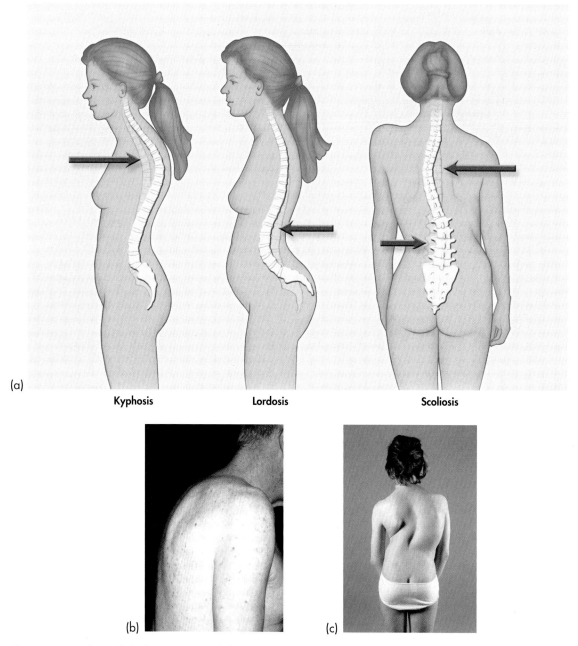

Figure 6.12 Spinal disfigurements. (a) Spinal disfigurements compared to healthy spinal curves. (b) Kyphosis. (*Source*: © Dr. P. Marazzi/Science Photo Library) (c) Scoliosis. (*Source*: © Medical Photo NHS Lothian/Science Photo Library)

Upper and lower extremities

The appendicular region consists of the arms and legs. Since these areas perform most of the body movement, the greatest number of sports-related injuries occurs here. See Figure 6.13, which shows the bones of the upper and lower extremities. These figures show the bony landmarks of the limbs and girdles, which will be beneficial in learning and understanding where the muscles discussed in Chapter 7 will attach.

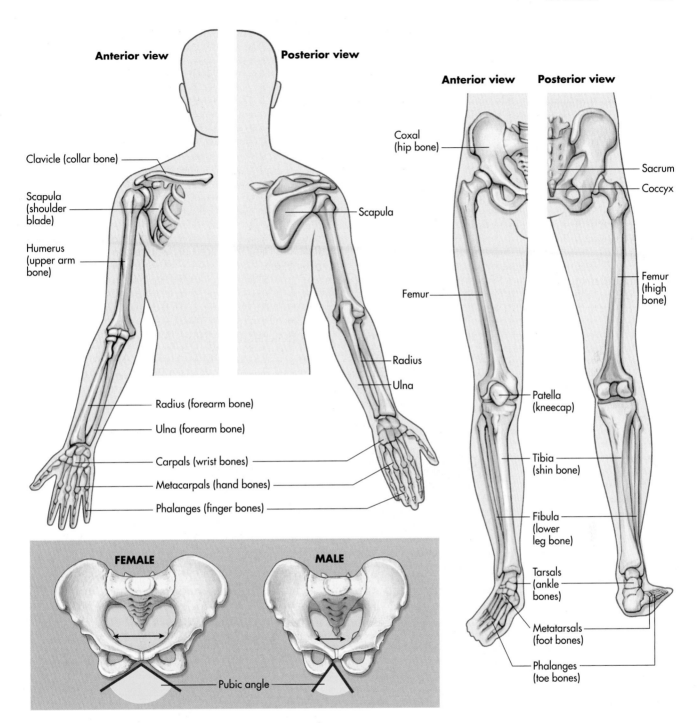

Figure 6.13 Bones of the upper and lower extremities.

Test your knowledge 6.3

Answer the following:

1. What is the difference between the axial skeleton and the appendicular skeleton?

2. Which of the following bones is considered to be part of the axial skeleton?
   a. humerus
   b. patella
   c. femur
   d. sternum

3. The number of vertebra in the thoracic region is
   a. 5
   b. 7
   c. 12
   d. 120

4. Describe the difference between flexion and extension.

Notice that the pelvic girdle in women is different than in men. Women have a greater pubic angle that facilitates childbirth and also a relatively broad girdle to support the extra weight of the child. This difference can be used to identify the sex of a skeleton, such as in a murder case or an archaeological find.

Common disorders of the skeletal system

All good things must come to an end, and the same can be said about the health of your skeletal system. In general, as the body ages, the cartilage and bones deteriorate. Although this is a natural process that we will all encounter, in some cases we can slow the process down.

As the body ages, the chemical composition of cartilage changes. The bluish tint and flexibility of young skeletal cartilage, changes to a more brittle and

Clinical application

Bone fractures

Chances are that you have broken a bone or know someone who has broken a bone in the past, but did you know that not all breaks are the same? Although there are a variety of fractures, the following are the more common ones. Please refer to the illustrations for further clarification.

A *hairline* fracture, which looks like a piece of hair on the X-ray, is a fine fracture that does not completely break or displace the bone. A *simple* or *closed* fracture is a break without a puncture to the skin. An individual in an accident who has a bone that is severely twisted may receive a *spiral* fracture. *Greenstick* fractures are incomplete breaks, which more often occur in children because they have softer, more pliable bones (like sapling branches) than adults (like seasoned twigs). If a bone is crushed to the point that it becomes fragmented or splintered, that is classified as a *comminuted* fracture. A fracture in which the bone is pushed through the skin is referred to as a *compound* or *open* fracture. These fractures are particularly nasty because deep tissue has the potential to be exposed to bacteria once the bone is set into place, and, hence, the chance for infection in addition to the break is increased. See Figure 6.14 for examples of common fractures.

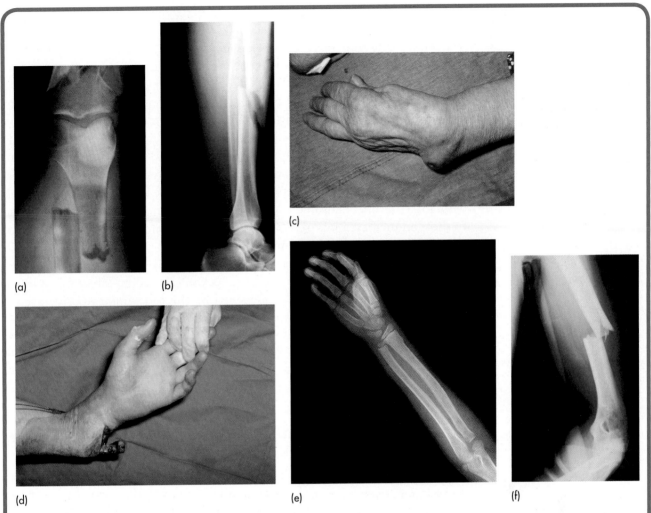

Figure 6.14 (a) Femur, AP view, comminuted fracture. (*Source:* © Doug Sizemore, Visuals Unlimited/ Science Photo Library) (b) Tibia, simple, transverse fracture. (*Source:* © HR Brauaz ISM/Science Photo Library) (c) Open fracture of the wrist. (*Source:* © Medical-on-Line/Alamy) (d) Displaced fracture of the distal radius. (*Source:* © Dr. M. A. Ansery/Science Photo Library) (e) X-ray of complete fracture of the radius. (*Source:* © Living Art Enterprises, LLC/Science Photo Library) (f) Fractured humerus. (*Source:* © Matt Meadows/Peter Arnold/Science Photo Library)

opaque yellow colour. Calcification or hardening of cartilage leads to brittleness. Articular cartilage, once it becomes brittle, doesn't function as well as young, healthy cartilage and can become arthritic. **Arthritis** is an inflammatory process of the joint or joints. The related tendons and ligaments also become less flexible, causing a decrease in the range of motion in joints.

Bone mass also changes with age. In our fifties, the skeleton begins to change: the breakdown of bone becomes greater than the formation of new bone. At the cellular level, the osteoclasts are tearing down more bone than the osteoblasts are forming. As a result, we see total bone mass beginning to gradually decrease. A microscopic examination of bones going through this process reveals increasing holes in the bone. This bone is lighter in weight and weaker than healthy bone, thereby making it more prone to breakage. Men appear to lose less than 25 per cent of their bone mass, whereas women experience a

osteoporosis (*OSS tee oh poor OH siss*)
  osteo = *bone*
  porosis = *condition of porous nature*

35 per cent loss on average. This condition of decreasing bone density, known as osteoporosis, is a serious problem.

Even though bone mass loss is a natural process of ageing, it can be slowed by a healthy lifestyle. It is important to consume the proper amount of dietary calcium to build strong bones in the first place. Proper calcium intake during the formative years, including the teenage years, and continued calcium consumption as the body ages, is crucial. Vitamin D is important because it allows your body to absorb ingested calcium from the digestive tract. As previously discussed, exercise (especially weight-bearing forms) also plays a vital role in developing and maintaining bones, so stay active. And, surprisingly, the excessive use of caffeine and cigarette smoking can reduce bone density. Individuals who, for a lifetime, consume two cups of black caffeinated coffee per day tend to show an increased loss of bone density, while cigarette smokers have shown a loss of 5 to 8 per cent bone mineral density.

While osteoporosis and arthritis are big concerns, there are many more potential disorders of bones and joints. As you can see in Tables 6.2 and 6.3, these disorders can generally be classified by the following causative agents: congenital, degenerative, nutritional, secondary disorders, infection, inflammation, trauma and tumours.

Table 6.2 Bone disorders

osteomalacia (*OSS tee oh ma LAY she ah*)
  malacia = *softening*

chondrosarcoma (*KON drow sar COE mar*)
  chondr = *cartilage*
  oma = *tumour*
  sarcoma = *tumour of connective tissue*

Classification	Example(s)
congenital disorders	abnormal curvature of the spine (kyphosis, lordosis, scoliosis), cleft palate, clubfoot
degenerative disorders	osteoporosis
infection	osteomyelitis
nutritional disorders	**osteomalacia** (vitamin D deficiency), rickets (vitamin D deficiency), scurvy (vitamin C deficiency)
secondary disorders	endocrine system dysfunction: gigantism, pituitary dwarfism
trauma	bruises, fractures
tumours	**chondrosarcomas**, myelomas, osteosarcomas

Table 6.3 Joint disorders

Classification	Example(s)
degenerative disorders	osteoarthritis
infection	gonococcal arthritis, rheumatic fever, septic arthritis, viral arthritis
inflammation	bursitis, arthritis
secondary disorders	immune system dysfunction: rheumatoid arthritis; metabolic dysfunction: gout
trauma	ankle and foot injuries, dislocations, hip fractures, knee injuries

Summary

- In addition to providing support and protection for the body, the skeleton also produces blood cells and acts as a storage unit for minerals and fat.
- The 206 bones of the skeleton can be classified according to their shapes: long bones, short bones, flat bones and irregular bones.
- Bone is covered with periosteum, which is a tough, fibrous connective tissue. In long bones, each bone end is called an *epiphysis*, and the shaft is called the *diaphysis*. The hollow region within the diaphysis is called the *medullary cavity* and stores yellow marrow.
- Compact bone is a dense, hard tissue that normally composes the shafts of long bones or is found as the outer layer of the other bone types. Spongy bone is different in that it contains irregular holes that make it lighter in weight and provides a space for red bone marrow, which produces red blood cells.
- Ossification is the formation of bone in the body. *Osteoprogenitor* cells are non-specialised cells that can turn into *osteoblasts*, which are the cells that actually form bones. *Osteocytes* are considered mature bone cells that were originally osteoblasts. Osteoclasts originate from a type of white blood cell called a *monocyte*, found in red bone marrow. Osteoclasts break down bone material and help move calcium and phosphate into the blood.
- A thin band of cartilage forms an epiphyseal plate (often referred to as the *growth plate*), and as long as it exists, the length and width of the bone will increase.
- Cartilage is a special form of dense connective tissue that can withstand a fair amount of flexing, tension and pressure and makes a flexible connection between bones, as between the breastbone and ribs. It also acts as a cushion between bones.

→

- Various types of joints join two or more bones and provide various types of movement. The point at which they join is called an articulation. Ligaments are very tough, whitish bands that connect from bone to bone to hold the joint together and can withstand heavy stress.

- The skeleton can be divided into two main sections. The axial skeleton includes the bones of the bony thorax, spinal column, hyoid bone, bones of the middle ear, and skull. The appendicular skeleton is the region of your appendages (arms and legs) as well as the connecting bone structures of the hip and shoulder girdles.

- As we age, the chemical composition of cartilage changes, causing it to become more brittle. Articular cartilage that ages or becomes injured can lead to arthritis, which is an inflammatory process of the joint or joints. Bone mass also gradually decreases with age, beginning in a person's fifties. Even though this is a natural process of ageing, it can be slowed by a healthy lifestyle.

Case study

A somewhat frail 76-year-old female (Maria) visits her General Practitioner for an annual check-up. Her social history shows she smokes a pack of cigarettes a day and is a heavy coffee drinker. She has had several fractured bones in the last five years that required medical attention. During initial examination, measurements show that the patient has lost approximately 2.5 cm of height over the past year. She has also lost several kilos but states she still wears the same size clothes.

1. What possible bone disease do you think she is exhibiting?

   _____

   _____

2. Describe the bone changes in this condition on a macro- and microcellular level.

   _____

   _____

3. What treatments and/or lifestyle changes would you suggest?

   _____

   _____

Review questions

Multiple choice

1. Your elbow is an example of which type of joint?
   a. hinge joint
   b. ball and socket joint
   c. gliding joint
   d. fibrous joint

2. The sternum is the correct medical term for which bone?
   a. shin bone
   b. breastbone
   c. shoulder blade
   d. collar bone

3. The end of a long bone is the:
   a. diaphysis
   b. epiphysis
   c. condyl
   d. fossa

4. As long as this exists, your bones will increase in length and width:
   a. facet
   b. ossification centre
   c. periosteum
   d. epiphyseal plate

5. The ageing process, excessive caffeine and cigarette smoking can each contribute to this bone disease:
   a. kyphosis
   b. osteoporosis
   c. osteomalacia
   d. osteosarcomas

6. The superior part of the sternum is known as the:
   a. xiphoid
   b. manubrium
   c. body
   d. clavicle

Fill in the blanks

1. Name three large appendicular bones: _____, _____ and _____.

2. List three places where cartilage is found in the body: _____, _____ and _____.

3. _____ is a liquid found in joints that keeps them lubricated.

4. The specialised cells that constantly rebuild bone are called _____.

5. These specialised cells are needed to tear down bone: _____.

Short answers

1. Describe the difference in function between tendons and ligaments.

   _____

   _____

   _____

   _____

2. List three functions of the skeletal system.

_____
_____
_____
_____

3. What happens to our skeletal system as we age?

_____
_____
_____
_____

4. What are the functions of cartilage?

_____
_____
_____
_____

Suggested activities

1. Using art paper, cut out, label and assemble bones to create a full-sized skeleton.

2. Act out various range-of-motion exercises and perform and identify different joint movements.

3. Visit the library and research how the human skeleton has evolved, also noting the differences between male and female skeletons, posture changes and changing shapes of various bones.

Answers

Answers to case study

1. Osteoporosis
2. Bone becomes lighter and weaker and microscopically the bone starts to develop more holes.
3. Stop smoking, reduce caffeine intake, take up regular exercise, increase calcium and vitamin D intake in diet – possible use of supplements.

Answers to multiple choice questions

1. a
2. b
3. b
4. d
5. b
6. b

Answers to fill in the blanks

1. Femur, tibia, fibula, humerus, radius, ulna.
2. Ears, nose, end of bones (as articular cartilage), rib cage.
3. Synovial fluid
4. Osteoblasts
5. osteoclasts

Answers to short answer questions

1. Tendons attach muscles to bones; ligaments hold two bone ends together in a synovial joint.
2. A framework for the human body; production of red blood cells; provides protection for organs; stores minerals e.g. calcium; and, along with the muscular system, allows for movement.
3. Bone mass decreases.
4. Cartilage makes a flexible connection between bones and can also act as a cushion between bones.

The muscular system

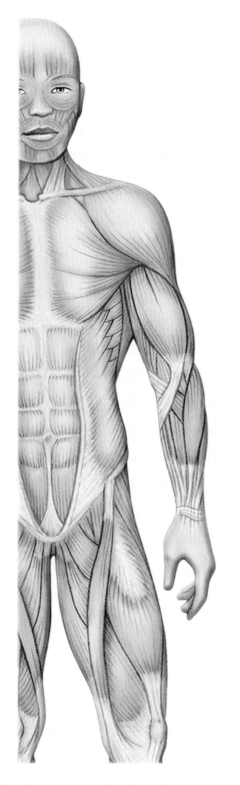

While the skeletal system provides the framework for the human body, the body also needs a system that allows movement, or locomotion, which is the job of the muscular system. The movement we are most familiar with is the use of our external muscles to walk, run or lift objects. However, movement is also required within the body. This internal movement occurs when food, air, waste products and body fluids such as blood must all be transported within our bodies. For example, if you drink some bad water, the smooth muscles in your digestive tract will rapidly pass it through your system to be expelled in the form of urgent diarrhoea. Different types of specialised muscles within the muscular system allow for both external and internal movement. This chapter defines and contrasts the different muscle types needed for external and internal body movement.

Learning objectives

At the end of this chapter, you will be able to:

◆ Differentiate between the three major muscle types
◆ Discuss the function of tendons and ligaments
◆ Explain the difference between voluntary and involuntary muscles
◆ Describe the various types of skeletal muscle movement
◆ Identify and explain the components of a muscle cell
◆ Describe the chemical activities required for muscle movement
◆ Contrast the activity of cardiac, smooth and skeletal muscle
◆ Discuss common disorders of the muscular system

Pronunciation guide

acetylcholine (*ah SEAT isle COE leen*)
actin (*ACT in*)
adenosine triphosphate (*ah DEN oh seen try FOSS fate*)
ataxia (*ah TAK see yah*)
atrophy (*AT row fee*)
diaphragm (*DYE uh fram*)
electromyography (*eh LEK troh my OG ruff ee*)
fibromyalgia (*fye bro my AL jar*)

flaccid (*FLASS id*)
flexion (*FLEK shun*)
glycogen (*GLY coh jin*)
Guillain-Barré syndrome (*gil ann barr ee SIN drome*)
hypertrophy (*high PER troh fee*)
intercalated discs (*in ter CAL ate ed*)
muscular dystrophy (*MUSS kew lar DISS troh fee*)
myalgia (*my AL jar*)

myasthenia gravis (*my ass THEE nee yah GRAV iss*)
myofibril (*my oh FYE bril*)
myosin (*MY oh sin*)
rigor mortis (*rig er MORE tiss*)
sarcomeres (*SAR koh meerz*)
sphincters (*SFINK ters*)
tetanus (*TET ah nuss*)
tonus (*TONE uss*)

Overview of the muscular system

Because of the numerous functions they must perform, muscles come in many shapes and sizes. The structure of the muscle matches its function, as you will shortly see.

Types of muscles

Muscle is a general term for all contractile tissue. The term muscle comes from the Latin word *mus*, which means 'mouse', because the movement of muscles looks like mice running around under our skin. The contractile property of muscle tissue allows it to become short and thick in response to a nerve impulse and then to relax once that impulse is removed. This alternate contraction and relaxation is what causes movement. Muscle cells are elongated and resemble fibres such as in clothing. Muscular tissue is constructed of bundles of these muscle fibres. These fibres are approximately the diameter of human hair. Under the direction of the nervous system, all of the muscles provide for motion of some type for your body.

The body has three major types of muscles: **skeletal**, **smooth** and **cardiac**. We begin with a general description and comparison of these three muscle types and then get more specific about each type.

Skeletal muscles are voluntary muscles, which means they are under conscious control and derive their name because they are attached to the bones of the skeletal system. The fibres in skeletal muscles appear to be striped and are therefore called *striated* (striped) muscle. These muscles allow us to perform external movements – running, lifting or scratching, for example. These are the muscles we try to develop through exercise and sports and also so we look good at the beach.

Unlike skeletal muscle, smooth muscle is involuntary and not under our conscious control. It is also called smooth muscle because it does not have the striped appearance of skeletal muscles. This involuntary muscle is found within certain organs, blood vessels and airways. Because it is the muscle of

Amazing body facts

Muscles

- Muscles make up almost half the weight of the entire body.
- The size of your muscles depends on how much you use them and how big you are. This is why ice skaters have large leg muscles.
- Individual elongated muscle cells can be up to 30 centimetres in length.
- At about the age of 40, the number and diameter of muscle fibres begin to decrease, and by age 80, 50 per cent of the muscle mass may be lost.

organs, it is sometimes called *visceral muscle*. Smooth muscle allows for the internal movement of food (*peristalsis*) in the case of the stomach and other digestive organs. Smooth muscle also facilitates the movement of blood by changing the diameter of the blood vessels (vasoconstriction and vasodilatation), and also the movement of air by changing the diameter of the airways found in our lungs.

The third type of muscle is the specialised cardiac muscle, which has a striated appearance. This muscle type is found solely in the heart. It makes up the walls of the heart and causes the heart to contract. These contractions cause the internal movement (circulation) of blood within the body. Fortunately, cardiac muscle, like smooth muscle, is an involuntary muscle. Imagine if we had to think each time in order for our heart to beat. Figure 7.1 contrasts the three types of muscles found within the body. We will now explore each of these types of muscles in further depth.

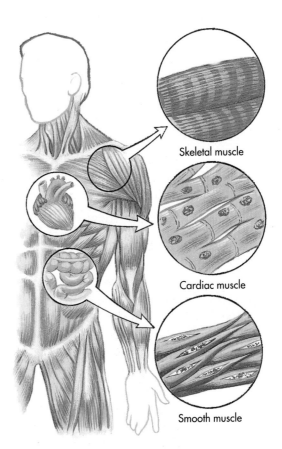

Skeletal muscle

Cardiac muscle

Smooth muscle

Figure 7.1 The three types of muscle: skeletal, cardiac and smooth.

Test your knowledge 7.1

Choose the best answer:

1. The biceps muscle is an example of a:
   a. smooth muscle
   b. cardiac muscle
   c. skeletal muscle
   d. involuntary muscle

2. Smooth muscle is found in all the following *except*:
   a. airways
   b. digestive system
   c. blood vessels
   d. heart

3. Which types of muscles are striated?
   a. smooth and cardiac
   b. cardiac and skeletal
   c. skeletal and smooth
   d. smooth only

Skeletal muscles

Skeletal muscles are attached to bones and provide movement for your body. Remember from Chapter 6 that **tendons** are fibrous tissues that usually attach skeletal muscle to bones and **ligaments** attach bone to bone. Note that some muscles can attach to a bone or soft tissue without a tendon. Such muscles use broad sheets of connective tissue called an *aponeurosis*. This type of connection is found, for example, in some facial muscles and the tongue.

Skeletal muscle is also known as voluntary muscle because its movement can be controlled by conscious thought. The numerous skeletal muscles found throughout the body are responsible for movement, maintaining our body posture, and heat generation. See Figure 7.2, which shows some of the major muscles found in the human body.

Skeletal muscles of specific body regions

Figure 7.2 shows the anterior and posterior major muscles. The following series of figures will provide you with greater detail.

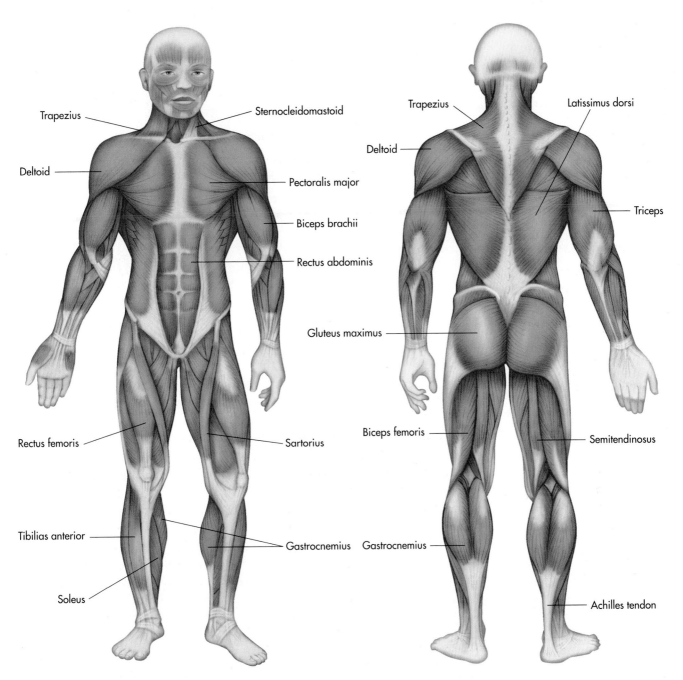

Figure 7.2 Anterior and posterior view of major muscles.

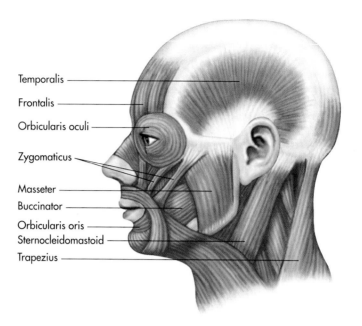

Figure 7.3 Skeletal facial muscles.

Facial skeletal muscles Please see Figure 7.3 which shows the facial skeletal muscles.

Anterior and posterior trunk skeletal muscles Now, take an in-depth look at the muscles of the anterior and posterior trunk of the body in Figure 7.4.

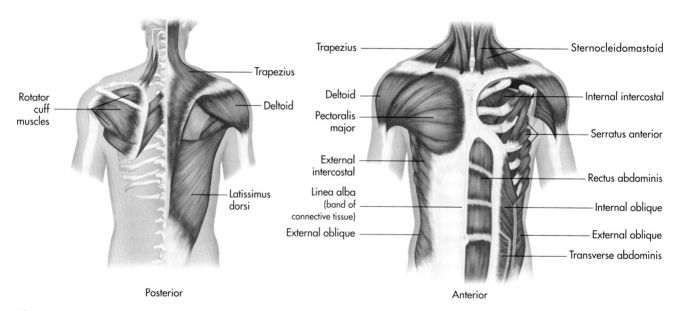

Figure 7.4 Skeletal muscles of the posterior and anterior trunk.

Skeletal muscles of the hand, arm and shoulder Moving out to the peripheral area of the body, we now zoom in on the skeletal muscles of the hand, arm and shoulder in Figure 7.5.

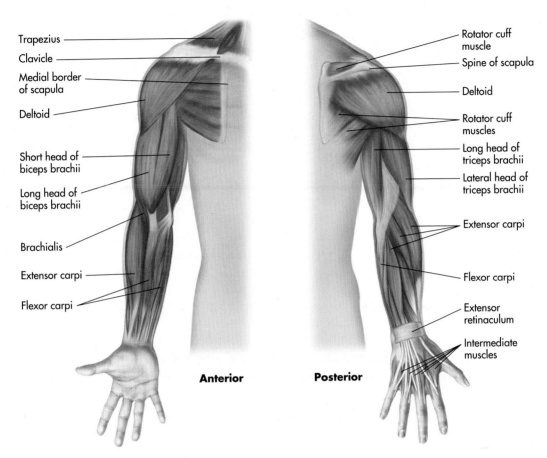

Trapezius

Clavicle

Medial border of scapula

Deltoid

Short head of biceps brachii

Long head of biceps brachii

Brachialis

Extensor carpi

Flexor carpi

Anterior

Rotator cuff muscle

Spine of scapula

Deltoid

Rotator cuff muscles

Long head of triceps brachii

Lateral head of triceps brachii

Extensor carpi

Flexor carpi

Extensor retinaculum

Intermediate muscles

Posterior

Figure 7.5 Skeletal muscles of the shoulder, arm and hand.

Skeletal muscles of the legs We finish our tour with the skeletal muscles of the hip and leg in Figure 7.6.

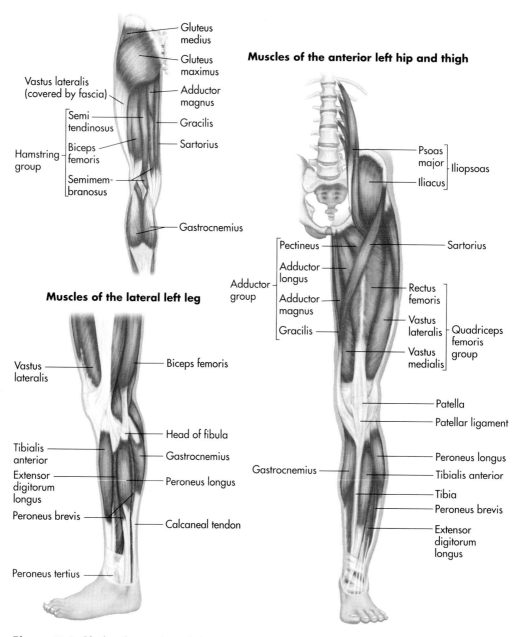

Figure 7.6 Skeletal muscles of the hip and leg.

Skeletal muscle movement

The body requires several different types of movement for various tasks. This movement is accomplished through the coordination of the contraction and relaxation of various muscles.

Contraction and relaxation

Movement of the body is a result of the contraction (shortening of the muscle fibres) of certain muscles and the relaxation of others. Consider the act of bending your arm so your fingers touch your shoulder. To really learn the concept, actually bend your arm and touch your fingers to your shoulder while resting your other hand on your biceps muscle. In order to do this, your forearm is drawn to your shoulder as a result of the contraction of your biceps. Did you feel the shortening and bulging of the biceps? Muscles, either by themselves or in muscle groups that cause movement, are known as **agonists** or **primary movers**.

agonist (*AGG on ist*)

The chief muscle causing the movement is the primary mover, and in this example it is the biceps muscle. Typically, as your muscle contracts, one of the bones will move (lower forearm) while the other (humerus) will remain stationary. The end of the muscle that is attached to the stationary bone is the **point of origin**, and in this example, it is at the shoulder area. The muscle end that is attached to the moving bone is the **point of insertion**; in this example, it is near the elbow (see Figure 7.7).

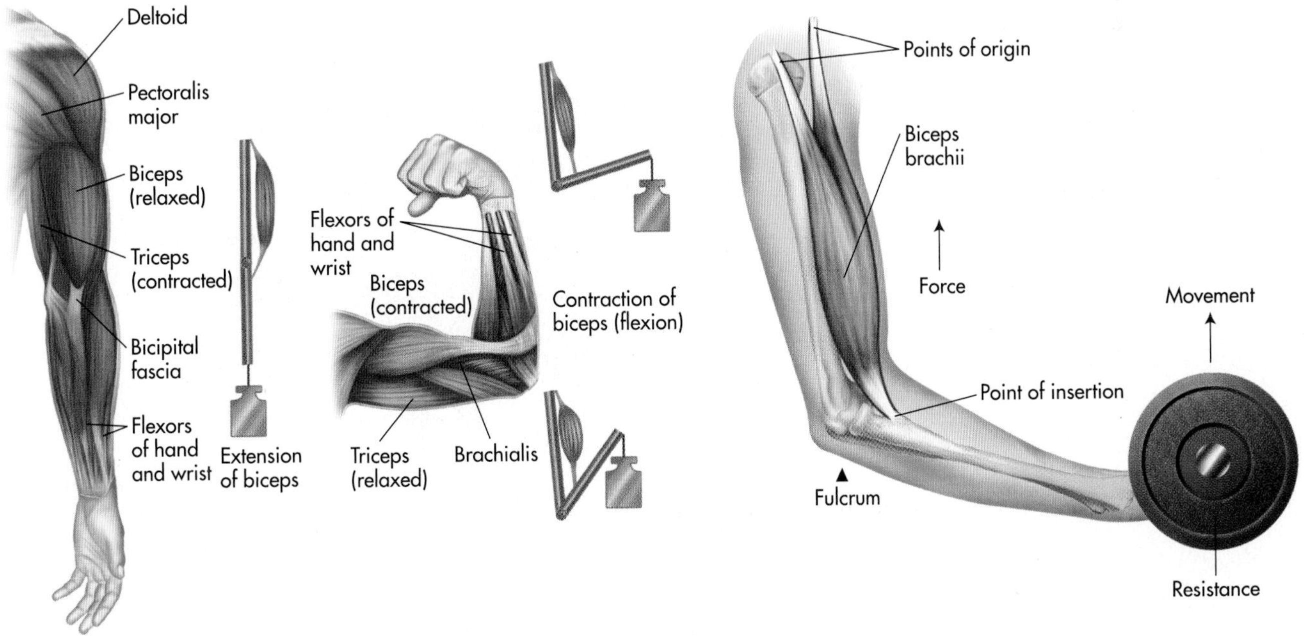

Figure 7.7 Coordination of antagonist muscles to perform movement.

synergistic (*sin er JISS tick*)

Other muscles can assist this movement, such as some of the muscles in the hands and wrist. These are called **synergistic** muscles because they assist the primary mover. To straighten out that same arm requires you to relax your biceps muscle and to contract the triceps muscles. Since these muscles cause movement in the opposite direction when they contract, they are called **antagonists**. This brings us to an important concept. All movement is a result of contraction of primary movers and relaxation of *opposing* muscles. In our example, you cannot forcefully contract the biceps muscles and straighten out your arm. Try it. Again, see Figure 7.7 for the illustration of the antagonist muscles of the biceps and triceps.

One very important skeletal muscle that controls our breathing is the **diaphragm**. This dome-shaped muscle separates the abdominal and thoracic cavities and is responsible for performing the major work of bringing air from the atmosphere into our lungs. Exactly how this process occurs is discussed in detail in Chapter 15, 'The respiratory system'. The diaphragm is unique in that it is under both voluntary and involuntary control. You don't have to think each time you breathe, but you can voluntarily change the way you breathe. Figure 7.8 shows the major muscle of breathing.

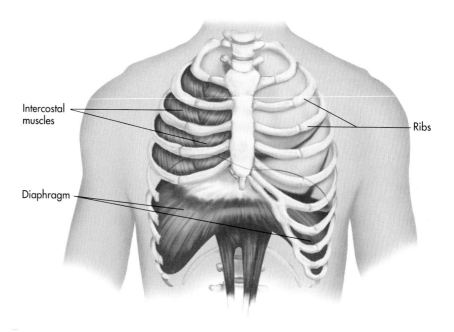

Intercostal muscles

Ribs

Diaphragm

Figure 7.8 The diaphragm: the major muscle of breathing.

Movement terminology

Certain terms are utilised to describe the direction of body movement. In Chapter 6, we discussed movement as it relates to joints in the skeletal system. In this chapter, we briefly discuss movement as it relates to muscles. **Rotation** describes circular movement that occurs around an axis. Rotation occurs, for example, when you turn your head from left to right or right to left. **Abduction** means to move *away* from the midline of the body. When you raise your arm to point when giving directions, you are performing abduction. **Adduction** occurs when you produce a movement that moves *toward* the midline of the body. When you bring your arm back down to your side from pointing, you are performing adduction.

abduction (*ab DUCK shun*)
   ab = *away, as in abduct or abnormal*

adduction (*ad DUCK shun*)
   ad = *toward*

extension (*ek STEN shun*)

flexion (*FLEK shun*)

Extension is a term used for *increasing* the angle between two bones connected at a joint. Extension is needed when you kick a football. In this situation, extension occurs when your leg straightens out during the kick. The muscle that straightens the joint is called the **extensor muscle**. **Flexion** is the opposite of extension. In this situation, you *decrease* the angle between two bones. Flexion occurs when you bend your legs to sit down. Flexion and rotation occur when you get your arm into position to arm wrestle. The muscle that bends the joint is called the **flexor muscle**. In this case a picture is worth a thousand words or at least the 124 words used to explain these concepts. Figure 7.9 illustrates these movements.

Applied science

Kinesiology

Kinesiology is the study of muscles and movement. A kinesiologist is someone who studies movement and can employ therapeutic treatment (kinesitherapy) by specific movements or exercises.

Test your knowledge 7.2

Give the correct body movement term for the following activities:

1. Looking right and left at a 'T' junction

   _____

2. Doing a split _____

3. Patting yourself on the back for labelling all these activities correctly _____

4. The first movement in lifting a weight

   _____

5. Returning the weight from the lifted position to your side _____

Flexion

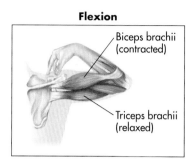

Biceps brachii
(contracted)

Triceps brachii
(relaxed)

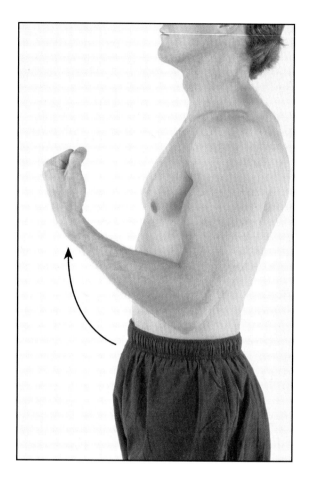

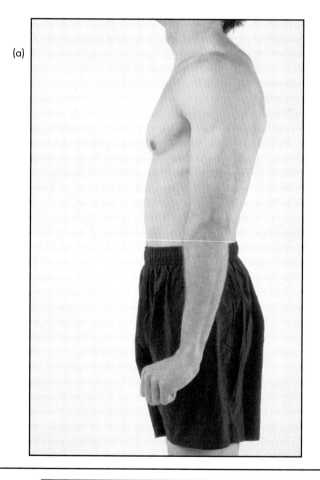

(a)

Extension

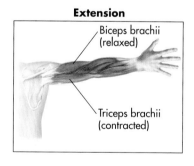

Biceps brachii
(relaxed)

Triceps brachii
(contracted)

(b)

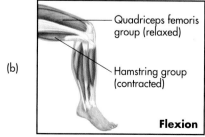

Quadriceps femoris
group (relaxed)

Hamstring group
(contracted)

Flexion

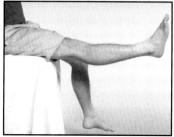

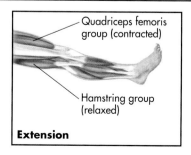

Quadriceps femoris
group (contracted)

Hamstring group
(relaxed)

Extension

Figure 7.9 The types of skeletal movement. (a) Flexion and extension of left forearm (*Source*: Mike Gallitelli/ Pearson Education Inc.) (b) Flexion and extension of the leg (*Source*: Michael Heron/Pearson Education Inc.)

Muscular movement at the cellular level

Exactly how is muscular contraction and relaxation accomplished? How does the muscle tissue cause a coordinated and smooth contraction? Let's look in more detail at how muscles work.

The functional unit of the muscle

We have talked about muscles on a macro 'or very large' scale. For example, how does that large biceps you've developed from working out contract when you touch your shoulder? Now, let's explore the make-up of the individual muscle fibre to learn exactly how this contraction takes place. As stated previously, muscle consists of elongated cells called muscle fibres, which can be up to 30 centimetres in length. Each muscle fibre contains functional units called **myofibrils**, and several of these myofibrils can be bundled together to form a muscle cell. The muscle myofibril can be considered a strand of metal, and these strands of metal can be put together to form a cable, which would be a muscle segment (see Figure 7.10(a)).

In order for contraction to take place, each fibre must possess many functional contractile units called **sarcomeres**. Each fibre has the ability to contract because of the make-up of the sarcomere. Each sarcomere unit has two types of thread-like structures called thick and thin myofilaments. The thick myofilaments are made up of the protein **myosin**, and the thin ones are primarily made up of the protein **actin**. The sarcomere has the actin and myosin filaments arranged in repeating units separated from each other by dark bands called **Z lines**, which give the striated appearance to skeletal muscle (see Figure 7.10(a) and (b)).

Note in Figure 7.10(c) that the contraction of a muscle causes the two types of myofilaments to slide toward each other, shortening each sarcomere and, therefore, the entire muscle. Picture a tube sliding within a tube, such as on a trombone. This sliding filament action and corresponding contraction requires that temporary connections, or crossbridges, be formed between the thick filament heads (myosin heads) and the thin filaments (actin) to pull the sarcomere together. Once the crossbridges form, the myosin heads rotate and pull the actin toward the centre of the sarcomere. When the sarcomere relaxes, the filaments return to their resting or relaxed position. Visualise a raised drawbridge, where cars cannot pass. In order to be functional, the cross-connection – or lowering of the drawbridge – must occur, similar to the crossbridges needed to be formed for a muscle contraction.

myofibril (*my oh FYE bril*)
myo = *muscle*

sarcomeres (*SAR koh meerz*)

myosin (*MY oh sin*)
actin (*ACT in*)

acetylcholine (*ah SEAT isle COE leen*)

Applied science

Interrelatedness of the neuromuscular system

Contraction of a skeletal muscle requires the coordination of both the muscular and nervous systems. The initiation of a skeletal muscular contraction requires an impulse from a motor neuron of the nervous system to trigger a release of a neurochemical transmitter called **acetylcholine** (ACh), which opens the sodium channels and sets the process of muscle contraction into motion. This all occurs at the neuromuscular junction. The nervous system's role in this action and the neuromuscular connection are fully explored in Chapters 9 and 10.

(a) MUSCLE SEGMENT

(b) MUSCLE SEGMENT WITH SARCOMERE

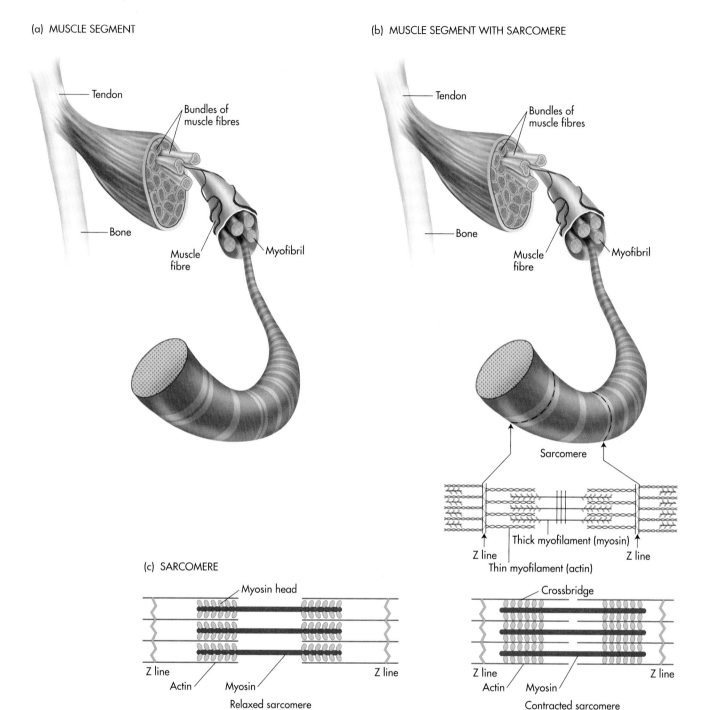

(c) SARCOMERE

Figure 7.10 (a) The muscle segment. (b) The muscle segment with sarcomere. (c) Relaxed and contracted sarcomeres.

Important ingredients: ATP and calcium

In the previous analogy of the drawbridge, consider that a toll must be paid in order for the bridge to lower and connect. The body's toll is the energy molecule **adenosine triphosphate** (ATP) and **calcium** (Ca^{++}), which are needed for contraction and relaxation. ATP provides the energy to help the myosin heads form and break the crossbridges with actin. When the muscle is relaxed, calcium is stored away from the actin and myosin in the **sarcoplasmic reticulum**, which is a specialised series of interconnecting tubules and sacs that surround each myofibril. When the muscle is stimulated, calcium is released from the sarcoplasmic reticulum and causes actin, myosin and ATP to interact, which causes the contraction. When calcium leaves the muscle and returns to the sarcoplasmic reticulum, the crossbridges are broken and the muscle relaxes.

When the nervous system tells the muscle to contract, the signal causes the muscle fibre to open what are called *sodium ion channels*. Sodium ions flow through these channels into the muscle fibres. This causes the muscle fibres to become excited and calcium is released from the sarcoplasmic reticulum. The calcium, now free in the cytoplasm of the muscle fibre, helps myosin bind with actin in the presence of ATP. The calcium is then pumped back into the sarcoplasmic reticulum for storage and the muscle now relaxes.

Have you ever heard of a dead body rising from a table or showing signs of movement? This may sound like the opening for a film about zombies or the 'undead'. Actually, it is a normal physiologic process called **rigor mortis** that can be explained by science and not by science fiction.

rigor mortis (*rig er MORE tiss*)

When a body dies, all the stored calcium cannot be pumped back into the sarcoplasmic reticulum. Therefore, excess calcium remains in the muscles throughout the body and causes the muscle fibres to shorten (contract) and stiffen the whole body. In addition, ATP is not present in a dead body to break the crossbridges – this is due to a lack of oxygen being made available to the cells. This stiffening process of the entire body is termed rigor mortis.

Test your knowledge 7.3

Fill in the blanks.

1. The region between two Z lines is called a _____.

2. The thick myofilament needed for muscle contraction is called _____.

3. The thin myofilament needed for muscle contraction is called _____.

4. The two ingredients needed for crossbridges to form and break are _____ and _____.

Visceral or smooth muscle

We've introduced the concept of smooth muscle earlier in this chapter; now let's take a closer look. **Visceral**, or **smooth**, **muscle** is found in the organs (except the heart) of your body, such as your stomach and other digestive organs, and in the blood vessels and bronchial airways. The ability for smooth muscle to expand and contract plays a vital role in many of the body's internal workings. For example, the vital sign blood pressure can be affected by whether the blood vessels get larger in diameter (**vasodilate**) or get smaller

vasodilate (*VAY so dye LATE*)

vasoconstrict (*VAY so con STRICT*)

in diameter (**vasoconstrict**). Vasodilatation can lead to decreases in blood pressure due to smooth muscle relaxation in the vessel that allows it to enlarge. The enlarged vessel has less resistance to flow, and the blood pressure therefore goes down. Conversely, vasoconstriction can cause increased blood pressure due to the smooth muscle contraction that restricts the blood vessel.

As another example, during an asthma attack, smooth muscles in the airways of the lungs constrict, making it difficult to get air in and out of the lungs. This is what causes the wheezing sound heard during an attack.

sphincters (*SFINK ters*)

A special type of smooth muscle, called **sphincters**, is found throughout your digestive system. These ring-shaped muscles act as doorways to let materials in and out by alternately contracting and relaxing. For example, the pyloric sphincter of the stomach acts like doors that open up to allow food from the stomach out into the small intestine. Have you ever swallowed a large amount of bread or stuffing and had it get stuck on the way down to your stomach? This is a painful reminder that there is a sphincter that must relax and open to allow food to enter your stomach. The muscles of the digestive system are discussed in greater depth in Chapter 16, 'The gastrointestinal system'.

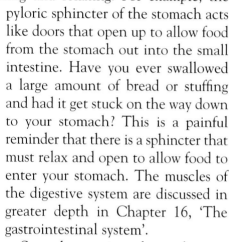

Learning hint

Smooth muscle regulation of blood pressure

In considering blood pressure, visualise a large motorway. If one lane is taken away (vasoconstriction), the same number of cars must now fit through one less lane, leading to traffic congestion (increase in pressure). If you open up another lane (vasodilate), you relieve some of this pressure.

Smooth, or visceral muscles, are involuntary muscles and do not contract as rapidly as skeletal muscles. Skeletal muscles, once stimulated, can contract 50 times faster than smooth muscle. Because of their slower activity and lower metabolic rate, smooth muscles receive only moderate amounts of blood. Once injured, smooth muscle rarely repairs itself and, instead, forms a scar.

Cardiac muscle

Cardiac muscle forms the walls of the heart. The contraction of cardiac muscle squeezes blood out of the chambers of the heart, causing the blood to circulate through the body. Cardiac muscle is involuntary muscle. Remember, this means that we don't have to consciously think about making our heart contract every time we need a heartbeat. Cardiac muscle fibres are somewhat shorter than the other muscle types. Since the heart must work constantly until you die, the cardiac muscles must receive a generous blood supply to get enough oxygen and nutrition, as well as to get rid of waste. In fact, cardiac muscle has a richer supply of blood than any other muscle in the body.

intercalated disc (*in ter CAL ate ed*)

The cardiac muscle fibres are connected to each other by **intercalated discs**. Because of this connection, as one fibre contracts, the adjacent one contracts, and so on. This is similar to the domino effect or the human 'Mexican' wave at a football stadium if done correctly. A wave of contraction occurs, allowing blood to be squeezed out of the heart and into the body. This directed wave is

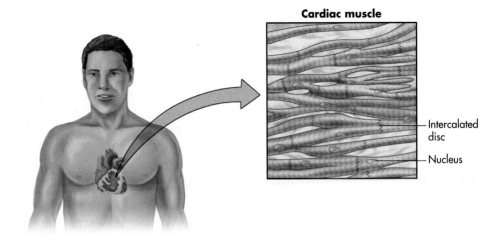

Figure 7.11 Heart and intercalated discs.

important for a full and effective emptying of the blood within the heart. See Figure 7.11.

Cardiac muscle does not regenerate after severe damage; this leads to tissue death such as occurs in a severe heart attack. If the blood supply going to the heart from the coronary arteries is blocked, cardiac muscle damage can occur, causing scarring of the heart. Scar tissue does not help the healthy muscles of the heart to contract. If the scarred area is extensive, the remaining cardiac muscle may not be sufficient to pump blood efficiently. An individual with scarred cardiac muscle may have a severely diminished cardiac output, which could lead to severe disability or even death.

Clinical application

Muscle tone

Have you ever had a plaster cast on for an extended period of time? When it is removed, the arm or leg is much smaller and weaker than the limb without the plaster cast. Why does this occur? Normally, all muscles exhibit muscle tone (**tonus**). Tonus is the partial contraction of a muscle with a resistance to stretching. Athletes who exercise regularly have increased muscle tone, making their muscles more pronounced. The muscle fibres in an athlete increase in diameter (**hypertrophy**) and become stronger. Hypertrophy refers to increased growth or development. When muscles are not used, they begin to lose their tone and become flaccid (soft and flabby). For example, if a patient is required to remain in bed (bedbound) for an extended period of time, his or her muscles waste away (**atrophy**) from the lack of use. One of the reasons patients are assisted to mobilise as soon as possible is to prevent atrophy from occurring. If skeletal muscle is damaged, it can regenerate itself. However, if the damage is extensive, then a scar forms.

hyper = *greater than normal*

trophy = *growth or development*

a = *without*

Muscular fuel

glycogen (*GLY coh jin*)

glucose (*GLOO kose*)

Muscle, like all tissue, needs fuel in the form of nutrients and oxygen in order to survive and function. The body stores a carbohydrate called glycogen in the muscle. Glycogen is always on reserve waiting to be converted to a usable energy source. When needed, the muscle can convert glycogen to glucose, which releases energy for the muscle to function. Muscles with very high demands (such as leg muscles) also store fat and use it as energy. The release of energy also produces heat, and this is why strenuous or prolonged exercises can overheat our bodies.

The higher-demand muscles not only use fat as an energy source, but have a much richer blood supply than do less demanding muscles. These muscles are needed for endurance, such as required by long-distance running. The richer blood supply carries extra oxygen to hardworking muscles, giving those muscles a darker colour.

Some muscles, such as those in the hand, have fewer heavy demands placed on them and need only a small supply of blood. These muscles utilise the local blood supply for glucose and the glycogen stored within. They therefore have a lighter colour. These muscles are faster but do not have the endurance capabilities that heavily used muscles have. Next time you take a long walk, keep pumping your hand. While the hand can move faster than the leg muscles, it will tire more quickly.

Another example can be found in chicken. Because its breast and wing muscles are not heavily used, those parts contain white meat. Its legs, however, endure constant use, and the meat is therefore dark. By contrast, a woodcock, a migratory bird that must fly long distances (endurance), has dark breast meat. Now you know why a chicken's breast meat is white. When is the last time you saw a chicken flying overhead?

Applied science

Maintaining a core body temperature

Not only do muscles produce movement, but they help maintain posture, stabilise joints and produce heat. Producing heat is important in maintaining the body core temperature. As the energy-rich ATP is used for muscle contraction, 75 per cent of its energy escapes as heat. This process helps to maintain body temperature by producing heat when muscles are utilised. This is why your temperature rises when exercising and also why you shiver when you are very cold. Shivering is your body's way of saying it is too cold and it needs to generate a lot of heat via many muscle contractions (shivering). In turn, this increases the body temperature.

Common disorders of the muscular system

myalgia (*my AL jar*)
fibromyalgia (*fye bro my AL jar*)
   algia = *pain*

Because there are so many muscles covering the entire body and they are constantly being used, many disorders occur within this system. Here are just a few examples. Myalgia means pain or tenderness in a muscle. Fibromyalgia may be one of the most common musculoskeletal disorders affecting women under age 40, but it is still not fully understood. Symptoms include aches, pains and muscle stiffness with specific tender points on anatomical regions of the body. The exact cause is unknown, and it is linked with other diseases such as chronic fatigue syndrome.

ataxia (*ah TAK see yah*)
  a = *without*
  tax/o = *coordination*
lysis = *destruction of*

itis = *inflammation of*
electromyography (*eh LEK troh my OG ruff ee*)
  electro = *electric*
  myo = *muscle*
  graphy = *graph*

muscular dystrophy (*MUSS kew lar DISS troh fee*)
  dys = *difficult*
  trophy = *growth or nourishment*
Guillain-Barré syndrome (*gil ann barr ee SIN drome*)

tetanus (*TET ah nuss*)

Ataxia is a condition in which the muscles are irregular in their actions or there is a lack of coordination. **Paralysis** is the partial or total loss of the ability of voluntary muscles to move. Sometimes it might be temporary; other times it might be permanent. A muscle that involuntarily suddenly and violently contracts for a prolonged period of time is said to have a **spasm** or **cramp**. A spasm can occur in a single muscle or in a muscle group. **Sprains** are tears or breaks in ligaments, while **strains** are tears or injury in muscles and tendons. A common running exercise-related inflammatory condition of the extensor muscles and surrounding tissues of the lower leg is **shin splints**.

A **hernia** occurs when there is a tear in the muscle wall and an organ of the body protrudes through that opening. **Tendonitis** is a condition in which tendons become inflamed. Muscular disorders can be diagnosed by **electromyography** (EMG), a test in which a muscle or group of muscles is stimulated with an electrical impulse. This impulse causes a muscle contraction. The strength of that muscle contraction is then recorded. Certain diseases can alter the strength of muscles.

Due to the close integration of the two systems, several diseases involve both the *nervous* system and the *muscular* system – hence, the term **neuromuscular** disease. **Myasthenia gravis** is a neuromuscular disease in which the patient exhibits gradually increasing profound muscle weakness. The first symptom of this disease is often the drooping of one or both of the upper eyelids. There is also progressive paralysis. Interestingly, tendon reflexes almost always remain. **Muscular dystrophy** is an inherited muscular disease in which muscle fibres degenerate and there is progressive muscular weakness. **Guillain–Barré syndrome** is a disorder of the *peripheral* nervous system that causes *flaccid* paralysis (limp muscles) and the loss of reflexes. Interestingly, the paralysis is usually *ascending*, meaning that it starts in the feet or lower extremities and progresses toward the head. Paralysis usually peaks within 10 to 14 days. Eventually, most patients return to normal, although it may take several weeks or months. **Tetanus**, on the other hand, creates rigid paralysis. With this disease, any type of minor stimulus can cause muscles to go into major spasm. The stimulus can be something as simple as a loud noise or turning on a light in a room. Tetanus is a result of toxins produced by bacteria found in the ground and can be spread by any type of puncture, not just the 'rusty nail' many were warned about when they were children. Smooth and cardiac muscle conditions are discussed in upcoming chapters.

Applied science

A useful application of a deadly toxin

Botulism is a potentially deadly disease caused by food poisoning with the *Clostridium botulinum* bacteria. Science has found a way to utilise the poison generated by this bacteria for medical and cosmetic treatment. Small amounts of botulinus toxin are injected into facial muscles to stop previously untreatable facial twitching. The toxin basically paralyses the muscles. The same toxin is used to treat wrinkles without the use of surgery and is known as Botox injections.

Summary

- The three main types of muscles are skeletal, smooth and cardiac.
- Skeletal muscle is striated, or striped, voluntary muscle that allows movement, stabilises joints and helps maintain body temperature.
- Smooth muscle is a non-striated involuntary muscle found in the organs of the body and linings of vessels; it facilitates internal movement within the body.
- Cardiac muscle is involuntary, striated muscle found only in the heart.
- All movement is a result of contraction of primary movers and relaxation of opposing muscles.
- Muscles usually attach to bones via tendons.
- Large muscles consist of many single muscle fibres comprised of myofibrils. The smallest functional contractile unit is called a sarcomere.
- Each sarcomere unit contains the two thread-like contractile proteins: myosin and actin.
- Muscles contract as the actin and myosin protein filaments, in the presence of ATP and calcium, form crossbridges that cause the filaments to slide past each other, thereby causing the muscle to contract or shorten.
- There is an interrelation between the nervous and muscular systems in which the motor neuron of the nervous system initiates the activity of muscle contraction through the release of a neurotransmitter.
- There are many common diseases and conditions of the muscles, and because the nervous system is so closely related, there are also many common neuromuscular diseases.

Case study

A 30-year-old patient complains of ascending flaccid paralysis that began with tingling in the toes and muscle weakness. This individual presented to Accident and Emergency after the leg weakness became so profound that he could barely walk, and now he notices his arms weakening. Loss of reflexes was also noted.

1. What disease do you think this is?

   _____

   _____

2. Knowing that this patient is losing the ability to use skeletal muscles, what life-threatening condition could occur?

   _____

   _____

3. What vital signs must you monitor?

   _____

   _____

4. Why is muscle atrophy a problem?

    _____

    _____

5. What areas of patient care must be addressed?

    _____

    _____

6. What is the likely prognosis?

    _____

    _____

Review questions

Multiple choice

1. Another name for voluntary muscle is:
    a. skeletal
    b. smooth
    c. cardiac
    d. non-striated

2. Which structure does *not* contain smooth muscle?
    a. blood vessels
    b. heart
    c. digestive tract
    d. bronchi

3. Most skeletal muscles attach to bones via:
    a. ligaments
    b. joints
    c. flexors
    d. tendons

4. The state of partial skeletal muscle contraction is known as:
    a. homeostasis
    b. muscle tone
    c. partialus contractus
    d. flexerus

5. Cardiac muscle:
    I.   is a voluntary muscle
    II. has intercalated discs to assist contraction
    III. regenerates after injury
    IV. lines the blood vessels

    a. I only
    b. I and II
    c. II only
    d. I, II, III, IV

Fill in the blanks

1. A sudden or violent muscle contraction is a _____.

2. Partial or total loss of voluntary muscle use is _____.

3. A tear in the muscle wall through which an organ can protrude is a _____.

→

4. The body stores a carbohydrate called _____ in the muscle; it can be converted to a usable energy source.

5. _____ means pain or tenderness in the muscle.

Short answers

1. List the three major muscle types and give an example of each.

    _____

    _____

    _____

    _____

2. Contrast the terms hypertrophy and atrophy and give an example of how each situation could occur.

    _____

    _____

    _____

    _____

3. Explain how vasoconstriction and vasodilatation affect blood pressure.

    _____

    _____

    _____

    _____

4. Identify two important substances required for muscle contraction and relaxation.

    _____

    _____

    _____

    _____

Suggested activities

1. Pick a major muscle group and discuss how your life would be different if that group could not function properly due to a disease or accident.

2. Bodybuilding requires an extensive knowledge of muscles and muscle groups. Demonstrate five different exercises and the different muscles they would develop.

3. Pair off with a partner and perform various muscle movements showing rotation, abduction, adduction, extension and flexion. See if your partner can accurately classify each motion.

Answers

Answers to case study

1. Guillain–Barré syndrome
2. Problems with breathing as a result of loss of use of the diaphragm.
3. Respiratory rate
4. The muscle is wasted, becomes weak and less easy to use.
5. Ensuring that there is minimal muscle wasting, and that the patient is reassured that the effects are likely to be temporary.
6. Good – the muscle paralysis peaks at about 10–14 days but eventually things return to normal – this may however take some weeks or months.

Answers to multiple choice questions

1. a
2. b
3. d
4. c
5. c

Answers to fill in the blanks

1. Cramp
2. Paralysis
3. Hernia
4. Glycogen
5. Myalgia

Answers to short answer questions

1. Smooth – bronchioles in the lungs
   Cardiac – heart
   Skeletal – biceps of upper arm
2. Atrophy – muscle wasting from prolonged immobility e.g. bedrest.
   Hypertrophy – increased muscle mass due to regular exercise.
3. When blood vessels dilate (open up) blood pressure drops.
   When blood vessels vasoconstrict (narrow) blood pressure goes up.
4. Calcium and adensoine triphosphate (ATP)

The integumentary system

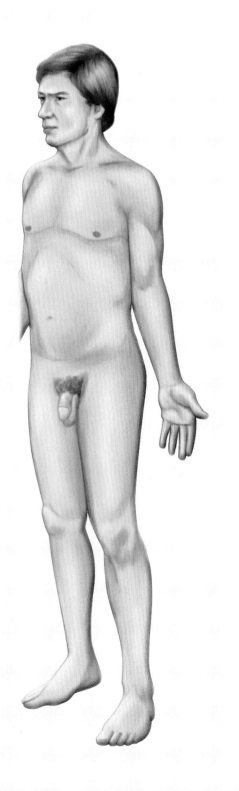

You learned in Chapter 6 that the skeletal system is like the framework of a building or house. But the framework is just one part of a building; the integrity of a house wouldn't last very long without tiles on its roof, walls of some sort, and windows, all of which help prevent the environment from doing damage to the main structures and inner workings. Like a house, the human body must be sheltered from the environment: that is the job of the integumentary system. Your skin forms a protective barrier to shield your body from the elements, guard against pathogens, and perform several other vital functions.

Think how important your skin is to your well-being. Without your skin, you would be unable to regulate your body temperature and would be uncomfortable in any environment. We paint our houses to protect them from the elements; likewise, we apply sun screen when we spend a day at the beach. While your skin is an important organ, there are several other accessory components such as nails, hair and glands involved, hence the name integumentary *system*.

Learning objectives

At the end of this chapter, you will be able to:

◆ Discuss the functions of the integumentary system
◆ List and describe the layers of the skin
◆ Explain the healing process of skin
◆ Describe the structure and growth of hair and nails
◆ Explain how the body regulates temperature through the integumentary system

Pronunciation guide

apocrine (*AP oh kreen*)
carotene (*CAR oh teen*)
eccrine (*EK kreen*)
epidermis (*ep ee DER miss*)
epithelial cells (*ep ee THEE lee yall*)
keratin (*CARE ah tin*)

keratinisation (*CARE ah tin eye ZAY shun*)
lesion (*LEE zhun*)
lunula (*LOON yoo lar*)
melanin (*MEL an in*)
melanocytes (*mel AN oh sights*)
pustule (*PUST yule*)

sebaceous gland (*sib AY shuss*)
sebum (*SEE bum*)
squamous cells (*SKWAY muss*)
stratum corneum (*STRAH tum core NEE yum*)
subcutaneous fascia (*sub kew TAY nee yuss FAY shee yah*)

System overview

In this section, we will look at the functions and parts of the integumentary system. This system is the protective covering of the body and is the most exposed system. Get ready for this chapter to 'get under your skin'.

Integumentary system functions

The integumentary system is comprised of the skin and its accessory components of hair, nails and associated glands. Your integumentary system performs several vital functions besides protecting you from an invasion of disease-producing pathogens. This system helps keep the body from drying out, acts as storage for fatty tissue necessary for energy, and, with the aid of some sunshine, produces vitamin D (needed to help your body utilise phosphorus and calcium for proper bone and teeth formation and growth). In addition, the skin provides sensory input (pleasant and unpleasant sensations involving pressure and temperature, for example) for your brain and helps regulate your body temperature.

The skin

Your skin is quite a large organ, easily weighing twice as much as your brain, approaching approximately 9 kg in an average adult. In fact, the skin is the largest organ. It covers an area of about 6.3 metres on an adult-sized body. A closer examination of a cross-section of skin reveals three main layers of tissue:

- epidermis
- dermis
- subcutaneous fascia (also called the hypodermis layer because it lies *under* the dermis)

As we discuss these three layers, please refer to Figure 8.1 for further clarification.

Epidermis

The epidermis is the layer of skin that we normally see. It is made up of five or six even smaller layers of stratified squamous epithelium. The epidermis is interesting for several reasons. First, it contains no blood vessels (**avascular**) – this is why a paper cut that only goes into the epidermis produces no bleeding – or nerve cells. Second, the cells on the surface of this layer are constantly

---

epidermis (*ep ee DER miss*)
   epi = upon
dermis (*DER miss*) true skin
subcutaneous fascia (*sub kew TAY nee uss FAY shee yah*)
   sub = under
   cutane/o = skin
   fascia = band

a = without

vascular = referring to vessels

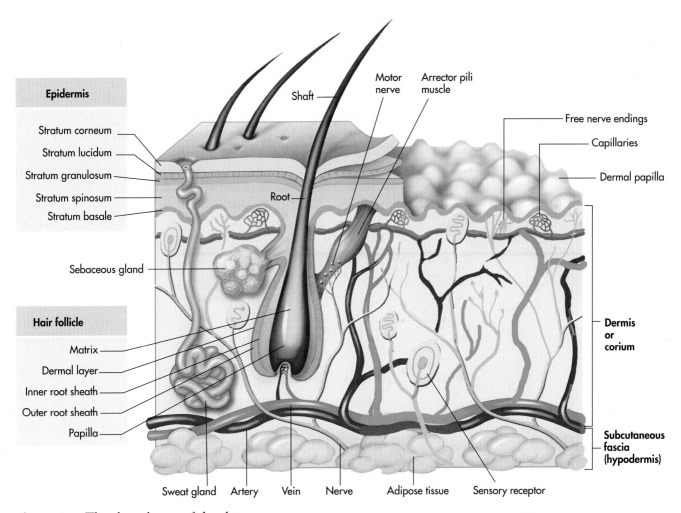

Figure 8.1 The three layers of the skin.

stratum corneum (*STRAH tum core NEE yum*)
stratus = *to spread out*
kerat/o = *hard or horny*

being shed, and are replaced with new cells that arise from the deeper region called the **stratum basale**, or basal layer of the epidermis, in a process that takes two to four weeks. In fact, the outermost surface of skin is actually a layer of dead cells called the stratum corneum, which are characteristically flat, scaly, **keratinised** (hardened) epithelial cells. The major cell type found in the epidermis is the keratinocyte which arises from the basal layer. As the cells push toward the surface, they die and fill up with the protein, keratin. Through everyday activities such as bathing, drying off and moving around, the body sloughs off *500 million* cells a day, equalling about 0.7 kg of dead skin a year! This continuous replacement of cells is very important because it allows your skin to quickly repair itself in cases of injuries.

melanocytes (*mel AN oh sights*)
melan/o = *black, extremely dark hue*
melanin (*MEL an in*)

Specialised cells called melanocytes are located deep in the epidermis and are responsible for skin colour. Melanocytes produce melanin, which is the actual substance that affects skin colour. An interesting note is that all people possess about the same number of melanocytes. The variations of skin colour are a result of the amount of melanin that is produced and how it is distributed. This is obvious when you are exposed to the ultraviolet rays of the sun. In order to protect your skin, melanocytes produce more melanin and, *voila!* you've got a tan. You will also note that in time you will develop a tan line demonstrating that the production of melanin occurs only as needed, in the areas where needed. Regardless of an individual's skin tone or colour, all skin

carotene (*CAR oh teen*)

types respond to sun exposure. For some of us, the melanin locates in patches on the skin, forming *freckles*. **Carotene**, which is another form of skin pigment, gives a yellowish hue to skin. Individuals with a pinkish hue derive that colour from the haemoglobin in their blood.

There are times when skin colour can indicate an underlying disease. Although carotene gives a yellowish hue to skin, that is normal. In a situation where liver disease exists, the body can't excrete a substance called *bilirubin*. As bilirubin builds up in the body, *yellow jaundice* occurs, giving the skin a deeper yellow colour. The changes in colour are not as apparent on individuals with darker skin, but the yellowish colour is easily seen in the whites of the eyes. Although bronze skin is associated with healthy outdoor living, individuals with a malfunctioning adrenal gland may have the same colour due to excessive melanin deposits in the skin (Addison's disease). Excessive bruising (black and blue marks called ecchymosis) could indicate skin, blood or circulatory problems as well as possible physical abuse.

Amazing body facts

More a bothersome fact of life

Sometimes the sebaceous glands within the skin become blocked. As a result, sebum stagnates and is exposed to air, drying it out. When this occurs, the sebum turns black, creating the infamous *blackhead*! To make matters worse, if that blackhead becomes infected, a *pimple* is formed (medically known as a *pustule*), usually right before your big date. Blackheads should not be squeezed because doing so can create craters at the site, not to mention interesting formations on the mirror! The best thing to do is keep those areas clean through gentle washing with soap and water and let nature take its course.

This brings us to the topic of cleaning your skin. Be careful not to wash too frequently with too hot water and/or too harsh soap, and avoid aggressive drying. This is important because excessive cleaning has the potential to dry out your skin and remove the antibacterial layer of sebum. In fact, aggressive washing with a soap that isn't pH balanced (the same acidity as your skin) can cause skin to lose its antibacterial abilities for 45 minutes after washing. As with most things in life, moderation is the key.

Dermis

The layer immediately below or inferior to, the epidermis layer is the thicker dermis layer. This layer of dense, irregular, connective tissue is considered the 'true skin' and contains the following:

- capillaries (tiny blood vessels)
- collagenous and elastic fibres
- involuntary muscles
- nerve endings
- lymph vessels (transport fluids from tissue to the blood system)
- hair follicles
- sudoriferous glands (sweat)
- sebaceous glands (oil)

Small 'fingers' of tissue project from the surface of the dermis and anchor this layer to the epidermal layer. Fingerprints, toe prints and other unique skin patterns also arise from this layer. Nerve fibres are located in the dermis so the body can sense what is happening in the environment. Because this layer also possesses blood vessels, this is where your blush comes from when you get embarrassed!

The collagenous and elastic fibres of this layer help your skin flex with the movements that you make. Without that ability, your skin would eventually tear. In addition to flexing, these fibres allow your skin to return to its normal shape when at rest. In older people and people regularly exposed to high levels of sunlight, the skin's firmness and ability to recoil to normal decreases. To better understand this process, try this experiment. With one of your hands resting palm down, take the thumb and index finger of your other hand and gently pinch and pull up the skin on the back of your resting hand. Let go and observe how quickly the skin recoils to normal. Try this experiment with an older person and observe how much slower his or her skin returns to normal. Another example of the skin's resilience due to the collagenous and elastic fibres is when it returns to normal after an injury that caused swelling.

apocrine (AP oh kreen)
eccrine (EK kreen)

There are two main types of sudoriferous or sweat glands: the **apocrine** and **eccrine** glands (see Figure 8.2). Apocrine sweat glands secrete at the hair

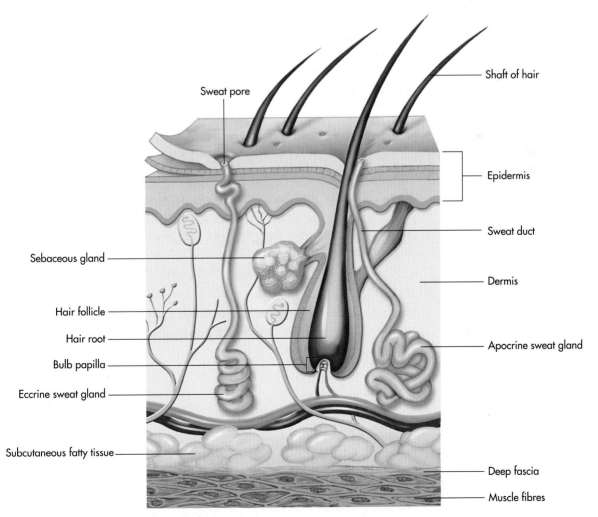

Figure 8.2 Sweat and sebaceous glands.

follicles in the groin and anal region as well as the armpits. These glands become active around puberty and are believed to act as a sexual attractant. Located all over your skin, eccrine glands are important in the regulation of body temperature. Eccrine glands are found in greater numbers on your palms, feet, forehead and upper lip. Your body has approximately 3,000,000 sweat glands! It is interesting to note that sweat by itself does not have a strong odour. However, if it is left on the skin, bacteria degrades substances in the sweat into chemicals that give off strong smells, commonly called body odours.

Sebaceous glands play an important role by secreting oil, or sebum, that keeps the skin from drying out. Because it is somewhat acidic in nature, it also helps destroy some pathogens on the skin's surface.

Subcutaneous fascia

hypo = *below, under*

lipo = *fat*

cyte = *cell*

Finally, the innermost layer of skin is the subcutaneous fascia, or **hypodermis**, which is composed of elastic and fibrous connective tissue and fatty tissue. Within this layer, **lipocytes**, or fat cells, produce the fat needed to provide padding to protect the deeper tissues of the body and act as insulation for temperature regulation. Fat is also necessary as an efficient store for energy. The hypodermis is also the layer of skin that is attached to the muscles of your body.

Test your knowledge 8.1

Complete the following:

1. List the three main layers of skin.

    a. _____

    b. _____

    c. _____

2. List four of the functions of your integumentary system.

    a. _____

    b. _____

    c. _____

    d. _____

Choose the best answer:

3. Which cells are responsible for your normal skin colour?

    a. monocytes

    b. osteocytes

    c. reticulocytes

    d. melanocytes

4. The two main types of sudoriferous glands are:

    a. apocrine and pelicine

    b. apotine and eccrine

    c. eccrine and apocrine

    d. sudacrine and melocrine

How skin heals

Just as storms can damage homes by high winds tearing off slates or damaging walls, lightning burning portions of the exterior, or damage caused by ice, everyone has had a skin injury of some type. To combat this, the body has developed ways to repair itself when injury threatens its first line of defence – the skin (see Figure 8.3).

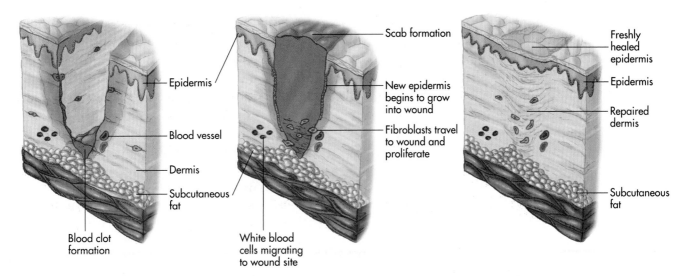

Figure 8.3 Wound repair.

fibro = *fibres, or fibrous tissue*
blast = *immature cellular development*

If blood vessels are damaged as a result of an injury to the skin, then initially blood clotting (haemostasis) occurs in order to form a clot to minimise blood loss from the injured blood vessels. The top part of the clot that is exposed to the air becomes a scab.

The first stage of wound healing is *inflammation*. This is usually indicated as five key signs at the area of the injury – pain, redness, swelling, heat and loss of function of the affected part of the body. These signs are as a result of a number of chemicals (such as histamine and substance P) being released from damaged skin cells causing vasodilation of local blood vessels. This leads to hyperaemia (increased blood flow) to the wounded area of skin – hence the redness and heat experienced. The hyperaemia also allows for an increase in the delivery of oxygen and nutrients (such as glucose) to the wound which will contribute to the healing process. Many of these blood vessels also become more leaky (increased permeability) and fluid leaks out of the blood vessels into the surrounding tissue spaces – this causes the swelling and contributes to the pain. Loss of function is usually due to the pain, but helps with healing as the affected area becomes rested. A number of leucocytes such as neutrophils and monocytes migrate to the injured area. They begin the process of removal of any foreign substances such as bacteria and dead cells.

The second stage is the *reconstruction* stage. By five to seven days post-injury, inflammation has settled and this second stage is now fully implemented. *Granulation* occurs – this involves the development of new dermal cells and tissue such as collagen **fibres** (made by fibroblasts) and new blood vessels (angiogenesis) to begin the formation of granulation tissue which will eventually become scar tissue. This develops from the base of the wound upwards. At the wound edges new epidermal cells are being produced and eventually these new cells will cover the granulation tissue and the edges of the wound will come together allowing wound contraction.

The third and final stage of wound healing is the *maturation* stage. Wound contraction has completed and the wound edges have all come together. As the wound is now almost healed there is less need for a blood supply and so this reduces. This leads to the pink granulation tissue becoming paler and looking more like scar tissue. As the epidermis is so thin this scar tissue can be seen through it. Over time the scar gradually becomes paler and smaller. The maturation stage can continue for over a year. The amount of scarring that occurs depends on the degree of tissue loss – a surgical incision for example produces a thin scar, but a burn wound is likely to cause a larger scar as more tissue has been injured.

Burns to the skin

Burns to the skin present special problems for healing. We naturally think that burns are caused by heat, and that is true. However, burns can also be caused by chemicals, electricity or radiation. When assessing the damage caused by burns, there are two factors to consider: the *depth* of the burn and the *size of the area damaged* by the burn.

The depth of a burn relates to the layer or layers of skin affected by the burn. A *first-degree burn* has damaged only the outer layer of skin, the epidermis. In this case there will be skin redness and pain, but no blistering. The pain usually subsides in about two to three days with no scarring. The damaged layer of skin usually sloughs off in about a week or so. Sunburn is a classic example of a first-degree burn.

Second-degree burns involve the entire depth of the epidermis and a portion of the dermis. Such burns cause pain, redness and blistering. The extent of blistering is directly proportional to the depth of the burn. Blisters continue to enlarge even after the initial burn. Excluding any additional complications such as infection, these blisters usually heal within 10 to 14 days, but burns reaching deeper into the dermis require anywhere from 4 to 14 weeks to heal. Scarring in second-degree burn cases is common.

Third-degree burns affect all three of the skin layers. Here the surface of the skin has a leathery feel to it and varies in colour: black, brown, tan, red or white. The victim will feel no pain because pain receptors are destroyed by third-degree burns. Also destroyed are the sweat and sebaceous glands, hair follicles and blood vessels. *Fourth-degree burns* are burns that penetrate to the bone.

Clinical application

Medicine delivery via the integumentary system

A variety of medicines can be delivered via the integumentary system. Medicines can be applied to adhesive patches that are placed on the skin where it is slowly absorbed into the bloodstream. These are called transdermal patches. Nicotine (for smoking cessation), nitroglycerin (for vasodilatation of blood vessels in the heart), birth control compounds, and pain medication, for example, can be delivered in this manner. If a more rapid response is required, the cardiac drug nitroglycerin can be placed under the tongue (*sublingually*) where it is rapidly absorbed into the bloodstream because of the high vascularisation of the mucosa in that area.

Of course, the other method of injecting drugs is a little more painful but very effective. This method is used when a drug can't be taken by mouth or the digestive system may alter the desired effects of the drug. Medication can be injected utilising a syringe and needle to deliver the medication either *subcutaneously* (under the skin) or *intradermally* (into the skin). Other routes of injection also include intramuscular (IM), intraspinal and intravenous (IV) routes.

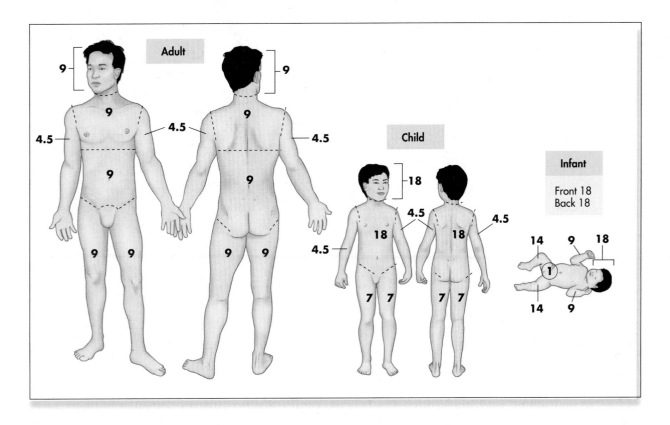

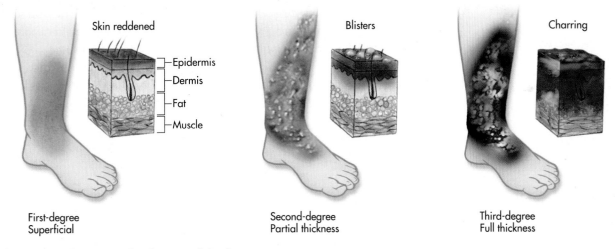

Figure 8.4 Assessing the degree of the burn.

A clinician can estimate the extent of the area covered by the burn by using the 'rule of nines'. As you can see in Figure 8.4, the body is divided into the following regions and given a percentage of body surface area value: head and neck, 9 per cent; *each* upper limb, 9 per cent; *each* lower limb, 18 per cent; front of trunk, 18 per cent; back of trunk and buttocks, 18 per cent; perineum (including the anal and urogenital region), 1 per cent. These regions can also be further divided for smaller burn areas, as you can see in the figure.

The clinical concerns for burn patients relate to the functions of the skin already discussed:

- bacterial infection
- fluid loss
- heat loss

Severe burns require healing steps at an intensity level that the body can't normally achieve on its own. Damaged skin must be removed as soon as possible to allow the process of skin grafting to begin. In this process, healthy skin is placed over damaged areas so it may begin to grow. Ideally, it is best to use the patient's own skin, known as **autografting**, because it generally eliminates the chances of tissue rejection. However, the destruction may be severe enough to require tissue from a donor (heterografting). Grafting usually requires repeat surgeries because large areas cannot be done all at once, and often the grafts don't 'take'. Other options may include growing sheets of skin tissue in a laboratory from cells of the patient or utilising synthetic materials that act as skin.

Nails

Specialised epithelial cells originating from the *nail root* form your nails (see Figure 8.5). As these cells grow out and over the *nail bed* (actually a part of the epidermis), they become keratinised, forming a substance similar to the horns on a bull that is the same protein that fills the cells of the stratum corneum. This process occurs as cells dry and shrink, are pushed to the surface, and become filled with the hard protein keratin. The **cuticle** is a fold of tissue that covers the nail root. The portion that we see is called the *nail body*. Nails normally grow about 1 millimetre every week. The pink colour of your nails comes from the vascularisation of the tissue under the nails, while the white half-moon shaped area, or lunula, is a result of the thicker layer of cells at the base.

auto = *self*

keratinised (*CARE ah tin ised*)
kerat/o = *hard or horny*

lunula (*LOON yoo lar*)
luna = *moon*

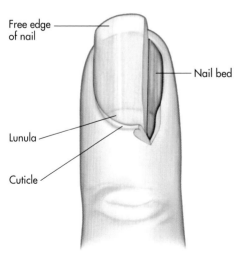
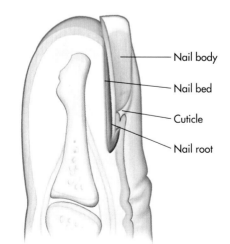

Figure 8.5 Structures of the fingernail.

Clinical application

Assessing peripheral perfusion

The pink colour of the nail bed is clinically significant in that it can aid in the assessment of perfusion (blood flow) to the extremities and can be a determinant of oxygenation. If you pinch one of your fingernails straight down with the thumb and index finger of your other hand for five seconds and release that pinch, you will note that the nail bed went from a blanched white colour back to pink in a matter of seconds. This shows good perfusion as a result of the blood rushing back into the nail bed. In cases of poor perfusion, it takes longer for the nail bed to 'pink-up'. Normally, it takes less than three seconds for the nail to return to pink from the blanched white state. If that time is greater than three seconds, perfusion to the extremities is considered sluggish. If the refill time is greater than five seconds, there is clearly an abnormal situation occurring.

Diabetes is a disease that causes a condition of reduced blood flow to the extremities known as **peripheral vascular disease** (**PVD**). PVD can cause an increase in the time required to re-perfuse the nail bed. Blood clots or vascular spasms can decrease blood flow and thereby extend refill time, as can hypothermia, which naturally constricts blood vessels in the periphery to conserve heat.

In addition, nail beds can change colours under certain conditions. For example, as the level of oxygen decreases in the tissue, the nail beds become bluish in tint.

Hair

Body hair is normal and served important purposes in our evolutionary past as well as today. Hair helps to regulate body temperature, as you will see in the next section, and it functions as a sensor to help detect things on your skin, such as bugs or cobwebs. Eyelashes help to protect our eyes from foreign objects, and hair in the nose helps to filter out gross particulate matter in the air that we breathe.

keratin (*CARE ah tin*)

follicle (*FOLL ickle*)

Hair is composed of a fibrous protein called **keratin** just like your fingernails and toe nails. The hair that you see is called the shaft, and the root extends down into the dermis to the **follicle** (see Figure 8.6). The follicle is formed by epithelial cells, which have a rich source of blood provided by the dermal blood vessels. As a result, cells divide and grow in the base of the follicle. As new cells continually form, the older cells are crowded out and pushed upward toward the skin's surface. As these old cells are pushed away from the blood source (which provides their nourishment), they die, becoming keratinised in the process. So, basically, the hair you see on the individual next to you is a bunch of dead cells. This isn't so bad considering that if your hairs were alive, your haircuts would be extremely painful with a good chance of your bleeding to death if you got more than a light trim. The old, popular belief that shaving or frequent cutting of hair makes it grow quicker or thicker is wrong. Neither shaving nor cutting does anything to affect the rate of cell growth at the base of those hair follicles.

sebaceous gland (*sib AY shuss*)
sebo = *tallow*
sebum (*SEE bum*)

If you look at Figure 8.6, you will note that there is a **sebaceous gland** associated with the hair follicle. The sebaceous gland secretes **sebum**, an oily substance that coats the follicle and works its way to the skin's surface. Sebum

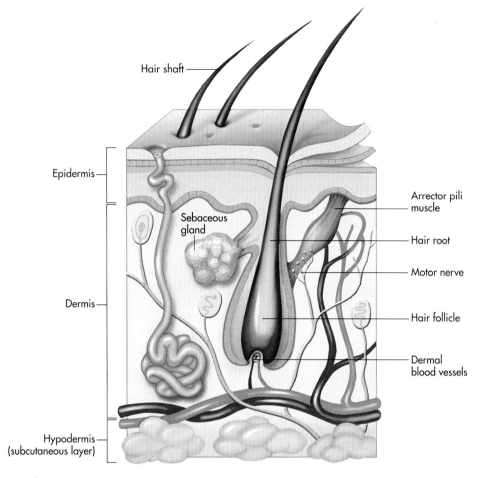

Figure 8.6 Diagram of a hair follicle.

Test your knowledge 8.2

Choose the best answer:

1. The blood clot that forms in a wound that is exposed to air becomes a:
   a. scar
   b. scab
   c. haematoma
   d. keloid

2. When assessing the skin damage caused by a burn, the two main factors to assess are:
   a. temperature and depth
   b. depth and odour
   c. area of damage and depth
   d. odour and colour

3. The white, half-moon shaped area of your fingernail is called:
   a. cuticle
   b. lunula
   c. keratin
   d. lingula

4. Subcutaneous fascia is another term for the:
   a. epidermis
   b. hypodermis
   c. corneum
   d. dermis

Applied science

Forensics and hair

An interesting point about hair is its ability to tell a pathologist if an individual ingested certain drugs or other substances, such as lead or arsenic. Trace amounts of ingested substances can become part of the hair's composition. As a result, analysis of a hair sample can reveal what and how long ago something was ingested. The longer the length of hair, the longer the record of what was consumed by that individual.

is somewhat antibacterial, so it aids in decreasing infections on your skin. Although it also waterproofs and lubricates the skin and hair, sebum production decreases with age. As a result, older individuals exhibit drier skin and more brittle hair.

Like skin colour, your hair colour is dependent on the amount and type of melanin you produce. Generally, the more melanin, the darker the hair. White hair occurs in the absence of melanin. Red hair is a result of an altered melanin that has iron in it.

You may be extremely envious of someone with a head full of curly hair or of someone with absolutely straight hair. Don't be envious of the person – just envy his or her hair shafts! Flat hair shafts produce curly hair, while round hair shafts produce straight hair.

The lifespan of hair is dependent on location. Eyelashes last around three to four months, while the hair on the scalp lasts for about three to four years.

Temperature regulation

The integumentary system plays a major role in the regulation of the body's temperature. It is amazing how we can stay in a relatively 'tight' range of body temperature while we do a variety of things in a variety of environments. Of course, clothes do help tremendously, but still it is critical to have a properly functioning temperature regulator like our integumentary system. This is accomplished through a complex series of activities.

Part of temperature regulation is accomplished by changes in the size of blood vessels in your skin. As your temperature rises, your body signals the blood vessels in your skin to get larger in diameter. The correct term is **vasodilatation**. This is the body's attempt to get as much 'hot' blood exposed to a cooler surrounding environment so the heat radiates away from the body. In addition, sweat glands excrete water (as well as some waste products such as nitrogenous wastes and sodium chloride) onto the skin's surface. As the water evaporates, cooling occurs. As long as you stay well hydrated you can produce sweat – this system works pretty well. Hydration is especially important before and during activities to avoid feeling thirsty. Thirst indicates the body is already dehydrating. It's pretty easy to dehydrate, especially when you realise that you can have up to 3000 sweat glands on a square inch of skin on your hands and feet and you can potentially excrete up to 12 litres of sweat in a 24-hour period. The risk of dehydration can be serious.

Conversely, if you were in a cold environment and needed to warm up, your blood vessels would become smaller in diameter, or **vasoconstrict**. This act forces blood away from the skin and back toward the core of the body where the heat is. The perfect visual explanation of vasoconstriction is a skinny little kid running around a pool on a chilly summer day, shivering, with blue lips. The lips are blue because of vasoconstriction. Rings have a tendency to slide

vasodilatation (*VAY so dye lat AY shun*)
vaso = *vessel*
dilatation = *enlargement*

vasoconstrict (*VAY so con STRICT*)

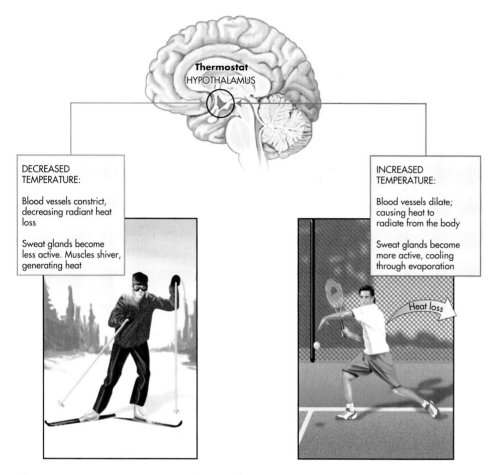

Figure 8.7 Integumentary regulation of body temperature.

off your fingers in cold weather more readily than in hot weather because of vasoconstriction.

Body temperature regulation is also aided by the hairs on your skin. Muscles in your skin called *arrector pili*, or erector muscles, are attached to your hairs, and when those muscles contract, they make your hairs stand erect. The constriction of those muscles shows up as goose bumps when you are chilled. When the hair stands up, pockets of still air are formed right above the skin, creating a dead air space that insulates the skin from the cooler surrounding environment. This response, however, does not have much effect on keeping the body warm. This is also how goose down (feathers) clothing works in protecting against cold weather. See Figure 8.7, which illustrates how your integumentary system regulates body temperature with the aid of your nervous system.

Common disorders of the integumentary system

There are whole sections of medical libraries dedicated to diseases of the skin. This section will provide you with some basic terms and diseases related to the integumentary system.

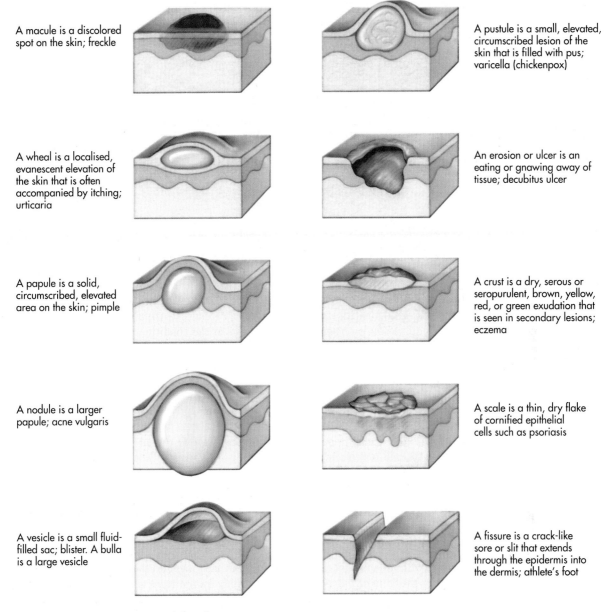

A macule is a discolored spot on the skin; freckle

A pustule is a small, elevated, circumscribed lesion of the skin that is filled with pus; varicella (chickenpox)

A wheal is a localised, evanescent elevation of the skin that is often accompanied by itching; urticaria

An erosion or ulcer is an eating or gnawing away of tissue; decubitus ulcer

A papule is a solid, circumscribed, elevated area on the skin; pimple

A crust is a dry, serous or seropurulent, brown, yellow, red, or green exudation that is seen in secondary lesions; eczema

A nodule is a larger papule; acne vulgaris

A scale is a thin, dry flake of cornified epithelial cells such as psoriasis

A vesicle is a small fluid-filled sac; blister. A bulla is a large vesicle

A fissure is a crack-like sore or slit that extends through the epidermis into the dermis; athlete's foot

Figure 8.8 Various types of skin lesions.

A **lesion** is a pathologically altered piece of tissue that can include a wound or injury or a single infected patch of skin. The colour of a lesion is usually different to that of normal skin. Figure 8.8 shows a variety of lesion types.

In addition to several types of skin lesions, there are many common pathologic conditions of the integumentary system. Table 8.1 describes some of these conditions. Please see Figure 8.9 to view various photos on types of integumentary conditions.

Table 8.1 Common pathological conditions of the integumentary system

Name of condition	Description
abrasion	A condition that results from mechanically scraping away a portion of the skin's layer(s). This may be a result of injury or a deliberate clinical procedure.
acne	Sebaceous glands over-secrete sebum which, along with the dead keratinised cells, clog the hair follicle. If the blocked follicle becomes infected with bacteria, pimples develop.
athlete's foot	This is a common fungal infection that occurs in areas of continuous moisture, such as between the toes, or on palms or fingers. Jock itch is a fungal infection of the groin area.
boil (furuncle)	Also known as a furuncle (*FOO rung kle*), this is an acute inflammatory* process involving either the subcutaneous layer of skin, a hair follicle or a gland.
cold sore	These watery vesicles are caused by the herpes simplex virus.
contusion (bruise)	A traumatic skin injury in which the skin is not broken but an injury still occurs. A contusion can present with pain, discoloration due to breakage of small blood vessels (capillaries), and swelling.
dermatitis	An inflammatory process that can be caused by a variety of irritants such as from plants or chemical sources. Patients with dermatitis can exhibit erythema (redness), papules, vesicles and crusty scabbing. Touching poison ivy or poison oak can lead to the condition called contact dermatitis.
eczema	A superficial form of dermatitis that exhibits redness, papules, vesicles and crusting.
hives (urticaria)	This disorder is a result of an allergic reaction and produces reddened patches (wheals) and itching (pruritis), which can be severe.
pressure ulcers (decubitus ulcers or bed sores)	These sores are a result of a lack of blood flow to skin that has had pressure applied to a bony prominence. This often occurs in bedridden patients who aren't turned often enough. Often called decubitus (*dee KYOO bih tus*) ulcers.
psoriasis	From the Greek word meaning 'to itch', psoriasis is a chronic, inflammatory skin condition that exhibits red, dry, crusty papules, which form circular borders over the affected areas.
scabies	An infectious and contagious disease caused by egg-laying mites, usually seen in children; causes severe itching.
shingles	This is a very painful inflammatory skin condition that also involves the nervous system. Shingles presents itself in the form of patches of vesicles mainly on the trunk of the body, but can be found on other body regions. This condition is caused by the virus herpes zoster. *Zoster* is from the Greek word meaning 'belt' and relates to the distribution of those patches around the trunk. The pain often remains for some time after the vesicles have healed.
skin cancer	There are a variety of skin cancers. Squamous cell carcinoma and basal cell carcinoma are the two most common. Basal cell carcinoma usually spreads locally and therefore can usually be successfully treated. Squamous cell carcinoma may develop deeper into tissue, but it usually doesn't spread. The most serious and least successfully treated skin cancer is malignant melanoma. This is a cancer that initially affects the melanocytes, which produce skin pigments. This cancer can spread throughout the body to various organs.

*Inflammation literally means to 'flame within'. The inflammatory response is tissue's reaction to injury where there is pain, heat, redness, swelling and loss of function of the affected area.

Figure 8.9 (*opposite*) Various types of integumentary conditions. (a) Urticaria (hives). (*Source*: © Scott Camazine/Phototake) (b) Malignant melanoma. (*Source*: © James Stevenson/Science Photo Library) (c) Erythema infectiosum (fifth disease). (*Source*: © Bubbles Photo Library/Alamy) (d) Acne. (*Source*: © Medical-on-Line/Alamy) (e) Poison ivy (dermatitis). (*Source*: © Doug Diamond/Alamy) (f) Herpes simplex. (*Source*: © Hercules Robinson/Alamy) (g) Burn, second-degree. (*Source*: © Dr. P. Marazzi/Science Photo Library)

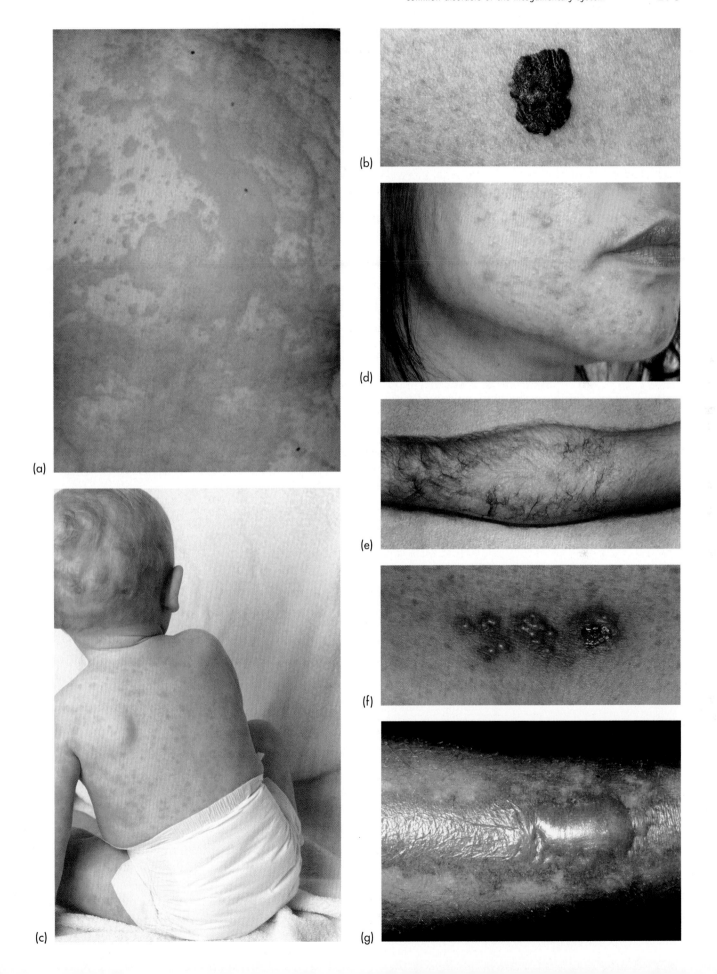

(a)

(b)

(c)

(d)

(e)

(f)

(g)

Summary

- Your skin is your largest organ.
- Your skin is an amazing organ that does the following:

  a. Acts as a barrier to infection both as a physical shield and through secretion of an antibacterial substance

  b. Acts as a physical barrier to injury

  c. Helps to keep the body from dehydrating

  d. Stores fat (yes, you do need some fat)

  e. Synthesises and secretes vitamin D with the help of sunshine

  f. Regulates body temperature

  g. Provides a minor excretory function in the elimination of water, salts and urea

  h. Provides sensory input

- The skin is composed of three layers: epidermis, dermis and subcutaneous fascia.
- Skin is not static; it constantly recreates itself.
- Various glands in the skin help moisturise, waterproof and control body temperature as well as excrete some waste products.
- The severity of burns to the skin is evaluated by the depth of the burn and the area that the burn covers.
- Nails are protective devices composed of dead material.
- Hair (also dead material) aids in controlling body temperature.

Case study

A 27-year-old female presents to her General Practitioner with complaints of red, itching and oozing skin for the past two days. Physical examination and history reveal the following: a well-nourished white female who is in otherwise good health, no known allergies, normal vital signs, pupils normal and reactive, reflexes good, breath sounds are normal. There are red wheals and itching on both legs, arms and torso. The patient stated that she changed her brand of washing powder recently.

1. What diagnosis is suggested by her history?

   _____

   _____

2. What types of treatment would you recommend?

   _____

   _____

Review questions

Multiple choice

1. The substance that is *mainly* responsible for skin colour is:
   a. melanin
   b. pigmentin
   c. carotene
   d. vitamin D

2. Whether you have naturally curly or straight hair is dependent on the shape of your:
   a. hair follicle
   b. hair shaft
   c. hair root
   d. sebaceous gland

3. The fibrous protein that makes up your hair and nails and fills your epidermal cells is called:
   a. carotene
   b. myelin
   c. keratin
   d. dermasene

4. In a cold environment, in order to maintain a core body temperature, peripheral blood vessels:
   a. vasodilate
   b. venospasm
   c. shiver
   d. vasoconstrict

5. The hair on your scalp can last:
   a. 3 to 4 months
   b. until your 40s
   c. 3 to 4 years
   d. until you have children

Fill in the blanks

1. The three main layers of skin are the _____, _____ and _____.

2. The two main types of sweat glands are the _____ and the _____ glands.

3. Sebaceous glands secrete an oily substance called _____.

4. For some individuals, melanin locates in small patches called _____.

5. Yellow jaundice, a condition associated with liver disease, occurs as a result of the build-up of _____.

Short answers

1. Discuss three functions of the integumentary system.

   _____

   _____

   _____

   _____

2. Discuss the functions of sebum.

   _____

   _____

   _____

   _____

3. Why is there an increased production of melanin when there is an increased sun exposure?

   _____

   _____

   _____

   _____

4. How does the integumentary system assist in the regulation of body temperature?

   _____

   _____

   _____

   _____

Suggested activities

1. Contact a local dermatologist to speak to your class on diagnosing and treating skin diseases.

2. Research the variety of medications used to treat acne and focus on the mode of action, benefits and side effects.

Answers

Answers to case study

1. Her history suggests urticaria caused by the change in washing powder.
2. Treatment can include changing the washing powder back to the original, anti-histamines, soothing lotions, e.g. calamine lotion.

Answers to multiple choice questions

1. a
2. b
3. c
4. d
5. c

Answers to fill in the blanks

1. epidermis; dermis; subcutaneous fascia
2. apocrine; eccrine
3. sebum
4. freckles
5. bilirubin

Answers to short answer questions

1. Any three from:
   - Protection
   - Storage for fatty tissue needed for energy
   - Produces vitamin D
   - Sensory input
   - Regulates body temperature

2. Sebum has an antibacterial function; it waterproofs the skin and lubricates it
3. Exposure to ultraviolet light results in increased production of melanin by melanocytes, and confers protection from the sun's rays
4. Changes blood flow to the skin (contracts if cold, dilates if hot); produces sweat (helps to cool the body)

The nervous system

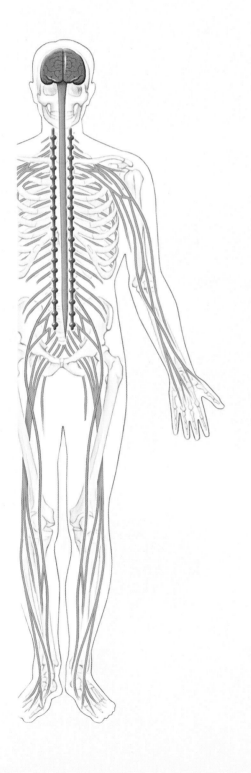

Part I

So far, we have seen the infrastructure, building blocks and support systems, of the city. Soon we will learn about transportation, protection and energy delivery systems. Like any good city, the body must have a control system, a system to monitor conditions, take corrective action when necessary, and keep everything running smoothly. Imagine what would happen if a traffic light network in a city were to suddenly fail. The control systems of the body are the nervous and endocrine systems, which receive help from your special senses. Like any control system, they have a large, complex job that is sometimes difficult to understand. They must keep track of everything that is happening in the body. Therefore, the nervous and endocrine systems are perhaps the most complex and vital systems in the body. To make the material more manageable, we have subdivided the nervous system into two separate chapters. In this chapter, we start at the bottom of the control hierarchy, the cells and the spinal cord. In Chapter 10 we focus on the higher-level control of the brain.

Learning objectives

At the end of this chapter, you will be able to:

◆ List and describe the components and basic operation of the nervous system
◆ Contrast the central and peripheral nervous systems
◆ Explain the relationship of the sensory system to the nervous system
◆ Define the parts and functions of the nervous tissue
◆ Describe the process of neuromuscular transmission
◆ Discuss the anatomy and physiology of the spinal cord
◆ List and describe various nervous system disorders of the nerves and spinal cord

Pronunciation guide

arachnoid mater (*ah RAK noyd MAY ter*)
astrocytes (*AST row SIGHTS*)
axon (*AK son*)
cerebrospinal fluid (*ser ree bro SPY nall*)
chemical synapse (*KEM ik all SIGH naps*)
commissures (*COM mih shores*)
dendrites (*DEN drights*)
dorsal root ganglion (*DOR sall root GANG lee yon*)

dura mater (*DURE ah MAY ter*)
ependymal cells (*ep PEN dye mall*)
epidural space (*ep ee DURE all*)
ganglia (*GANG lee yah*)
glial cells (*GLY all*)
gyri (*JIE ree*)
meninges (*men IN jeez*)
microglia (*my crow GLY ah*)
myelin (*MY lin*)
neuroglia (glial cells) (*new ROH gly ah*)
nodes of Ranvier (*ran vee AYE*)

oligodendrocytes (*olly go DEN drow sights*)
pia mater (*PEE ah MAY ter*)
plexus (*PLEK suss*)
Schwann cells (*shwon*)
somatic nervous system (*soh MAT ik*)
subarachnoid space (*sub ah RAK noyd*)
subdural space (*sub DURE all*)
sulcus (*SULL cuss*)
vesicle (*VEE sickle*)

Organisation

We begin this chapter with an overview of the entire system to show how all the components are interrelated. Let's start with the basic operations.

The parts and basic operation of the nervous system

The organisation of the nervous system can be compared to a computer. Information is input into the computer by various means ('senses'): keyboard, mouse, microphone, internet connection, and so on. The main components of the computer's 'brain' are the hard drive ('long-term memory'), random access memory ('short-term memory') and central processing unit ('thinking' and 'decision-making'). The computer's output (control of muscles and glands) – its 'interaction with the world' – exits via ports and cables to printers, displays, speakers and other devices. There is also a 'default' or return to normal setting (homeostasis), which is similar to an internal diagnostic procedure in the computer. There is only one nervous system, but typically, we refer to the brain and spinal cord as the **central nervous system** (or **CNS**) and everything outside of the brain and spinal cord, which represent the input and output pathways, as the **peripheral nervous system** (or **PNS**). Figure 9.1 is a schematic of the organisation of the branches of the nervous system.

The nervous system's 'input devices' comprise the **sensory system**. Your senses sample the environment and bring the information to the nervous system, as explored further in Chapter 11. Everything that can possibly be measured about your body and the world around you (both consciously and subconsciously) is measured by your sensory system. The sensory information goes into the nervous system where it is handled by the brain and spinal cord. The brain and spinal cord (the hard drive, random access memory and processor) combine the input with other kinds of information, compare it to information from past experiences, and make decisions about how to respond to the information. Once the brain and spinal cord decide what response is required to deal with the new information, the output side is activated. The output side carries out the orders from the brain and spinal cord. The output side, often

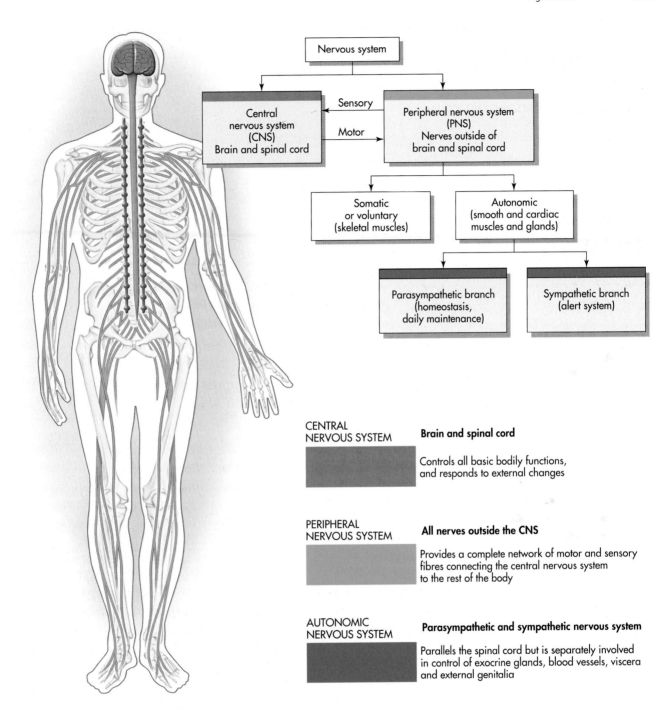

Figure 9.1 Organisation of the nervous system.

moto = *movement*

somatic nervous system *(soh MAT ik)*

called the **motor** or **efferent system**, carries orders to all three types of muscles (smooth, skeletal and cardiac) and to the body's glands, telling them how to respond to the new information. The motor system is divided into two different branches: the somatic nervous system, which controls skeletal muscle and usually voluntary movements, and the **autonomic nervous system**, which controls smooth and cardiac muscle in your organs and also glands. Autonomic output is involuntary and not under conscious control.

The autonomic nervous system is further divided into two branches: **para-sympathetic** and **sympathetic**. The parasympathetic branch, often called 'resting

Amazing body facts

Faster than a speeding bullet

Well, not literally! Your nervous system must respond very quickly to stimuli. Think about how fast you pull your hand away from a hot plate or step on the brake when something runs in front of your car. Nerve impulses can move very quickly. Some neurons have speeds as fast as 100 metres per second. That's roughly 200 miles per hour. Bullets, on the other hand, can travel 3000 miles per hour! In some diseases, for example diabetes mellitus, the speed of the nerve impulse is slowed or deflected (almost a ricochet) and so it takes longer for the impulse to reach the brain to say, for example, that the temperature of the water is too hot (see common disorders of the nervous system). This results in a delay in behavioural change (i.e. removing the hand from the hot water) and so tissue damage occurs. Of course, this sequence can occur in many diseases where damage to the nervous system has occurred (e.g. spina bifida, some spinal cord injuries, etc.).

and digesting', deals with normal body functioning such as digesting food or emptying the urinary bladder, while the sympathetic branch is the body's alert system, commonly known as the 'fight-or-flight' response system. A further subdivision is the **enteric nervous system**. This consists of an extensive network of axons and neuron cell bodies (collectively called plexuses) within the wall of the digestive tract. The enteric nervous system can control the digestive tract independently of the central nervous system. It is considered to be part of the autonomic nervous system because of the parasympathetic and sympathetic neurons, which contribute to the plexuses.

The organisation of the nervous system is easier to understand if we look at a real-life event. You drive over to your friend's house. As you open her gate, a large dog bounds down the front steps barking and snarling at you. Your sensory system gathers the following information about the new stimulus: a large unfriendly dog; you are far from the protection of the car; nobody is coming to help. The information goes into your spinal cord and brain, and several decisions are made. You are in danger and something must be done. Your brain and spinal cord send directions via your autonomic nervous system to your organs to gear up for action. Your heart rate, blood pressure and respiration rate rise. You begin to sweat. More blood is delivered to your skeletal muscles and heart in order to get you fully ready to respond. This is all involuntary, meaning you cannot consciously control it. Your nervous system readies your skeletal muscles to help you to run away. This fight-or-flight response is discussed later in further depth. If you can control your fear, you back slowly away from the situation. If you are scared witless, you run from the garden as fast as you can. Either way, you can hopefully escape the danger with your skin (if not your dignity) intact.

Each part of the nervous system has separate but unmistakably connected roles to play in assessing a situation and responding to it. The nervous system is active 24 hours a day, 7 days a week, for your entire life. Some situations, like the unfriendly dog, or a sudden drop in blood pressure, or a terrible car accident, may be life-threatening. Others, like an ant crawling across your foot or a pencil rolling under the desk just out of reach, are simply annoying. Everything that happens in your world is monitored and responded to by your nervous system. It's a truly Herculean task.

Nervous tissue

Like all organs, the components of the nervous system are made up of tissue. But unlike other systems, the nervous system contains no epithelium, connective tissue or muscle tissue. Nervous tissue is made up of two different types of cells: neuroglia (or glial cells) and neurons.

Neuroglia

neuroglia (*new ROH gly ah*)
glial cells (*GLY all*)

The **neuroglia** or **glial cells** (meaning nerve glue) are specialised cells in nervous tissue that allow it to perform nervous system functions. In the CNS, neuroglia are the most abundant cells, accounting for more than half of the brain's weight. They are the major supporting cells in the CNS with many functions such as phagocytosis, formation of the permeability barrier between the blood and neurons (the blood–brain barrier), production of cerebrospinal fluid and formation of myelin sheaths. There are four types of glial cells, each with a unique function and structure. **Astrocytes** are metabolic and structural support cells that are star-shaped where the extensions spread out and can cover the surfaces of blood vessels, neurons and the pia mater (discussed later). Astrocytes have a role in regulation of the extracellular composition of brain fluid, releasing substances that promote formation of tight junctions (see Chapter 3). It is the endothelial cells and their tight junctions that form the blood–brain barrier, which protects neurons from some toxic substances in the blood but allows exchange of nutrients and waste products. **Microglia** are specialised macrophages that remove debris by phagocytosis. **Ependymal cells** line the ventricles of the brain and central canal of the spinal cord, doing the job of epithelial cells. Specialised ependymal cells and blood vessels form the choroid plexuses. The choroid plexuses secrete cerebrospinal fluid. **Oligodendrocytes** make a lipid insulation called **myelin**, described in more detail a bit later. In the PNS, there are only two types of glial cells: **Schwann cells**, which make myelin for the PNS, and **satellite cells**, which are support cells providing nutrients to neuron cell bodies. Figure 9.2 shows a cutaway view of the spinal cord of the CNS showing the four types of neuroglial cells found there.

astrocytes (*AST row SIGHTS*)
astro = *star*
cyte = *cell*

microglia (*my crow GLY ah*)
ependymal cells (*ep PEN dye mall*)

oligodendrocytes (*olly go DEN drow sights*)
oligo = *few*
dendro = *branches*
myelin (*MY lin*)
Schwann cells (*shwon*)

Neurons

The glial cells do all the support activities for the nervous system, such as lining and covering cavities and supporting and protecting structures. None of the glial cells, however, are capable of measuring the environment, making decisions or sending orders. All of the control functions of the nervous system must be carried out by a second group of cells called **neurons** or **nerve cells**. Neurons are rather bizarre-looking cells, often with many branches and what appears to be a tail (see Figure 9.3). Each part of a neuron has a specific function. The neuron body's main function is that of cell metabolism. Its **dendrites** receive information from the environment or from other cells. The **axon** generates and sends signals to other cells. Those signals leave the cell and travel down the axon until they reach the **axon terminal**, which then connects to a receiving cell. This combination of axon terminal and receiving cell is called a **synapse**. If the receiving cell is a skeletal muscle cell, then this particular synapse is called the *neuromuscular synapse* or *junction*.

Neurons can be classified by how they look (structure) or what they do (function). From a structural point of view, neurons can have either one axon

dendrites (*DEN drights*)
axon (*AK son*)

synapse (*SIGH naps*)

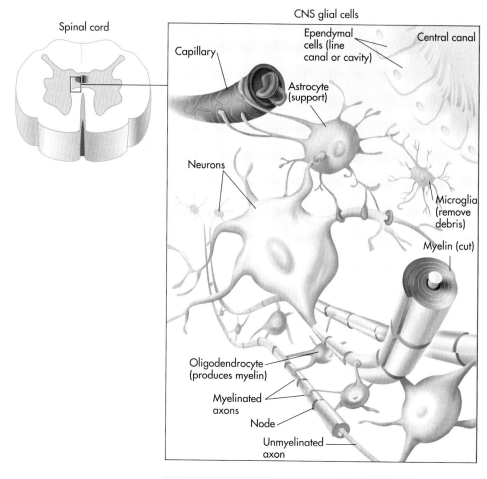

Figure 9.2 Glial cells and their functions.

and one dendrite (bipolar), one axon and many dendrites (multipolar), or one process that splits into a central and a peripheral projection (unipolar). Classified by function, input neurons are known as **sensory** or **afferent neurons** and output neurons are known as **motor** or **efferent neurons**. Neurons that carry information between neurons are called interneurons (*inter* means between) or **association neurons**.

interneurons (association neurons)
(in ter NEW rons)

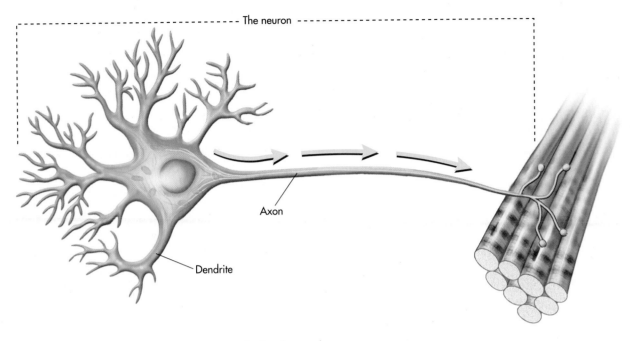

Figure 9.3 A neuron connecting to a skeletal muscle.

Test your knowledge 9.1

Choose the best answer:

1. Which cells are the support cells in the nervous system?

   a. neurons

   b. neuroglia

   c. epithelium

   d. all of the above

2. The output side of your nervous system is called:

   a. sensory

   b. motor

   c. central

   d. exit

3. The part of the nervous system that integrates and processes information is known as the:

   a. PNS

   b. PBS

   c. CNS

   d. CIA

4. The lipid insulation of nervous tissue is called:

   a. glia

   b. myelin

   c. Schwann cells

   d. astrocytes

Action potentials

A cell that is not stimulated or excited is called a resting cell and is said to be *polarised*. It has a difference in charge across its membrane, being more negative on the inside than on the outside. When that cell gets stimulated (excited), gates in the cell membrane spring open. When these gates, called sodium gates, open, they allow sodium ions (Na^+) to travel across the cell

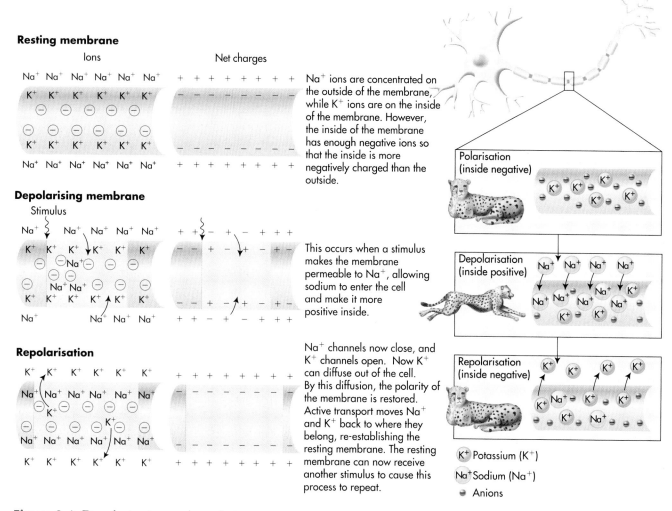

Figure 9.4 Depolarisation and repolarisation.

membrane. These sodium ions are positively charged, so when they go into the cell, the cell becomes more positive. A cell that is more positive than resting is called **depolarised**. In less than a millisecond, the gates on the sodium channels shut, just like the automatic doors at the supermarket shut by themselves. Then other gates open for potassium channels. Potassium (K⁺), which is also positive, leaves the cell, taking its positive charges with it. The inside of the cell becomes more negative again, eventually returning to rest. This is **repolarisation** (see Figure 9.4). Sometimes a cell overshoots and becomes more negative than when it is at rest. Then the cell is **hyperpolarised**. Eventually the cell will return to a resting state when the sodium exits and the potassium re-enters the cell. It should be noted that the cell is unable to accept another stimulus until it repolarises (returns to its resting state), and this time period during which it cannot accept another stimulus is called the **refractory period**. This whole series of permeability changes within the cell and the resultant changes of the internal and external charges to carry the impulse down the axon is called the **action potential**.

If you drink too much water, i.e. several litres all in one go, the concentrations of electrolytes in the extracellular fluids (ECF) can become 'diluted' which results in an inability of cells to pump Na^+/K^+ across the cell membrane and therefore the cell is unable to depolarise (and repolarise) correctly. This can result in a dysfunctional nervous system and eventual death.

How neurons work

Now that we know the basic structure of neurons, let's look closer at how they function.

Excitable cells

A neuron is a kind of cell called an *excitable cell*. Excitable cells can change their charges and therefore transmit information or effect muscular contraction. Each time charged particles flow across a cell membrane, a tiny electrical current is generated. (Electricity is just the movement of charges from one place to the other.) All three types of muscle cells are excitable cells, as are many gland cells. Because neurons are excitable cells, it makes sense that the way neurons send and receive signals is via tiny electrical currents. (This is also one of the reasons electrocution can cause nervous system damage. It literally shorts out the electrical pathways in the neurons, which causes the heart to stop beating.)

How, you might ask, can cells carry electricity? Cells contain substances called electrolytes, in different concentrations (see Table 9.1). Positively charged electrolytes are called cations whereas negatively charged electrolytes are called anions. Cells are 'bathed' in an electrolyte-rich fluid (interstitial fluid) which contains different concentrations of the same electrolytes.

An inflow of sodium into the cell and an outflow of potassium from the cell results in an action potential being generated. Normally, potassium and sodium would not be able to move into or out of the cell in sufficient quantities to make the cell excited, so there is a 'pump', called the Na^+-K^+-ATP pump, embedded in the cell membrane. This is responsible for the high concentration of potassium and low concentration of sodium in the cell. The 'pump'

Table 9.1

Ions (electrolytes)	Intracellular concentration (mmol/l)	Extracellular concentration (mmol/l)
Cations		
potassium (K^+)	148	5
sodium (Na^+)	10	142
calcium (Ca^{++})	<1	5
Anions		
Proteins	56	16
chloride (Cl^-)	4	103

Applied science

Fugu

The puffer fish, fugu, is considered a delicacy in Japan because if you eat it, you could die! Fugu can be served only by specially certified chefs trained to prepare the fish so it is safe to eat. Puffer fish contain a poison, tetrodotoxin (TTX), in their tissues that blocks sodium channels, preventing sodium from entering cells. Cells exposed to TTX cannot depolarise. Thus, neurons cannot fire action potentials. People who consume improperly prepared fugu become paralysed. Symptoms develop as ascending paralysis within 15 minutes to 20 hours post ingestion. If untreated, death can result in four to six hours.

moves more sodium ions out of the cell ($3Na^+$ compared to $2K^+$) resulting in a net transfer of positive charge out of the cell, which contributes to the resting membrane potential. The negative charge inside the cell is further enhanced by the presence of negatively charged proteins inside the cell.

It seems hard to believe, but cells are like miniature batteries, able to generate tiny currents simply by changing the permeability of their membranes. Perhaps the plot of the *Matrix* films in which humans are used by machines as batteries is not so far-fetched?

Local potentials

Neurons can use their ability to generate electricity to send, receive and interpret signals. Let's look at an example of how this works. You are hammering a picture hanger into the wall of your newly painted bedroom, and you hit your thumb. The blow from the hammer stimulates the dendrites in your thumb, and sodium gates open. Sodium flows into the dendrites, which become depolarised. If you hit your thumb softly, the cell is stimulated only a little, only a few gates open, and the cell does not depolarise very much. (The pain isn't too bad, either.) If you hit your thumb really hard, more gates open and the cell depolarises much more. (It hurts a lot more, too.) This phenomenon is known as a **local potential**. In a local potential, the size of the stimulus determines the excitement of the cell. A big stimulus causes a bigger depolarisation than a small stimulus. Many sensory cells work via local potentials. That's often how your CNS tells the size of the environmental change.

The dendrites carry the depolarisation to the sensory neuron cell body. The cell body takes that information and generates an action potential, if the stimulus is big enough. One difference between action potentials and local potentials is that action potentials are 'all-or-none', which means that the action potential, once it starts, will always finish and will always be the same size. You either have one or you don't. There are no small action potentials or big ones as there are with local potentials. Action potentials are always the same. Once an action potential is formed, it travels down the axon from the cell body to the terminal. This movement is called *impulse conduction*.

Clinical application

Multiple sclerosis

Multiple sclerosis (MS) is a disorder of the myelin in the CNS. Patients with multiple sclerosis have many areas where the myelin has been destroyed. In areas without myelin, impulse conduction is slow or impossible. Imagine an electrical wire with many bare spots or with no insulation. These bare or exposed parts can 'short out' and prevent the current from passing further down the wire. Symptoms of multiple sclerosis differ from patient to patient depending on where the myelin damage occurs. Patients may have disturbances in vision, balance, speech or movement. MS is more common in women than in men and is diagnosed most often in people under the age of 50.

Impulse conduction

The speed of impulse conduction along an axon is determined by two characteristics: the presence of myelin and the diameter of the axon. Myelin is a lipid insulation or sheath formed by the oligodendrocytes in the CNS and the Schwann cells in the PNS. In preserved brains, the myelinated axons look white. Unmyelinated parts of the CNS, like cell bodies, look grey. Therefore, what we call 'white matter' is typically made of axons, and what we call 'grey matter' is made of cell bodies and dendrites. The cell membranes of oligodendrocytes or Schwann cells are wrapped around an axon like a bandage. Between adjacent glial cells are tiny bare spots called **nodes of Ranvier** (see Figure 9.5). When an axon is wrapped in myelin, we call the axon myelinated. Myelin is essential for the speedy flow of action potentials down axons. In an unmyelinated axon, the action potential can only flow down the axon by

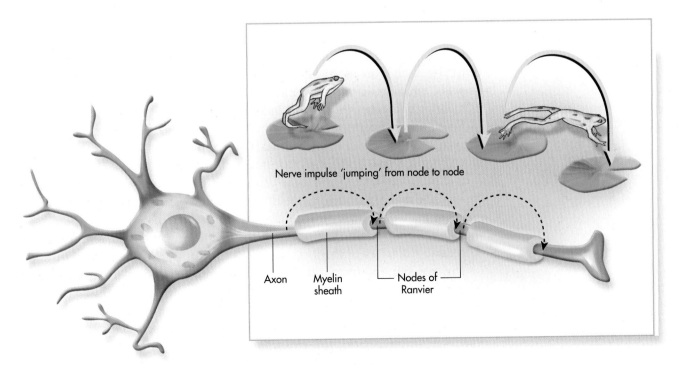

Nerve impulse 'jumping' from node to node

Axon   Myelin sheath   Nodes of Ranvier

Figure 9.5 Impulse conduction via myelinated axon.

depolarising each and every millimetre of the axon. Every single sodium chan-nel must open, and every single potassium channel must open. It is a relatively slow process. In a myelinated axon, only the channels at the nodes must open for the action potential to flow down the axon. Myelin prevents ions from passing through channels, because ions are water soluble and myelin is a lipid (charged particles cannot get through lipids). The action potential therefore 'jumps' down the axon from node to node (nodal transmission) rather than creeping along the entire length of the axon. It works the same way as if you try to walk across the floor heel to toe, never missing a spot and then jump across the floor in large leaps. Which is faster?

The diameter of the axon also affects the speed of action potential flow. Think about moving through two different pipes, say on a playground or obstacle course. One pipe is half a metre in diameter. To move through the pipe, you must crawl. The other pipe is 2 metres in diameter. Most of you can stand in this pipe with room to spare. Which trip would be faster? Obviously, it's the one through the larger diameter pipe. You don't have to crawl, you don't get caught on the side; you could even run through the big pipe. For ions, the axon is essentially a pipe. The wider the axon, the faster the ions flow. The combination of myelination and large diameter makes a huge differ-ence in speed. Small, unmyelinated axons have speeds as low as 0.5 metres per second, while large-diameter, myelinated axons may be as fast as 100 metres per second. That's 200 times faster.

How synapses work

In order for neurons to communicate, there must be some way for a message to be sent from one neuron to the other. This communication occurs at the synapse. An action potential is generated and flows toward the axon terminal.

Chemical synapses

When an impulse arrives at the axon terminal, the terminal depolarises and calcium gates open. Calcium flows into the cell. When calcium flows in, it triggers a change in the terminal. There are tiny sacs in the terminal called **vesicles**, which release their contents from the cell via exocytosis when cal-cium flows in. These vesicles are filled with molecules called **neurotransmitters**.

These neurotransmitters are used to send the signal from the neuron across the synapse to the next cell in line. The neurotransmitter binds to the cell receiving the signal and causes gates to open or close. Some neurotransmitters excite the receiving cell and some calm it down. In the case of the hammered thumb, the neurotransmitter would be released in your spinal cord and would excite a pain neuron. The receiving neuron would be stimulated and take the information about your thumb to your brain, where you would register the pain.

The last step in the transfer of information is clean-up. The neurotrans-mitter must be taken away from the synapse or it will continually bind to the receiving cell. This clean-up is accomplished through an inactivator. For example, if the neurotransmitter chemical is acetylcholine (ACh), it will be inactivated by **acetylcholinesterase**, an enzyme that breaks down ACh. This type of information flow, using neurotransmitters, is called a **chemical syn-apse**, because chemicals (neurotransmitters) are used to carry the information

vesicle (*VEE sickle*)

neurotransmitter (*new row TRANZ mitt er*)

ase = *to break down*

chemical synapse (*KEM ik all SIGH naps*)

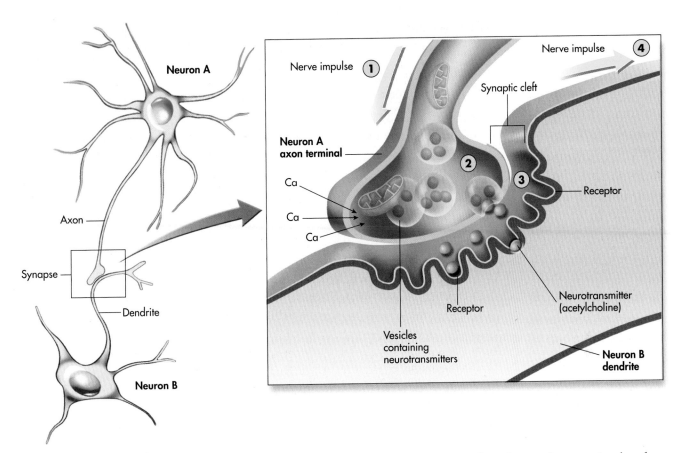

Figure 9.6 The chemical synapse. Step 1: the impulse travels down the axon. Step 2: vesicles are stimulated to release neurotransmitter (exocytosis). Step 3: the neurotransmitter travels across the synapse and binds with the receptor site of the post synaptic cell. Step 4: the impulse continues down the dendrite.

from one cell to another. See Figure 9.6 for the steps in chemical synaptic transmission.

Our understanding of chemical synapses has led to several breakthroughs for treating mental illness. Many medications on the market today are designed to modify synapses. Selective serotonin re-uptake inhibitors (SSRIs) are good examples. These medications prevent the clean-up of the neurotransmitter serotonin from synapses, thereby increasing the effects of serotonin on the receiving cell. Many antidepressants and anti-anxiety medications are SSRIs. When the neurotransmitter noradrenaline is released into the synaptic cleft, most of it is transported back into the presynaptic terminal, where it is repacked ready for reuse. Monoamine oxidase (MAO) is an enzyme that inactivates noradrenaline. Table 9.2 contains other examples of clinically important neurotransmitters.

Electrical synapses

Some cells do not need chemicals to transmit information from one cell to another. These synapses are called **electrical synapses**. The cells in an electrical synapse can transfer information freely because they have special connections called **gap junctions**. Such connections can exist between many types of excitable cells. They are found, for example, in the intercalated discs between cardiac muscle fibres and ensure coordinated contraction of the heart.

Table 9.2 Selected common neurotransmitters

Neurotransmitter	Location	Function	Comments
acetylcholine	CNS* and PNS*	generally excitatory but is inhibitory to some visceral effectors	Found in skeletal neuromuscular junctions and in many ANS* synapses. Alzheimer's disease is associated with a decrease in acetylcholine-secreting neurons
Catecholamines			
norepinephrine	CNS and PNS	may be excitatory or inhibitory depending on the receptors	Found in visceral and cardiac muscle. Cocaine and amphetamines increase the release and block the re-uptake of norepinephrine
epinephrine	CNS and PNS	may be excitatory or inhibitory depending on the receptors	Found in pathways concerned with behaviour and mood
serotonin	CNS	generally inhibitory	Found in pathways that regulate temperature, sensory perception, mood, onset of sleep. In schizophrenia there are elevated levels of serotonin
dopamine	CNS and ANS	excitatory or inhibitory	Destruction of dopamine-secreting neurons if found in Parkinson's disease
histamine	CNS	generally inhibitory	Histamine is thought to be involved in arousal from sleep
Amino acids			
gamma-aminobutyric acid (GABA)	CNS	generally inhibitory	Medications which increase GABA are used in epilepsy treatment
glycine	CNS	generally inhibitory	The poison strychnine inhibits glycine receptors, resulting in muscle contractions and convulsions
glutamate and aspartate	CNS	excitatory	Important in learning and memory
nitric oxide	CNS (and adrenal glands)	excitatory	Stimulating nitric oxide release is used in the phosphodiesterase type 5 inhibitors (sildenafil citrate – Viagra, tadalafil – Cialis, and vardenafil – Levitra)
Neuropeptides			
endorphins and encephalins	CNS and PNS	generally inhibitory	Inhibit release of sensory pain neurotransmitters
Substance P	CNS	generally excitatory	Substance P is a neurotransmitter associated with pain transmission

*CNS = central nervous system; PNS = peripheral nervous system; ANS = autonomic nervous system.

The neuromuscular junction

One chemical synapse, the **neuromuscular junction**, is particularly important to the function of your motor system, as we mentioned briefly in Chapter 7, 'The muscular system'. The neuromuscular junction is a specialised synapse between somatic (voluntary) motor neurons and the skeletal muscles they innervate. The surface of the muscle is studded with sodium channels that are ligand-gated. A ligand-gated channel is one which opens or closes when a molecule binds to the receptor that is part of the channel, like a key fitting into a lock. In the case of skeletal muscles, the ligand, or key, is the neurotransmitter acetylcholine. Acetylcholine is released from the terminal of a motor neuron and binds to the surface of skeletal muscle, opening sodium channels and causing the skeletal muscle to be depolarised. This depolarisation leads to the changes described in Chapter 7, and the muscle contracts. Like all chemical synapses, the neuromuscular junction must be cleaned up at the end of transmission. The enzyme responsible for cleaning up the synapse is called acetylcholinesterase.

Clinical application

Bioterrorism

If acetylcholinesterase activity is prevented, acetylcholine continually stimulates the muscle, eventually paralysing it. Some insecticides and the nerve gas Sarin, which killed/injured several hundred people during a terrorist attack in a Tokyo subway tunnel in 1995, are acetylcholinesterase inhibitors. They kill by causing paralysis of skeletal muscles, including the diaphragm, the most important respiratory muscle. Victims of Sarin or overexposure to organophosphate insecticides may die from respiratory arrest.

Test your knowledge 9.2

Choose the best answer:

1. Which ions move *into* neurons during the action potential?
   a. potassium
   b. calcium
   c. sodium
   d. acetylcholine

2. Which of the following is all-or-none?
   a. local potential
   b. action potential
   c. chemical synapse
   d. impulse conduction

3. The molecules used to send signals across synapses are called:
   a. hormones
   b. ions
   c. neurotransmitters
   d. messengers

4. When you hit your thumb with a hammer, what happens next in your nervous system?
   a. synaptic transmission
   b. a local potential
   c. an action potential
   d. resting potential

5. What is another name for myelinated axons?
   a. grey matter
   b. dura mater
   c. white matter
   d. duzitmater

Anaesthesia

As we can see from Table 9.2, some neurotransmitters are inhibitory. These effects are used clinically to reduce anxieties in patients who, for example, are going to undergo some form of intervention or surgical intervention. If we look at GABA, we find that it is an inhibitory neurotransmitter, and so enhancing it with a medication called a benzodiazepine results in reduction in anxiety; sedation; induction of sleep; and reduction in muscle tone. These effects can be desirable in someone who may be anxious about the forthcoming surgical procedure, and so a benzodiazepine is often given as part of a pre-medication (or 'pre-med'), before patients go to the anaesthetic room.

Spinal cord and spinal nerves

So far we've discussed the nervous system at the cellular and tissue level. In addition, we've described how impulses are transmitted in the nervous system. Now let's focus on the main highway that these impulses travel along.

Learning hint

Directional terms

In the spinal cord, directional terms – anterior, posterior, dorsal and ventral – are very important because function is linked to location. It is easy to get confused by the use of anterior and posterior for some structures and dorsal and ventral for others. Just keep in mind that, in humans, anterior is ventral and posterior is dorsal. The names are interchangeable for all the structures except the spinal roots. The spinal roots are *always* called dorsal and ventral.

External anatomy

The **spinal cord** is located in a hollow tube running inside the vertebral column from the *foramen magnum* to the second lumbar (L2) vertebra. You can think of your spinal cord as a very sophisticated neural information superhighway. It is divided into 31 segments, each with a pair of **spinal nerves**. The spinal cord segments and the spinal nerves are named for their corresponding vertebrae. The spinal cord ends at L2 in a pointed structure called the *conus medullaris*. Hanging from the conus medullaris is the ***cauda equina*** (literally 'horse's tail'). The cauda equina is the group of spinal nerves (L2 to the coccygeal (Co) nerves) dangling loosely in a bath of **cerebrospinal fluid** (CSF). The spinal cord has two widened areas, the cervical and lumbar enlargements, which contain the neurons for the limbs. See Figure 9.7.

Meninges

The CNS, both brain and spinal cord, are surrounded by a series of protective membranes called the **meninges**. The purpose of the meninges is to cover the delicate structures of the brain and spinal cord. In essence, they help to set up layers that act as cushioning and shock absorbers for the brain and spinal cord.

cauda = *tail*

equina = *horse*

cerebrospinal fluid (*ser ree bro SPY nall*)

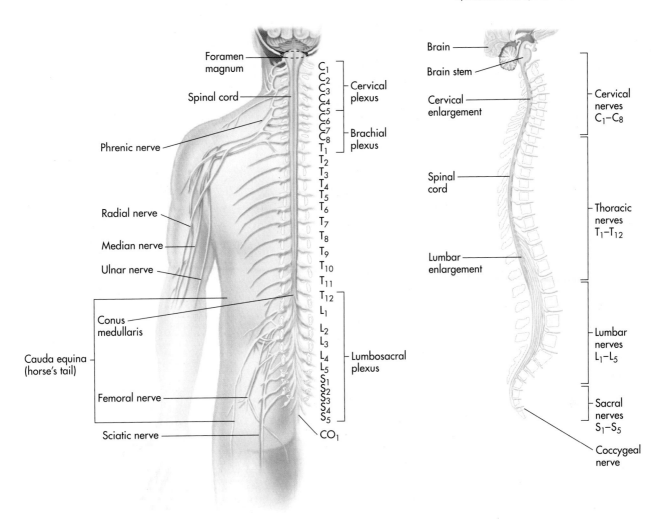

Figure 9.7 The spinal cord.

dura mater (*DURE ah MAY ter*)

arachnoid mater (*ah RAK noyd MAY ter*)

mater = *mother; the three layers formed by the meninges literally mean 'hard mother' (dura mater), 'soft mother' (pia mater), and 'spider mother' (arachnoid layer)*

pia mater (*PEE ah MAY ter*)

The meninges form three distinct layers. The outer layer of thick, fibrous tissue is called the **dura mater**. The middle layer, a wispy, delicate layer resembling spider webs, is the **arachnoid mater**. This layer, which is composed of collagen and elastic fibres, acts as a shock absorber. In addition, it can also transport dissolved gases and nutrients as well as chemical messengers and waste products. The third, innermost layer, which is fused to the neural tissue of the CNS, is the **pia mater**. This layer contains blood vessels that serve the brain and spinal cord.

Clinical application

Epidural anaesthesia

Often, during labour or in preparation for a Caesarean section, a woman receives an **epidural**. An epidural is an injection of local anaesthetic into the epidural space. The anaesthetic is usually delivered via a catheter (a small tube). Ideally, epidural anaesthesia allows a woman to continue to participate actively in the birth without severe labour pains. Epidural injections of steroids are sometimes prescribed for patients with chronic lower back injuries to relieve pain and inflammation.

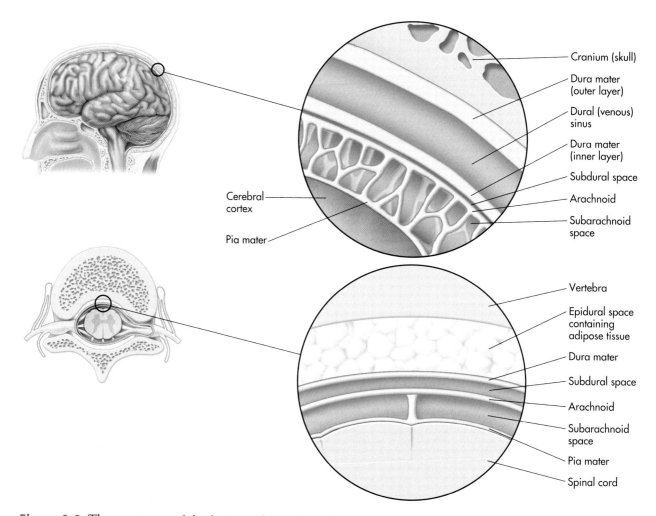

Cranium (skull)
Dura mater (outer layer)
Dural (venous) sinus
Dura mater (inner layer)
Subdural space
Arachnoid
Subarachnoid space

Cerebral cortex
Pia mater

Vertebra
Epidural space containing adipose tissue
Dura mater
Subdural space
Arachnoid
Subarachnoid space
Pia mater
Spinal cord

Figure 9.8 The meninges of the brain and spinal cord.

epidural space (*ep ee DURE all*)
  epi = *on top*
subdural space (*sub DURE all*)
  sub = *under*
subarachnoid space (*sub ah RAK noyd*)

A series of spaces is associated with the meninges. Between the dura and the vertebral column is a space filled with fat and blood vessels called the **epidural space**. Between the dura mater and the arachnoid mater is the **subdural space**, which is filled with a tiny bit of fluid. Between the arachnoid mater and the pia mater is the large **subarachnoid space**, filled with cerebrospinal fluid, acting as a fluid cushion for the CNS. These three membranes and their fluid-filled spaces, together with the bones of the skull and vertebral column, form a strong protection system against CNS injury (see Figure 9.8).

Internal anatomy of the spinal cord

fissure (*FISH yer*)
sulcus (*SULL cuss*)

The spinal cord is divided in half by a ventral median fissure and a dorsal median sulcus. A **fissure** is a deep groove on the CNS surface while a **sulcus** is a shallow groove on the CNS surface (see Figure 9.9). The interior of the spinal cord is then divided into a series of sections of white matter **columns** and grey matter **horns**. There are three types of horns. The dorsal horn is involved in sensory functions, the ventral horn in motor functions, and the lateral horn in autonomic functions. The horns are the regions where the neurons have their cell bodies.

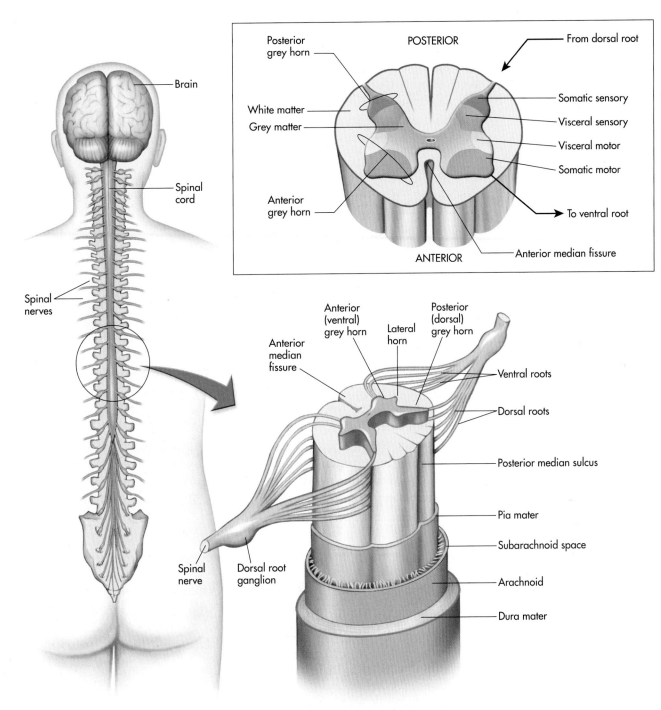

Figure 9.9 Internal anatomy of the spinal cord.

The columns also have a ventral, dorsal and lateral aspect. These columns act as nerve tracts, pathways, or axons, running up and down the spinal cord to and from the brain. Think of the columns as communication wires, part of a vast network, transporting information to the appropriate parts of the system. This is the neural information superhighway. (We discuss the functions of these regions later in the chapter.)

In addition, the spinal cord has several other features. The **commissures**, grey and white, connect left and right halves of the cord so the two sides of the CNS can communicate. (So the right hand does usually know what the

commissures (*COM mih shores*)

ganglion (*GANG lee yon*)

left hand is doing even if it doesn't always seem like it.) The central canal is a cavity in the centre of the spinal cord that is filled with cerebrospinal fluid. The **spinal roots**, projecting from both sides of the spinal cord in pairs, fuse to form spinal nerves. (The roots are the entry points and exit points of the neural superhighway.) The dorsal root, with the embedded **dorsal root ganglion**, a collection of sensory neurons, carries sensory information, while the **ventral root** carries motor information. Refer to Figure 9.9.

Spinal nerves

Nerves are the connection between the CNS (brain and spinal cord) and the world outside the CNS. Nerves are therefore part of the PNS. Isn't it amazing that the brain, which is totally encased in darkness, can receive and interpret the nerve messages from the PNS to allow us to see the wonderful world around us? All nerves consist of bundles of axons, blood vessels and connective tissue. Nerves run between the CNS and organs or tissues, carrying information into and out of the CNS. The nerves connected to the spinal cord are called *spinal nerves*. There are 31 pairs of spinal nerves, each named for the spinal cord segment to which they are attached. Since each spinal nerve is a fusion of dorsal and ventral roots, spinal nerves carry both sensory and motor information. A nerve that carries both types of information is called a **mixed nerve**. All spinal nerves are mixed. When spinal nerves leave the vertebral column, they can go through a number of different pathways to reach the peripheral tissues. Spinal nerves from the thoracic spinal column project directly to the thoracic body wall without branching. Spinal nerves from the cervical, lumbar and sacral regions of the spinal cord go through complex branching patterns, recombining with nerves from other spinal cord segments before projecting to peripheral structures. These complex branching patterns are called **plexuses**. See Figure 9.10.

Clinical application

A matter of centimetres

Did you know that the difference between being able to breathe on your own after a spinal cord injury and being dependent on a ventilator is literally a matter of centimetres? It's true. One of the nerves that projects from the cervical plexus is called the phrenic nerve. This nerve is the motor nerve for your diaphragm, your main breathing muscle. If the spinal cord is damaged below the cervical plexus, say in the high thoracic region, the phrenic nerve can receive signals from the brain and send them to the diaphragm and the injured person is able to breathe. However, if the damage to the spinal cord is between the brain and the cervical plexus, the path from the brain to the phrenic nerve is blocked and the signals can't get to the diaphragm from the brain. The diaphragm is paralysed, and the person cannot breathe on his or her own. The difference is a matter of a few centimetres.

Reflexes

Reflexes are the simplest form of motor output. Reflexes are generally protective, keeping you from harm. They are involuntary, and usually the response

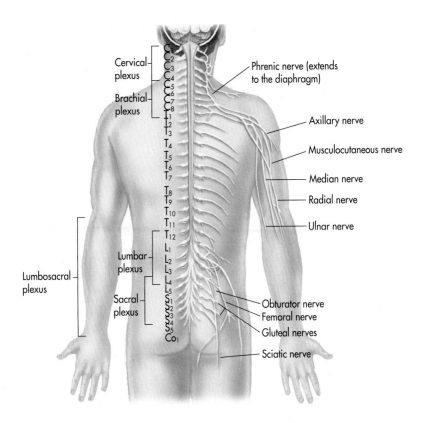

SPINAL NERVE PLEXUSES				
PLEXUS	**LOCATION**	**SPINAL NERVES INVOLVED**	**REGION SUPPLIED**	**MAJOR NERVES LEAVING PLEXUS**
Cervical	Deep in the neck, under the sternocleidomastoid muscle	C_1–C_4	Skin and muscles of neck and shoulder; diaphragm	Phrenic (diaphragm)
Brachial	Deep to the clavicle, between the neck and the axilla	C_5–C_8, T_1	Skin and muscles of upper extremity	Musculocutaneous Ulnar Median Radial Axillary
Lumbosacral	Lumbar region of the back	T_{12}, L_1–L_5, S_1–S_4	Skin and muscles of lower abdominal wall, lower extremity, buttocks, external genitalia	Obturator Femoral Sciatic Pudendal

Figure 9.10 Spinal cord plexuses.

is proportionate to the stimulus. Some familiar reflexes are the *withdrawal* reflex, activated for example, when you pound your thumb with a hammer or touch a hot stove; the *vestibular* reflex, which keeps you vertical; and the *startle* reflex, which causes you to jump at loud sounds. The amazing thing about reflexes is that they can often occur without your brain being involved. For many reflexes, only the spinal cord is necessary.

The most common example of a reflex is the patellar tendon (knee-jerk) reflex. You may have experienced this reflex at the doctor's surgery. The doctor taps your knee with a hammer and your leg kicks, seemingly against your will. What exactly is going on? When the doctor taps your knee, the hammer gently tugs on a tendon that is connected to your quadriceps muscles. The quads are gently tugged. This causes them to lengthen slightly. This length change is received by a sensory neuron and transmitted back to the spinal cord. The

Test your knowledge 9.3

Choose the best answer:

1. The segments of the spinal cord are named for:
   a. the bones of the skull
   b. the vertebrae
   c. their function
   d. none of the above

2. This layer of the meninges is fused to the surface of the CNS:
   a. dura mater
   b. pia mater
   c. arachnoid mater
   d. whatsa mater

3. The ventral root carries _____ information, and the dorsal root carries _____ information.
   a. motor, motor
   b. motor, sensory
   c. sensory, sensory
   d. sensory, motor

4. What is the term for axon pathways carrying information up and down the spinal cord?
   a. horns
   b. roots
   c. columns
   d. ganglia

sensory neuron synapses with a motor neuron in the ventral horn. The motor neuron sends a signal to the quadriceps to stop the stretch. How do the quads stop the stretch? They shorten (contract). When a muscle shortens, it causes a movement. The shortening of the quadriceps muscles extends your knee, causing your leg to kick.

This action is totally reflexive (no pun intended). You can't stop yourself from kicking, even if you try. You do not have to think about kicking. The action happens without your brain. Only the spinal cord is necessary.

Common disorders of the nervous system

Peripheral neuropathy encompasses a number of disorders involving damage to peripheral nerves. Because peripheral nerves are involved in sensory, motor and autonomic function, the symptoms of peripheral neuropathy vary greatly among patients. Symptoms include muscle weakness and decreased reflexes, numbness, tingling, paralysis, pain, difficulty controlling blood pressure, abnormal sweating and digestive abnormalities. Non-genetic neuropathy can be grouped into three broad categories: systemic disease, trauma, and infection or autoimmune disorders. Trauma, such as falls or automobile accidents, causes mechanical injury to nerves. Nerves may be severed, crushed or bruised. Systemic disorders that cause peripheral neuropathy include kidney disorders, hormonal imbalance, alcoholism, vascular damage, repetitive stress (like carpal tunnel), chronic inflammation, diabetes, toxins and tumours. Infections such as shingles, Epstein–Barr virus, herpes, HIV, Lyme disease and polio cause peripheral neuropathy. Guillain-Barré syndrome is an acute form of peripheral neuropathy. Some forms of neuropathy are inherited.

Even though the spinal cord is well protected by the bones and meninges, trauma can cause damage to the delicate neural tissue. The spinal cord may be partially or completely severed, crushed or bruised. Bruises to the spinal cord may resolve with time and rehabilitation, but a severed or crushed spinal cord is a permanent injury. Spinal cord injury usually results in paralysis and sensory loss below the injury, and the extent of the paralysis is related to the location of the spinal cord injury. Patients with injuries to the cervical spinal cord are quadriplegics, paralysed in all four limbs. Some quadriplegics, with damage very high in the cervical spinal cord, have paralysed diaphragms and cannot breathe on their own. Patients with injuries in the thoracic spinal cord and lower have paraplegia. They can move their arms but their legs are paralysed.

Guillain-Barré syndrome (GBS) is a paralysis caused by inflammation of peripheral nerves. Patients develop, over variable periods of time, weakness and ascending paralysis of the limbs, face and diaphragm. Some patients may have a mild form of Guillain-Barré syndrome, but those with severe disease must be kept on a ventilator until the paralysis resolves. The cause of the disease is not known, though many patients develop Guillain-Barré syndrome after a viral infection. Other evidence suggests that autoimmune attack of peripheral myelin may be to blame. There is no effective treatment except supportive care. Fortunately, the disorder is usually temporary. Many patients need rehabilitation after their PNS recovers.

Myasthenia gravis is an autoimmune disorder in which the immune system attacks and destroys acetylcholine receptors at the neuromuscular junction. Motor neurons continue to release acetylcholine, but the receptor number is reduced, so motor neurons cannot communicate with skeletal muscles. Eye muscles are typically the first muscles affected, but some patients initially experience difficulty chewing, swallowing or talking. The disorder, like most autoimmune disorders, is progressive, though the course of the disease varies widely among patients. Treatment for myasthenia gravis includes acetylcholinesterase inhibitors, corticosteroids, immunosuppressant drugs and plasma exchange. In a few patients, the disease disappears spontaneously.

Botulism is a form of paralysis caused by toxins produced by the bacterium *Clostridium botulinum*. Botulism can be caused by ingesting the toxin in food and can result from wound infections. The bacteria grow most commonly in improperly prepared canned food, especially home-canned food. The toxin keeps neurotransmitters from being released at the neuromuscular junction, causing paralysis. Initial symptoms include vision disturbances, slurred speech, dry mouth and muscle weakness. If left untreated, paralysis will spread to limbs and respiratory muscles. Botulism can be treated by administration of antitoxin and supportive care. Botulism is a rare disorder.

Injuries to the brain can be primary (direct) injuries or secondary, caused by infection or cerebral oedema, for example. A head injury, where there is sudden acceleration or deceleration, can result in the brain 'hitting' the inside of the skull. Due to the cerebrospinal fluid in which the brain is cushioned, the brain then 'bounces' backwards, causing further damage to the 'back' of the brain. The injuries are referred to as coup and contrecoup contusions, where the coup is the site of the impact, and the contrecoup is the injury caused by the rebound. During this type of head injury, blood vessels in the brain can become damaged, and bleed into the space surrounding the skull, and nerve tracts and other structures in the brain can become torn or bruised. Consciousness may be lost, and it is useful to know the estimated time of loss

Learning hint

The big picture

Often on a journey, you cannot see or appreciate the big picture. For example, if you are in a portion of a city, it can be quite impressive, but when you fly over the same city, the 'whole picture' becomes clear and spectacular. This chapter on the nervous system has given you a picture of the nervous tissue and how nerve transmission occurs. In addition, this chapter has begun to develop the role of the spinal cord. The next chapter focuses on the brain and then pulls everything together so you can see the big picture.

of consciousness. In a mild head injury, there may be only a momentary loss of consciousness; a moderate head injury is characterised by a longer period of unconsciousness, whereas a severe head injury may mean that consciousness is lost for a substantial period of time. Clearly the duration of the loss of consciousness equates to a poorer prognosis.

Meningitis is an infection, usually from viruses or bacteria, of the meninges, the lining of the brain and spinal cord. Bacterial meningitis is a potentially fatal infection. The bacteria first infect the upper respiratory tract and then travel to the meninges. High-risk groups include the elderly, people with suppressed immune systems, very young children, and university students who live in halls of residence. Patients who survive bacterial meningitis often have severe neurological impairment, including deafness and severe brain damage.

The relationship between the muscular system and nervous system can be seen in the condition of **carpal tunnel syndrome** (CTS), which occurs with an inflammation and swelling of the tendon sheaths surrounding the flexor tendon of the palm. This is a result of repetitive motion such as typing on a keyboard. As a result of this inflammation and swelling, the median nerve is compressed, producing a tingling sensation or numbness of the palm and first three fingers.

Summary

- The nervous system is the body's computer. It has a sensory (input) system, an integration centre, the CNS, and a motor (output) system. The input and output nerves are in the PNS, and the brain and spinal cord are the CNS.

- The tissue of the nervous system is made up of two types of cells: neurons, which send, receive and process information; and neuroglia, which support the neurons.

- Neurons are excitable cells. They do their jobs by carrying tiny electrical currents caused by changes in cell permeability to certain ions.

- These tiny electrical currents can be all-or-none (action potentials), can change depending on the size of the stimulus (local potentials), can travel down axons (impulse conduction), or can be used to transmit information from one cell to another (synapses).

- The cavities in your brain are known as ventricles, and the spinal cord cavity is the central canal. These cavities are part of an elaborate protection system for your CNS and are filled with cerebrospinal fluid.

- Your CNS is surrounded by a three-layered membrane system: dura mater, arachnoid mater and pia mater, collectively known as the meninges. Cerebrospinal fluid is also contained in the space between the arachnoid and pia mater.

- The spinal cord has 31 segments, each with a pair of spinal nerves. The spinal nerves are a part of the peripheral nervous system.
- The spinal nerves are made of a pair of spinal roots. The ventral root is integral to motor function, and the dorsal root is integral to sensory function. Spinal nerves are mixed: they carry both sensory and motor information.
- A series of tracts run up and down the spinal cord to and from the brain. The tracts going toward the brain carry sensory information to the brain. The tracts coming from the brain toward the spinal cord carry motor information from the brain.

Case study

David was crossing the road outside his home when he was struck by a fast-moving car. His head hit the windscreen, making a 'target' indentation on the windscreen. He lost consciousness for several minutes and appears to have paralysis of the left side of his body.

1. What type of brain injury do you think he has?

   _____

   _____

2. How severe is it?

   _____

   _____

3. What type of treatments should be given?

   _____

   _____

Review questions

Multiple choice

1. The input side of your nervous system is known as:
   a. motor
   b. sensory
   c. association
   d. all of the above

2. Neurons with a central and peripheral projection are known as:
   a. unipolar
   b. bipolar
   c. multipolar
   d. northpolar

3. During depolarisation, _____ ions move _____ a neuron.
   a. K^+, out of
   b. K^+, into
   c. Na^+, out of
   d. Na^+, into

4. The ventral root of the spinal cord is:
   a. sensory
   b. motor
   c. association
   d. none of the above

5. Spinal nerves carry what kind of information?
   a. sensory
   b. motor
   c. mixed
   d. vertebral

Fill in the blanks

1. The speed of impulse conduction is determined by _____ and _____.

2. _____ potentials are all-or-none.

3. The spinal cord has white matter _____ and grey matter _____.

4. _____ fluid is contained in the _____ space between the arachnoid mater and pia mater.

5. A _____ is an involuntary, protective movement that is sometimes generated without the brain.

Short answers

1. Explain the changes in a neuron during an action potential.

2. List the steps in chemical synaptic transmission.

3. List the three main spinal nerve plexuses.

4. List the types of neuroglia and their functions.

Suggested activities

1. With a partner, test your knee-jerk reflexes using reflex hammers. Try to control the reflex. Can you? Try it with your eyes closed. What does that say about reflexes?

2. Grab a friend and explain the action potential and chemical synapse to her or him. Can you do it?

Answers

Answers to case study

1. Concussion
2. Concussion is a mild form of traumatic brain injury whose severity is dependent on how long consciouness was lost and whether there is any amnesia.
3. Monitoring and rest.

Answers to multiple choice questions

1. b
2. a
3. d
4. b
5. c

Answers to fill in the blanks

1. the presence of myelin; the diameter of the axon
2. Action
3. columns; horns
4. Cerebrospinal; subarachnoid
5. reflex

Answers to short answer questions

1. When a cell gets stimulated (excited), gates in the cell membrane spring open. Sodium gates open allowing sodium ions (Na^+) to travel across the cell membrane. These sodium ions are positively charged, so when they go into the cell, the cell becomes more positive. A cell that is more positive than resting is called **depolarised**. In less than a millisecond, the gates on the sodium channels shut. Then other gates open for potassium channels. Potassium (K^+), which is also positive, leaves the cell, taking its positive charges with it. The inside of the cell becomes more negative, eventually returning to rest, called **repolarisation**.

2. Calcium flows into the cell, triggering a change in the axon terminal. These terminals are called vesicles and these contain neurotransmitters. The neurotransmitters are used to pass the signal from one cell to the next. The neurotransmitter is removed from the synapse otherwise it would continually bind to the receiving cell. The chemicals that remove neurotransmitters are inactivators, for example, acetylcholine is inactivated by acetylcholinesterase.

3. Cervical (located in the neck under the sternocleidomastoid muscle); brachial (deep to the clavicle, between the neck and the axilla); and lumbrosacral (lumber region of the back).

4. Neuroglia: Astrocytes – metabolic and structural support cells that regulate extracellular composition of brain fluid, releasing substances that promote formation of tight junctions. Microglia – specialised macrophages that remove debris by phagocytosis. Ependymal cells – line the ventricles of the brain and central canal of the spinal cord; some ependymal cells form the choroid plexuses. Oligodendrocytes – make myelin. In the peripheral nervous system there are two types of neuroglia – Schwann cells (which make myelin) and satellite cells which are support cells, providing nutrients to neuron cell bodies.

The nervous system

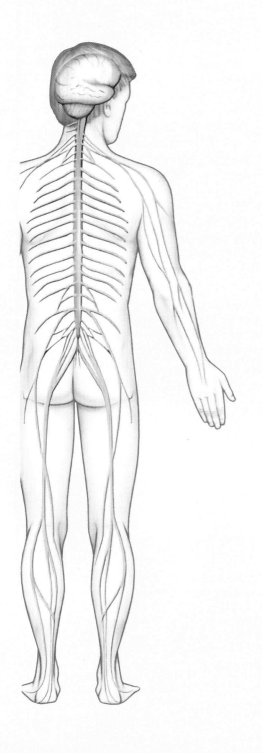

Part II

In this chapter, we focus on the main controller of the nervous system: the brain. Then we pull everything together to see how all the pieces of the puzzle form the big picture. In a city, there are all kinds of roads that lead to all sorts of places. However, the flow of traffic must be precisely controlled, or chaos and accidents would occur. The control system of the roads includes such things as traffic signs and traffic lights, roundabouts and toll booths/bridges. The brain is the control system of the nervous system 'superhighway', attempting to keep everything running smoothly.

Learning objectives

At the end of this chapter, you will be able to:

◆ Organise the hierarchy of the nervous system
◆ Locate and define the external structures and their corresponding functions of the brain
◆ Locate and define the internal structures and their corresponding functions of the brain
◆ Describe the sensory functions of the brain with related structures
◆ Describe the motor functions of the brain with related structures
◆ Contrast the parasympathetic and sympathetic branches of the autonomic nervous system
◆ Discuss some diseases of the nervous system

Pronunciation guide

anterior commissure (*an TEE ree ur KOM i sure*)

basal nuclei (*BAY sal new klee ie*)

cerebellum (*ser eh BELL um*)

cerebrum (*ser EE brum*)

corpus callosum (*KOR pus ka LOH sum*)

corticobulbar tract (*KOR ti coe BUL bar*)

corticospinal tract (*KOR ti coe SPY nal*)

diencephalon (*dye en KEFF ah lon*)

fornix (*FOR niks*)

gyri (*JIE ree*)

hypothalamus (*high poh THAL a muss*)

limbic system (*LIM bik*)

medulla oblongata (*me DULL lah ob long GAR ta*)

occipital lobe (*ok SIP it all*)

parietal lobe (*par EE tall*)

pineal body (*pin EE all*)

spinocerebellar tract (*SPY no ser eh BELL ar*)

spinothalamic tract (*SPY no thal A mic*)

subarachnoid space (*sub ah RAK noyd*)

sulcus (*SULL cuss*)

thalamus (*THAL a muss*)

The brain and cranial nerves

The brain and cranial nerves represent the major control system of the nervous system. The brain acts as the main processor and director of the entire system. The cranial nerves leave the brain and go to specific body areas where they receive information and send it back to the brain (sensory), and the brain sends back instructions as to the appropriate response (motor). Let's again begin with an overview and then get more specific.

Overall organisation

Note

The scarecrow on The Wizard of Oz 'thought' he needed a brain. As you work through this chapter, can you 'think' of several reasons why you knew all along he had a brain?

At the top of the spinal cord, beginning at the level of the foramen magnum and filling the cranial cavity, is the brain. Just like a supermarket needs to be organised into sections such as vegetables, meats and dairy, etc., the brain can be divided into several anatomical and functional sections. We will talk about each section separately and then describe the interactions between each of the parts of the brain and the spinal cord.

The brain's external anatomy

Let's start by looking at the outside of the brain, and then we will zoom in on the internal structures. From the outside, you can see that the brain consists of a **cerebrum**, **cerebellum** and **brainstem** (see Figure 10.1).

cerebrum (*ser EE brum*)

cerebellum (*ser eh BELL um*)

Cerebrum

The cerebrum, which is the largest part of the brain, is divided into two **hemispheres**, right and left, by the longitudinal fissure, and divided from the cerebellum ('little brain') by the transverse fissure. The surface of the cerebrum is not smooth, but is broken by ridges (**gyri**) and grooves (**sulci**),

gyri (*JIE ree*)

sulci (*SULL kee*)

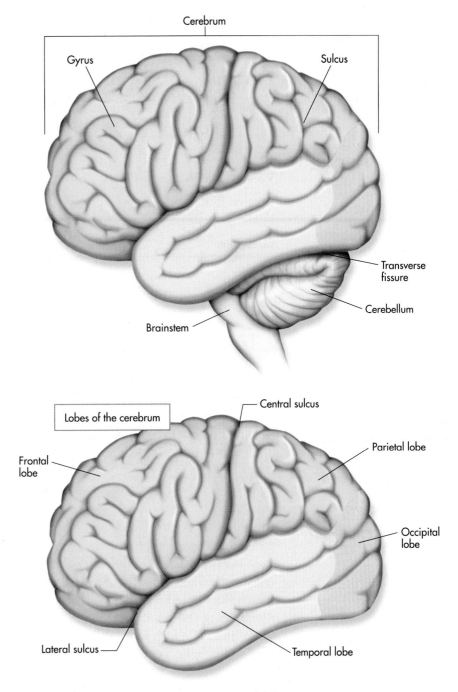

Figure 10.1 External brain anatomy and lobes.

collectively known as *convolutions*. These convolutions serve a very important purpose by increasing the surface area of the brain, yet allowing it to be 'folded' into a smaller space. Most of the sulci are extremely variable in their locations among humans, but a few are in basically the same place in every brain. These less variable sulci are used to divide the brain into four large sections called **lobes**, much like the major departments in a supermarket are separated by aisles.

Table 10.1 Cerebral lobes and cerebellum

Structure	Major functions
Cerebral lobes	
frontal lobe	motor function, behaviour and emotions, memory storage, thinking, smell
parietal lobe	body sense, perception, taste and speech
occipital	vision
temporal lobe	hearing, taste, language comprehension, integration of emotions
cerebellum	the 'little brain'; sensory and motor coordination and balance

Amazing body facts

Why not a smooth brain?

The surface of the brain is folded and rippled into a series of convolutions. These convolutions allow lots of brain surface area to fit into a very small space. If you could lay the convolutions flat, they would be the size of a pillowcase.

parietal lobe (*par EE tall*)

occipital lobe (*ok SIP it all*)

contra = *against, opposite*

lateral = *side*

Note

Does contralateral control mean that if the left brain controls the right hand, and the right brain controls the left hand, then only left-handed people are in their right minds?

medulla oblongata (*me DULL lah ob long GAR tah*)

The lobes (see Table 10.1) are named for the bones in the skull that cover them, and they occur in pairs, one in each hemisphere. The most anterior lobes, separated from the rest of the brain by the central sulci, are the **frontal lobes**. The frontal lobes are responsible for motor activities, conscious thought, and speech. Posterior to the frontal lobes are the parietal lobes. The parietal lobes are involved with body sense perception, primary taste and speech. Posterior to the parietal lobes are the occipital lobes, which are responsible for vision. The parietal and occipital lobes are separated by the parieto-occipital sulcus. The most inferior lobes, separated by the lateral fissures, are the **temporal lobes**, which are involved in hearing and integration of emotions. Again, see Figure 10.1, which shows the lobes of the brain and the sulci that separate them. There is a section of the brain, the **insula**, deep inside the temporal lobes, which is often listed as a fifth lobe, but it is not visible on the surface of the cerebrum. The insula helps coordinate autonomic functions. Much of the information coming into your brain is **contralateral**. That is, the left side of the body is controlled by the right side of your cerebrum, and the right side of the body is controlled by your left brain.

Cerebellum

The cerebellum lies dorsally from under the occipital lobe of the cerebrum and plays an important role in sensory and motor coordination and balance. Its surface is also convoluted like that of the cerebrum. From its external appearance, it is easy to see why anatomists consider the cerebellum the 'little brain'.

The brainstem

The brainstem (see Table 10.2) is a stalk-like structure inferior to and partially covered by the cerebrum. It is divided into three sections. The **medulla**

Table 10.2 The brainstem

Structure	Function
midbrain	relays sensory and motor information
pons	relays sensory and motor information; role in breathing
medulla oblongata	regulates vital functions of heart rate, blood pressure, breathing, and reflex centre for coughing, sneezing, swallowing and vomiting

oblongata is continuous with the spinal cord. The medulla is responsible for impulses that control heartbeat, breathing and the cardiovascular system, muscle tone and therefore blood pressure. The **pons** is just superior to the medulla oblongata and connects the medulla oblongata and the cerebellum with the upper portions of the brain. The pons plays a role in breathing. The **midbrain**, which is the most superior portion of the brainstem, acts as a two-way conduction pathway to relay visual and auditory impulses (see Figure 10.2). The brainstem receives sensory information and contains control systems for vital functions such as blood pressure, heart rate and breathing. The brain controls the vital functions of life, and patients with severe brain injuries with an intact brainstem can continue in a vegetative state as long as they are nutritionally supported. This condition is called a persistent vegetative state (PVS).

The brain, like the spinal cord, is covered with protective membranes called *meninges*. Again, refer to Figure 10.2. The meninges of the brain are continuous with the spinal cord meninges. A potentially fatal condition is an infection of the meninges called **meningitis**, which can rapidly spread and affect the brain and spinal cord through this common covering.

meningitis (*men in JYE tiss*)
itis = *inflammation of*

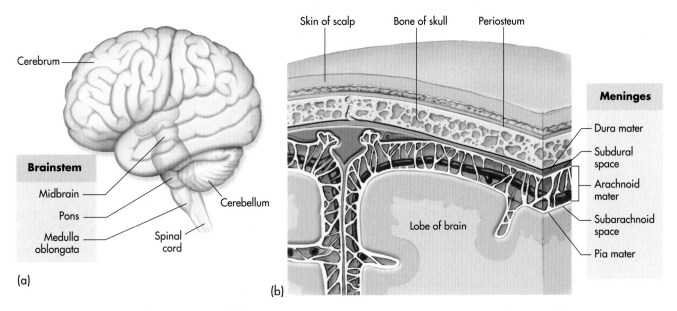

Figure 10.2 (a) The brain stem and (b) meninges.

Internal anatomy of the brain

The inside of the brain has, like the spinal cord, white matter and grey matter, and hollow cavities containing cerebrospinal fluid. Unlike the spinal cord, however, the white matter is surrounded by the grey matter in the brain. (Remember, in the spinal cord, the white matter surrounds the grey matter.) The layer of grey matter surrounding the white matter is called the **cortex**. In the cerebrum it is called the cerebral cortex, and in the cerebellum it is called the cerebellar cortex. Throughout the brain there are deep 'islands' of grey matter surrounded by white matter. These islands are called **nuclei**.

The fluid-filled cavities in the brain are called ventricles, and they are continuous with the central canal of the spinal cord and the subarachnoid space of both the brain and the spinal cord. These ventricles allow for the circulation of cerebrospinal fluid throughout the brain. The lateral ventricles (ventricles 1 and 2) are in the cerebrum; the third ventricle is in the diencephalon, a region between the cerebrum and brainstem; and the fourth ventricle is in the inferior part of the brain between the medulla oblongata and the cerebellum (see Figure 10.3).

The cerebrum

The inside of the cerebrum reflects its external anatomy (see Figure 10.4). The lobes – frontal, temporal, parietal and occipital – are clearly visible. On either side of the central sulcus are two gyri named for their locations: the **precentral gyrus**, anterior to the central sulcus, and the **postcentral gyrus**, posterior to

ventricles (*VEN trik alls*)
subarachnoid space (*sub ah RAK noyd*)

gyrus (*JIE russ*)

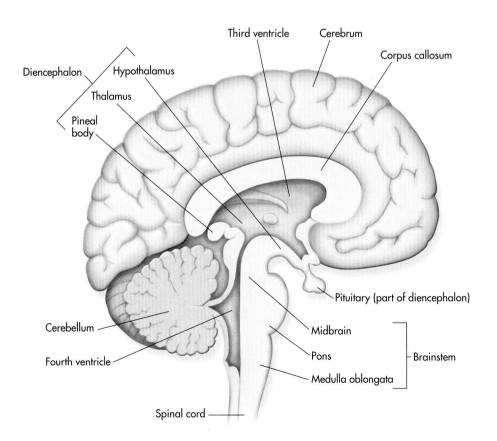

Figure 10.3 Internal anatomy of the brain.

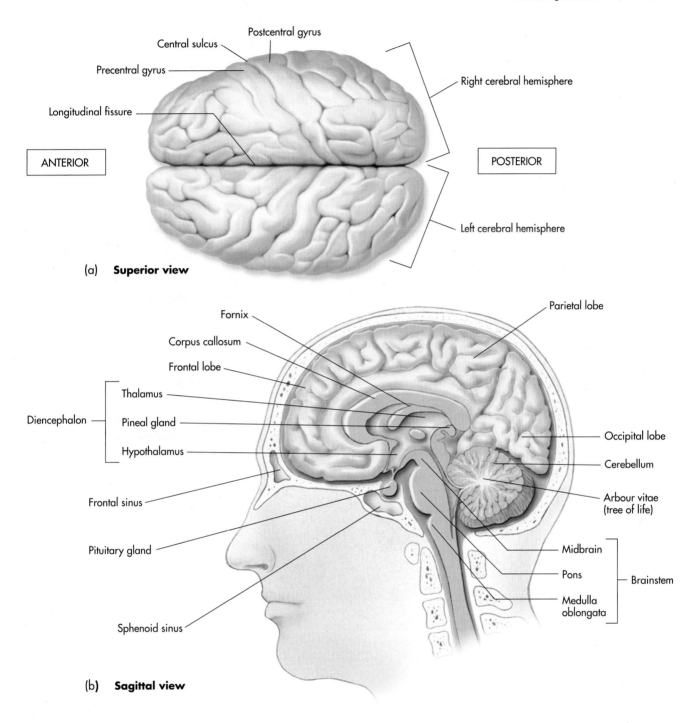

(a) **Superior view**

(b) **Sagittal view**

Figure 10.4 Superior (a) and sagittal (b) views of the brain.

corpus callosum (KOS puss ka LOH sum)

the central sulcus. The importance of these gyri will become obvious when we discuss function of the brain later in the chapter. The right and left hemispheres are connected by a collection of white matter surrounding the lateral ventricles, called the **corpus callosum**. This connection allows for cross-communication between the right and left sides of the brain. Many of our day-to-day activities, like walking or driving, require both sides of the body, and therefore both sides of the brain, to be well coordinated. Imagine walking if your legs were acting independently!

Clinical application

Cerebrospinal fluid circulation and hydrocephalus

The ventricles of the brain, the central canal of the spinal cord, and the subarachnoid space, surrounding both parts of the CNS, are all filled with cerebrospinal fluid (CSF). The CSF is filtered from blood in the ventricles by tissue called the choroid plexus. Produced at a rate of 500 millilitres daily, CSF made in the lateral ventricles flows through a tiny opening into the third ventricle and then through another opening into the fourth ventricle. From the fourth ventricle, CSF flows into the central canal of the spinal cord and the subarachnoid space. CSF is returned to the blood via special 'ports' (the arachnoid villi) between the subarachnoid space and blood spaces in the dura mater (the dural sinuses).

The balance of CSF made and CSF reabsorbed by the blood is very important. The brain is a very delicate organ captured between the liquid CSF and the bones of the skull. If there is too much CSF, pressure inside the skull rises, eventually crushing brain tissue. This condition, in which there is too much CSF, is called hydrocephalus (literally 'water in the head'). Hydrocephalus can be caused by blockage of the narrow passages between the ventricles due to trauma, by birth defects or tumours, or by decreased reabsorption of CSF due to subarachnoid bleeding. Hydrocephalus may be treated by medication, but the most common treatment is insertion of a shunt, a tube that drains the extra CSF into a patient's neck vein.

The diencephalon

diencephalon (*dye en KEFF ah lon*)

thalamus (*THAL ah muss*)

hypothalamus (*high poh THAL ah muss*)

pineal body (*pin EE yall*)

Inferior to the cerebrum is a section of the brain that is not visible from the exterior, called the diencephalon. The diencephalon consists of several parts: the thalamus, hypothalamus, pineal body and the **pituitary gland** (see Table 10.3). The hypothalamus, pituitary and pineal body represent an interface with the endocrine system (the other control system), covered in Chapter 12. The diencephalon contains the third ventricle and a number of nuclei that are part of the basal nuclei and limbic system. Specific nuclei of the hypothalamus are responsible for controlling hormone levels, hunger and thirst, body temperature, and the sleep–wake cycle, and for coordinating the flow of information around the brain.

Learning hint

Grey vs. white matter

The grey matter is composed of the cell bodies of the neurons. White matter is composed of axons, which are surrounded with myelin. Remember from the previous chapter that myelin allows messages to be transmitted faster.

Table 10.3 Diencephalon

Structure	Function
thalamus	relays and processes information going to the cerebrum
hypothalamus	regulates hormone levels, temperature, water-balance, thirst, appetite and some emotions (pleasure and fear); regulates the pituitary gland and controls the endocrine system
pineal body	responsible for secretion of melatonin (body clock)
pituitary gland	secretes hormones for various functions (explained in Chapter 12)

The cerebellum

The external similarities between the cerebellum and cerebrum are also obvious internally (see Figure 10.4). The cerebellum has a grey matter cortex and a white matter centre, known as the arbour vitae (tree of life). The cerebellum also has nuclei that coordinate motor and sensory activity. Essentially, the cerebellum fine-tunes voluntary skeletal muscle activity and helps in the maintenance of balance.

Cranial nerves

In order for the CNS to function, it must be connected to the outside world via nerves of the PNS. We have already seen that the spinal cord is connected to the outside via spinal nerves. The brain also has nerves to connect it to the outside, aptly named **cranial nerves** (see Figure 10.5). Cranial nerves are like spinal nerves in that they are the input and output pathways (PNS) for the brain, just as the spinal nerves are the pathways for the spinal cord. However, that is really where the similarities end. You should remember that there are 31 pairs of spinal nerves and that all of them are mixed nerves: they carry both sensory and motor information because they are formed by a combination of dorsal and ventral roots. There are far fewer cranial nerves – only 12 pairs. All but the first two pairs arise from the brainstem. Cranial nerves are not *all* mixed as are spinal nerves. Some cranial nerves are mainly sensory nerves, providing input; some are mainly motor nerves, directing activity; and some are mixed. Cranial nerves are much more specialised than spinal nerves and are named based on their speciality. Some cranial nerves carry sensory and motor information for the head, face and neck, while others carry visual, auditory, smell or taste sensation. (Table 10.4 lists cranial nerves and their functions.)

Learning hint

Mnemonic devices

A mnemonic device is a tool used to help you memorise long lists. It can be very useful in anatomy. To make a mnemonic device, take the first letter of each part of the list you are trying to memorise and make it into a sentence. An example for the cranial nerves is: **O**n **O**ld **O**lympus' **T**owering **T**op, **A** **F**riendly **V**iking **G**rew **V**ines **A**nd **H**ops.

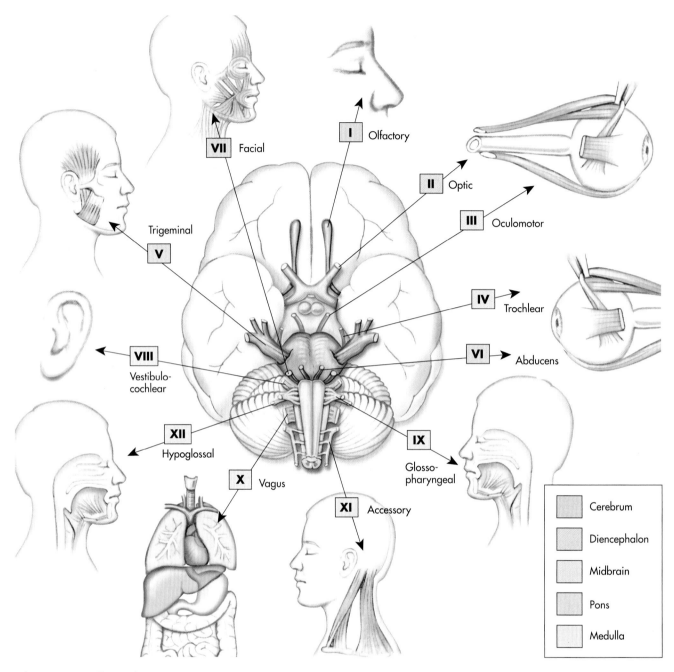

Figure 10.5 Cranial nerves.

Table 10.4 Cranial nerves and functions

Nerve	Function
olfactory (I)	sensory (smell)
optic (II)	sensory (vision)
oculomotor (III)	mixed, chiefly motor for eye movements
trochlear (IV)	mixed, chiefly motor for eye movements
trigeminal (V)	mixed, chiefly motor for face
abducens (VI)	motor fibres for lateral eye movement
facial (VII)	mixed, for lower face, throat and mouth; sensory impulses for taste from the tongue
vestibulocochlear (VIII)	sensory, hearing and balance
glossopharyngeal (IX)	mixed, motor for throat muscles; sensory for taste
vagus (X)	mixed, motor for autonomic heart, lungs, viscera; sensory for viscera, taste buds, and so on
accessory (XI)	mixed, chiefly motor; motor and sensory for larynx, soft palate, trapezius, and sternocleidomastoid muscles
hypoglossal (XII)	mixed, chiefly motor for tongue muscles; sensory, same as motor

Test your knowledge 10.1

Choose the best answer:

1. Which of the following is *not* part of the brainstem?
   a. pons
   b. medulla oblongata
   c. midbrain
   d. diencephalon

2. Deep islands of grey matter are known as:
   a. hemispheres
   b. gyri
   c. nuclei
   d. nerves

3. Which of the following is *not* found in the diencephalon?
   a. thalamus
   b. pineal body
   c. postcentral gyrus
   d. hypothalamus

4. This lobe contains the primary visual cortex.
   a. frontal
   b. temporal
   c. occipital
   d. parietal

5. The cerebrum is divided into right and left hemispheres by the:
   a. central sulcus
   b. transverse fissure
   c. corpus callosum
   d. none of the above

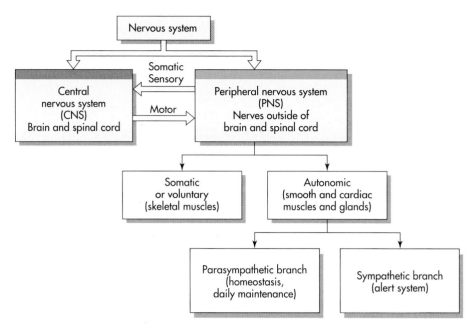

Figure 10.6 Nervous system flow chart highlighting the areas thus far explained.

The big picture: integration of brain, spinal cord and PNS

So how does the brain work with the spinal cord and peripheral nervous system? Let's begin to put the pieces together by revisiting the overall organisational chart of the nervous system. To serve as our guide or map to keep our location clear, see Figure 10.6.

The somatic sensory system

somatic (soh MAT ik)

The **somatic sensory system** provides sensory input for your nervous system. We have already mentioned the parts of your brain that are dedicated to your special senses: vision (occipital lobes), hearing (temporal and parietal lobes), taste (parietal lobes) and smell (temporal lobes). We have not yet talked about the sense of touch, called **somatic** **sensation**. Somatic sensation allows you to feel the world around you. Somatic sensation includes not just fine touch, which allows you to tell the difference between a peach and a nectarine or a golf ball and a ping pong ball, but also crude touch, vibration, pain, temperature and body position. While your special senses are all carried on cranial nerves, information for somatic sensation comes into *both* the brain and the spinal cord. Ultimately, for you to attach meaning to the sensation, the sensory information must get to your brain.

Let's start with the spinal cord. When we talked about reflexes, we saw somatic sensory information come into the spinal cord via the dorsal root and synapse with a motor neuron in the ventral horn. Let's say you place your hand on a hot iron. The sensory neuron carries pain information to your spinal cord. The motor neuron then activates the muscle, allowing you to respond to the stimulus *immediately* and pull your hand away to minimise injury. The spinal cord further carries the sensory information to your brain via tracts in the white matter of the spinal cord, so you feel the pain associated with your action and wonder why you ever put your hand on the hot iron in the first place.

Because there are so many variations of 'touch' information that must come in from all parts of the body, different neural highways are needed to more effectively take that information to the brain. Three pathways on either side of the spinal cord, the **dorsal column tract**, the spinothalamic tract and the **spinocerebellar tract**, carry somatic sensory information to the brain from all parts of the skin, joints and tendons.

- The dorsal column tract carries fine touch and vibration information to the cerebral cortex.
- The spinothalamic tract carries temperature, pain and crude touch information to the cerebral cortex.
- The spinocerebellar tract carries information about posture and position to the cerebellum.

The axons transport information to specific parts of the somatic sensory cortex that correspond to parts of the body. This route can be mapped out, as in Figure 10.7. The neurons in the somatic sensory cortex are the neurons that allow conscious sensation. You feel an insect crawling on your arm because the 'arm' neurons in your somatic sensory cortex are stimulated by the insect.

Somatic sensory association area: how do we interpret touch?

There is another area of the cerebral cortex that allows *understanding* and *interpretation* of somatic sensory information. It is located in the parietal lobe

spinothalamic tract (*SPY no thal A mic*)

spinocerebellar tract (*SPY no ser eh BELL ar*)

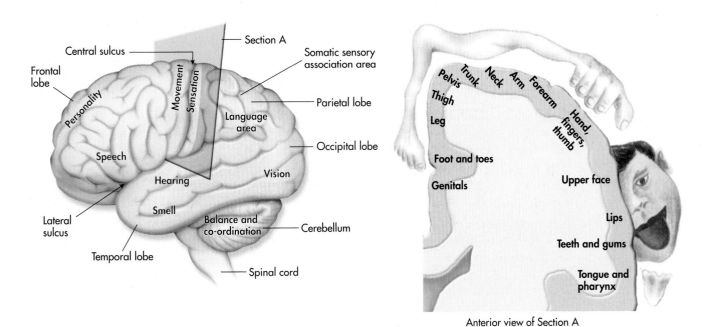

Anterior view of Section A

Figure 10.7 Primary somatic sensory area. Notice the size of the body parts is proportional to the amount of the sensory input provided. For example, the hands provide much more sensory input due to touch than your neck would and as a result the area devoted to the neck is much smaller. Remember that the map is contralateral.

just posterior to somatic sensory cortex and is known as the somatic sensory association area.

The somatic sensory system works on a kind of hierarchy, with the sensory neurons in the spinal cord and brainstem (brainstem neurons transport information in the same fashion as the spinal sensory neurons, but without going to the spinal cord first) collecting information and passing it to areas in the thalamus, cerebellum and cerebral cortex for processing. The *actual* understanding of complex sensory input happens only *after* the information is passed to the somatic sensory cortex and somatic sensory association area.

Let's go back to hitting your thumb with a hammer. The pain neurons, with bodies in the dorsal root ganglion, are depolarised and send signals to the spinal cord via the spinal nerve and dorsal root. The neurons synapse with motor neurons, causing you to pull your hand away from the stimulus. These same pain neurons synapse with the spinothalamic tract in your spinal cord and simultaneously send a pain signal to your brain. The signal goes first to the thalamus and other parts of the diencephalon, causing physiological symptoms, like sweating and increased heart rate. The pain then continues along the pathway to your somatic sensory cortex. There, the exact location of the pain becomes apparent to you. Now you know *where* the pain is but still may not recognise it as pain. The *understanding* occurs when the information is integrated with sensation in the somatic sensory association area. Now you know that you have hurt yourself! All this receiving of information and processing and associated actions happen almost instantaneously.

The big picture: the motor system

The motor system is also a hierarchy, working in parallel with the somatic sensory system. However, now information is moving in the opposite direction, from brain to spinal cord. See Figure 10.8 to show our current progress on the nervous system organisational chart.

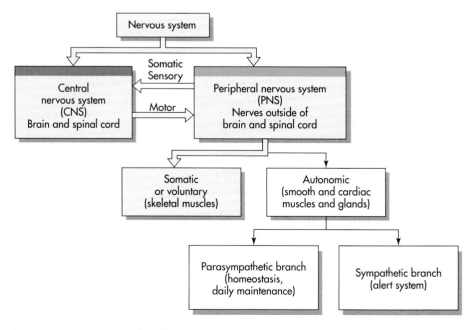

Figure 10.8 Progress thus far on the nervous system flow chart.

The somatic motor system controls voluntary movements under orders from the cerebral cortex. In the frontal lobe are the pre-motor and prefrontal areas, which *plan* movements. The plan from these two areas is sent to the primary motor cortex. The primary motor cortex is located in the precentral gyrus, in the frontal lobe, just anterior to the somatic sensory cortex just discussed. The primary motor cortex should seem familiar. Just like the somatic sensory cortex, the motor cortex has a map of the body. The size of the map in the motor cortex is proportional to the amount of movement control. Therefore, the hands and tongue have larger maps than the trunk or forearms.

Subcortical structures

basal nuclei (*BAY sal new klee EYE*)
limbic system (*LIM bic*)

Deep in the cerebrum are several areas of grey matter, which are surrounded by white matter and are known as **nuclei**. The nuclei in the cerebrum can be part of the **basal nuclei**, which is a motor coordination system, or part of the **limbic system**, which controls emotion and mood. We are concerned with motor neurons now, so the basal nuclei are our focus.

The plan for movement leaves the motor cortex and connects with neurons in the thalamus, which is located in the diencephalon. The thalamus, basal nuclei and cerebellum are part of a complicated *motor coordination loop*. Here, the movement must be fine-tuned, posture and limb positions are taken into account, other movements are turned off, and movement and senses are integrated. This loop is fundamental. Without the coordination loop among the subcortical structures (subcortical: 'under the cortex'), movements would be at the very least jerky and inaccurate. Some movements would be impossible. Those with Parkinsonism have a disorder of one of the basal nuclei and are unable to start new movements or turn off other movements. They can also have difficulty walking and swallowing and usually have an uncontrollable tremor when sitting still.

Spinal cord pathways

corticospinal tract (*KOR ti coe SPY nal*)
corticobulbar tract (*KOR ti coe BUL bar*)

After the movement, information is processed by the thalamus basal nuclei and cerebellum; it moves to the spinal cord and brainstem via the **corticospinal** and **corticobulbar tracts** and several other tracts. The corticospinal and corticobulbar tracts from your motor cortex are direct pathways, while others coming from subcortical structures are considered indirect pathways. The axons from all pathways synapse on motor neurons in the ventral horn. These motor neurons connect to skeletal muscles via the cranial nerves (in the brainstem) or the ventral roots and spinal nerves (spinal cord), sending orders to the skeletal muscles to carry out the planned movement or coordinating ongoing movements. (Remember, these neurons communicate with the muscles via the neuromuscular junction and therefore release the neurotransmitter acetylcholine across the synapse.)

A second function of the motor tracts is fine-tuning of reflexes. These tracts inhibit reflexes, making them softer than they would be if they had no influence from the brain.

Let's go back again to hitting your thumb with the hammer. After you hit your thumb, the first motor activity is a withdrawal reflex. You pull your

Amazing body facts

Size matters

Check out the size of the primary motor cortex dedicated to each body part in Figure 10.9. Does the size of the map make sense given the size of the body parts? Not really: the hands, lips and head have very large maps, while the legs and arms have very small maps compared to each body part's size.

What determines map size? What do the hand and lips have in common that makes them different from the arms and legs? They are required to perform more finely coordinated movements, such as speaking and handwriting (or typing), and therefore have more motor output. If you aren't convinced, think about how much harder it would be to type with your feet than with your hands. This map can be drawn as a homunculus (little man) with huge hands, lips and head (see Figure 10.9).

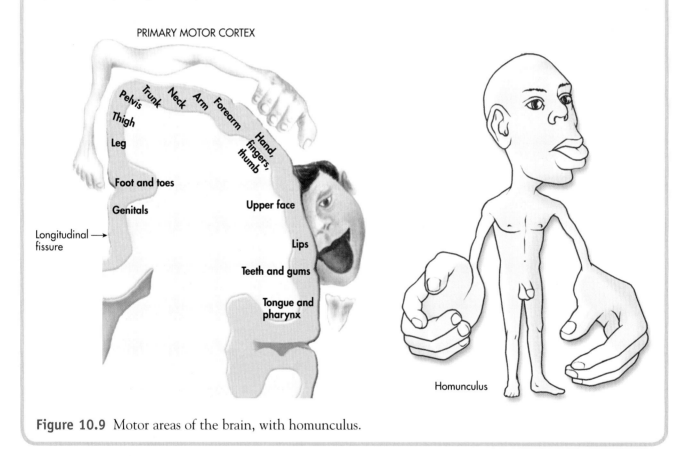

Figure 10.9 Motor areas of the brain, with homunculus.

thumb away from the painful stimulus. As we mentioned before, this is not a voluntary movement. It isn't planned. Only your spinal cord motor neurons are necessary for withdrawal. However, after the initial withdrawal, some jumping up and down and perhaps some cursing, you stop and think. What do you do next? You look at your thumb. That is a planned movement. It requires coordination between your motor system, to uncover your thumb and move it into your visual field, and your visual system to look at your thumb. You walk to the kitchen, open the freezer, get out the ice, wrap it in a cloth and make an ice pack. Then you may make a drink and walk back into the living room, nursing your wounds. All of this activity requires motor planning and coordination. You must reach for the freezer door and open it accurately. You must stay upright with respect to gravity. You must open a plastic bag and put ice in it. None of this happens without careful motor planning and coordination.

The role of the cerebellum

We have really glossed over the important function of the cerebellum. The cerebellum has both motor and sensory inputs and outputs from the cerebral cortex, the thalamus, the basal nuclei and the spinal cord. The cerebellum gets information about the *planned* movement and the *actual* movement and *compares* the plan to the actual. If the plan and the actual do not match, the cerebellum can adjust the actual movement to fit the plan. The function of the cerebellum is subtle and still a bit of a mystery, but without the cerebellum, movements would be inaccurate at best.

Test your knowledge 10.2

Choose the best answer:

1. The size of the map for a particular body part in the precentral gyrus is determined by:
   a. the amount of fine motor control
   b. the size of the body part
   c. the importance of the body part
   d. all of the above

2. The cerebellum compares:
   a. shapes of objects
   b. planned movement to actual movement
   c. textures
   d. position of objects in space

3. Which of the following spinal cord pathways carries pain and temperature information?
   a. dorsal column
   b. spinothalamic
   c. corticospinal
   d. thermostatic

4. The _____ sends direct messages to ventral horn motor neurons, while the _____ are involved in indirect pathways.
   a. thalamus, basal nuclei
   b. primary sensory cortex, thalamus
   c. cerebellum, primary motor cortex
   d. primary motor cortex, basal nuclei

The autonomic nervous system

The peripheral nervous system is divided into two systems: the somatic system, which controls skeletal muscles, and the autonomic system, which controls involuntary muscles. These muscles include the smooth muscles found in structures such as the blood vessels and airways, and cardiac muscle found in the heart. Glands are also controlled by this system. The autonomic system, then, controls physiological characteristics such as blood pressure, heart rate, respiration rate, digestion and sweating.

The neurons for the autonomic system, like the somatic motor neurons, are located in the spinal cord and brainstem, and release the neurotransmitter acetylcholine. This is where the similarity ends. The autonomic motor neurons are all located in the lateral horn rather than the ventral horns, and unlike the somatic motor neurons, autonomic neurons leaving the spinal cord do not connect directly to muscles. Instead, they make a synapse in a ganglion outside of the CNS. A **ganglion** is a group of nerve cell bodies outside of the CNS. You can think of this as a junction box where the signal can be passed on to the next part of the circuit. Then a second motor neuron, called a *post-ganglionic neuron*, connects to the smooth muscle or gland.

ganglion (*GANG lee yon*)

The sympathetic branch

para = *near or around*

The autonomic nervous system is divided into two subdivisions: the **sympathetic** and the **parasympathetic** divisions (see Figure 10.10). The sympathetic division controls the flight-or-fight response. It is charged with getting your

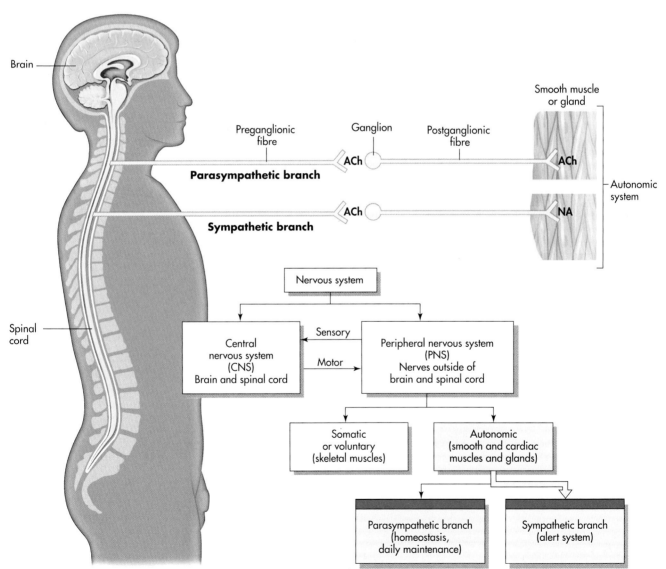

Figure 10.10 General representation of autonomic nervous system. ACh = acetylcholine and NA = noradrenaline.

body ready to expend energy. Sympathetic effects include increased heart rate, increased blood pressure, sweating and dry mouth, all the symptoms of an adrenaline rush. This part of the autonomic system was responsible for your racing heart, rapid breathing, and intense sweating when confronted with the snarling dog. Your heart pumped more blood to your muscles, and your lungs took in more oxygen, both of which got you 'up' to either fight or, most likely, flee. Another sympathetic response is dilation of your pupils to help you see the situation at hand much better. Sympathetic output is pretty strong when you hit your thumb with the hammer, too.

The *preganglionic neurons* for the sympathetic system are located in the thoracic and first two lumbar segments of the spinal cord. They are called the *thoracolumbar division*. The preganglionic neurons, which secrete acetylcholine, synapse with the postganglionic neurons in the sympathetic ganglia. The ganglia for the sympathetic division form a pair of chain-like structures that run parallel to the spinal cord. These are called *paravertebral ganglia*. The postganglionic neurons release the neurotransmitter *noradrenaline* (norepinephrine). One of the most important effects of the sympathetic system is its stimulation of the adrenal medulla to release the hormone *adrenaline* (epinephrine), the chemical that causes that familiar adrenaline rush by circulating in the bloodstream.

The parasympathetic branch

If you have an accelerator pedal, you also need a brake pedal. The parasympathetic division is often called 'resting and digesting' because it has the opposite effect of the sympathetic division. The parasympathetic division is responsible for maintenance of everyday activities. It also helps bring you back down to normal from a sympathetic response. Parasympathetic effects include decreased heart rate, respiration and blood pressure and increased digestive activity, including salivation and even stomach rumbling. The parasympathetic system allows you to calm down after you get the ice pack on your thumb or the dog is safely shut in another room.

The neurons of the parasympathetic system are in the brainstem and the sacral spinal cord and therefore are called the *craniosacral division*. They, too, release acetylcholine. The preganglionic neurons synapse with postganglionic neurons in the parasympathetic ganglia. Parasympathetic ganglia are located near the organs. Postganglionic neurons release the neurotransmitter acetylcholine. (Remember, acetylcholine excites skeletal muscle but can either inhibit or excite smooth and cardiac muscle.)

Putting the autonomic system all together

Let's revisit the snarling dog. When you encounter him, the sympathetic nervous system is stimulated and the impulses are sent from the thoracolumbar region of the spinal cord to preganglionic fibres, which release the neurotransmitter acetylcholine. Acetylcholine combines with receptors on the postganglionic neurons which generate an action potential and carry the impulse to the target area where noradrenaline is released. If this occurs at the heart (cardiac muscle), the rate and force of contraction will supply more blood for the fight-or-flight response.

Once the emergency is over, the sympathetic pathway will not be as active, and the noradrenaline will be metabolised so you don't remain in that stimulated state. The parasympathetic system now takes over to return your system to normal. This is another example of homeostasis in which normal heart rate is regulated by a balance between these two systems.

Other systems

Two other systems that do not have a single, specific location but are found within the brain are the limbic and reticular systems.

The limbic system

The limbic system is a series of nuclei in the cerebrum, diencephalon and superior brainstem. These nuclei are involved in mood, emotion and memory. One area helps attach emotion to movement. Another coordinates emotion and your sense of smell. Still another is responsible for storing and retrieving memories.

The reticular system

The reticular system is a diffuse network of nuclei that is responsible for 'waking up' your cerebral cortex. The reticular system activity is vital for the maintenance of conscious awareness of your surroundings. When your alarm clock wakes you in the morning, your reticular system is responsible for nudging your cortex out of slumber. General anaesthesia inhibits the reticular system, rendering surgical patients unconscious and unresponsive to any form of stimuli. Injury due to ischaemia, mechanical damage or drugs can damage the reticular system and lead to coma.

Common disorders of the nervous system

Paralysis is the inability to control voluntary movements. Paralysis can be *spastic* or *flaccid*. **Spastic paralysis** is characterised by muscle rigidity or increased muscle tone (*hypertonia*) and overactive reflexes (*hyperreflexia*). In spastic paralysis, the muscles are rigid and the reflexes overactive because of decreased communication from the brain to the ventral horn motor neurons in the spinal cord. Muscles contract randomly, and reflexes do not have any control signals from the brain. Stroke, head injuries and spinal cord injuries can cause spastic paralysis.

Flaccid paralysis is characterised by floppy muscles (decreased muscle tone, or *hypotonia*) and decreased reflexes (*hyporeflexia*). In flaccid paralysis, the damage is to the spinal nerves. Impulses cannot get to the muscles from the motor neurons. Therefore, the muscles are floppy and reflexes are absent. Flaccid paralysis can be caused by peripheral injury or disorders like polio and Guillain-Barré syndrome.

Cerebral palsy is a collection of movement disorders that are not progressive and that occur in young children. Patients with cerebral palsy exhibit signs of classic spastic paralysis. The disorder is caused by improper development of or damage to the motor system of the brain. Symptoms of cerebral palsy can range from minor motor loss to significant motor deficits, including the inability to walk or speak. Patients with cerebral palsy may have significant developmental delays or blindness or may have average or above-average intelligence.

The condition known as a stroke, or **cerebrovascular accident** (CVA), is caused by the disruption of blood flow to a portion of the brain due to either haemorrhage or blood clot. If the oxygen is disrupted for long enough, brain tissue will die. The symptoms of stroke vary depending on the location of the stroke. Muscle weakness, paralysis or the lack of sensation due to stroke is contralateral (remember, the right hemisphere controls the left side of the body). Strokes can also rob patients of the ability to speak and can cause blindness and destroy memory. Symptoms of stroke appear suddenly. Some patients have a series of minor strokes (almost like tiny earthquakes before the big one) with minor, temporary symptoms before they have a major stroke. These are called *transient ischaemic accidents*, or TIAs.

A **subdural haematoma** is a pool of blood between the dura mater and arachnoid mater, in the subdural space. Subdural haematomas are usually caused by head injuries. A blow to the head ruptures tiny blood vessels in the skull, causing them to bleed into the space. If the haematoma is large, or growing, the condition can cause irreversible brain damage and death. Haematomas cause damage by increasing pressure inside the skull. As the blood collects in the subdural space, pressure increases. The skull cannot expand and blood cannot be compressed as pressure rises, so the increasing pressure can cause severe damage to the delicate brain tissue. Some subdural haematomas resolve themselves, and some can be treated with medication. However, large or rapidly growing subdural haematomas must be treated by surgery.

Huntington's disease is a progressive genetic disorder causing deterioration of neurons in the basal nuclei and eventually of the cerebral cortex. The disease begins with mood swings and memory disturbances, writhing movements of the hands and face, or clumsiness. Eventually, the disease causes difficulty swallowing, speaking and walking, as well as memory loss, psychosis and loss of cognitive function. There is no cure for Huntington's disease, and most patients die from accidental injuries, infections or other complications. There is a genetic test to determine if a person carries the gene for the disease. Offspring of parents who have the disease have a 50 per cent chance of inheriting the gene for Huntington's disease. A carrier of the gene will eventually develop the disease.

Summary

- The nervous system is your body's computer system, its information superhighway. Without the nervous system, people could not sample the environment, make decisions or respond to stimuli. The nervous system handles millions of pieces of information every minute, making sure that every system in the body is working properly and correcting any problem that occurs.

- The brain is a hierarchical organ. It is divided into compartments (lobes), each with very specific functions. The brain also has nerves attached to it, called cranial nerves. There are 12 pairs of cranial nerves, and they can be sensory, motor or mixed.

- The cerebrum controls your conscious movement and sensation. Beneath the cerebrum are the diencephalon, brainstem and cerebellum. Each part plays important roles in coordinating sensory and motor information for the cerebrum. Other parts of the brain, called association areas, allow you to make connections between different types of sensory information and to compare current experience to memories.

- The somatic sensory cortex is in the postcentral gyrus of the parietal lobe. Sensory information from the spinal cord tracts eventually ends up in this part of the cortex. When the information arrives there, you become aware of your sense of touch.

- There are motor and sensory maps of the body in the cerebral cortex. Orders for voluntary movements originate in the primary motor cortex in the precentral gyrus of the frontal lobe and travel down the spinal cord via direct spinal cord tracts. Subcortical structures coordinate this information via indirect tracts.

- The nervous system also controls involuntary movement via a part of the system known as the autonomic nervous system. The sympathetic division controls the fight-or-flight response, and the parasympathetic division controls day-to-day activities.

Case study

Sarah finds her elderly father lying at the bottom of the stairs. He is semiconscious. He is paralysed on his right side, but he seems to be able to feel that side of his body. At the hospital, he is diagnosed with a cerebrovascular accident (or stroke).

1. What part of his brain is damaged?

2. How can you tell?

Review questions

Multiple choice

1. One of the following brain parts is *not* subcortical. Which one?
   a. hypothalamus
   b. medulla oblongata
   c. precentral gyrus
   d. pineal body

2. This cranial nerve controls the abdominal visceral organs.
   a. olfactory (I)
   b. trigeminal (V)
   c. vestibulocochlear (VIII)
   d. vagus (X)

3. The size of the map of each body part in the postcentral gyrus is determined by the:
   a. sensitivity of the body part
   b. size of the body part
   c. importance of the body part
   d. fine motor control of the body part

4. The sympathetic nervous system:
   a. causes decreased heart rate
   b. has ganglia near the organs
   c. has ganglia near the spinal cord
   d. all of the above

5. This part of the brain contains the body's set points and controls most of its physiology, including blood pressure and hunger level.
   a. thalamus
   b. hypothalamus
   c. amygdala
   d. hippocampus

Fill in the blanks

1. The _____ are nuclei that coordinate motor output.

2. The occipital lobe is responsible for this sensation: _____.

3. Emotion, mood and memory are controlled by this collection of nuclei: _____.

4. This portion of the brainstem has vital nuclei for respiration and the cardiovascular system: _____.

Short answers

1. List the differences between cranial and spinal nerves.

   _____

   _____

   _____

   _____

→

2. List the differences between the sympathetic and parasympathetic nervous systems.

_____

_____

_____

_____

Suggested activity

1. Find the parts of the brain on a diagram or model.

Answers

Answers to case study

1. The left side of the brain has been damaged due to an interruption of blood flow.
2. Control of the body is contralateral, the left side of the brain controls the right side of the body.

Answers to multiple choice questions

1. c
2. d
3. a
4. c
5. b

Answers to fill in the blanks

1. basal ganglia
2. Sight/vision
3. Limbic system
4. Medulla oblongata

Answers to short answer questions

1. Most cranial nerves carry mixed and sensory information (the exceptions are the optic, olfactory and vestibulocochlear); motor nerves carry mixed motor and sensory information.
2. Sympathetic nervous system controls the active/fight-or-flight responses; whereas the parasympathetic nervous system controls homeostasis.

The special senses

As we stroll through a city, we take in many sights, sounds and smells. As we look around we may see many different buildings and types of people. The city is also noisy, full of car horns blaring, brakes squealing and people talking. We might smell delicious food wafting from restaurant windows or the stench of rubbish. Our special senses receive all of this input and send it to the brain for interpretation so we can understand and appreciate what is happening around us. These special senses are highly integrated with the nervous system, enabling us to respond quickly and thereby protecting us from harm. For example, we need to *see* that oncoming speeding car as we step away from the kerb and *hear* the blaring horn in order to respond by quickly stepping back to the safe confines of the pavement. See how sensory input can determine motor response?

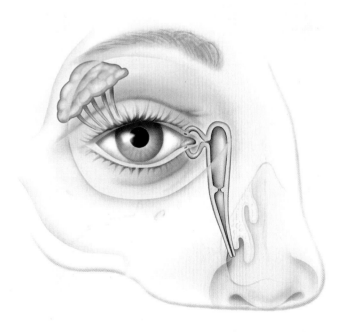

Learning objectives

At the end of this chapter, you will be able to:

◆ Differentiate general and special senses
◆ Describe the internal and external anatomy and functions of the eye
◆ Describe the internal and external anatomy and functions of the ear
◆ Discuss the process involved with the senses of taste, smell and touch
◆ Contrast the types of pain and the pain response
◆ Explain several common disorders of the eye and ear

Pronunciation guide

amblyopia (*am bly OH pee ah*)
aqueous humour (*AY kwee uss HUW mer*)
auricle (*AW rih kl*)
cataract (*KAT ah rakt*)
cerumen (*seh ROO men*)
ceruminous glands (*seh ROO men us*)
choroid (*COR royd*)
ciliary muscles (*SILL ee air ee*)
cochlea (*COCK lee ah*)
conjunctiva (*kon JUNK tye vah*)
endolymph (*EN doe limf*)
Eustachian tubes (*yoo STAY shun*)

external auditory meatus (*AW dih tor ee mee YAY tuss*)
glaucoma (*gl OW KOH mah*)
gustatory sense (*GUSS ta TOR ee*)
hyperopia (*HIGH per OH pee yah*)
incus (*ING kuss*)
labyrinth (*LAB er inth*)
lacrimal apparatus (*LAK rim all app ah RAY tuss*)
malleus (*MALL ee uss*)
Ménière's disease (*MANY erz*)
myopia (*my OH pee ah*)
ossicle (*OSS ickle*)

otitis media (*oh TYE tiss ME dee ah*)
perilymph (*peri LIMF*)
photo pigment FOE toe PIG ment
pinna (*PINN ah*)
presbyopia (*PRESS bee OH pee ah*)
sclera (*SKLAIR ah*)
stapes (*STAY peez*)
tactile corpuscles (*KOR puss els*)
tinnitus (*tin IT uss*)
tympanic membrane (*tim PAN ik*)
vestibule chamber (*VESS ti bule*)
vitreous humour (*VITT ree uss HUW mer*)

The different senses

Our body senses allow us to experience all aspects of life. They are truly remarkable sensors through which we see, hear, smell, taste and feel the world around us.

Our senses monitor and detect changes in the environment and send this information from the receptor to the brain via sensory or afferent neurons. The brain interprets the information and in many circumstances makes the appropriate motor or efferent response.

Traditionally, we are taught that we possess five senses: vision, hearing, smell, taste and touch. However, there are more areas of sensory input into the brain. What about pain and pressure sensations? How do we 'feel' hot and cold? How does our body sense position and balance (equilibrium)? What about feelings of hunger and thirst? These, too, are senses that are very important to our survival.

The senses of sight (eyes), sound and equilibrium (ears), taste (tongue), and smell (nose) are referred to as our *special senses*. These senses are found in a well-defined region of the body. However, other senses, called *general senses*, are scattered throughout various regions of the body. These include the sensations of heat, cold, pain, nausea, hunger, thirst and pressure or touch.

cutaneo = *skin*

visceral = *pertaining to organs*

The senses can be further broken down. For example, the receptors of the skin are called the **cutaneous** senses and include touch, heat, cold and pain. The **visceral** senses include nausea, hunger, thirst, and the need to urinate and defecate.

Before finishing this discussion of the various senses, we should also mention a final, more controversial sense. By 'reading your mind', we see you already identified it as extrasensory perception or ESP. Notice that this means senses outside the normal sensory perceptions. While there is still debate over whether this phenomenon exists, we simply 'know' this chapter will be an 'eye-opening' experience for you. We just hope the puns aren't stimulating your visceral senses and making you nauseous!

Note

While it may not be an actual sense, have you ever heard of common sense? What do you think of the statement that 'common sense is not all that common nowadays'?

Sense of sight

The eye has many similarities to a camera. The light rays from the images photographed with a camera pass through the small opening (comparable to the pupil) and through the transparent lens (lens of the eye) where the rays are then focused on a photoreceptive film (retina). The shutter of a camera opens and closes at various speeds to adjust the amount of light exposure on the film. The shutter (iris) must allow just the right amount of light to enter and focus properly on the film for a clear image. A camera may be packed away in a suitcase or thrown in a car and therefore needs protection from being dropped or exposed to a harsh environment. You will soon see how the external structures of the eye help protect it from injury, much like a camera case protects a camera. In addition, the camera lens must be kept clean to insure a clear picture. The lacrimal glands that secrete tears help to perform this function. Now that you have this analogy to 'give you the big picture', let's explore the specific structures and functions of the eye.

The external structures of the eye

The **orbit** is a cone-shaped cavity formed by the skull that houses and protects the eyeball. This cavity is padded with fatty tissue that cushions and protects the eye from injury and has several openings through which nerves and blood vessels can pass. The eyeball is connected to the orbital cavity with six short muscles that provide support and allow rotary movement so you can see in all directions. Also protecting the eye is a pair of movable folds of skin, commonly called eyelids, which contain eyelashes to help prevent gross particles from entering. The eyelashes act as a sensor to cause rapid closure as a foreign object approaches the eyeball. The eyelids close over the eye much like the lens cover of a camera to protect it from intense light, foreign particles or impact injuries. The eyelids also contain sebaceous glands that secrete the oily substance *sebum* onto the eyelids to keep them soft, pliable, and a little sticky to trap particles.

conjunctiva (kon JUNK tye vah)

lacrimal apparatus (LAK rim all app ah RAY tuss)

   A protective membrane called the **conjunctiva** lines the exposed surface of the eyeball and acts as a protective covering for the exposed eye surface. Each eye has a **lacrimal apparatus** that produces and stores tears. The lacrimal apparatus includes the **lacrimal gland** and its corresponding ducts or passageways that transport the tears. The lacrimal glands (exocrine because their secretion of tears go outside the body) produce the tears needed for constant cleansing and lubrication which are spread over the eye surfaces by blinking. Our eyes are constantly tearing, but they do not overflow because excess tears drain into the nose via two small holes in the inner corner of the eye. However, when we cry, the excess runs down our cheeks and more drains into our nose, causing it to run. The tears also act as an **antiseptic** to keep the eyeballs free of germs. Please see Figure 11.1 for the structures involved with tearing.

The internal structures of the eye

The globe-shaped eyeball is the organ of vision and is separated into two chambers of fluid that help to protect the eye. These 'fluids of the eye' are

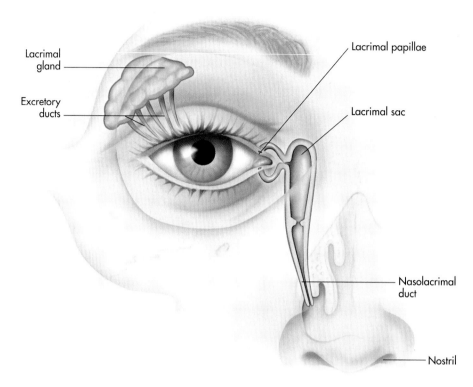

Lacrimal gland

Excretory ducts

Lacrimal papillae

Lacrimal sac

Nasolacrimal duct

Nostril

Figure 11.1 Lacrimal structures of the eye.

aqueous humour (*AY kwee uss HUW mer*)

vitreous humour (*VITT ree uss HUW mer*)

sclera (*SKLAIR ah*)

choroid (*COR royd*)

retina (*RETT in ah*)

ophthalm/o = *eye*

iris (*EYE riss*)

pupil (*PEW pill*)

ciliary muscles (*SILL ee air ee*)

called *humours*. The **aqueous humour** ('watery' humour) bathes the iris, pupil and lens and fills the anterior and posterior chambers of the eye. The second humour is called the **vitreous humour** and is a clear, jelly-like fluid that occupies the entire eye cavity behind the lens.

The eyeball has three layers. Please see Figure 11.2 for the layers and internal structures of the eye. These layers are the **sclera**, **choroid** and **retina**. The sclera is the outermost layer and is a tough, fibrous tissue that serves as the protective shield we commonly call the 'whites of the eye'. The sclera contains a specialised portion called the **cornea**, which is transparent to allow light rays to pass into the eye. The cornea has a curved surface that allows it to bend the entering light waves to focus them on the surface of the retina.

The middle layer, or choroid, is a highly vascularised (rich blood supply) and pigmented region that provides nourishment to the eye. This layer also contains the **iris** and the **pupil**. The iris is the coloured portion of the eye that controls the size of the opening (pupil) where light passes into the eye. The iris is a sphincter, which means it can relax or contract, thereby making the centre opening, or pupil, larger or smaller depending on light conditions. In low light, the iris relaxes, causing the pupil to dilate and thereby allowing more light into the eye for a better image.

The third and innermost layer is the retina. This area contains the nerve endings that receive and interpret the rays of light into images. Located behind the pupil is the **lens**, which is surrounded by **ciliary muscles**. These muscles can alter the shape of the lens, making it thinner or thicker to allow the incoming light rays to focus on the retinal area. This process is called *accommodation*, which basically combines changes in the size of the pupil

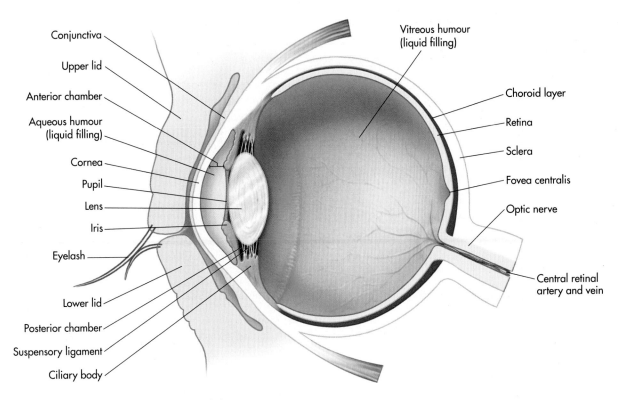

Figure 11.2 Internal structures of the eye.

Amazing body facts

Why we can see in the dark

Why are there two types of photoreceptors? Just like a video camera, the eyes are exposed to various lighting conditions. In low light, video camera images are black and white, while in good light, the images become fully 'colourised'. The rods of the eye are the photoreceptors for dim light and provide black and white images in dark conditions. The cones function in bright light for colour vision. The retina of each eye contains about 3 million cones and 100 million rods.

optic = *pertaining to the eye*
photopigment (*FOE toe PIG ment*)

and the lens curvature to make sure the image converges in the same place on the retina and therefore is properly focused.

The retina is a delicate membrane that continues posteriorly and joins to the **optic** *nerve* (*cranial nerve II*). It contains two types of light-sensing receptors called **rods** and **cones**. The rods are active in dim light and do not perceive colour, while the cones are active in bright light and do perceive colour. These receptors contain **photopigments** that cause a chemical change when light hits them. This chemical change causes an impulse to be sent to the optic nerve and then to the brain where the impulse is interpreted, and we 'see' the object. This interpretation occurs in the visual part of the cerebral cortex located in the occipital lobe.

In summary, light rays enter the eye and pass through the conjunctival membrane, cornea, aqueous humour, pupil, lens and vitreous humour and are focused on the retina. Here the photoreceptors in the retina cause an impulse to be sent to the optic nerve, which carries it to the brain for the interpretation we call vision. See if you can trace the pathway light takes in Figure 11.2. Refer to Table 11.1, which summarises the major structures and their corresponding functions of the eye.

Table 11.1 Structures and functions of the eye

Organ or structure	Primary function
Orbit	cone-shaped cavity that contains the eyeball; padded with fatty tissues, the orbit has several openings for nerves and blood vessels to pass through
Eye muscles	six short muscles that provide support and rotary movement
Eyelids	movable folds of skin containing eyelashes to protect eye from intense light, foreign particles and impact injuries
Conjunctiva	protective membrane that lines the exposed surface of eyeball
Lacrimal apparatus	includes the lacrimal gland that produces tears that lubricate and cleanse the eye and the corresponding ducts or passageways to transport the tears
Eyeball	globe-shaped organ of vision
Sclera	outermost layer known as the 'white' of the eye; contains the transparent curved cornea, which bends outside light rays to focus them on the surface of the retina
Choroid	the middle layer that has rich blood vessels and pigmentation to prevent internal reflection of light rays; also contains the iris and pupil
Iris	coloured portion of eye that controls the size of the opening (pupil) where light passes into the eye
Pupil	the opening through which light passes into the eye
Retina	innermost layer that contains the nerve endings that receive and interpret the rays of light for vision
Lens	located behind the pupil, the lens is controlled by ciliary muscles that shape the lens by thinning or thickening the lens to allow light to focus on the retinal surface; this process, called accommodation, combines changes in pupil size and the lens curvature to ensure the light rays focus properly on the retina

Test your knowledge 11.1

Choose the best answer:

1. Which of the following is *not* a layer of the eye?
   a. photo optic
   b. sclera
   c. choroid
   d. retina

2. Which layer of the eye contains the iris and the pupil?
   a. cornea
   b. sclera
   c. choroid
   d. retina

3. The _____ is the coloured portion of the eye that controls the opening, or _____, where light passes through.
   a. retina, pupil
   b. pupil, iris
   c. sclera, pupil
   d. iris, pupil

The sense of hearing

Our ears are responsible for our sense of hearing and our sense of balance so we don't fall and get hurt. We can 'hear' your heart pounding with anticipation, so let's explore the specific structures and functions of the ear.

Structures of the ear

The ear is responsible for hearing and maintaining our equilibrium, or sense of balance. We hear by receiving sound vibrations usually via the air (unless we are under water) and translating them into an interpretable sound via the eighth cranial nerve. The ear can be separated into three divisions: the **external ear**, the **middle ear** or **tympanic** cavity, and the **inner ear** (also called the labyrinth.) See Figure 11.3 for the structures of the ear.

tympanic (*tim PAN ik*)
labyrinth (*LAB er inth*)

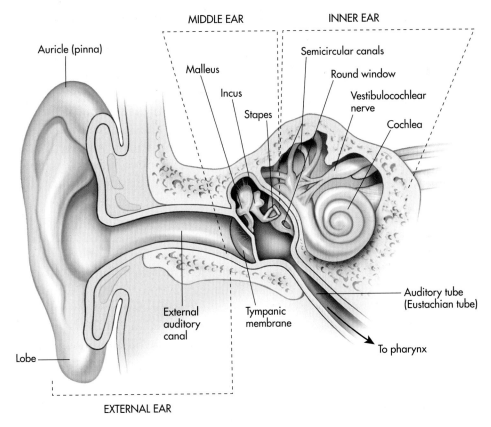

Figure 11.3 Structures of the ear.

Applied science

Sound conduction

Many people are under the false impression that sound travels best through air because this is the medium we are immersed in. However, sound is transmitted due to molecular collisions and is better transmitted where the molecules are closer together, such as in a liquid or solid medium. This is why whales can talk to each other up to two miles apart under water. This is also why trapped underground miners tap on the wall instead of shouting for help.

pinna (*PINN ah*)

auricle (*AW rih kl*)

external auditory meatus (*AW dih tor ee mee YAY tuss*)

ceruminous glands (*seh ROO men us*)

cerumen (*seh ROO men*)

tympanon = *Greek word for drum*

ossicle (*OSS ickle*)

malleus (*MALL ee yus*)

incus (*ING kuss*)

stapes (*STAY peez*)

Eustachian tubes (*yoo STAY shun*)

cochlea (*COCK lee ah*) *Latin for snail shell*

vestibule chamber (*VESS ti bule*)

perilymph (*peri LIMF*)

The external ear

The external ear is the outer projection, the part we can see. It also includes the canal (where we put cotton swabs despite the warning label) leading into the middle ear. The projecting part is called the **pinna** or **auricle**, which collects and directs sound waves into the **auditory canal** or **external auditory meatus**. The canal contains earwax, called **cerumen**, which is secreted by the **ceruminous glands** to lubricate and protect the ear. At the end of the canal is the **eardrum**, or **tympanic membrane**, where the external ear ends and the middle ear begins. Don't you wish there were just ONE term everyone agreed upon for these structures!

The middle ear

The middle ear, or tympanic cavity, is basically a space that contains the three smallest bones of your body. The three bones, or **ossicles**, are joined so they can amplify the sound waves the tympanic membrane receives from the external ear (sound travels best through a solid). Once amplified, the sound waves are transmitted to the *fluid* contained in the internal ear. Once again, this is another example of sound waves being transmitted more efficiently though a liquid medium than in air.

The bones of the ears are named according to their shapes. The first ossicle attached to the tympanic membrane is the **hammer**, or **malleus**. The **anvil**, or **incus**, is attached to the hammer. Finally, the **stirrup**, or **stapes**, connects to a membrane called the **oval window**. The oval window begins the internal ear and carries the amplified vibrations from the tympanic ossicles. During transmission, the sound or vibrations can be amplified as much as 22 times their original level.

Also contained within the middle ear are the **Eustachian tubes**. These tubes allow the air pressure on either side of the eardrum to be equalised. The tubes connect the nose and throat to the middle ear. Therefore, they transmit both the outside atmospheric pressure through the nose and throat opening and the inner ear pressure where they are located to allow for an equalisation of pressure between the atmosphere and the middle ear. This equalising of pressure between the middle ear and external atmosphere allows the eardrum to freely vibrate with incoming sound waves. Sudden pressure changes, such as caused by flying in an aeroplane, can affect this area. This is why, when flying, you are instructed to chew gum or swallow so the inner ear can better sense and adjust to the rapidly changing outside atmosphere via the Eustachian tubes.

The inner ear

The oval window membrane is the portal into the inner ear or labyrinth. This area comprises three separate, hollow, bony spaces that form a complex maze of winding and twisting channels. Since another name for a maze is labyrinth, this area can also be called the bony labyrinth. The three areas are the **cochlea**; the **vestibule chamber**, which houses the internal ear; and the **semicircular canals**.

The cochlea is the bony spiral or snail shell-shaped entrance to the internal ear connected to the oval window membrane (see Figure 11.4). The cochlea contains fluid called **perilymph**, which helps to transmit the sound through this area. The sound is then transmitted to the back of the maze, which

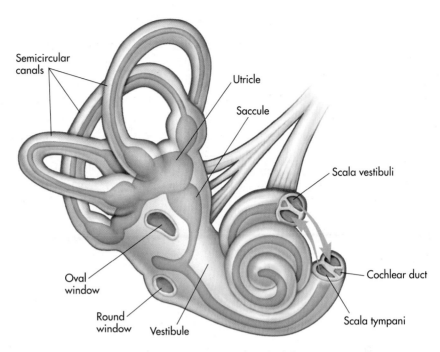

Figure 11.4 The internal structure of the cochlea.

endolymph (*EN doe limf*)

vestibulocochlear (*VESS tib you low COCK lee are*)

contains another fluid called endolymph. Here the sound is carried to tiny hair-like receptors that are stimulated and conduct the signal to the brain via the **acoustic** or **vestibulocochlear nerve (cranial nerve VIII)**. Table 11.2 lists the major structures and functions of the ear.

In summary, sound waves enter the external canal and vibrate the eardrum or tympanic membrane in a process called *sound conduction*. The middle ear then amplifies the sound through the respective ossicles. This process is called *bone conduction* of sound. The last ossicle (stapes) vibrates and causes a gentle pumping against the oval window membrane. This causes cochlear fluid to vibrate small hair-like nerves found in an area called the *Organ of Corti*. As a result of the vibrating sensory cells (hair-like nerves), a nerve impulse is sent to the temporal lobe of the brain where it is interpreted as sound, a process called *sensorineural conduction*.

Low-intensity sound waves, similar to a clock ticking, send vibrations that cause the sensory cells to move in waves that are interpreted by the brain as that 'tick' sound. In extreme cases where intense sound waves are produced, such as from a gun blast, it is believed that the vibrations are so great that they may knock over the hair-like cells, much like an earthquake knocks over tall trees. Repeated assaults can lead to permanent hearing damage. Therefore, it is a 'sound' investment to wear proper hearing protection around loud noises.

The ear is also responsible for your sense of balance or equilibrium. The semicircular canals process sensory input related to equilibrium. They contain nerve endings or receptors in the form of hair cells. The semicircular canals are three loops within the inner ear that help to maintain balance. Like the cochlea, they are filled with endolymph fluid, and each canal duct contains a sensory receptor. This fluid moves when you change body position. The movement is picked up by the sensory receptor, which triggers a nerve impulse to travel to the brainstem and the cerebellum. Here the impulse is interpreted as body position to help maintain muscle coordination and body equilibrium.

Table 11.2 Structures and functions of the ear

Organ or structure	Function
External ear	
Auricle, or pinna	cartilaginous projection that collects and directs sound waves into the auditory canal much like a satellite dish collects transmissions from space
Auditory canal, or external auditory meatus	canal that contains earwax, or cerumen, secreted by the ceruminous glands that lubricate and trap foreign particles
Eardrum, or tympanic membrane	membrane that separates the external and middle ear
Middle ear	
Ossicles	three small bones (hammer-, anvil- and stirrup-shaped) that help amplify and transmit sound
Eustachian tubes	allows for equalisation of external (atmospheric) and internal (within the middle ear) pressure on the tympanic membrane so the eardrum can freely vibrate with incoming sound
Inner ear, or labyrinth	
Cochlea	bony, snail shell-shaped entrance to the internal ear containing perilymph fluid, which helps to transmit sound
Semicircular canals	three canals containing endolymph fluid, which transmits positional changes to tiny, hair-like receptors that are stimulated and conduct the signal to the brain via the acoustic or vestibulocochlear nerve (eighth cranial nerve) to help maintain balance

Test your knowledge 11.2

Choose the best answer:

1. The structure that marks the end of the external ear and the beginning of the middle ear is called the
   a. pinna
   b. hammer
   c. tympanic membrane
   d. labyrinth

2. The structures of the ear that are important for balance are the:
   a. ceruminous glands
   b. incus, malleus and stapes
   c. semicircular canals
   d. Eustachian tubes

Provide the synonymous terms for each of the following:

3. auricle _____

4. earwax _____

5. malleus _____

6. anvil _____

7. stirrups _____

Other senses

Other senses also help us to interpret the world around us. These include the senses of taste, smell and touch.

Taste

gustatory sense (*GUSS ta TOR ee*)

The sense of taste is referred to as the **gustatory sense**. Taste receptors are located in the tongue and are called **taste buds**. Notice in Figure 11.5 that the four tastes of *sweet*, *sour*, *salty* and *bitter* are located in different regions of our tongue. Two additional tastes have been discovered, *umami* and *water*. *Umami* has been included with the traditional four because it is the distinct taste of glutamates, which cannot be duplicated by the combination of any of the other four tastes. Water may be considered to be tasteless, but there are specific water receptors located in the pharynx. Quite often, taste preferences change with the body's need, which is why, for example, pregnant women may crave a variety of food throughout their pregnancy. The refinement of food taste is primarily dependent on the sense of smell.

Smell

The sense of smell arises from the receptors located in the olfactory region or the upper part of the nasal cavity (see Figure 11.6). We 'sniff' in order to bring

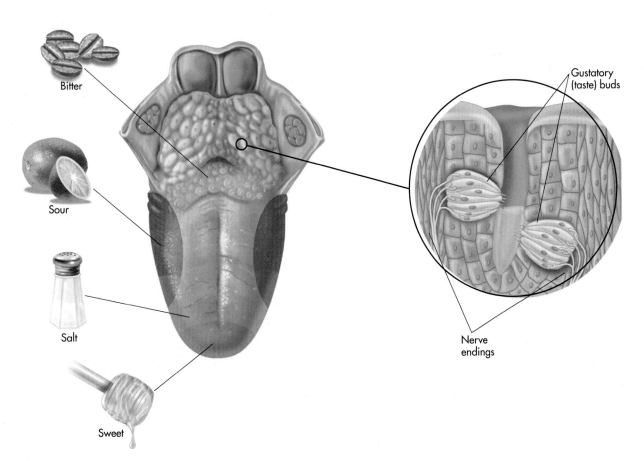

Figure 11.5 The sense of taste.

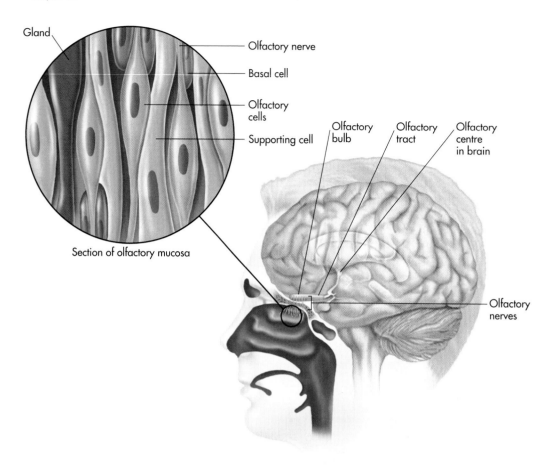

Gland

Olfactory nerve

Basal cell

Olfactory cells

Supporting cell

Section of olfactory mucosa

Olfactory bulb

Olfactory tract

Olfactory centre in brain

Olfactory nerves

Figure 11.6 The sense of smell.

smells up to this area, where they can be interpreted. Remember taste and smell are closely related, which is why we can't taste foods when we have a severe head cold. Pleasant food odours also initiate digestive enzymes, so when you smell that cinnamon apple pie baking, your mouth really may water in anticipation. Not all people have a sense of smell. Those without this sense are said to be anosmic.

Touch

tactile corpuscles (*KOR puss els*)

Touch receptors are small, rounded bodies called **tactile corpuscles** located in skin and especially concentrated in the fingertips. They are also located on the tip of the tongue. It is interesting to note that even when a patient is anaesthetised and there are no pain sensations, patients are still conscious of pressure through these deep touch sensors.

Temperature sensors are also found in the skin. The body has separate heat and cold receptors. These receptors may cause an interesting phenomenon called **adaptation** to occur. Continued sensory stimulation causes the sensors to desensitise or adapt. For example, if you are continually exposed to the cold, the receptors adjust so you won't feel so cold. That is why an outside temperature of 50°F (10° celsius) seems warm after a long cold winter, but that same temperature seems cold after a long hot summer. Another example is when you first enter a hot bath, it may feel extremely hot. After a few seconds, it doesn't feel so hot, yet the actual temperature of the water has not yet changed.

Pain is a very important protective sense. It is the body's way of drawing attention to a particular danger, such as the 'hitting your thumb with a hammer' example used earlier in the book. Pain is the most widely distributed sense, found in skin, muscle, joints and internal organs. Pain receptors are merely branches of nerve fibres called *free nerve endings*.

There are even different types of pain. *Referred pain* originates in an internal organ yet is felt in another region of the skin. For example, liver and gallbladder disease often cause pain in the right shoulder. *Phantom pain* can result from an amputated limb: an individual can feel pain in an arm or leg he or she no longer has.

Clinical application

Heat and cold therapy

Heat and cold therapy are used for a variety of injuries. They rely on physiologic changes in the body in response to either heat or cold. For example, heat relaxes muscles and dilates vessels, thereby bringing more blood flow to the site of injury. Cold therapy constricts blood vessels and minimises the amount of blood and swelling at a site.

anaesthesia (*an ess THEE zee ah*)
an = *without*
aesthesia = *feeling*

Pain receptors do not adapt as heat and cold receptors do. Pain is felt for as long as the stimulus that is causing it remains or unless a person is under **anaesthesia**. An interesting debate is whether or not some people have higher or lower thresholds of pain. See Figure 11.7 for the senses of touch.

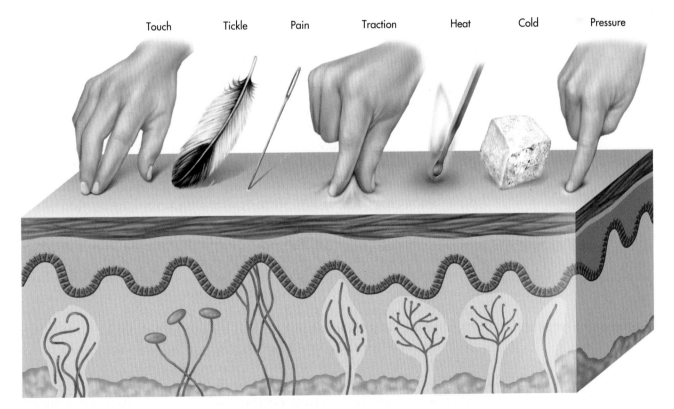

Figure 11.7 The sense of touch.

Common disorders of the eye and ear

conjunctivitis (*kon JUNK tih VYE tiss*)

Conjunctivitis is an inflammation of the membrane that lines the eye. This condition can either be acute or chronic and is caused by a variety of irritants and pathogens. The acute phase is commonly called *pink eye* and is a highly contagious form caused by a bacterium.

cataract (*KAT ah rakt*)

A cataract is a condition in which the lens loses its flexibility and transparency, and light cannot easily pass through the clouded lens. It has been shown that increased exposure to sunlight may speed up the development of cataracts. Untreated, this condition can lead to blindness. It is interesting to note that cataract surgery was one of the earliest recorded surgical procedures, dating back to ancient Greece.

glaucoma (*glaw KOH mah*)

Glaucoma can also lead to blindness. It is caused by increased pressure in the fluid of the eye, which interferes with optic nerve functioning. This is a tragic loss because glaucoma can be readily diagnosed and treated.

The eye can have several defects that impair vision. The eyes may have difficulty focusing on near or far objects because the light rays are not focusing properly on the retina. Hyperopia (farsightedness) occurs when the eye cannot focus properly on nearby objects. As the ciliary muscles age, they weaken and pupil size is decreased, reducing the amount of light coming into the retina. Presbyopia is farsightedness that occurs with age, usually between 40 and 45 years. The lens becomes stiff and yellowish. Such age-related changes make it difficult for older adults to focus and make them more sensitive to glare, which can impair their night-time driving abilities. Myopia (nearsightedness) causes objects at a distance to appear blurred. Amblyopia, or lazy eye, usually occurs in childhood. Here poor vision in one eye is caused by the abnormal dominance of the other eye, which does most of the work. Most people have a dominant eye. To find yours, look at a small, distant object with both eyes open. Extend both arms with both palms facing that object. Slowly bring your hands together so there is a small opening formed in the space between the thumbs and the index finger. Now locate your distant object in that opening with both eyes open. Now alternately close each eye to determine which eye still sees that object. This is your dominant eye. Generally speaking, right-handed people have right-dominant eyes.

hyperopia (*HIGH per OH pee yah*)

presbyopia (*PRESS bee YOH pee yah*)
  presby = *old*
  opia = *refers to vision*

myopia (*my OH pee yah*)
amblyopia (*am bly OH pee yah*)

In addition, the eyes can be used to help diagnose a variety of diseases. For example, a yellow tint to the conjunctiva (jaundice) may indicate a liver disease. A neurological assessment called PERLA, which stands for *pupils equal, reactive to light and accommodation*, can be used to assess brain injury. The **rapid eye movement**, or **REM**, stage of sleep is measured during sleep studies and helps to diagnose sleep disorders.

otitis media (*oh TYE tiss ME dee yah*)
  oto = *ear*
  itis = *inflammation*
  media = *middle*

Otitis media is an infection of the middle ear, usually caused by a bacterium or virus, and is frequently found in infants and young children. It is commonly associated with an **upper respiratory infection**, or **URI**, such as a cold. By examining the structure of the ear, can you see how a sinus infection can spread to an ear infection and vice versa?

labyrinthitis (*LAB er inth EYE tiss*)

Labyrinthitis is an inflammation of the inner ear and usually is caused by high fevers. Labyrinthitis can cause **vertigo**, which is a feeling of dizziness or whirling in space. (If you are not sure what vertigo is, watch the Alfred Hitchcock classic movie of the same name. Not only does it clearly demonstrate vertigo, but it is a great mystery film.) Ménière's disease is a chronic

Ménière's disease (*MANY erz*)

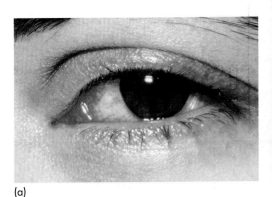
(a)

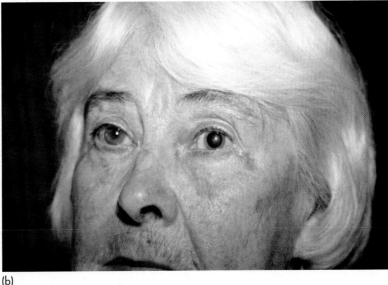
(b)

Figure 11.8 Some common eye disorders. (a) Conjunctivitis ('pink eye') (*Source*: © Medical-on-Line/Alamy) (b) Cataract of right eye (*Source*: © SHOUT/Alamy)

condition that affects the labyrinth and leads to progressive hearing loss and vertigo.

Deafness can be either partial or complete and is caused by a variety of conditions, ranging from inflammation and scarring of the tympanic membrane to auditory nerve and brain damage. Finally, **tinnitus** is a ringing sound in the ears which, according to superstition, means someone is talking about you. Clinically, it can occur as a result of chronic exposure to loud noises, Ménière's disease, some medications, wax build-up or various disturbances to the auditory nerve. Figure 11.8 shows some common eye disorders.

tinnitus (*tin IT uss*)

Summary

- The senses of sight (eyes), sound and equilibrium (ears), taste (tongue) and smell (nose) are called special senses. The body feels other sensations, such as touch, heat, cold and pain, which are called general senses.

- The eye is very similar to a camera, with lens cover (eyelids), opening (pupils), shutter (iris), lens (eye lens) and photoreceptive film (retina).

- Light rays enter the eye and pass through the conjunctival membrane, cornea, aqueous humour, pupil, lens and vitreous humour, and are focused on the retina. The photoreceptors in the retina cause a chemical impulse to be sent to the optic nerve, which carries it to the brain for the interpretation we call vision.

- The ear has three major divisions: the external, middle and inner ear. The ear is the organ for hearing and maintaining our sense of balance.

- Sound waves enter the external canal and vibrate the eardrum, or tympanic membrane. The middle ear then amplifies the sound through the respective tiny bones, or ossicles. The last ossicle (stapes) vibrates and causes a gentle pumping against the oval window membrane. This causes cochlear fluid (perilymph) to move and vibrates tiny, hair-like neurons, which transmit an impulse to the hearing centres in the brain where the sound is interpreted.

- The semicircular canals are responsible for maintaining body balance.
- Our sense of taste, or gustatory sense, has traditionally been thought to consist of sweet, sour, salty and bitter; a fifth taste, umami, has recently been distinguished as its own category. The sixth taste is, perhaps controversially, water. The sense of taste originates on taste buds on the tongue and is closely associated with the sense of smell.
- The sense of smell arises from the olfactory region of the nose.
- The sense of touch allows perceptions of pain, temperature, pressure, traction and the sensation of being 'tickled'.

Case study

Paul is a 40-year-old male who works for a local council. He presents complaining of tinnitus and vertigo. He complains that his hearing is getting progressively worse and he is having dizzy spells and nausea.

1. Describe the patient's complaints in your own words.

2. What possible disease is present?

3. What part of the ear is affected and why?

Review questions

Multiple choice

1. The part of the eye that allows for varying amounts of light onto the retina is the:
    a. lens
    b. humour
    c. iris
    d. optic nerve

2. The photopigment structures responsible for the ability to see colours are:
    a. cones
    b. rods
    c. iris
    d. pupil

3. The incus is found in the:
    a. inner ear
    b. middle ear
    c. external ear
    d. region of South America

4. What is the correct descending order for the media through which sound travels, with the most efficient conductor of sound listed first:
    a. liquid, air, solid
    b. solid, liquid, air
    c. air, liquid, solid
    d. they are all equal

5. Another word for the sense of taste is:
    a. olfactory
    b. vertigo
    c. mastication
    d. gustatory

Fill in the blanks

1. The two functions of the auditory system are _____ and _____.
2. The three ossicles of the ear are the _____, _____ and _____.

Short answers

1. What are the five basic tastes?

   _____

   _____

2. How does the body protect the eyes?

   _____

   _____

3. How is an eye like a camera?

   _____

   _____

Suggested activity

1. The sense of sight is also important to the interpretation of our sense of taste. Blindfold participants and ask them to identify similar products with varying tastes, such as different types of fizzy drink or fruit pastilles, and see if they can correctly identify the flavour.

Answers

Answers to case study

1. He is complaining of ringing in the ears and a feeling of dizziness or whirling in space.
2. Ménière's disease.
3. The inner ear (or labyrinth) is affected because there is disturbance to the auditory nerve.

Answers to multiple choice questions

1. c
2. a
3. b
4. b
5. d

Answers to fill in the blanks

1. Hearing and balance
2. Stapes, malleus and incus

Answers to short answer questions

1. Sweet, Sour, Salty, Bitter, Umami (and possibly water)
2. Iris (limits light entry); eyelids protect from intense light
3. Lenses to focus; pupil is the opening (like a shutter); image captured on the retina (like a camera film).

The endocrine system

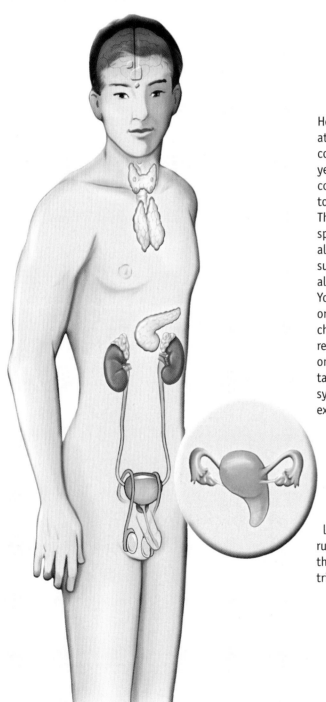

Here we are again, talking about control. We have already looked at one of the control systems: the complex structure of cells and connections known as the nervous system. Now we will look at yet another control system: the endocrine system. These two control systems may seem like separate systems, but they are totally interconnected and always monitor each other's activities. The nervous system collects information and sends orders with a speed that is truly mind boggling. While the endocrine system also collects information and sends orders, it's a slower, more subtle control system. The endocrine system's orders to the body also last much longer than those made by the nervous system. You might think of the endocrine system as sending 'standing orders', which are orders meant to be obeyed indefinitely unless changed by another set of orders. The orders change subtly on a regular basis, but their intention is constant. The nervous system, on the other hand, issues orders that are to be obeyed instantaneously but are used for short-term situations. The endocrine system demands organs to 'carry on' while the nervous system expects them to respond immediately.

Suppose we spent the day at a funfair riding roller coasters. Afterwards, en route to our next destination, we were forced off the road because of a near miss with a lorry. In both cases – when the roller coaster ride is over and the lorry is long gone – our legs still shake, our heart continues to race, and our blood pressure remains elevated, even though we are no longer in danger. We call such lingering effects the 'adrenaline rush'. These lingering effects are not due to continued activity of the nervous system, but rather to endocrine activity deliberately triggered by the autonomic nervous system.

Learning objectives

At the end of this chapter, you will be able to:

- ◆ Discuss the functions of the various endocrine glands
- ◆ Describe the purpose and effects of hormones within the body
- ◆ Discuss the process of homeostatic control
- ◆ Differentiate between hormonal and humoral control
- ◆ Explain common diseases of the endocrine system

Pronunciation guide

adrenal cortex (*ad REE nal KOR teks*)
adrenal medulla (*ad REE nal meh DULL lah*)
endocrine (*EN doh kreen*)
epinephrine (*EP ee NEFF rinn*)
homeostasis (*hoh mee oh STAY siss*)

hypothalamus (*high poh THAL ah mus*)
norepinephrine (*NOR ep ee NEFF rinn*)
oxytocin (*OCK see TOH sin*)
parathyroid gland (*PARA THIGH royd*)

pineal gland (*pin EE all*)
pituitary (*pi TURE it air ee*)
prolactin (*proh LAK tinn*)
testes (*TESS teez*)
thymus (*THIGH muss*)

Organisation of the endocrine system

The endocrine system has many organs that secrete a variety of chemical substances. Let's begin by looking at the basic organisation of the system.

Endocrine organs

endocrine (*EN doh kreen*)
endo = *into*
crine = *to secrete*

The endocrine system is a series of organs and glands (see Figure 12.1) in your body that secrete chemical messengers called hormones directly *into* your bloodstream. In contrast, exocrine glands (such as sweat and salivary glands) produce secretions that exit via a duct onto an epithelial surface.

We have already discussed some of the endocrine glands, such as the hypothalamus, pituitary and pineal glands, because they are part of the nervous system and provide a link *between* the two control systems. We visit some of the other endocrine organs later when we journey through the urinary, reproductive and digestive systems. Many of these glands, like the hypothalamus and pancreas, have multiple functions.

It may seem like an overwhelming task to learn all the endocrine glands and their associated hormones, and therefore we will begin our discussion with a concise overview to lay a foundation upon which to build. These concepts are reinforced and expanded. For now, see Table 12.1, which lists the wide variety of functions of endocrine organs. (The hypothalamus and pituitary will be discussed in greater depth under 'The major endocrine organs' later in this chapter.)

Amazing body facts

Lesser-known endocrine glands

Did you know that many organs, such as the heart, small intestines, stomach and placenta, can also secrete hormones? These and many other organs are not listed as endocrine organs because their primary jobs are focused on other tasks, like pumping blood, storing and digesting food, or nourishing an embryo. But the hormones secreted by other organs are still an important part of the body's control systems. We will learn more about them when we discuss the body's other systems.

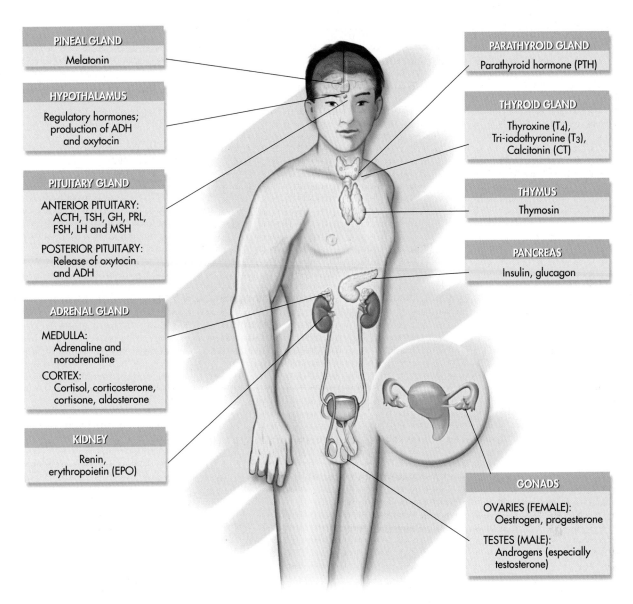

Figure 12.1 The endocrine glands and their hormones.

Hormones

hormone (*HOR moan*)

The chemical messengers released by endocrine glands are called **hormones**. We have already seen one type of chemical messenger, the neurotransmitter. Neurotransmitters are released by neurons at chemical synapses. They diffuse across the synapse, a very tiny space, to a cell on the other side and bind to that cell. They are cleaned up quickly, so their effects are localised and short lived. Hormones, on the other hand, are released into the bloodstream and travel all over your body. Some hormones can affect millions of cells simultaneously. Their effects last for minutes or even hours or days. Many hormones are secreted constantly, and the amount secreted changes as needed. To help clarify the similarities and differences between a neurotransmitter and hormone, see Table 12.2.

Table 12.1 Endocrine organ functions

Endocrine organ	Hormone released	Effect
hypothalamus	(see Table 12.3)	controls pituitary hormone levels
pineal	melatonin	believed to regulate sleep
pituitary	(see Table 12.3)	controls other endocrine organs
thyroid	thyroxine, tri-iodothyronine	controls cellular metabolism
	calcitonin	decreases blood calcium
parathyroid glands	parathyroid hormone	increases blood calcium
pancreas	insulin	lowers blood glucose
	glucagon	raises blood glucose
adrenal glands	adrenaline, noradrenaline	flight-or-fight response
	adrenocorticosteroids	many different effects
ovaries/testes	oestrogen, progesterone	control sexual reproduction
	testosterone	secondary sexual characteristics

Table 12.2 Comparison of neurotransmitters and hormones

Neurotransmitters	Hormones
chemical messengers	chemical messengers
bind to receiving cell	bind to receiving cell
control cell excitation	control cell activities
released by neurons	released by neurons, glands or organs
released at chemical synapse	released into bloodstream
intended target very close	travel to distant target
effects happen quickly (less than a second)	effects take time (seconds or minutes)
effects wear off quickly (few seconds)	effects long lasting (minutes or hours)
affects a few cells at a time	can affect many cells

How hormones work

Like neurotransmitters, hormones usually work by binding to receptors on the surface of target cells. When hormones bind to the outside of the cell, they can have several different effects, either changing cellular permeability or sending the target cell a message that changes enzyme activity inside the cell. Thus, the target cell changes what it has been doing, usually by making a new protein or turning off a protein it has been making. This is known as the second messenger system – as the hormone binds to the surface receptor it

Test your knowledge 12.1

Choose the best answer:

1. What chemical, when secreted into the bloodstream, controls the metabolic processes of target cells?

    a. neurotransmitter

    b. secretion

    c. hormone

    d. ligand

2. Steroid hormones are very powerful because they:

    a. are hormones

    b. are medicine

    c. interact directly with DNA

    d. are secreted outside the body

3. Which of the following is true of hormones?

    a. They last a short time.

    b. They are fast acting.

    c. They affect distant targets.

    d. They leave the body.

activates a cascade of events inside the cell. But hormones may bind not only to sites on the outside of the cell, like neurotransmitters, but also to sites inside the cell.

One special class of hormones, **steroids**, is particularly powerful because steroids can bind to sites inside cells. Steroids are lipid molecules that can pass easily through the target cell membrane. These hormones, then, can interact directly with the cell's DNA, the genetic material, to change cell activity. This is known as direct gene activation. These hormones are carefully regulated by the body because of their ability, even in very small amounts, to control target cells. See the section 'The adrenal glands' for further discussion of steroid hormones and the dangers of taking steroids.

Control of endocrine activity

Many endocrine organs are active all the time. The amount of hormone they secrete changes as the situation demands, but unlike neurons, the cells in the endocrine organs often secrete hormones continuously. How is the activity controlled? How do the organs know how much hormone to secrete?

Homeostasis and negative feedback

In order to understand how the endocrine system is controlled, we first have to revisit the concept of **homeostasis** discussed in Chapter 1 (see Figure 12.2). Recall that many of the chemical and physical characteristics of the body have a standard level, or **set point**, that is the ideal level for that particular value. Blood pressure, blood oxygen, heart rate and blood sugar, for example, all have 'normal' ranges. Your control systems, nervous and endocrine, work to keep the levels at or near ideal. There is a way for your body to measure the variable, a place where the 'ideal' level is stored, and a way for the body to correct levels that are not near ideal. For example, neurons measure your

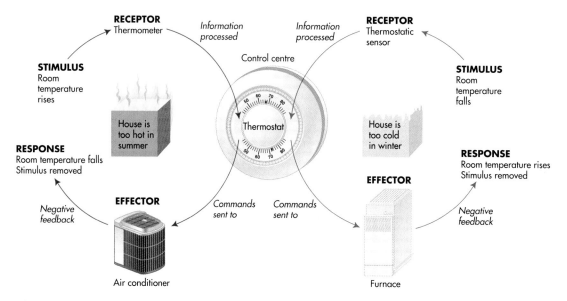

Figure 12.2 Homeostasis is analogous to regulation of temperature via a thermostat.

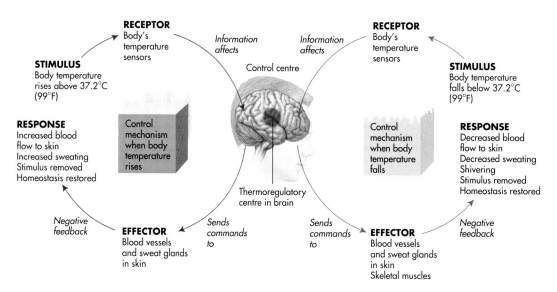

Figure 12.3 Homeostasis and negative feedback as related to control of body temperature.

body temperature. The hypothalamus stores the set point. If your temperature falls below the set point temperature, the hypothalamus causes shivering to produce additional heat. If body temperature rises above the set point, the hypothalamus causes sweating.

If any of the body's dozens of homeostatic values become seriously disrupted, the control systems work to bring them back to the set point. This process is called **negative feedback** (see Figure 12.3). Most of you are familiar with negative feedback in real life. When you fill your car with petrol, there is a sensor in the nozzle that turns *off* the flow of petrol when the tank is full. That is negative feedback. The petrol is flowing, the tank is filling, and when the goal is reached, the petrol stops flowing. In the body, negative feedback counteracts a change. As blood pressure rises, for example, your body works to

Amazing body facts

Fever

When you get an infection, sometimes your body temperature rises above normal (set point). The medical term for a patient with a fever is *febrile*. A fever is the deliberate raising of your body temperature set point by your hypothalamus. In an attempt to inhibit the growth of micro-organisms that have invaded your body, the hypothalamus resets your body temperature set point to a higher level. Initially, the new set point is higher than your normal temperature. Your body thinks you are cold and takes steps to raise your body temperature to the new set point. Now you have a fever. Your head aches as blood vessels dilate, your heart rate and blood pressure rise. You feel miserable, and you have your hypothalamus to blame. (Actually it's the fault of the invading bacteria or virus the hypothalamus is trying to defeat.)

bring it down to 'normal', the set point. If blood pressure falls, your body works to raise it back to the set point. Hormones work in the same way. If hormone levels rise, negative feedback turns off the endocrine organ that is secreting the hormone.

The body is also capable of **positive feedback**, which increases the magnitude of a change. Certain processes in the body are positive feedback mechanisms; these include blood clotting, breast feeding and giving birth.

Sources of control of hormone levels

Hormone levels can be controlled by the nervous system (neural control), by other hormones (hormonal control), or by body fluids such as the blood (humoral control).

Clinical application

Childbirth and positive feedback

A good example of necessary positive feedback is the continued contraction of the uterus during childbirth. When a baby is ready to be born, a signal, not well understood at this time, tells the hypothalamus to release the hormone **oxytocin** from the posterior pituitary. Oxytocin increases the intensity of uterine contractions. As the uterus contracts, the pressure inside the uterus caused by the baby moving down the birth canal increases the signal to the hypothalamus: more oxytocin is released and the uterus contracts harder. As pressure gets higher inside the uterus, the hypothalamus is signalled to release more oxytocin, and the uterus contracts even harder. This cycle of ever-increasing uterine contractions due to ever-increasing release of oxytocin from the hypothalamus continues until the baby is born.

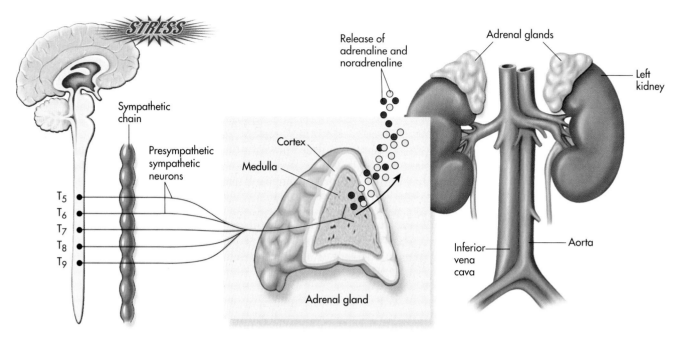

Figure 12.4 Sympathetic control of adrenal gland.

Neural control

There are three basic ways in which endocrine organs function to maintain hormone levels and body function. Some hormones are directly controlled by the nervous system. For example, the adrenal glands receive signals from the sympathetic nervous system. When the sympathetic nervous system is active (remember the near-miss), it sends signals to the adrenal glands to release adrenaline (epinephrine) and noradrenaline (norepinephrine) as hormones, prolonging the effects of sympathetic activity (see Figure 12.4).

Hormonal control

Other hormones are part of a hierarchy of hormonal control in which one gland is controlled by the release of hormones from another gland higher in the chain, which is controlled by another gland's release of hormones yet higher in the chain. Orders are sent from one organ to another. This is very similar to a relay race at an athletics event where the baton is handed from one runner to the next, hopefully in a smooth manner to send the baton to the finish line. Negative feedback controls the flow of orders via hormones from one part of the 'chain of command' to the other. For example, the hypothalamus has control over the pituitary, which has control over the adrenal gland, which secretes the hormone cortisol. Increased cortisol secretion is one way that the body copes with stress, and as cortisol levels rise in the blood, further release of hormones at the hypothalamus is depressed (see Figure 12.5).

Humoral control

Still other endocrine organs can directly monitor the body's internal environment by monitoring the body fluids, such as the blood, and respond

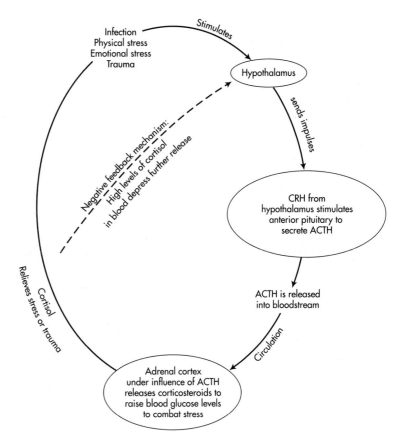

Figure 12.5 Hormonal control of adrenal gland. ACTH = adrenocorticotropic hormone; CRH = corticotropin-releasing hormone.

accordingly. *Humoral* pertains to body fluids or substances, and therefore control through body fluids is called *humoral control*. For example, the pancreas secretes insulin in response to rising blood glucose, as shown in Figure 12.6.

Test your knowledge 12.2

Choose the best answer:

1. Which of the following is *not* a way that hormone levels are regulated?

    a. negative feedback

    b. chain of command

    c. positive feedback

    d. direct control by nervous system

2. The 'ideal' value for a body characteristic is called the:

    a. set point

    b. average

    c. goal

    d. feedback point

3. _____ feedback enhances a change in body chemistry:

    a. negative

    b. positive

    c. regular

    d. cyclic

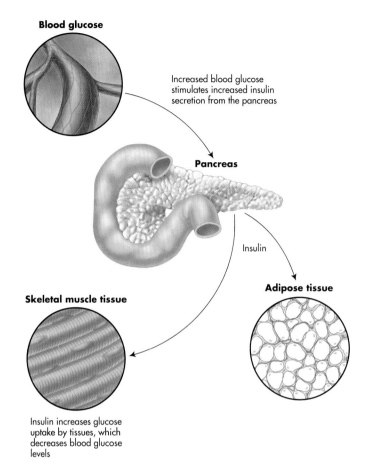

Blood glucose

Increased blood glucose stimulates increased insulin secretion from the pancreas

Pancreas

Insulin

Adipose tissue

Skeletal muscle tissue

Insulin increases glucose uptake by tissues, which decreases blood glucose levels

Figure 12.6 Humoral control of blood sugar levels.

The major endocrine organs

The endocrine system has several organs, and each has specific tasks within the body. The messages to carry out these tasks are related to the hormones they release.

The hypothalamus

hypothalamus (*high poh THAL ah mus*)

We considered the hypothalamus when we looked at the central nervous system. Located in the diencephalon, this gland is an important link between the two control systems, nervous and endocrine (please see Figure 12.7). The hypothalamus controls much of the body's physiology, including hunger, thirst, fluid balance and body temperature, to name only a few of its functions. The hypothalamus is also, in part, the 'commander-in-chief' of the endocrine system, because it controls the pituitary gland and therefore most of the other glands in the endocrine system. Table 12.3 lists the hypothalamic and pituitary hormones.

The pituitary (or hypophysis)

pituitary (*pi TURE it air ee*)

The pituitary, also a part of the diencephalon, is commonly known as the 'master gland', indicating its important role in control of other endocrine

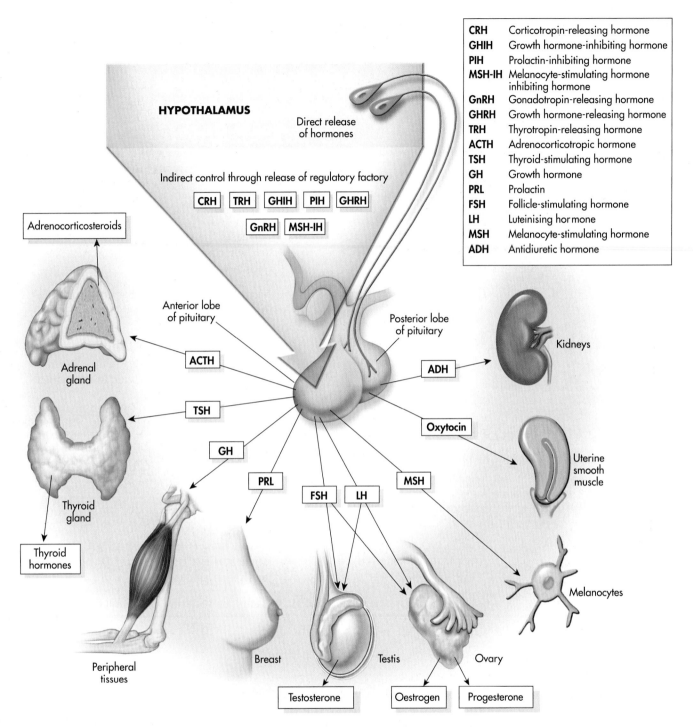

CRH	Corticotropin-releasing hormone
GHIH	Growth hormone-inhibiting hormone
PIH	Prolactin-inhibiting hormone
MSH-IH	Melanocyte-stimulating hormone inhibiting hormone
GnRH	Gonadotropin-releasing hormone
GHRH	Growth hormone-releasing hormone
TRH	Thyrotropin-releasing hormone
ACTH	Adrenocorticotropic hormone
TSH	Thyroid-stimulating hormone
GH	Growth hormone
PRL	Prolactin
FSH	Follicle-stimulating hormone
LH	Luteinising hormone
MSH	Melanocyte-stimulating hormone
ADH	Antidiuretic hormone

Figure 12.7 The hypothalamus, anterior and posterior pituitary glands, and their targets and associated hormones.

glands. However, that name is misleading because the pituitary gland rarely acts on its own. The pituitary acts only under orders from the hypothalamus. If the hypothalamus is the 'commander-in-chief', the pituitary is a high-ranking soldier who carries out the orders.

The posterior pituitary (or neurohypophysis)

The pituitary is split into two segments: the *posterior* pituitary and the *anterior* **pituitary**. The posterior pituitary is an extension of the hypothalamus.

Table 12.3 Selected hypothalamic and pituitary hormones

Hormone	Function
Hypothalamus	
growth hormone-releasing hormone (GHRH)	increases the release of growth hormone from the pituitary gland
growth hormone-inhibiting hormone (GHIH)	decreases the release of growth hormone from the pituitary gland
corticotropin-releasing hormone (CRH)	increases the release of adrenocorticotropic hormone from the pituitary gland
gonadotropin-releasing hormone (GRH)	increases the release of luteinising hormone and follicle-stimulating hormone from the pituitary gland
thyrotropin-releasing hormone (TRH)	increases the release of thyroid-stimulating hormone from the pituitary gland
Posterior pituitary	
antidiuretic hormone (ADH)	dilutes blood and increases fluid volume by increasing water reabsorption in the kidney
oxytocin	increases uterine contractions
Anterior pituitary	
growth hormone (GH)	increases tissue growth
thyroid-stimulating hormone (TSH)	increases secretion of thyroid hormones
adrenocorticotropic hormone (ACTH)	increases steroid secretion from adrenal gland
prolactin	increases milk production
luteinising hormone (LH)	stimulates ovaries and testes for ovulation and sperm production
follicle-stimulating hormone (FSH)	oestrogen secretion and sperm production

antidiuretic hormone (*AN tee dye yoo RET ik*)

oxytocin (*OCK see TOH sin*)

Hypothalamic neurons, specialised to secrete hormones instead of neurotransmitters, extend their axons through a stalk to the posterior pituitary. Using the posterior pituitary as a sort of launch pad, these neurons secrete two hormones: antidiuretic hormone (**ADH**) (also called vasopressin) and oxytocin. Both of these hormones are secreted from the posterior pituitary, but they are made by the hypothalamus.

ADH does exactly what its name suggests. A diuretic is a chemical that increases urine production, so an antidiuretic decreases urine production. The effect of ADH, then, is to decrease fluid lost via the kidneys, increasing body fluid volume. ADH is secreted when the hypothalamus senses decreased blood volume or increased blood osmolarity (more solids suspended in blood). ADH circulates through the bloodstream and targets the kidneys specifically, causing them to absorb more water. It is very important in long-term control

Learning hint

Hormone names

Most hormones are named according to where they are secreted or what they do. If you learn the meanings of their names, you can usually tell something about the hormone. For example, growth hormone stimulates cells to grow. **Prolactin** increases milk production. Even a complicated hormone name like adrenocorticotropic hormone can be picked apart fairly easily. *Adreno* refers to the adrenal gland, *cortico* refers to the cortex, and *tropic* means change. Therefore, adrenocorticotropic hormone is a hormone that changes the activity (in this case increases) of the adrenal cortex. Also keep in mind that most hormones are known by their abbreviations, for obvious reasons. Adrenocorticotropic hormone, for example, is abbreviated ACTH, which is much easier to say and write.

prolactin (*proh LAK tinn*)
pro = *for*
lactin = *milk*

Amazing body facts

ADH, alcohol and coffee

Drinking too much alcohol can lead to several unpleasant consequences, not the least of which is the development of a hangover. The symptoms of a hangover are due to many side effects from alcohol consumption, but the most important may be that alcohol turns off ADH. The more alcohol you drink, the less ADH you secrete and the more dehydrated you become. That makes you thirsty, so if you drink some more alcohol you secrete even less ADH and become even more dehydrated. (This is another example of a vicious cycle.) The more alcohol you drink, the more you urinate and the more dehydrated you become. This is part of the reason why consuming too much alcohol can make you miserable the following morning. The caffeine in coffee (and tea!) also has the same effect in inhibiting ADH. That's why you make frequent trips to the toilet if you pull an 'all-nighter' studying for your A&P using coffee or caffeine-containing drinks to stay awake! Too much caffeine will even make you feel pretty miserable.

of blood pressure, especially during dehydration. (For more information on ADH, see Chapter 17 'The urinary system'.)

The second hypothalamic hormone released from the posterior pituitary is oxytocin. Oxytocin is important in maintaining uterine contractions during labour in women and is involved in milk ejection in nursing mothers. Oxytocin's function in males is unknown.

The anterior pituitary (or adenohypophysis)

The anterior pituitary is also controlled by the hypothalamus but is an endocrine gland in its own right. The anterior pituitary makes and secretes a number of hormones, under hormonal control of the hypothalamus, that control other endocrine glands. (Growth hormone and prolactin are exceptions to this rule.) Refer back to Table 12.3 and Figure 12.7 for a list of hypothalamic and pituitary hormones. We discussed this relationship previously when we talked about hormonal control. The hypothalamus secretes a hormone that controls hormone secretion by the anterior pituitary, which usually controls the secretion of hormones by another endocrine gland. The hormone levels are controlled by negative feedback to both the pituitary and the hypothalamus.

Clinical application

Stature disorders

Stature disorders are those disorders that result in *well*-below-average height (called dwarfism) or *well*-above-normal height (called giantism or gigantism). Some of these disorders are caused by abnormalities in skeletal development or nutritional deficiencies. However, growth hormone (GH) problems are often implicated. If GH secretion is insufficient during childhood, children do not grow to 'standard' height. This type of dwarfism results in stunted adult height. However, if GH deficiency is diagnosed before closure of the growth zones of the long bones, it can be treated with GH injections. Children treated with GH injection attain full height. On the other end of the spectrum are those who secrete too much GH. If the over secretion happens during childhood, people get extremely tall. (Robert Pershing, the tallest man according to the *Guinness Book of World Records*, was 8 feet 11.1 inches tall – 2.72 metres.) Gigantism causes many health problems. The body gets so big that it cannot support itself. Surgery and medication are the only treatments. If GH over secretion begins after a person has stopped growing (bone closure), he or she does not get any taller, but the tissues of the hands, feet, face and many internal organs continue to grow out of control, causing pain and organ dysfunction. Most over secretion of GH is caused by non-cancerous pituitary tumours.

Test your knowledge 12.3

Choose the best answer:

1. Oxytocin:
   a. is secreted by the anterior pituitary
   b. decreases uterine contractions
   c. is released by the posterior pituitary
   d. is a way to get more oxygen to your toes

2. The _____ is controlled by hormones from the hypothalamus, while the _____ actually secretes hypothalamic hormones.
   a. posterior pituitary, posterior pituitary
   b. anterior pituitary, anterior pituitary
   c. anterior pituitary, posterior pituitary
   d. posterior pituitary, anterior pituitary

3. This gland, under orders from the hypothalamus, releases hormones that control other endocrine glands.
   a. adrenal gland
   b. anterior pituitary
   c. thyroid gland
   d. pancreas

4. Which gland does ACTH control?
   a. adrenal gland
   b. anterior pituitary
   c. thyroid gland
   d. pancreas

The thyroid gland

thyroid (*THIGH royd*)

The **thyroid** gland, located in the anterior portion of your neck, is a butterfly-shaped organ. It is responsible for secreting the hormones thyroxine (T_4) and tri-iodothyronine (T_3), under orders from the pituitary gland (see Figure 12.8).

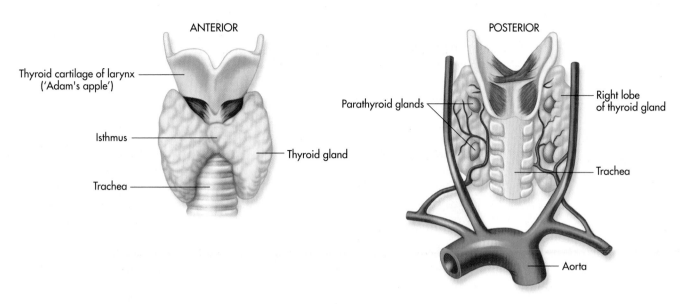

Figure 12.8 The thyroid and parathyroid glands.

Thyroxine and tri-iodothyronine contain iodine and control cell metabolism and growth. The thyroid gland also secretes a third hormone, calcitonin, which decreases blood calcium by stimulating bone-building cells. Thyroxine and tri-iodothyronine are generally referred to as 'thyroid hormones' and are of great clinical importance. Overproduction (hyperthyroidism) or underproduction (hypothyroidism) can cause a variety of clinical symptoms, because the level of these hormones is essential in controlling growth and metabolism of body tissues, particularly in the nervous system. The importance of these hormones is so great that table salt contains added iodine to ensure that people get enough iodine in their diets to make thyroid hormones.

Clinical application

Hyperthyroidism

Jo was a healthy 25-year-old teacher starting her first job when she began to have strange symptoms. Her heart sometimes beat so quickly that it frightened her, she would sweat copiously, and she was always hungry. Initially, she ignored these changes, attributing them to the stress of moving away from home for the first time, but then more alarming symptoms appeared. Typically rather laid back, Jo became irritable and restless. She began to have trouble sleeping and couldn't concentrate. Soon she could barely focus long enough to read the newspaper or watch a 30-minute sitcom. Her thoughts became so scattered she thought she might be losing her mind. Frightened, she made an appointment to see her GP. Testing confirmed that she did have a problem, but she was not losing her mind. Jo had Graves' disease, a disorder that causes the thyroid gland to secrete too much thyroid hormone.

parathyroid (*PARA THIGH royd*)

The thyroid gland has two small pairs of glands embedded in its posterior surface. These glands are called the parathyroid **glands**, and they produce **parathyroid hormone (PTH)** which regulates the levels of calcium in the bloodstream. If calcium levels get too low, the parathyroid glands are stimulated to release PTH, which stimulates bone-dissolving cells and thereby releases needed calcium in the bloodstream. Again, see Figure 12.8.

The thymus gland

thymus (*THIGH muss*)

The thymus **gland** is located in the upper thorax and plays an important function in the immune system. It produces a hormone called thymosin, which helps with the maturation of white blood cells during childhood to fight infections. This gland is further discussed in Chapter 14, 'The lymphatic and immune systems'.

The pineal gland

pineal gland (*pin EE all*)
melatonin (*mel ah TOH nin*)

The tiny pineal gland is found in the brain, and its full function still remains unknown. However, it has been shown to produce the hormone melatonin, which rises and falls during the waking and sleeping hours. It is believed this hormone is what triggers our desire to sleep by peaking at night and causing drowsiness.

The pancreas

pancreas (*PAN kree yass*)

The pancreas is largely responsible for maintaining blood sugar (glucose) levels at or near a set point. The normal clinical range for blood glucose levels is 4–8 mmol/l (70–105 mg/dl (milligrams per decilitre)). The pancreas can measure blood glucose, and if the blood glucose is high or low, the pancreas releases a hormone to correct the level. To understand the importance of the pancreas, let's go back to the chapter on cells. Why does it matter how much glucose is in your blood? Why devote an organ, and a pretty big one at that, to controlling blood sugar? There are two reasons why blood glucose is important. Too much glucose floating around in your blood causes many problems with the fluid balance of your cells. Recall that if the concentration of the fluid outside a cell is high in solutes, the cell will lose water to the surroundings. If the solutes are low outside the cell, the cell will fill with water and eventually can even burst. Obviously, that's a serious problem. It does not matter if the solutes are salts or glucose, the result is the same. Blood glucose must therefore be maintained at a certain level for cells to neither gain nor lose water. Why else is glucose important? Glucose is vital for cellular respiration. Cellular respiration is needed to get energy, by making adenosine triphosphate (ATP) so cells need to have enough glucose that they can make sufficient ATP to carry out their daily activities.

The pancreas makes two hormones that control blood glucose: *insulin*, which most of you have heard of before, and *glucagon*. Insulin, the hormone that is missing or ineffective in diabetes mellitus, removes glucose from the blood by directing the liver to store excess glucose and by helping glucose to get inside the cells so it can be used to make ATP. (Remember that glucose, a carbohydrate, is large and water-soluble, so it cannot get into cells by itself.) Insulin is released from the beta cells of the pancreas. When would insulin be secreted by the pancreas: when blood glucose is high or when blood glucose is low? Because insulin removes glucose from the blood, it lowers blood sugar, so it is released when blood sugar is high (hyperglycaemia), for example, straight after a meal.

Glucagon does the opposite of insulin. Glucagon puts glucose into the bloodstream mainly by directing the liver to release glucose which has been stored in the liver in the form of glycogen. Glucagon is released typically several hours after a meal to prevent blood glucose from dropping too low (hypoglycaemia). Glucagon is released from the alpha cells of the pancreas. These two hormones control blood glucose very tightly in healthy humans (see Figure 12.9). Other hormones, like the adrenal hormone cortisol, also aid in the control of blood sugar.

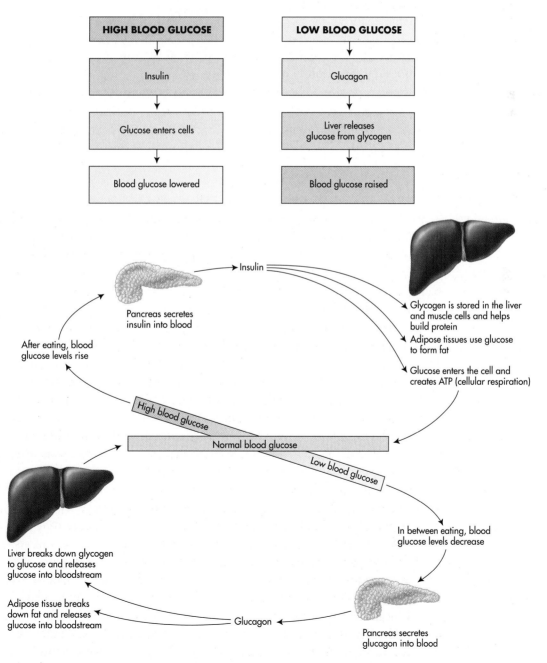

Figure 12.9 Control of blood glucose by pancreatic hormones.

Clinical application

Diabetes mellitus

Diabetes mellitus is a condition characterised by abnormally high blood glucose (hyperglycaemia). Insulin-dependent diabetes mellitus (referred to as Type 1 diabetes mellitus) is caused by the destruction of the insulin-producing cells of the pancreas. Patients with Type 1 diabetes mellitus do not produce enough insulin. They are always dependent on daily insulin injections. Non insulin-dependent diabetes mellitus (referred to as Type 2 diabetes mellitus) is caused by insensitivity of the body's tissues to insulin. Patients with Type 2 diabetes mellitus can often be treated with a carefully controlled diet and weight-loss regimen, but may require insulin and other medication to help control blood glucose levels. In both types of diabetes, abnormally high blood glucose must be resolved. If blood glucose remains high, the kidneys work overtime to secrete the excess sugar. Increased urination and dehydration are the most obvious symptoms. But the stress of trying to get rid of the excess blood sugar eventually causes kidney damage. In addition, if insulin is not effective, glucose cannot get into cells. Cells must have glucose in order to make ATP. If cells can't get glucose and can't make ATP, they will look for other sources of energy. Untreated diabetics often lose weight as their body searches for other energy sources. Often, their blood becomes increasingly acidic as waste products from abnormal cell metabolism accumulate in the bloodstream. The changes in blood chemistry lead to tissue and organ damage. Left untreated, diabetes mellitus may lead to coma and death.

The adrenal glands

adrenal (ad REE nal)

adrenal cortex (ad REE nal KOR teks)

adrenal medulla (ad REE nal meh DULL lah)

The adrenal glands are a pair of small glands that sit on top of your kidneys. The adrenal glands are split into two regions: the adrenal cortex, an outer layer, and the adrenal medulla, the middle of the gland. (Note that it is best to be specific when talking about the adrenal cortex, since your cerebrum has a cortex, too.)

The adrenal medulla

epinephrine (EP ee NEFF rinn)

norepinephrine (NOR ep ee NEFF rinn)

The adrenal medulla releases two hormones: **adrenaline** (also known as epinephrine) and **noradrenaline** (norepinephrine). These hormones increase the duration of the effects of your sympathetic nervous system. (Remember your friend's snarling dog.) Cells of the adrenal medulla receive the neurotransmitter noradrenaline (it is both a hormone and a neurotransmitter, depending on where it is released) from the sympathetic nervous system. The neurotransmitter triggers the release of noradrenaline and adrenaline into the bloodstream, increasing heart rate, blood pressure and respiration rate, and giving you sweaty palms and dry mouth. (Not a great combination.) Again, the effects of the hormones last much longer than the effects of the neurotransmitter.

The adrenal cortex

The adrenal cortex, on the other hand, makes steroid hormones known collectively as *adrenocorticosteroids* (steroids in the adrenal cortex). The adrenal cortex releases steroid hormones under the direction of the anterior pituitary. Many of these steroid hormones are so important that a decrease in their production could be fatal relatively quickly. Some of these hormones regulate electrolyte (salt) and fluid balance, others regulate blood sugar, and still others control cell metabolism, growth and immune system function.

Clinical application

Prednisone

Prednisone (hydrocortisone) is clinically important in the treatment of inflammation, organ transplant rejection and immune disorders, but prescription steroids are a double-edged sword. These medications are so powerful that they can cause dangerous side effects, such as bone density loss, weight gain, hair growth, fat deposits and delayed wound healing. Long-term use of steroids can also cause the development of diabetes mellitus. These drugs, even when taken for a short time, cannot be discontinued suddenly. Patients must be weaned off their medication slowly. Why? When a patient takes a steroid medication, the adrenal gland decreases steroid production in response. (Remember negative feedback: as hormone levels rise, hormone secretion decreases.) Therefore, if the medication is removed suddenly, patients are left with a severe hormone deficiency, which could be fatal. Their adrenal gland must be given some time to 'gear up' to secrete hormones at the appropriate level. It is no surprise, then, that taking steroid medications for the purpose of increasing athletic performance is prohibited by amateur and professional sports organisations.

The gonads

The chief function of the gonads – the testes in males and the ovaries in females – is to produce and store gametes; eggs in females, and sperm in males. However, the gonads also produce a number of sex hormones that control reproduction in both males and females, including testosterone in males and oestrogen and progesterone in females. For more details on the hormones produced by the gonads, see Chapter 18.

Common disorders of the endocrine system

Anabolic steroids are a class of steroid molecules that cause large increases in muscle mass when compared to working out without steroids. Some athletes use anabolic steroids to enhance performance or to get big muscles much faster than they would without steroids. Because steroid hormone levels are so tightly controlled by the body, anabolic steroids have a number of side effects. (Think back to the discussion of prednisone at therapeutic levels; abuse levels are much higher.) Men abusing steroids may experience changes in sperm production, enlarged breasts and shrinking of their testes. Women may experience deepening of the voice, decreased breast size and excessive body hair growth. Steroid abuse may lead to cardiovascular diseases and increased cholesterol levels. Many steroids suppress immune function, and because steroid use is illegal, many abusers expose themselves to hepatitis B and HIV when sharing needles. Steroid abuse has also been linked to increased aggressive behaviour. All major professional and amateur athletic organisations ban the use of steroids.

Hashimoto's disease is a form of hypothyroidism caused by an autoimmune attack on your thyroid gland. For unknown reasons, the immune system begins to attack the cells in the thyroid, causing inflammation and damage to the gland. This damage eventually leads to decreased production of thyroid hormones – hypothyroidism. In addition, the thyroid may swell, causing pain and difficulty swallowing. Hashimoto's disease, like many autoimmune disorders, is most common in women between 30 and 50 years old. It can be treated by taking thyroid hormones daily.

Graves' disease is also an immune disorder that affects the thyroid, but in this case the immune system stimulates the thyroid, resulting in hyperthyroidism and bulging eyes. Treatment for Graves' disease involves decreasing the activity of the thyroid with medication. If medication does not work, the thyroid may be removed, and the patient must take thyroid hormones.

A **phaeochromocytoma** is a tumour of the adrenal gland that causes the gland to secrete excess adrenaline. The symptoms of these tumours are what you might expect: an adrenaline rush. Patients experience severe headaches, excessive sweating, racing heart, anxiety, abdominal pain, heat intolerance and weight loss. Phaeochromocytomas are not generally cancerous, but they must be removed or the effects of excess adrenaline will be fatal.

Addison's disease is caused by insufficient production of the adrenocorticosteroid cortisol. The deficiency causes weight loss, muscle weakness, fatigue, low blood pressure and excessive skin pigmentation. Aldosterone may also be deficient. Many cases of Addison's disease are autoimmune. Addison's disease is treated with hormone replacement.

Cushing's syndrome is caused by over secretion of cortisol. Symptoms include upper body obesity, round face, easy bruising, weakened bones, fatigue, high blood pressure and high blood sugar. Women may have excess facial hair and irregular periods; men may have decreased fertility and decreased sex drive. Cushing's syndrome may be a side effect of medical use of steroids, like prednisone, or may be due to pituitary tumours, lung tumours, adrenal tumours, or one of several genetic disorders. Treatment depends on the underlying cause of the disorder. Typically, the cause of the excess production must be removed, and patients generally must take hormone replacement. Figure 12.10 shows examples of some common endocrine disorders.

(a)

(b)

Figure 12.10 Examples of endocrine disorders. (a) A six-year-old child with congenital hypothyroidism. (*Source*: © John Paul Kay/Peter Arnold Inc./Science Photo Library) (b) A patient with gigantism. (*Source*: © Bettina Cirrone/Science Photo Library)

Summary

- The endocrine system works together with the nervous system to regulate the activities of all the body systems. The endocrine system is linked to the nervous system but works very differently. The endocrine system secretes hormones that act very slowly on distant targets. Their effects are long lasting.

- Some organs, like the pancreas and the thyroid gland, function mainly to release hormones. However, many other organs, like the heart and stomach, can also release hormones. They aren't considered endocrine glands because hormone release isn't their primary role.

- Most hormones act on cells by binding to external receptors, causing changes in enzyme activity inside the target cell.

- Hormone levels are controlled largely by negative feedback. When hormone levels rise, signals are transmitted to the endocrine organ releasing the hormone, telling the organ to decrease the amount of hormone released. Hormone levels will then decrease. The optimal level of the hormone is called the set point. If the signal brings a hormone back to set point, the action is called negative feedback. If the signal causes the hormone to get further away from set point, the action is positive feedback.

- Hormone levels can be regulated by three mechanisms: changes in the body's internal environment, control by hormones released by another endocrine gland, and direct control by the nervous system.

- The hypothalamus, a part of the diencephalon, controls much of the endocrine system by controlling the pituitary gland. The pituitary gland has two parts: the posterior pituitary, which is part of the hypothalamus and actually secretes hypothalamic hormones (ADH and oxytocin), and the anterior pituitary, which secretes several different hormones under the influence of hormones from the hypothalamus. The hormones secreted by the anterior pituitary typically control other endocrine glands (growth hormone is an exception).

- Several other endocrine glands have important control functions. The thyroid gland secretes the iodine-containing hormones tri-iodothyronine (T_3) and thyroxine (T_4), which control growth and cellular metabolism.

- The pancreas secretes two main hormones: insulin, which lowers blood glucose, and glucagon, which raises blood glucose. Diabetes mellitus is caused by a decrease in insulin secretion or decreased sensitivity to insulin. Very high blood glucose is the result.

- The adrenal glands are split into two parts. The adrenal medulla is an extension of the sympathetic nervous system, releasing adrenaline and noradrenaline as hormones during fight-or-flight response. The adrenal cortex releases different adrenocorticosteroid hormones, which control reproduction, inflammation, tissue growth and immunity.

Case study

Bill is a 20-year-old man who presents to Accident and Emergency with the following symptoms:

- recent weight loss
- generalised weakness
- excessive thirst and urine output

A blood glucose measurement shows a level of 17 mmol/l and the urinalysis shows an acidic urine.

1. What type of diabetes does he have?

2. What organs will be affected if he is not properly diagnosed and treated?

3. What treatment and lifestyle suggestions would you give?

Review questions

Multiple choice

1. ADH stands for:
   a. antidiuretic hormone
   b. androdoginin hormone
   c. American department of health
   d. all-diglyceride hormone

2. The 'master gland' is the:
   a. adrenals
   b. pituitary
   c. Graves'
   d. pancreas

3. The thymus gland's main function is for:
   a. reproduction
   b. growth
   c. immunity
   d. RBC levels

4. The pineal gland is located in/on the:
   a. kidneys
   b. brain
   c. thorax
   d. abdomen

5. Glucagon performs the opposite action of:
   a. glucose
   b. insulin
   c. ATP
   d. WBCs

Short answers

1. Compare and contrast neurotransmitters and hormones.

2. List the sources of control of hormone levels.

3. Explain negative feedback and its role in controlling hormone levels.

4. Discuss why the use of anabolic steroids is outlawed for performance enhancement.

5. What is the difference between neural control and humoral control of endocrine glands?

Matching

1. Match the hormone or neurotransmitter on the left with the description on the right

   _____ ADH

   _____ Insulin

   _____ Glucagon

   _____ Oxytocin

   _____ Adrenaline

   _____ Thyroxine

   _____ Prolactin

   _____ ACTH (adrenocorticotropic hormone)

   _____ TSH

   _____ growth hormone

   A. decreases blood sugar

   B. increases thyroid hormone secretion

   C. regulates cell metabolism

   D. increases steroid release

   E. increases uterine contractions

   F. decreases urine output

   G. prolongs sympathetic response

   H. stimulates tissue growth

   I. increases blood sugar

   J. increases milk production in females

Suggested activities

1. Review the hormones secreted by each endocrine organ and their functions. Make up some cards with the endocrine gland on the front and the hormones it produces on the back. Pick a partner and quiz each other. Once you and your partner can match the gland to the hormones, make up another set of cards with the hormone on the front and its activity or function on the back, and again quiz each other.

2. Review the endocrine diseases listed in this chapter. Make up some cards with the signs and symptoms on the front and the actual disease printed on the back. In a small group or with a partner, try to stump each other in determining the correct disease diagnosis.

Answers

Answers to case study

1. Type 1 diabetes mellitus
2. Cardiovascular, neurological, renal, visual, as well as most other systems
3. He needs to have his diagnosis confirmed (initially by fasting glucose measurement); he needs to eat a balanced diet and ensure that insulin is administered pre-meals. Regular blood glucose monitoring (as well as regular monitoring of glycated haemoglobin-HbA1C).

Answers to multiple choice questions

1. a
2. b
3. c
4. b
5. b

Answers to short answer questions

1. Neurotransmitters are released by neurons at chemical synapses, diffusing across the synapse. Their effects are localised and short-lived. Hormones are released into the bloodstream, can affect multiple cells simultaneously and can last for minutes, hours or days.
2. stimulus, receptor, control centre, effector, response.

3. If the homeostatic mechanisms become disrupted, the control systems bring them back to the set point; if hormone levels rise, negative feedback turns off the endocrine organ that is secreting the hormone.
4. Anabolic steroids mimic the effects of testosterone and dihydrotestosterone, increasing protein synthesis, resulting in a build-up of cellular tissue (anabolism) in muscles. Long-term exposure to anabolic steroids can result in increased low density lipoprotein (the harmful cholesterol), hypertension, liver damage and can change the structure of the left ventricle of the heart.
5. Neural control is hormones controlled by the nervous system whereas humoral control relates to control by body fluids.

Answers to matching

ADH – F
Insulin – A
Glucagon – I
Oxytocin – E
Adrenaline – G
Thyroxine– C
Prolactin – J
ACTH – D
TSH – B
Growth Hormone – H

The cardiovascular system

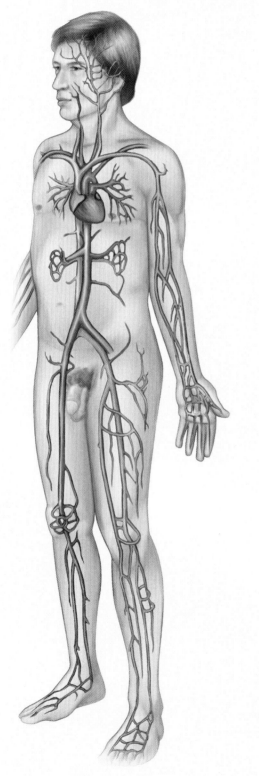

Historically, city waterways, such as rivers and canals, have been used as a way of transporting food and supplies to people. These very same waterways also had industrial by-products dumped into them. This is very similar to the cardiovascular system, where nutrients and oxygen are transported to the cells in the body, and carbon dioxide and other waste products of cells' metabolism are removed. The cardiovascular system is also a lot like a hot water heating system. The boiler has a pump (heart) to circulate the hot water (blood) through the piping system (vessels) to deliver the much needed heat to every room throughout the building. So now let's get right to the 'heart' of the matter!

Learning objectives

At the end of this chapter, you will be able to:

◆ Identify the structures and functions of the cardiovascular system
◆ Trace blood flow through the chambers of the heart and the blood vessels
◆ Explain the structure and function of the coronary circulation of the heart
◆ Describe the contraction of the heart and the conduction system
◆ Differentiate between arteries, veins and capillaries
◆ List the major components of blood and their functions
◆ Discuss the importance of blood typing
◆ Explain the process of blood clotting
◆ Describe various cardiovascular diseases

Pronunciation guide

agglutinate (*a GLUE tin ate*)
albumin (*ALB you men*)
anaemia (*a NEE mee yah*)
aneurysm (*AN you rizm*)
arterioles (*ar TEE ree olls*)
arteriosclerosis (*ar tee ree oh skler ROW sis*)
atrioventricular node (*ay tree oh vehn TRIK yoo lahr*)
atrium; atria (*AY tree um; AY tree ah*)

autorhythmicity (*aw to rith MISS city*)
basophils (*BAY soh fills*)
cor pulmonale (*KOR pull mun AH lee*)
diastole (*dye ASS toe lee*)
embolus (*EM boh luss*)
endocardium (*EN doh KAR dee um*)
eosinophils (*EE oh SIN oh fills*)
erythrocytes (*eh RITH roh sights*)
haemophilia (*HEE moh FILL ee ah*)
haemostasis (*HEE moh STAY siss*)

inotropism (*EYE no TROPE izm*)
ischaemia (*iss KEE mee ah*)
polycythaemia (*poll ee sigh THEE mee ah*)
prothrombin (*pro THROM bin*)
systole (*SISS toh lee*)
thrombocytes (*THROM boh sights*)
tunica externa (*TUE nik ah ex TERN ah*)
tunica interna (*TUE nik ah in TERN ah*)
venules (*VEN yules*)

System overview

cardio = *heart*

vascul/o = *vessels*

The major components of the **cardiovascular** system include the *heart* (which is the organ that pumps blood through the system), *blood* (a form of connective tissue that has a fluid component called *plasma* and a variety of cells and substances) and *blood vessels* (a network of passageways to transport the blood to and from the body's cells).

Blood vessels can be further classified. Vessels that carry blood *away* from the heart are called *arteries*. These main vessels branch out into ever smaller vessels called **arterioles**, which eventually become *capillaries*. Capillaries are where the exchange of nutrients, gases and waste products occurs at the cellular level. Capillaries are also the transition vessels where blood begins its trip back to the heart through ever-merging vessels, the tiniest of which are called **venules** that form the larger *veins*. To view this transitional region along with the cardiovascular system, see Figure 13.1.

arterioles = *small arteries*

venules = *small veins*

In general, veins differ from arteries not only because veins bring blood back to the heart but because the blood now is *oxygen-poor* (contains less than the normal arterial amount of oxygen) and has a higher level of carbon dioxide and other waste products of cellular metabolism. Veins also have thinner walls than arteries, are more numerous, and have a larger capacity to hold blood.

Applied science

Colour-coded blood

Notice in Figure 13.1 that blood is depicted as being blue or red depending on the area in which it is located. When the blood contains high amounts of oxygen (oxygen-rich), it causes a chemical change within the red blood cell. This occurs in the blood of most arteries. As a result of that change, the arterial blood turns a bright red colour. As the oxygen is being delivered to the tissues in the body, the blood begins to turn darker red. This occurs mostly in the veins that have lower levels of oxygen (oxygen-poor) due to their delivery of oxygen to the tissues. You can see the darker veins within your wrist or arms. However, when you cut a vein, the blood is no longer darker red because it immediately mixes with the oxygen in the atmosphere and becomes bright red.

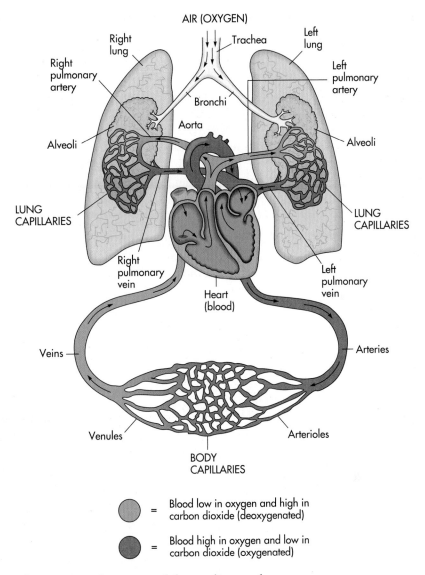

Figure 13.1 Overview of the cardiovascular system.

Learning hint

Arteries or veins?

Remembering what arteries and veins do can be confusing. One easy way to remember is that arteries take blood *away* from the heart. Both words start with *a*. Obviously, then, veins have to bring blood back to the heart.

Incredible pumps: the heart

Let's begin with the main organ of the cardiovascular system, the heart. While it is often described as the 'pump' of the cardiovascular system, you will soon see that it is actually *two* pumps working together.

General structure and function

The heart is a specially shaped muscle about the size of your fist, containing a series of chambers that move blood throughout the body. The heart is surrounded by a double layer serous membrane called the pericardium. The outer layer is a tough fibrous connective tissue called the parietal pericardium. This helps to anchor the heart to surrounding structures, for example the diaphragm. The inner layer, the visceral pericardium, is fused to the heart surface and there is a potential cavity between the layers called the pericardial cavity. The visceral pericardium is also known as the epicardium. The middle layer of the heart wall, the myocardium, is made of cardiac muscle. The heart is lined by epithelium, called the endocardium. See Figure 13.2.

The heart's location is slightly left of the centre of your chest and above your diaphragm. As strange as it may initially appear, the **base** of the heart is proximal to your head, while the **apex** of the heart is distal. Although the heart is a single organ, it is easier to understand its function if you think of it as *two separate pumps working together*. The right side of the heart is responsible for receiving blood and sending it to the lungs to pick up oxygen and get rid of carbon dioxide. The left side of the heart receives blood from the lungs and pumps it around the body. Before we explore the heart, it's important to review that blood returning to the heart travels through **veins** (thus, venous blood) and blood travelling away from the heart travels through **arteries** (thus, arterial blood).

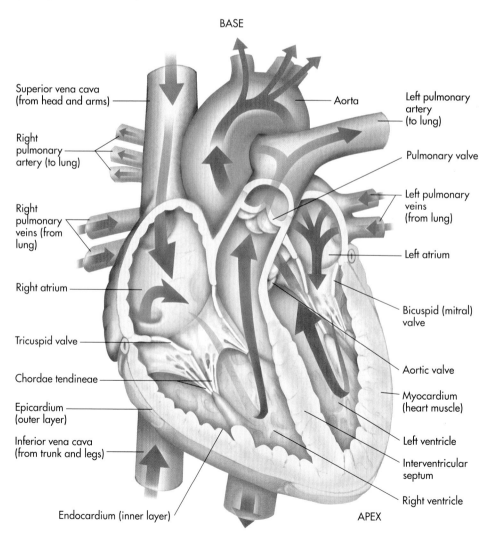

Figure 13.2 The anatomy of the heart.

Opening the heart reveals four chambers. The chambers of the right side of the heart are separated from the chambers of the left side of the heart, so there is no mixing of blood from one side to the other. The wall that separates the two smaller chambers is called the *interatrial* **septum**; while the wall between the two larger chambers is called the *interventricular* **septum**. When looking at Figure 13.2, you will notice the small chamber in the upper left quadrant of the picture. That is the *right* **atrium** (remember, locations are based on the *patient's* perspective). This is a collecting chamber where blood is returned to the heart after its trip through the body. The two large veins that bring the blood to the right atrium are the **superior vena cava** (blood from the head, neck, chest and upper extremities) and **inferior vena cava** (blood from the trunk, organs, abdomen, pelvic region and lower extremities). Once the blood is collected, it drains through a one-way valve to a larger chamber in the right side called the *right* **ventricle**. That one-way valve is called the right **atrioventricular valve**, or **AV valve**. This is also called the **tricuspid** valve because the valve is formed with three cusps.

septum = wall (SEP tum)

atrium (AY tree um)

atrioventricular (ay tree oh vehn TRIK yoo lahr)
tri = three

Cardiac cycle

The movements of the heart, called the cardiac cycle, can be divided into two phases called systole and diastole. Usually when discussing heart movement we refer to ventricle activity.

When the right ventricle is full of blood, the heart contracts (**systole**). Because the tricuspid valve is a one-way valve, as the right ventricular pressure increases, the valve shuts so blood cannot return back into the right atrium. As the pressure increases, the blood has to go somewhere. Now the only way for the blood to travel is through the *pulmonary semilunar valve* to the pulmonary trunk, which divides into the left and right **pulmonary arteries**.

systole (SISS toh lee)

Each pulmonary artery goes to its respective lung and branches down into ever smaller vessels to the point where they become capillaries that form a network around each air sac in the lungs. This is where the blood gives up one of the waste products of metabolism by cells (carbon dioxide) and picks up a fresh supply of oxygen from the lungs. These capillaries containing freshly oxygenated blood converge into increasingly larger vessels until they form the left and right pulmonary veins. Figure 13.3 illustrates the blood flow through the heart.

The pulmonary veins meet and pour their contents into the *left* atrium. Once the left atrium is filled, blood flows through the left AV valve and into the *left* ventricle. When the left ventricle is full, the heart contracts

Learning hint

Locating the tricuspid

Remember the tricuspid valve is on the *R*ight side of the heart.

Amazing body facts

Pulmonary arteries and veins

True to what we have told you, the pulmonary arteries take blood away from the heart. What is amazing is that they are the *only* arteries in the body that carry deoxygenated blood away from the right heart to travel to the lungs to become oxygenated. The pulmonary veins collect this oxygenated blood and return it to the left side of the heart and therefore are the only veins to carry oxygenated blood.

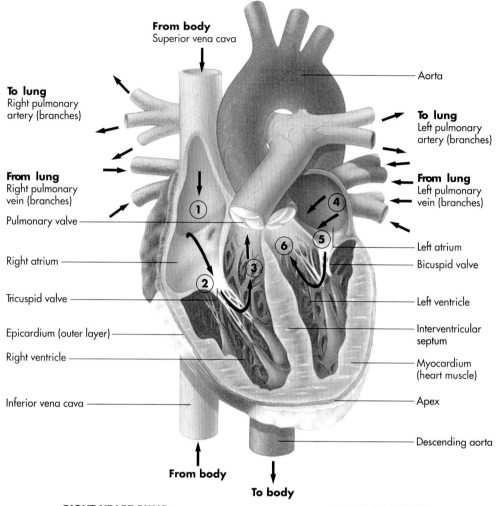

From body
Superior vena cava

Aorta

To lung
Right pulmonary
artery (branches)

To lung
Left pulmonary
artery (branches)

From lung
Right pulmonary
vein (branches)

From lung
Left pulmonary
vein (branches)

Pulmonary valve

Left atrium

Right atrium

Bicuspid valve

Tricuspid valve

Left ventricle

Epicardium (outer layer)

Interventricular
septum

Right ventricle

Myocardium
(heart muscle)

Inferior vena cava

Apex

Descending aorta

From body

To body

RIGHT HEART PUMP

1. Deoxygenated blood returns from the upper and lower body to fill the right atrium of the heart, creating a pressure against the atrioventricular (AV) or tricuspid valve.

2. This pressure of the returning blood forces the AV valve open and begins filling the ventricle. The final filling of the ventricle is achieved by the contracting of the right atrium.

3. The right ventricle contracts, increasing the internal pressure. This pressure closes the tricuspid valve and forces open the pulmonary valve, thus sending blood toward the lung via the pulmonary artery. This blood will become oxygenated as it travels through the capillary beds of the lung and will then return to the left side of the heart.

LEFT HEART PUMP

4. Oxygenated blood returns from the lung via the pulmonary vein and fills the left atrium, creating a pressure against the bicuspid valve.

5. This pressure of returning blood forces the bicuspid valve open and begins filling the left ventricle. The final filling of the left ventricle is achieved by the contracting of the left atrium.

6. The left ventricle contracts, increasing the internal pressure. This pressure closes the bicuspid valve and forces open the aortic valve, causing oxygenated blood to flow through the aorta to deliver oxygen throughout the body.

Figure 13.3 The functioning of the heart valves and blood flow.

(squeezes) again. The ventricular pressure increases, forcing the left AV valve shut and ejecting the blood out of the left ventricle through the aortic semilunar valve to the ascending aorta, sending it on its way throughout the body. Two more familiar names for the left AV valve are bicuspid (formed by only two leaves or cusps) or **mitral** valve. Now the ventricles and atria can rest (**diastole**) as the atria fill with blood before they squeeze another load of blood into the ventricles. It is important to note that the chambers of the heart fill during diastole, or the relaxation phase, and eject blood during systole, or the contraction phase.

Some points to remember are that

- both atria fill at the same time
- both ventricles fill at the same time
- both ventricles eject blood at the same time when the heart contracts.

Did you ever get shouted at or get angry with the *other* person for squeezing the toothpaste tube in the middle? What's the big deal? When you eat an ice pop, do you squeeze it in the middle or from the bottom and work the contents up to your mouth? These two examples are important to visualise because the heart has to contract in a certain way to make sure that all of the blood is squeezed out during each contraction. In order to do this, the contraction begins at the apex and travels upward. As you trace the flow of blood in Figure 13.3, you will see how efficient this is.

If you further examine the heart illustration, you will notice that the walls of the atria are thinner than the ventricular walls. This is because higher pressures are generated in the ventricles to move blood. You should also note that the walls of the left ventricle are thicker than the walls of the right ventricle. When you think about it, this makes total sense. The right ventricle has to pump blood only a short distance through the vasculature of the lungs and back to the heart. The left ventricle, on the other hand, has to pump all of the blood throughout the body and back to the right atrium. The resistance of all those blood vessels in the body is six times greater than the resistance of the lung **vasculature** (network of blood vessels).

In the Amazing body facts box, you learned that the heart muscles have a blood-rich environment. In Figure 13.4, you can see that a portion of the newly oxygen-enriched blood is diverted from the aorta by the right and left **coronary arteries**. These arteries continuously divide into smaller branches, forming a web of interconnections known as **anastomoses**, which enable the heart muscle to constantly and fairly consistently receive a rich supply of blood. It is interesting to note that regular aerobic exercise can increase the density of these blood vessels that supply the heart. The number of anastomoses also increases, as does

anastomoses (*An AST owe mow sees*)

Amazing body facts

How does it keep going and going and going?

One of the amazing things about your heart is that it continues to beat day after day without your even thinking about it. Here is an experiment. Previously, we said that your heart is about the size of your fist. Let's pretend that your fist is your heart. To mimic the pumping action of your heart, open your fist with your fingers fully extended. Now make a tight fist. Continue this action of fully opening and tightly closing your hand for the next *60 seconds*. Chances are good that your hand will feel like it's ready to fall off of your arm! If this is how your hand feels after 60 seconds, how does your heart constantly beat approximately 100,000 times and move approximately 7000 litres of blood each day for decades and decades? Think back to Chapter 7 on muscles. Cardiac muscle cells have specialised connections called *intercalated discs*. These connections, along with associated pores, provide an efficient connection with adjacent muscle cells so electrical impulses, ions and various small molecules can readily travel throughout the heart, allowing a smooth contraction from one area of the heart to another. The vasculature of the heart takes approximately 5 per cent of oxygenated blood from each heartbeat to ensure there is a blood-rich environment so plenty of oxygen and nutrients are available to the heart muscle itself.

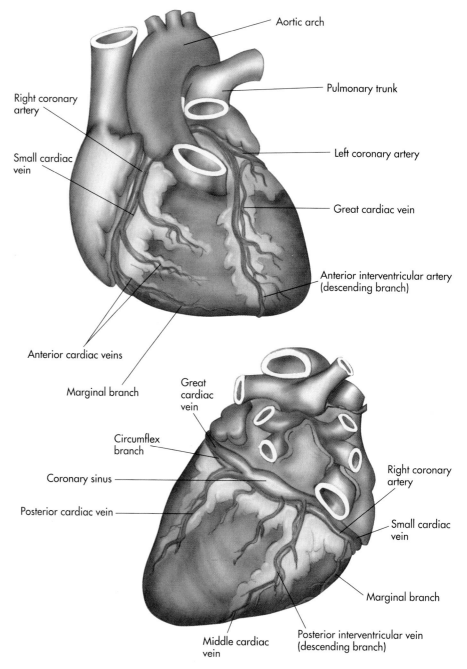

Figure 13.4 Coronary circulation.

their number of locations. This is important for people with a blockage of a small coronary artery. They have an increased survival rate because the blood now has alternative routes to travel, which helps to prevent heart muscle damage.

The right coronary artery provides blood for the right ventricle, the posterior portion of the interventricular septum, and inferior parts of the heart. The left coronary artery provides blood to the left lateral and anterior walls of the left ventricle and to portions of the right ventricle and interventricular septum.

Test your knowledge 13.1

Choose the best answer:

1. Which blood vessels carry blood away from the heart?
   a. capillaries
   b. venules
   c. veins
   d. arteries

2. How many chambers are found in the human heart?
   a. one
   b. two
   c. three
   d. four

3. Which blood vessel type is involved in the exchange of oxygen and nutrients with the tissues of the body?
   a. arterioles
   b. sphincters
   c. arteries
   d. capillaries

Complete the following:

4. The chamber responsible for pumping blood to the body's various organs is the _____.

5. Which side of the heart pumps blood to the lungs?
   _____.

The conduction system (the electric pathway)

Cardiac muscles don't always rely on nerve impulses or hormones to contract. In fact, they can contract on their own. This unique ability is known as **autorhythmicity**. The problem with this ability is that uncontrolled *individual* contractions would prohibit the heart from contracting effectively. This potential problem is solved through the use of specialised cardiac cells that create and distribute an electrical current that causes a *controlled* and *directed* heart contraction.

autorhythmicity *(aw to rith MISS city)*

Nodal cells (also called **pacemaker cells**) are specialised cells that not only create an electrical impulse but create these impulses at a regular interval. These cells are connected to each other and to the conducting network, which we discuss soon. Nodal cells are divided into two groups. The main group of pacemaker cells is found in the wall of the right atrium, near the entrance of the superior vena cava. This collection of pacemaker cells forms the **sinoatrial** node, or **SA node**. The SA node generates an electric impulse at approximately 70 to 80 impulses per minute. There is a second collection of pacemaker cells located at the point where the atria and the ventricles meet. This collection forms what is called the **atrioventricular node**, or **AV node**. The cells in the AV node generate an electric impulse at a rate of 40 to 60 beats per minute. The bundle of His also generates an electric impulse but this is at a rate of 30 beats per minute.

sinoatrial *(sigh no AY tree all)*

So, which one dictates how fast the heart beats? Think of the SA node as the power station; the AV node as a substation that supplies electricity via an electrical grid for a small city and the bundle of His as the local supply to your house. The SA node sends its impulse to the AV node for distribution before the AV node can send its own. However, *if* the SA node cannot generate an impulse, the AV node takes control and sends out impulses that result in a slower heartbeat, but a heartbeat nonetheless. If the AV node cannot generate an impulse. The bundle of His will generate a heartbeat. However, the heart

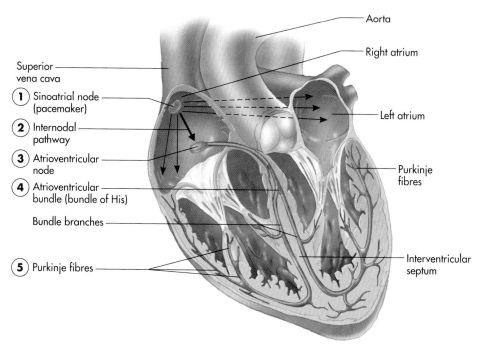

1. The sinoatrial (SA) node fires a stimulus across the walls of both left and right atria, causing them to contract.

2. The stimulus arrives at the atrioventricular (AV) node.

3. The stimulus is directed to follow the AV bundle (bundle of His).

4. The stimulus now travels through the apex of the heart through the bundle branches.

5. The Purkinje fibres distribute the stimulus across both ventricles, causing ventricular contraction.

Figure 13.5 Conduction system of the heart.

rate from the bundle of His is incompatible with life. Figure 13.5 shows the conduction system of the heart.

'So, how come my heart rate isn't always 70–80 beats per minute like when I exercise or get frightened?' you ask. It's true that the SA node sets the heart rate when the body is at rest, but several influences can increase or decrease heart rate. The *autonomic nervous system* (both the sympathetic and parasympathetic divisions) has direct connections to the SA and AV nodes as well as to the myocardium. The sympathetic division can release neurotransmitters that increase heart rate and the force of the contraction (**inotropism**). To counteract this, the parasympathetic division, through the **vagus nerves**, releases a neurotransmitter that can decrease both the pulse rate and force of contraction.

In addition, ions, hormones and body temperature can alter heart rate. For example, as body temperature increases, so does the rate and force of contractions because of the increased metabolic rate of cardiac muscle cells. Conversely, as the body cools down below normal temperature, the rate and force of cardiac contractions decrease. Adrenaline (epinephrine) is a hormone that has the same effects as the sympathetic nerves, so it increases heart rate. Electrolytes are important too, especially when there is an imbalance of too

inotropism (*EYE no TROPE izm*)

many or too few specific ones. Low sodium or potassium can alter heart activity, as can abnormal levels of calcium. Low potassium can lead to a weak heartbeat, while high levels of calcium can prolong heart muscle contractions to the point where the heart can stop beating. These are just a few examples of what can happen when electrolyte levels are outside of normal range. Age, gender, a history of exercise or lack thereof, all can impact on your heart rate. Normally, the resting heart rate for a female is 72 to 80 beats per minute, while the average resting heart rate for males is 64 to 72 beats per minute.

a = *without*

dys = *bad or difficult*

Clinical application

ECGs

Because the myocardial contraction is initiated and continues because of an electrical impulse, that charge can actually be detected on the surface of the body. This surface detection of the electric impulse travelling through the heart can be recorded by using an *electrocardiograph*, which records an *electrocardiogram* (ECG). See Figure 13.6.

The normal ECG has three distinct waves that represent specific heart activities. The *P wave* is the first wave on the ECG and represents the impulse generated by the SA node and depolarisation of the atria right before they contract. The next wave is called the *QRS complex* (a combination of Q, R and S waves). It represents the depolarisation of the ventricles that occurs right before the ventricles contract. The ventricles begin contracting right after the peak of the R wave. Due to the greater muscle mass of the ventricles compared to the atria, this wave is greater in size than the P wave. The final wave is the *T wave*, which represents the repolarisation of the ventricles where they are at rest before the next contraction. 'Aha,' you say, 'where is the repolarisation of the atria?' It occurs during the QRS complex but is usually overshadowed by the ventricles' activity. In the recording of a healthy heart, there are set ranges for the height, depth and length of time for each of the waves and wave complexes. Changes in those parameters, or the addition of other abnormal types of waves, known as **cardiac arrhythmias** or **dysrhythmias**, can indicate health problems that involve the heart.

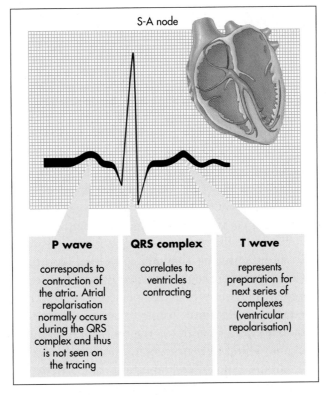

Figure 13.6 Typical ECG tracing.

So far, we have our two generators and regulators of electricity. We can see how electricity is moved in the form of electric lines along the road. In the heart, the movement of the electric impulses is done by specialised *conducting* cells. This power grid of electric distribution has to be set up so the following actions can occur:

1. First, the right and left atria contract together and *before* the right and left ventricles.
2. Then, the two ventricles must contract together.
3. But the direction of the wave of ventricular contraction has to be from the *apex* to the *base* of the heart. This ensures that all of the blood is squeezed out of the ventricles (remember the tube of toothpaste?).

So let's retrace the electrical 'wiring', as illustrated in Figure 13.5. Once an electric impulse is generated at the SA node, several pathways composed of conducting cells transmit that impulse to the AV node. A slight signal delay allows for the atria to fill with blood before contraction occurs. Once this charge reaches the AV node, it continues its journey through the **AV bundle**, also known as the **bundle of His** (which sounds like 'hiss'). Travelling down the interventricular septum, the AV bundle eventually divides into the *right bundle branch* and the *left bundle branch*. These branches spread across the inner surfaces of both ventricles. Finally, another type of specialised cells called *Purkinje cells* carry the impulse to the contractile muscle cells of the ventricles. And so the contraction begins at the apex, and the wave of contraction smoothly continues up the ventricles, squeezing out all of the blood.

Blood

Now that we understand the pump, let's talk about what exactly the heart pumps. Blood is a fluid form of connective tissue that is responsible for three very important functions.

Blood *transports* oxygen from the lungs, nutrients and fat cells from the digestive system, and hormones from endocrine glands to the approximately 75 *trillion* cells in the body. On the return trip from those cells, blood transports carbon dioxide and other waste products that were formed from metabolic activities of the cells to the kidneys, lungs and other organs for removal from the body. Blood also transports important plasma proteins made by the liver and various electrolytes, for example sodium and potassium.

Blood helps to *regulate* a variety of levels in the body to maintain homeostasis by ensuring that **pH** (levels of acidity or alkalinity) and **electrolyte** (ion) values are within normal parameters for proper cell functioning. Blood helps to regulate body temperature by absorbing heat generated by skeletal muscles, spreading it throughout the rest of the body. Conversely, blood can radiate excess heat out of the body through the skin. Blood can take in or give up more fluid to help regulate the fluid balance of the body.

Finally, but no less important, blood helps to *protect* us from invasion and infection by pathogens and toxins. This is done by specialised **white blood cells** (often shortened to **WBCs**) and special proteins called **antibodies**.

Amazing body facts

What's in a drop of blood?

In one *drop* of blood, you will find *5 million* red blood cells; *250,000* to *500,000* platelets; and *7500* white blood cells. Don't forget that there are a lot of other substances, such as plasma proteins, nutrients, oxygen, carbon dioxide, hormones and electrolytes, in there too. This helps to make blood *five times thicker* than water. With a life expectancy of approximately 120 days, new red blood cells are created to replace old ones at the rate of *2 million each second!* Even more amazing is that the total surface area of all the red blood cells in your body is greater than the surface area of a football field!

The amount of blood in the body depends on an individual's size and gender. Normally, the body contains between 4 and 6 litres of blood, which accounts for 7 to 9 per cent of total body weight.

Blood composition

Although we can classify blood as a connective tissue, it is important to understand all the components that make up blood. When blood is separated by a centrifuge, the *major* components are **plasma** and what we call **formed elements**. The centrifuge is a machine that spins a test tube of blood at a very fast rate. Due to the spinning force of the centrifuge, the heavier components, like the formed elements, are forced to the bottom of the tube and the lighter component (plasma) is displaced to the top of the tube (see Figure 13.7).

Plasma is the yellowish, straw-coloured liquid that comprises about 55 per cent of the blood's volume and contains about 100 different substances dissolved within. So, if your total blood volume is 5 litres, you have about 2.75 litres of plasma. While plasma is about 90 per cent water, nutrients, salts and a small amount of oxygen are also dissolved into the plasma for transport to the body's cells. Hormones and other cell activity-regulating substances are found in plasma. **Plasma proteins** are an important group of dissolved substances that include **albumin**, which aids in keeping the correct amount of water in the blood; fibrinogen, which is a substance needed for blood clotting; and globulins, which form antibodies that protect us from infection.

Formed or solid elements include the following:

1. red blood cells (RBCs), or **erythrocytes**
2. white blood cells (WBCs), or **leucocytes**, which can be further classified as basophils, eosinophils, lymphocytes, monocytes and neutrophils
3. thrombocytes (also known as platelets), which aid in clotting.

albumin (*ALB you men*)

erythrocytes (*eh RITH roh sights*)
   erythro = *red*
   cytes = *cells*
leucocytes (*LOO koh sights*)
   leuco = *white*
thrombocytes (*THROM boh sights*)
   thrombo = *clotting*

haemopoiesis (*HEE mow poy EE siss*)
   haemo = *blood*
   poiesis (**poietic**) = *making or producing*

Red blood cells

Lacking a nucleus, and therefore unable to divide to form new cells, red blood cells are created by the red bone marrow through a process called **haemopoiesis** and are similar in shape to a doughnut. Red blood cells perform two crucial functions. With the aid of an iron-containing red pigment called **haemoglobin**, red blood cells transport oxygen from the lungs to the cells in the body. In addition, they help to transport some carbon dioxide, a by-product of cellular metabolism, from the cells to the lungs for removal from the body.

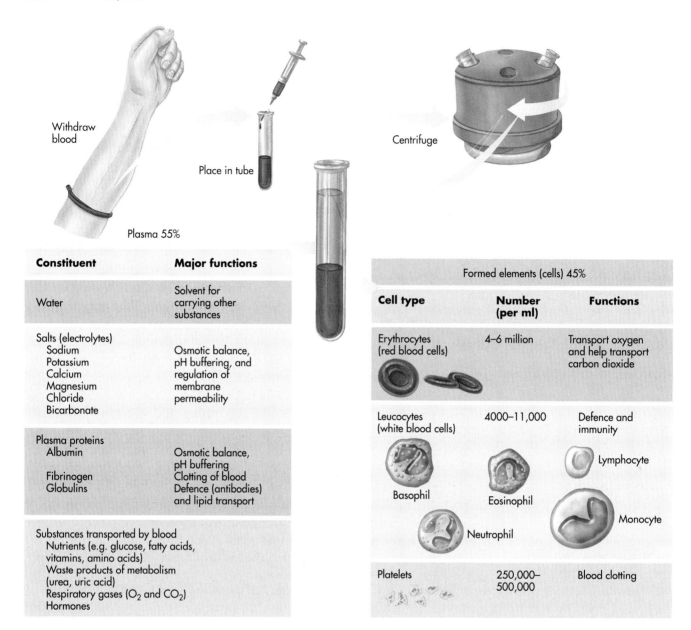

Figure 13.7 Composition of blood.

Red blood cells have a tough life and only last about 120 days, and so new red blood cells are made all the time. However, if we lose red blood cells (perhaps through blood loss following injury or surgery) this will result in a potential deficit of oxygen-carrying capability in the blood. This reduction in oxygen results in erythropoeitin being released from the kidneys and the liver. This acts on red bone marrow to increase red blood cell production.

White blood cells

There are several types of white blood cells. Polymorphonuclear granulocytes originate from red bone marrow. Also originating from bone marrow but

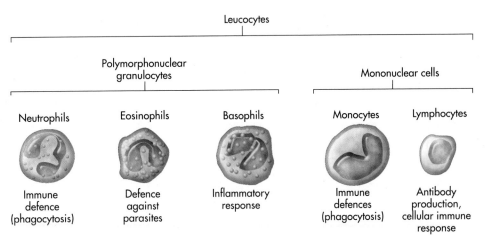

Figure 13.8 Functions of white blood cells.

maturing in lymphoid and myeloid tissues (mononuclear cells), leucocytes are our guardians from invasion and infection (see Figure 13.8).

The types of white blood cells that compose the polymorphonuclear granulocyte group are neutrophils, eosinophils and basophils. Neutrophils are the most aggressive white blood cells in cases where bacteria attempt to destroy tissue. **Phagocytosis** is the process in which neutrophils surround and ingest the invader and attempt to destroy it by utilising cell particles called **lysosomes** that release powerful enzymes. As an infection occurs, the body produces a higher than normal number of neutrophils. Eosinophils are utilised to combat parasitic invasion and a variety of body irritants that lead to allergies. Basophils are involved with allergic reactions by enhancing the body's response to irritants that cause allergies. In addition, basophils are important because they secrete the chemical **heparin**, which helps to keep blood from clotting as it courses through blood vessels.

The types of white blood cells that comprise the mononuclear cell group are monocytes and lymphocytes. Monocytes are found in higher than normal amounts when a chronic (long-term) infection occurs in the body. Like neutrophils, monocytes destroy invaders through phagocytosis. Even though it takes longer for monocytes to arrive on the scene of infection, their numbers are greater than neutrophils and therefore they destroy more bacteria. Lymphocytes are unique cells that protect us from infection. Instead of utilising phagocytosis, they are involved in a process that produces antibodies that inhibit or directly attack invaders. Fighting infection and foreign invaders is discussed, along with the lymphatic system, in Chapter 14.

Thrombocytes

Thrombocytes, also known as **blood platelets**, are the smallest of the formed elements and are responsible for the blood's ability to **clot**. In addition, thrombocytes can release a substance called *serotonin*, which can cause smooth muscle constriction and decreased blood flow. Both these actions can help to minimise bleeding.

---

phagocytosis (*FAG oh sigh TOH siss*)
  phago = *to eat*
  osis = *process*
lysosomes (*LIE soh soams*)
  lyso = *destruction*
  som(a) = *body*

Test your knowledge 13.2

Choose the best answer:

1. Which cell type does not belong in this group?

   a. erythrocyte

   b. lymphocyte

   c. monocyte

   d. eosinophil

2. What substance found in red blood cells is responsible for oxygen transport?

   a. gobuloglobin

   b. haemoglobin

   c. gammaglobulin

   d. cytoplasm

3. Which portion of an ECG tracing represents ventricular depolarisation?

   a. T wave

   b. QRS complex

   c. P wave

   d. SA node

Complete the following:

4. This important plasma protein helps to maintain fluid balance of your blood: _____.

5. The main pacemaker of the heart is the _____.

Blood groups and transfusions

Not all human blood is the same. A person in need of a blood transfusion cannot be given blood from a randomly selected donor. Incompatibility of blood groups is due in part to antigens. An antigen is a protein on cell surfaces that can stimulate the immune system to produce antibodies, which fight foreign invaders. Antigens are typically foreign proteins introduced into the body through wounds, blood transfusions, and so on. Because they are not 'native' to the body, they are called 'non-self' antigens. They differ from our own 'self-antigens' that exist on the cell membrane of every cell in the body. The chain of events that occurs between antigens and antibodies is called the antigen–antibody reaction, which is the basis for the immune response, as you will see in Chapter 14. Antibodies often react with the antigens that caused them to form, and the antigens stick together, or **agglutinate**, in little clumps. Although there are at least 50 different antigen types found on the surface of a red blood cell, our main focus is on the A, B and Rh antigens.

Everybody has only *one* of the following blood groups:

   group A

   group B

   group AB

   group O

Group A blood is very common. Approximately 42 per cent of the United Kingdom (UK) population has this blood group. A represents a specific type of antigen that is found on the cell membrane of each red blood cell in the body of a person with blood group A. Since that person was born with blood group A, no antibodies were created to fight it, so there are no anti-A

agglutinate (*a GLUE tin ate*)

antibodies in his or her plasma. However, blood group A plasma *does* contain anti-B antibodies.

Group B red blood cells possess B antigens and the plasma contains anti-A antibodies. Apparently, these two blood groups don't like each other! The percentage of people in the UK with blood group B is 8 per cent.

Group AB, however, tries to get along with both A and B. Its red blood cells have *both* of the A and B antigens on the cell surface with neither A nor B antibodies in the plasma. Blood group AB is quite rare in the UK population, with only 3 per cent of the population having this blood group.

Not to be outdone, group O red blood cells have *no* A or B antigens on the cell surface, but its plasma contains *both* A and B antibodies. Blood group O is very common in the UK, with 47 per cent of the population having this blood group.

Frequency of the blood groups within different populations varies. Blood group B is more common in people with an Asian background, with about 25 per cent of Asian people in the UK having blood group B. Blood group AB is rare in the UK (only 3 per cent); however, in Japan and China 10 per cent of the population have this blood group.

This information is important to know if there is a need to transfuse blood. If the donor's antigens and the antibodies in the blood recipient's plasma agglutinate, serious harm and even death can occur.

If a *donor* gives blood that contains no A or B antigens, agglutination by anti-A and/or anti-B antibodies in the *recipient's* blood is prevented. Since group O doesn't have A or B antigens, it can be given to anyone, so a donor with group O blood is a **universal donor**. Since group AB doesn't contain plasma anti-A antibodies or anti-B antibodies, it can't clump with any donated blood that contains A or B antigens. Because of this, a Group AB person is labelled a **universal recipient**. See Figure 13.9, which shows blood groups with matching antigen and antibodies and which recipient blood would safely match the donor's blood.

Are you with us so far? Good. Now there is one more thing we need to add: the **Rh factor**. Based on a discovery of a special blood antigen first found in the blood of Rhesus monkeys, it was discovered that approximately 85 per cent of the UK population possess the Rh antigen in their blood. Individuals with this antigen in their blood are said to be **Rh-positive**. Conversely, those without this antigen are **Rh-negative**. When typing an individual's blood, the term *Rh* is eliminated, so an individual would be O-positive or AB-negative, for example. A problem arises when there is a Rh-positive father and a Rh-negative mother. If their first baby inherits the father's Rh-positive blood trait, the mother will develop anti-Rh antibodies (remember the foreign invader scenario?). This baby will be okay, but any future baby born to these parents may be attacked by the anti-Rh antibodies of the mother *IF* that baby has the Rh-positive trait in its blood.

Amazing body facts

Transfusions and artificial blood

Currently, there is much research into the development of artificial blood or blood substitutes. The main objective is to develop a blood form that can be universally used for all humans without the need to match specific blood groups. Other objectives include making a substitute that is free of blood-borne diseases, has the ability to be rapidly infused, has increased oxygen-carrying capacity, and has an extended shelf-life. Some artificial blood is made by chemically altering and refining natural blood components, and some artificial blood is synthetic in nature.

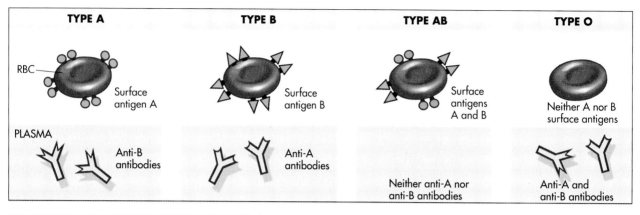

Recipient's blood		Reactions with donor's blood			
RBC antigens	**Plasma antibodies**	**Donor type O**	**Donor type A**	**Donor type B**	**Donor type AB**
None (Type O)	Anti-A Anti-B				
A (Type A)	Anti-B				
B (Type B)	Anti-A				
AB (Type AB)	(None)				

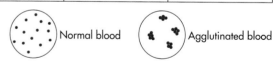

Figure 13.9 Blood groups and results of donor and recipient combinations.

Blood vessels: the vascular system

So far, we have a pump and some fluid. We now need a way to transport the blood away from and then back to the heart. The arteries and veins we discussed previously now come into play.

Structure and function

Initially, blood leaves the heart through the aorta, which branches into large vessels called *arteries*. Arteries divide into smaller and smaller ones as they

arterioles (*ar TEE ree oll*)

spread out through the body. The smallest form of artery is the **arteriole**. Arterioles feed the capillaries that form the capillary beds in your body's tissues. Here is where fresh oxygen and nutrients are supplied to the cells of your body and carbon dioxide, along with other waste products, is picked up by the blood for removal.

Blood from the capillary beds begins its return trip to the heart by draining into tiny veins called venules. Venules combine into veins, which eventually combine into the great veins (superior and inferior vena cavae) that empty back into your heart.

tunica interna (*TUE nik ah in TERN ah*)

For most blood vessels, the walls are composed of three layers, often called coats or tunics. See Figure 13.10. The **tunica interna** is the innermost layer and is composed of a thin, tightly packed layer of *squamous epithelial* cells over a layer of loose connective tissue. The compacting of the epithelial cells provides a smooth surface so blood can easily pass through. The next layer is thicker and is composed mainly of smooth muscle and elastic tissue and

tunica media (*TUE nik ah mee DEE ah*)

collagen. This middle layer is called the **tunica media**. By contracting or relaxing those muscles, this layer actually controls the diameter of the vessels to meet certain blood flow needs of the body at a given time. Your sympathetic nervous system determines these needs and signals the vessels to **vasoconstrict** (decrease the inner diameter or lumen) or **vasodilate** (increase the inner diameter or lumen) as needed. This change in diameter changes blood pressure (**BP**); as the vessels dilate, BP decreases. Constriction leads to

tunica externa (*TUE nik ah ex TERN ah*)

BP increases. The outermost, or external, coat is the **tunica externa**. Its job is to provide vessel support and protection, so it is composed of mostly fibrous tissue.

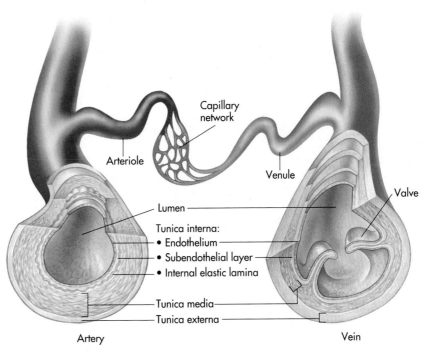

Figure 13.10 Blood vessels and the capillary connection.

Clinical application

Taking a blood pressure

Blood pressure is the pressure exerted by the blood on the walls of the vessels as it travels through them. It is measured in millimetres of mercury – mmHg.

An important diagnostic test is the measurement of arterial blood pressure (BP). This is done with a stethoscope and a sphygmomanometer. As shown in Figure 13.11, an inflatable cuff is placed around the arm, above the elbow, so that when the cuff is inflated, it squeezes the brachial artery shut. Your stethoscope is placed over the brachial artery in the proximity of the patient's elbow. The cuff can then be inflated by repeatedly squeezing the bulb while listening with the stethoscope.

It is first necessary, however, to determine the point at which the cuff can stop being deflated.

Before measuring the actual blood pressure, the cuff can be inflated whilst feeling the radial pulse on the same arm. As the cuff is inflated, the brachial artery will eventually close and at this point it will no longer be possible to feel the pulse. The pressure at this point should be noted. The blood pressure can now be measured and this time the cuff can be inflated to approximately 30 mmHg above the pressure obtained when the radial pulse disappeared.

You can read the pressure by one of three ways, depending on the device you are using. Some devices use a column filled with mercury (thus the pressure unit of measure mmHg) connected to the cuff via a hollow tube. Other devices use a round pressure gauge, and others use a digital readout (Figure 13.11). All of them measure pressure in units of mmHg.

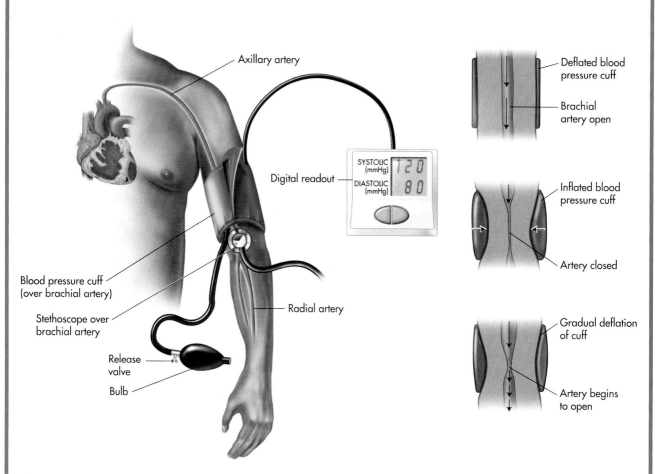

Figure 13.11 Blood pressure measurement.

Once the pressure of the cuff is 30 mmHg above the point where the pulse disappeared, the release valve is opened *slightly* so the cuff slowly deflates as you listen to the brachial artery. As the cuff pressure decreases to slightly below systolic pressure (pressure when the heart contracts), the sound of blood being pushed through the artery by the heart can be heard. That is the peak systolic pressure, or top number of the BP reading.

As the cuff pressure decreases, the sound of the pulse decreases and then disappears. That point is where the cuff pressure is equal to the arterial pressure when the heart is at rest (diastole). This is the bottom number in the BP reading. One-twenty over eighty (120/80) has traditionally been accepted as a normal BP for healthy young adults. Recently, lower values are being considered as more in line with a healthy value. Time and continued study will tell which will be the accepted value.

The sounds heard through the stethoscope when taking blood pressure are known as the Korotkoff Sounds (named after their discoverer). They are divided into five phases. The first sound heard (Phase 1) correlates to the systolic pressure equalling the cuff pressure as it is slowly deflated, and is a sharp tapping sound. As the cuff continues to be deflated the sounds change as the artery slowly opens and cuff pressure decreases until Phase 4 is reached. At Phase 4 the sounds become more muffled and in Phase 5 they cease altogether – this is the diastolic pressure.

The capillaries

The differences in the structure of the blood vessels vary depending on their job. Arteries possess much thicker walls than veins. As we said earlier, this makes sense because arteries have to deal with higher pressures. In fact, larger arteries contain complete sheets of elastic tissue, *elastic laminae*, in their middle walls to help deal with increased pressure. Veins can possess thinner walls because the pressure on them is lower than in arteries, but the inside opening, known as the **lumen**, is larger than in arteries. In addition, the larger veins of the body, especially in the legs, contain valves that prevent blood from flowing backward. Remember, the venous side is lower in pressure. Another means that the body has to help move venous blood toward the heart is through the use of skeletal muscle. The relaxation and contraction of skeletal muscles that surround veins – particularly in the legs – help to 'milk' the blood toward the heart.

Clinical application

Regulation of blood pressure

Blood pressure is controlled by length and diameter of the vessel (peripheral resistance) the viscosity of the blood (how thick the blood is) and the amount of blood pumped by the heart (cardiac output – CO). Cardiac output is a function of heart rate and the amount of blood pumped with each contraction (stroke volume – SV). Stroke volume is influenced mainly by blood volume. For example, increased fluid volume, increased heart rate, and increased peripheral resistance would lead to an increased blood pressure.

Typically in an adult, SV is about 70 ml, and so with a resting heart rate of about 70 beats per minute CO would be 4,900 ml/min but conventionally CO is recorded in litres/minute so CO becomes 4.9 l/min (CO = SV × HR). This is approximately equivalent to the total blood volume passing through the heart once every minute.

Finally, we have the capillaries, which are composed of only the tunica interna. With a diameter of only 0.008 millimetres (slightly larger than the diameter of a single red blood cell), this wall is only one cell in thickness so oxygen and nutrients can easily move into the tissue cells while carbon dioxide and wastes can move into the blood. This is important because, even at rest, metabolising tissue cells require approximately 250 millilitres of oxygen while producing almost 200 millilitres of carbon dioxide *every minute*. Dozens of capillaries form a web, or network, of vessels called a **capillary bed**. As you can see in Figure 13.12, capillary beds are composed of two types of blood vessels: a **vascular shunt**, which is a main road connecting the arteriole to the venule, and **true capillaries**, which make the actual exchanges with tissue cells. True capillaries can be considered the on-ramps and off-ramps to and from the vascular shunt. In Figure 13.12, you will notice a group of structures called **precapillary sphincters**. These structures are composed of smooth muscle and act as toll booths, either allowing blood to flow through or stopping blood flow when they contract. If the blood flows through, it travels through the true capillaries and to cells of the tissue. If the blood is stopped at the precapillary sphincters, then the blood travels through the vascular shunt.

Tissue fluid formation and reabsorption

As well as facilitating oxygen and nutrient movement into the tissues, the structure of the capillaries also allows the movement of water from the blood into the tissues and then to the cells to maintain hydration. This is enabled due to two pressures – osmotic pressure and hydrostatic pressure.

- Osmotic pressure – this pressure is present as a result of the existence of plasma proteins in the blood and its effect is to draw fluid back into the capillaries.

- Hydrostatic pressure – this is the blood pressure in the capillaries and tends to force fluid and solutes out of the capillaries into the interstitial spaces.

As you can see in Figure 13.12, a capillary bed has an arterial end and a venous end. At the arterial end the hydrostatic pressure exceeds the osmotic pressure and so fluid is forced out of the capillaries. At the venous end the osmotic pressure exceeds the hydrostatic pressure and therefore fluid in the interstitial spaces is drawn back into the capillaries. If this did not take place and the fluid remained in the interstitial spaces, then oedema (accumulation of fluid in the interstitial spaces) would occur. Changes in either of the two pressures may well lead to oedema. If osmotic pressure decreased below the hydrostatic pressure, due to a drop in the level of plasma proteins (caused by malnutrition/burns), then fluid could not be drawn back into the capillaries at the venous end and oedema would occur. Should the hydrostatic pressure increase as a result of hypertension, then fluid movement into the interstitial space would be increased and this may also compromise fluid return into the capillaries at the venous end.

Should there be excessive fluid in the interstitial spaces then one of the roles of the lymphatic capillaries is to pick up and return this fluid back to the circulation. Once the fluid enters the lymphatic system it is called 'lymph'. See Figure 13.16 for the relationship between the capillaries of the circulatory system and the capillaries of the lymphatic system.

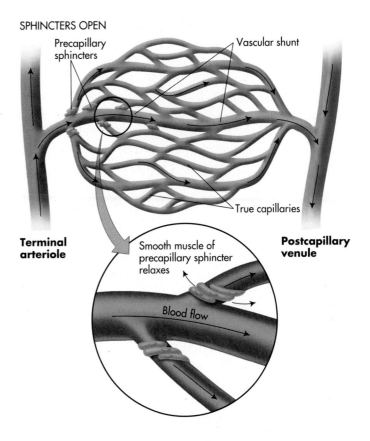

SPHINCTERS OPEN

Precapillary sphincters

Vascular shunt

Terminal arteriole

Postcapillary venule

True capillaries

Smooth muscle of precapillary sphincter relaxes

Blood flow

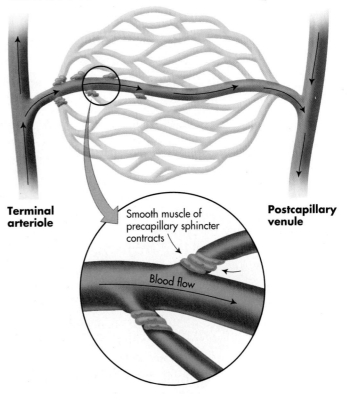

SPHINCTERS CLOSED

Terminal arteriole

Postcapillary venule

Smooth muscle of precapillary sphincter contracts

Blood flow

Figure 13.12 Capillary beds and sphincters.

Blood clotting

As we have discussed, the cardiovascular system is a closed and pressurised system. Imagine what could happen if a leak or a break in the system occurred. You have probably seen cars along the road with leaky radiators. The water leaks out and the car goes nowhere. A similar condition can occur in your body; if enough blood is lost, you won't go anywhere and will die. Thanks to several substances in your blood, some leaks or breaks that occur can be stopped. Haemostasis (the stoppage of blood) is accomplished through a chain of events shown in Figure 13.13.

Damage to the innermost wall of a vessel exposes the underlying collagen fibres. Platelets that are floating around in the blood begin attaching to the rough, damaged site. The attached platelets release several chemicals that draw more platelets to that site, creating a *platelet plug*. These platelets also release *serotonin*, which causes blood vessels to spasm, thereby decreasing blood flow to that area. Within approximately 15 seconds from the time of the initial injury, the actual **coagulation** (clotting) of blood begins. With the help of calcium ions and 11 different plasma proteins (clotting factors), a chain reaction starts. One of the clotting proteins, prothrombin, which is produced by the liver with the help of vitamin K, converts to thrombin. Thrombin transforms fibrinogen, which is dissolved in the blood, into its insoluble, hair-like form called **fibrin**. Fibrin forms a net-like patch at the injury site and snags more blood cells and platelets, and within three to six minutes, a clot is

haemostasis (*HEE moh STAY siss*)

prothrombin (*pro THROM bin*)
thrombin (*THROM bin*)
fibrinogen (*fye BRINN oh jenn*)

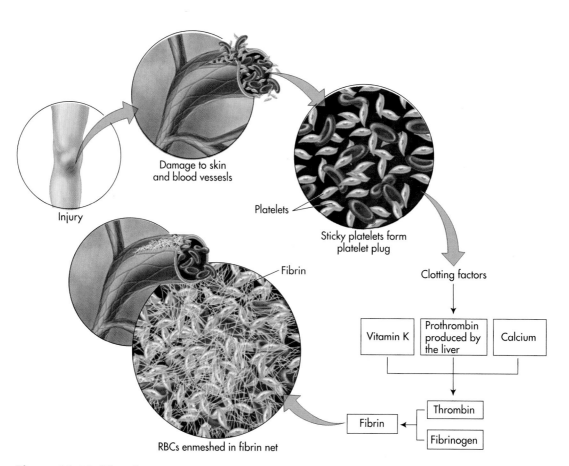

Injury

Damage to skin
and blood vessesls

Platelets

Sticky platelets form
platelet plug

Clotting factors

Fibrin

Vitamin K | Prothrombin produced by the liver | Calcium

Thrombin

Fibrin

Fibrinogen

RBCs enmeshed in fibrin net

Figure 13.13 The clotting process.

created. Once the clot is formed, it eventually begins to retract and, as a result, pulls the damaged edges of the blood vessel together. This allows the edges to regenerate the necessary epithelial tissue to make a permanent repair over time. Once the clot has outlived its usefulness, it is dissolved. Again, see Figure 13.13.

thrombus (*THROM buss*)

Clinical application

Clotting gone bad

The chain reaction that causes a clot must be stopped when it has accomplished its purpose, or else clotting would continue throughout the vascular system. However, there are times when unwanted clotting occurs. A rough surface on an otherwise smooth lumen of a blood vessel may allow platelets to begin 'sticking' there, forming a type of clot called a **thrombus**. A thrombus that forms in the vascular system of the heart can partially or totally block blood flow to a portion of the heart, resulting in a coronary thrombosis, which can cause a heart attack. The degree of blockage along with the heart area affected determines the severity of the attack. If allowed to increase in size as more blood cells attach to it, total blockage of blood flow can happen.

Another scenario is the potential for a portion, or several portions, of the thrombus to break off and flow through the circulatory system like an iceberg at sea. This floating thrombus, called an **embolus**, is not a problem until it travels down too narrow a blood vessel and becomes lodged, partially or totally blocking downstream blood flow. A cerebral embolus would affect blood flow to the brain, causing a stroke; a pulmonary embolus would lodge in the lung region and affect your ability to get oxygen into your blood, as you will see in Chapter 15.

Blood that is not travelling through the vessels at the rate it should, can also lead to unwanted clot formation. People who are bedridden, take long plane, bus or car journeys, or are immobile for extended periods of time are susceptible to thrombus formation. It also appears that women who smoke and use oral contraceptives and individuals on some types of chemotherapy are at a higher risk for clot formation.

Substances that decrease the blood's ability to coagulate, such as aspirin or heparin, help prevent unwanted clotting. Once a clot forms, 'clot busters' such as the drug streptokinase are given to regain proper blood flow.

Test your knowledge 13.3

Choose the best answer:

1. The universal donor blood type is:
   a. O-positive
   b. AB-positive
   c. Rh-positive
   d. B-positive

2. The smallest form of arteries are:
   a. arterules
   b. capillaries
   c. arterioles
   d. vessicles

3. A type of unwanted blood clot is a:
   a. bolus
   b. thrombus
   c. omnibus
   d. schoolbus

Complete the following:

4. List the three layers commonly found in a blood vessel:
   a. _____.
   b. _____.
   c. _____.

Common disorders of the cardiovascular system

The following are several types of cardiovascular conditions. This section discusses heart problems, blood vessel problems and blood disorders.

Pump problems

So far, we have discussed how a healthy heart works. However, certain events can affect the efficiency of the heart's pumping action. Remember that this is a *two-pump* system. What may damage one pump may not always damage the other pump.

Let's look at the right side of the heart first. Right-sided heart failure is a potentially serious condition in which the right-side pump can't move blood as efficiently as it should. This is a result of the heart muscles working harder than they normally do. As with any muscle that you exercise, heart muscle also becomes larger. In this case, the muscles on the right side of the heart become too large and can no longer efficiently pump blood. Disease conditions such as polycythaemia, in which the blood is thicker than normal and is harder to pump, or blood vessels in the lungs that constrict more than normal, making it harder to push blood through them, can cause the heart muscles to work harder. Because these two conditions are related to certain lung diseases, it is no surprise that 85 per cent of the patients with chronic obstructive pulmonary disease develop right-sided heart failure.

In right-sided heart failure the left side of the heart is pumping normally. As the left side pumps blood through your body and back to the right side, the right side cannot take all of the returning blood to pump it to the lungs. As a result, the blood begins to back up. The vessels in the body are flexible and can expand a little to take that extra volume of blood, so extended neck veins may be seen. Certain organs can hold more blood than usual, so engorgement of the liver and spleen may occur. As the volume of the blood increases in the vessels so does the pressure and eventually there is leakage of fluid from the vessels into the surrounding tissues. There is a gravity-dependent movement of the fluid and so swollen ankles, feet and/or hands can occur.

Now, let's consider left-sided heart failure. The healthy right pump pushes blood through the vasculature of the lungs on its way to the left-side pump. If the left side can't keep up with the blood being delivered to it, the blood backs up into the lungs, increasing the pressure in those blood vessels. Once that pressure reaches a certain point, fluid leaks out of the vessels and into the lung tissue. Pulmonary oedema is the term for fluid that forms in the lungs and causes difficulty breathing. See Figure 13.14.

Sometimes there is a problem with one or more of the heart valves that seal off the chambers during contraction. There are two possible types of problems: either the passageway through the valve is too small (stenosis) and restricts sufficient blood flow, or the passageway is too large and blood squirts backward into the chamber on contraction (valvular insufficiency). A problem that can occur in either case is the increased tendency to form clots in the damaged valve area. Such clots can detach, flow through the blood vessels, and cause a

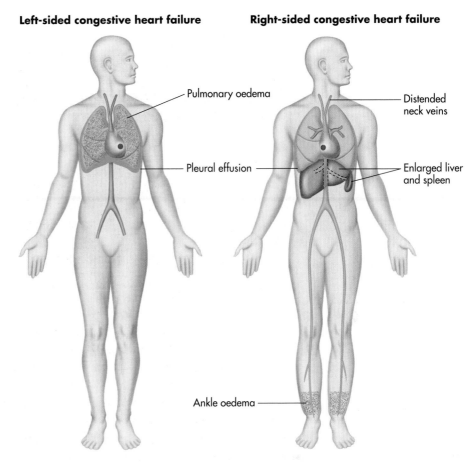

Figure 13.14 Left-sided and right-sided congestive heart failure.

pulmonary embolus in the lungs or can travel to the brain and cause a stroke. There are also potential problems with the small, specialised muscles called the papillary muscles that attach to the undersides of the cusps and contract when the ventricles contract so the cusps don't flap back up into the atria. The failure of the papillary muscles to properly contract will allow blood to flow backward into the atrium instead of flowing forward.

Vessel problems

A common problem with blood vessels occurs to some degree for all of us as we age. **Arteriosclerosis**, also known as *hardening of the arteries*, is a result of the thickening of the intima, which causes the involved vessels to become less flexible. Blood vessels in this condition have a tendency to rupture. Since these vessels are less flexible and can't readily accommodate increases in blood volume, the body is more susceptible to high blood pressure.

Normally, blood vessels have a smooth inner lining, which promotes efficient blood flow by decreasing resistance. **Atherosclerosis** is a potentially life-threatening condition in which fatty deposits, called **plaque**, build up on the inner lining of blood vessels. As a result, blood flow can become greatly restricted or totally blocked. The fatty material that makes up plaque is composed mostly of **cholesterol**. Interestingly, all blood vessels are susceptible to

arteriosclerosis (*ar tee ree oh skler ROW sis*)
  sclerosis = *hardening*

atherosclerosis (*ath er oh skler ROW sis*)
  athero *fatty or porridge like*
  sclero = *hardness*

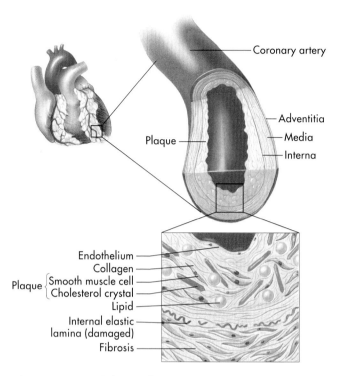

Figure 13.15 Atherosclerosis.

atherosclerosis, but the aorta and coronary arteries seem particularly suscept-ible to developing this condition. Cerebral arteries and those in the legs can also be affected (see Figure 13.15).

The largest arteries in the body are the femoral arteries, that carry blood to your legs, and are approximately 8–9 mm in diameter. The carotid arteries, that carry blood to the brain, are 5–6 mm in diameter, and the coronary arteries are approximately 3–4 mm in diameter.

If blood flow is restricted through one or more coronary arteries, the heart muscle may become oxygen-starved and cause tissue death. This would be a **myocardial infarction** (MI), or heart attack. If there is blockage of blood flow to the brain, a cerebral vascular accident (CVA) or stroke can occur. Poor blood flow to the arteries in the legs can lead to intermittent claudication – pain on walking. Reduced blood flow to tissues that leads to tissue *injury* but not tissue *death* is called tissue ischaemia.

Although the coronary arteries are some of the smaller arteries in the body, there are smaller arteries, for example the pudendal artery. This artery is approximately 1 mm in diameter and supplies the penis with blood. Like any artery in the body, it can be affected by atherosclerosis and this manifests as erectile dysfunction (ED). ED is the inability to gain or maintain an erection of sufficient strength for sexual activity, so ED can be a 'silent' indicator of atherosclerosis and therefore heart disease. The average time between onset of ED and first cardiac event is three and a half years. Therefore any man presenting with erection problems needs to be screened for cardiovascular disease.

Heredity seems to be one factor for atherosclerosis, and atherosclerosis is also a common complication of diabetes. It is interesting that many medical

infarct (*in FARKT*)

ischaemia (*iss KEE mee ah*)

professionals feel that diabetes should be classified as a cardiovascular disease. Diet and lifestyle also seem to predispose some individuals to atherosclerosis.

The main risk factors for atherosclerosis are considered to be male gender, age, diabetes mellitus, hypertension, smoking and hyperlipidaemia.

Arteries are designed to handle the increased pressures generated by a beating heart. In some individuals, those walls may not always be able to maintain their integrity. An **aneurysm** is a localised weakened area of blood vessel wall that may have been caused by a congenital defect, disease or injury. There also appears to be a familial tendency for abdominal aneurysms, which often present no warning symptoms. The danger is when the aneurysm expands to the point that it ruptures, causing haemorrhaging. If it is a major artery, an individual can bleed to death in a matter of minutes. If the aneurysm is detected on an X-ray or ultrasound, surgical intervention can remedy the situation. Early detection is important.

aneurysm (AN you rizm)

Blood problems

Secondary polycythaemia is a condition in which chronic low levels of oxygen (due to lung disease or living in high altitudes) cause the body to produce more than normal amounts of erythrocytes to transport more efficiently the smaller amounts of available oxygen. Primary polycythaemia does the same thing but can be caused by bone marrow cancer.

Anaemia is a blood condition in which there is a less than normal number of red blood cells or there is abnormal or deficient haemoglobin. Anaemia can be a result of bone marrow dysfunction, low levels of iron or vitamins, or the improper formation of red blood cells. Individuals with anaemia share these common symptoms:

anaemia (ah NEE mee ah)

- pale skin tone (*pallor*)
- pale mucous membrane and nail beds
- fatigue and muscle weakness
- shortness of breath
- chest pains in some heart patients due to decreased levels of oxygen being supplied to the heart.

Sickle cell anaemia is an inherited condition in which red blood cells and haemoglobin molecules do not form properly. The resultant red blood cells are crescent- or sickle-shaped and have a tendency to rupture. As they are destroyed, the body is stimulated to produce greater numbers of red blood cells to replace them. Unfortunately, at that high production rate, the blood cells cannot mature fast enough. The ruptured cells also clog up smaller blood vessels. Clogged vessels combined with the increased thickening of the blood from excess red blood cells and cell parts lead to increased clotting and an impaired ability to carry oxygen.

There are several blood problems involving white blood cells. **Leukaemia**, usually due to bone marrow cancer, is a condition in which a higher than normal number of white blood cells are produced. You might think this would be a good thing; however, the problem is that the white blood cells produced are immature and therefore ineffective in protecting the body from infection. **Leucocytosis** also exhibits as a situation in which there is a higher than

Clinical application

Heart attacks

A true heart attack occurs when there is an insufficient supply of blood from the coronary artery to the tissues of the heart. This could be a result of plaque build-up in the arteries decreasing flow or a piece of that plaque breaking off and occluding the artery. A clot that forms and blocks the artery can be another scenario. If the decreased blood flow is sufficient to kill heart tissue, the condition is called an acute myocardial infarction (or AMI). You may think that the classic heart attack is when the victim clutches his or her chest in extreme pain and falls over. However, most heart attacks start out slowly with little or no pain (often called a *silent MI*) and may progress over a few hours, days or even weeks, showing only subtle signs.

Symptoms that can be indicative of an MI are centrally located chest pain, chest heaviness or vague discomfort; pain in the left shoulder or shoulder blade, neck and jaw (where it mimics a toothache), radiating down the left arm; nausea, heartburn, weakness, or a clammy, sweaty feeling. Shortness of breath and/or dizziness can also be a warning sign. Women often exhibit 'non-traditional' signs and symptoms, such as pain in the shoulder blade or jaw, and such symptoms are missed as indicators of a heart attack, often with tragic results.

Another big problem is that the victim often goes into denial, trying to explain away the symptoms as the result of some other problem, such as indigestion, from eating too much or food that 'doesn't agree with me' or feeling that it's a gallbladder problem, or a pulled muscle. This can cost the patient valuable time in the treatment for a heart attack. Indeed, the first hour is when much of the heart damage occurs.

Emergency assistance should be obtained as quickly as possible if a heart attack is suspected. Research shows that an individual who is experiencing a heart attack should also immediately chew and swallow an aspirin tablet, preferably a baby aspirin or a plain, non-enteric coated adult aspirin (provided they are not allergic to them!). The anti-coagulating ability of aspirin helps to keep blood flowing through the coronary vasculature, hopefully decreasing the amount of damage to the heart muscle. Most heart attacks are survivable *if* you act quickly and seek treatment immediately. It is better to be safe than sorry!

haemophilia (*HEE moh FILL ee yah*)

thrombocytopenia (*THROM boh sigh TOH PEE nee ah*)

normal number of white blood cells. In this case, the cause is often an infection that is being fought. **Leucopenia** is a condition in which the number of white blood cells is lower than normal. This can be a result of drugs that suppress their production, such as corticosteroids and anti-cancer agents. Chronic infections can also wear the body down to the point that it cannot produce the necessary numbers of white blood cells.

Sometimes, there is a problem with the ability of blood to clot properly. Haemophilia is a general term used to describe inherited blood conditions that prohibit or slow down the blood's ability to clot. Thrombocytopenia is a condition in which there are fewer than normal circulating platelets. If the platelet count is low enough, even normal movement can lead to bleeding. This condition can be caused by liver dysfunction, Vitamin K deficiency, radiation exposure or bone marrow cancer.

The lymphatic connection

The lymphatic system, the topic of Chapter 14, has an important relationship with the cardiovascular system (see Figure 13.16). The lymphatic system runs parallel to the cardiovascular system and has three major responsibilities:

1. Helps to maintain the body's fluid balance by returning interstitial fluid to the venous side of the cardiovascular system.

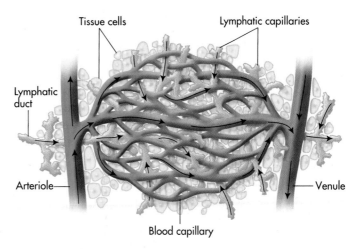

Figure 13.16 Relationship of the lymphatic system to the cardiovascular system.

2. Assists the cardiovascular system in distributing nutrients and hormones that may not easily enter the blood system directly, and assists in the removal of waste products from tissues.
3. Helps prevent infection and disease by utilising lymphocytes.

Summary

- The cardiovascular system is a closed, pressurised system.
- Like a system of rivers and canals, the cardiovascular system is responsible for transportation of oxygen, hormones and nutrients to the tissues of the body and for removing the by-products of metabolism by the cells.
- The cardiovascular system also helps maintain proper fluid balance of the body and assists in the control of body temperature.
- The cardiovascular system is a major player in the body's defence from infection.
- The heart is an organ that is actually two pumps working together to move blood.
- The heart's right pump moves blood collected from the body to the lungs, where oxygen is loaded, and carbon dioxide is removed to be exhaled by the lungs.
- The heart's left pump takes the freshly oxygenated blood and pushes it through the body so that tissue cells can be kept healthy.
- Arteries carry blood away from the heart.
- Veins carry blood to the heart.
- Capillaries are blood vessels with walls the thickness of only one cell, which readily allows for the transfer of oxygen and nutrients to the tissues in the body.
- This thinness of capillary walls also allows for waste products of the cells' metabolism to be picked up by the blood for removal.
- The major components of blood are plasma, erythrocytes (red blood cells, the main transporters of oxygen), leucocytes (white blood cells, protectors from infection) and platelets (aid in the clotting of blood).

Case study

A 55-year-old male (David) goes to his GP complaining about vague chest pains for the past several days. He said that he had a pain when out climbing a hill when on holiday abroad, but attributed that pain to being dehydrated. A quick patient history revealed the following: ex-smoker since the age of 18; height of 5ft 11 inches (1.82 m); weight 14 stone 4 lb (92.08 kg); lives with a partner, eats well but consumes more than the recommended daily alcohol limits. Although he does some walking, he drives to and from work and generally has a sedentary lifestyle and complains of occasional chest pains.

1. What disease process do you think this individual may be experiencing?

2. What examinations might be needed?

3. What suggestions would you give to the patient and his doctor?

Review questions

Multiple choice

1. A condition in which one side or the other of the heart cannot pump efficiently enough to overcome vascular resistance, resulting in excessive leakage from the vessels into tissues, is:
   a. heart failure
   b. cardiac tamponade
   c. vesiculitis
   d. atherosclerosis

2. Plaque deposits in blood vessels are composed mostly of:
   a. platelets
   b. cholesterol
   c. fibrin
   d. haem

3. A localised weakness in the walls of a blood vessel is called a(n):
   a. aneurysm
   b. altruism
   c. embolism
   d. stint

4. Which of the following symptoms would *not* normally be related to anaemia?
   a. shortness of breath
   b. fatigue
   c. increased urine output
   d. pallor

5. A general term to describe an inherited blood-clotting disorder is:
   a. haemopoiesis
   b. haemophilia
   c. haemoglobin
   d. polycythaemia

Fill in the blanks

1. List the three main responsibilities of the lymphatic system:
   a. _____.
   b. _____.
   c. _____.

2. Decreased blood flow to cardiac muscle that only *injures* the tissue creates a condition known as _____.

3. _____ is an important vitamin that is needed for the proper clotting of blood.

4. _____ is a term used for the dividing wall between the ventricles.

Short answers

1. Why is the direction of the wave of contraction so important in the heart?

   _____
   _____
   _____
   _____

2. Describe the flow of blood beginning at the right atrium and ending at the aorta.

   _____
   _____
   _____
   _____

3. Provide one reason why the proper amount of iron is so important in your diet.

   _____
   _____
   _____
   _____

4. Why can polycythaemia potentially cause a heart problem?

   _____
   _____
   _____
   _____

Suggested activities

1. Research and present to your group all the useful components of a donated unit of blood.

2. With a colleague take each other's blood pressure and pulse while sitting in a chair, lying down and immediately after walking up a flight of stairs. Compare your results with your classmates'. Discuss any similarities or differences.

Answers

Answers to case study

1. He has atherosclerosis affecting his coronary arteries and has experienced an episode of angina pectoris.
2. He needs to have an angiogram (an investigation to view the narrowing of the coronary arteries), a review of his blood profile (lipids as well as markers for myocardial infarction [troponin levels]), and an electrocardiogram (ECG). If his coronary arteries are narrowed, he may need a cardiac stent (a tube) that maintains the patency of the artery, or an angioplasty, where the fat in the artery is pushed into the arterial wall.
3. He needs to reduce his alcohol and saturated fat intake, and increase the amount of exercise he takes.

Answers to multiple choice questions

1. a
2. b
3. a
4. c
5. b

Answers to fill in the blanks

1. Maintains fluid balance by returning interstitial fluid to the venous side of the cardiovascular system; assists the cardiovascular system in distributing nutrients and hormones; helps prevent infection.
2. ischaemia.
3. Vitamin K.
4. Interventricular septum.

Answers to short answer questions

1. To ensure that the atria then ventricles contract in sequence. This ensures that blood is pumped from the atria to the ventricles, and then out into the pulmonary artery/aorta.
2. Blood enters the right atrium and flows into the right ventricle. When the sinoatrial node is activated, the remaining blood in the atria is forced into the right ventricle. From there the blood passes through the pulmonary semi-lunar valve into the pulmonary artery (the only artery in the body to carry deoxygenated or oxygen-poor blood) when the electrical impulse reaches the end of the Purkinje fibres (when the ventricles contract). Blood cannot pass back into the atria because the chordae prevent the tricuspid valve from opening. From the pulmonary artery, the blood goes to the right and the left lung where it releases carbon dioxide and takes on oxygen. The oxygen-rich blood returns to the heart via the pulmonary veins (the only veins in the body to carry oxygen-rich blood). The blood flows through the left atria into the left ventricle. From the left ventricle the blood exits via the aortic semi-lunar valve into the aorta.
3. It helps to make red blood cells.
4. Polycythaemia is a condition where there is excess cells in the blood. The thickness (viscosity) of the blood increases resulting in the myocardium working 'harder' to pump the 'thicker' blood around the body.

The lymphatic and immune systems

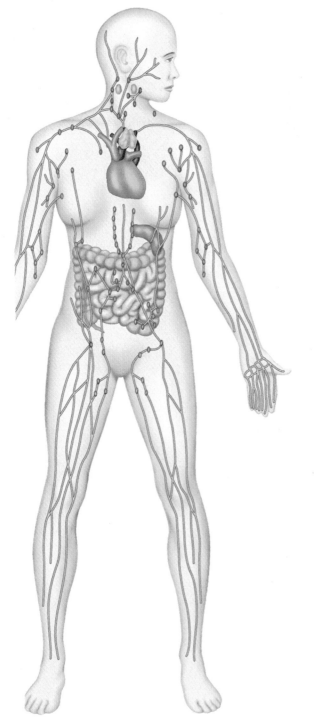

So far, we have considered control systems, transport systems and infrastructure. We have seen how each separate system works together to allow the body to function as an integrated unit. However, like all cities, the body must be protected. Cities have police, ambulance and fire stations. On a larger scale, countries have an army, navy and air force bases. Similarly, your body has the immune and lymphatic systems with a variety of protective mechanisms and cells each performing specific duties. These systems help to protect the body from pathogens that can produce disease. Without your immune and lymphatic systems, your life would be a very short one – the first exposure to a potential pathogenic organism would seriously affect the health of your body with disastrous consequences.

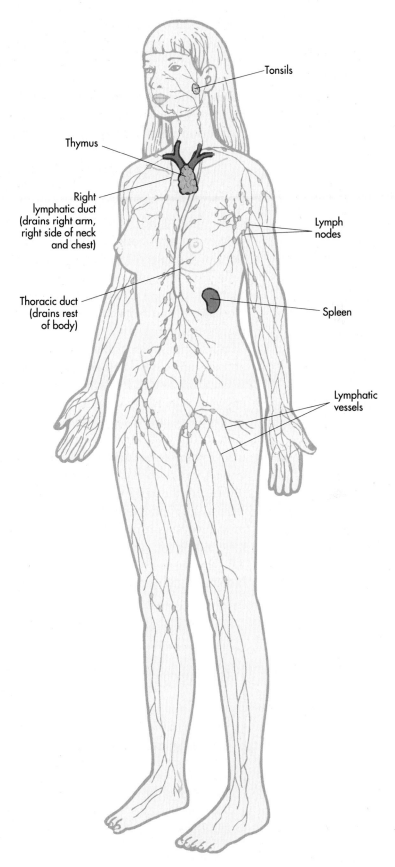

Figure 14.3 The spleen and thymus.

The fluid that moves from blood capillaries into tissues is thereby filtered through the white blood cells in the lymph nodes where pathogens are removed and destroyed and recycled back into the cardiovascular system.

As previously mentioned, there are two larger collections of lymphatic tissue, known as lymph organs. These lymph organs are the spleen and the thymus. While they are not strictly lymph nodes and are not part of lymph circulation, they are so similar to lymph nodes that they are classified as part of the lymph system.

The **spleen** (see Figure 14.3) is a spongy organ in the upper left quadrant of the abdomen. It is structurally similar to lymph nodes but instead of having lymphatic sinuses, the spleen has blood sinuses. Surrounded by the blood sinuses are islands of white pulp containing lymphocytes and islands of red pulp containing both red blood cells and white blood cells. One of the functions of the spleen is to remove and destroy old, damaged or fragile red blood cells. Can you guess the second function of the spleen, given its anatomy? Since it is similar to a lymph node, the spleen also filters pathogens from the bloodstream and destroys them in the same way that lymph nodes filter pathogens from lymph.

While the spleen is a very important organ, it is not one of the vital organs and in some cases, such as trauma, may need to be surgically removed. While its removal in children can severely compromise their ability to ward off disease, the spleen's removal in an adult has much less effect. One reason for this is that as we age, the body becomes better at fighting off infections that it has seen in the past, but new invaders present more of a challenge when we are first exposed to them as children. Because of the spleen's rich supply of blood vessels, injury to the spleen can often cause internal bleeding.

thymus (*THIGH muss*)

The **thymus** is a soft organ located between the aortic arch and sternum (again, see Figure 14.3). The thymus is very large in children because of all the new infections it must be ready to fend off. It gets smaller or even disappears in adults as the immune system fully matures in its ability to fight infection. The thymus is packed with lymphocytes, which mature into a type of white blood cell called a T lymphocyte. The thymus also secretes a hormone called thymosin, which stimulates the maturation of T lymphocytes in lymph nodes.

Clinical application

Cancer stages

When patients are diagnosed with cancer, they are often told they have a certain 'stage' of cancer. Cancer is staged and prognosis is determined by the amount of metastasis (spread). The lymphatic system is amazingly efficient at capturing and transporting pathogens from connective tissue to lymph nodes for destruction. Unfortunately, this ability can also be a liability when cancerous cells develop in one part of the body and use the lymphatic system to hitch a ride to distant areas of the body. Once the cancerous cells make their way from their point of origin into lymphatic capillaries, and if the lymphocytes in the lymph nodes do not overpower the tumour cells, cancer cells can easily move around the body, invading many different areas simultaneously.

Cancers that are diagnosed after they have already spread are much more likely to be fatal than cancer that is treated before cells have a chance to spread. This is why screening for certain types of cancer is so important and often why lymph nodes are removed for study around surgical sites of cancer, for example breast cancer. The earlier a patient is diagnosed, the better his or her chances of beating the disease. Though most types of cancer have specific staging criteria, cancer stages generally follow this pattern:

- Stage 1: no spread from origin
- Stage 2: spread to nearby tissues
- Stage 3: spread to nearby lymph nodes
- Stage 4: spread to distant tissues and organs

Stage 4 cancers are often terminal.

Test your knowledge 14.1

Choose the best answer:

1. Any organism that invades your body and causes disease is known as a:
   a. bacteria
   b. fungus
   c. virus
   d. pathogen

2. Which of the following areas do *not* have large numbers of lymph nodes?
   a. cervical
   b. axillary
   c. abdominal
   d. adrenal

3. Cancer that has entered the lymph nodes is in this stage:
   a. first
   b. second
   c. third
   d. centre

4. Which cells are housed in lymph nodes?
   a. red blood cells
   b. white blood cells
   c. platelets
   d. lymph nodules

5. The thoracic duct of the lymphatic system empties into this blood vessel.
   a. right subclavian vein
   b. left subclavian vein
   c. aorta
   d. hepatic portal vein

6. The function of the thymus is to:
   a. remove pathogens from blood
   b. destroy damaged red blood cells
   c. both a and b
   d. none of the above

The immune system

If the lymphatic system can be considered a transport and storage system for the body's defence systems, then the components of the immune system are the weapons and soldiers that can be called in to protect a city in times of extreme need. The immune system is a series of cells, chemicals and barriers that protect the body from invasion by pathogens. Some of the weapons are active, some are passive, some are inborn, others change with experience. Together they form a system that is remarkably good at keeping the body free of infection.

Antigens and antibodies

antigen (*AN tea jenn*)

As you should remember from our discussion of blood types in Chapter 13, cells have molecules on the outer surface of their membranes to distinguish whether they are friend or foe. These molecules are called antigens. Each human being has his or her own unique cell surface antigens, as do all other living things, including bacteria, viruses, animals and plants. The presence of these unique fingerprints or antigens allows the immune system to distinguish between cells that are naturally yours and cells that are not. This ability, called **self-recognition** and **non-self-recognition**, is at the heart of immune system function.

Antigens are like the identity codes sent out by aeroplanes. Air traffic controllers and fighter jets on patrol over a no-fly zone depend on the identity codes to tell which aircraft are friendly. Antigens do the same for your immune system. A well-functioning immune system ignores your antigens (self) and attacks other antigens (non-self). We discuss this in more depth later in this chapter.

As part of its defence system, the body can make proteins that bind to antigens, eventually leading to their destruction. Can you remember these proteins from the discussion of blood types? Yes! They are **antibodies**, one of the most potent weapons in the body's defensive arsenal. Therefore, antibodies are called into action when a foreign antigen invades the body. Antibodies are also known as **immunoglobulins** or **Igs** and there are five classes of them – IgM, IgA, IgD, IgG and IgE. Each class is found in slightly different regions of the body and has different functions (see Table 14.1).

Innate versus adaptive immunity (specific versus non-specific immunity)

The immune system defends the body on two fronts, by **innate immunity** and **adaptive immunity**. Innate immunity is the first line of defence against invasion. Innate immunity, as the name suggests, is the body's inborn ability to fight infection. Innate immunity prevents invasion, or, if pathogens do get inside, innate immunity recognises the invasion and takes steps to stop the infection from spreading. However, innate immunity can only recognise that something is not you, it can't identify the invaders. Innate immunity *cannot* improve with experience and, because it does not recognise specific pathogens, it cannot 'remember' an infection the body has encountered before.

Innate immunity consists of a collection of relatively crude mechanisms for defending the body from infection, sort of like building a wall around a city or having metal detectors at the airport. Walls can keep out some invaders and metal detectors can tell if someone might be carrying a metallic weapon, but neither can respond to specific threats. (There is a world of difference between car keys and pocket knives, but metal detectors cannot tell them apart. That's why you have to take your keys out of your pocket when you go through a

Table 14.1 Immunoglobin classes

Class	Generalised structure	Where found	Biological function
IgD		Virtually always attached to B cell	Believed to be cell surface receptor of immunocompetent B cell; important in activation of B cell
IgM	J chain	Attached to B cell; free in plasma	When bound to B cell membrane, serves as antigen receptor; first Ig class released to plasma by plasma cells during primary response; potent agglutinating agent; fixes complement
IgG		Most abundant antibody in plasma; represents 75% to 85% of circulating antibodies	Main antibody of both primary and secondary responses; crosses placenta and provides passive immunity to fetus; fixes complement
IgA	J chain	Some (monomer) in plasma; dimer in secretions such as saliva, tears, intestinal juice and milk	Bathes and protects mucosal surfaces from attachment of pathogens
IgE		Secreted by plasma cells in skin, mucosae of gastrointestinal and respiratory tracts, and tonsils	Binds to mast cells and basophils, and triggers release of histamine and other chemicals that mediate inflammation and certain allergic responses

metal detector.) Other parts of innate immunity are like weapons of mass destruction, indiscriminately killing pathogens and healthy tissue alike.

Innate immunity is backed up by a platoon of mechanisms that specifically target invaders, can remember invaders from previous encounters and therefore prepare for future invasions, and can improve their responses with experience. These mechanisms are known as *adaptive immunity* because the mechanisms 'learn' and change each time they are engaged. The components

Test your knowledge 14.2

Choose the best answer:

1. Cell surface molecules that can be used to identify cells are called:

   a. antibodies

   b. antigens

   c. antihistamines

   d. antibiotics

2. Proteins that bind to antigens are called:

   a. binding proteins

   b. receptors

   c. hormones

   d. antibodies

3. This type of immunity has no memory and is not specific:

   a. adaptive

   b. acquired

   c. innate

   d. non-specific

of adaptive immunity can be trained as an elite fighting force for particular pathogens. Their goals are 'surgical strikes' targeting particular invaders and sparing as much of the healthy body tissue as possible.

It is tempting to think of these two parts of immunity as separate entities. However, the two work closely together. Innate immunity prepares the way for adaptive immunity, weakening some pathogens and stimulating components of adaptive immunity. Adaptive immunity in turn further stimulates innate immunity. It is through the mutual cooperation of both innate and adaptive immunity attacking the pathogen on two fronts that invaders can be removed from the body.

Components of the immune system

We often think the immune system begins in the blood with the white blood cells. However, physical barriers exist to act as a first line of defence to attempt to stop the infective agents from getting into the body in the first place. This is much like concrete traffic barriers you see in front of public buildings.

Barriers

Anything that prevents invaders from getting inside your body prevents infection. Therefore, your body has many barriers located in the places where invaders are most likely to gain entrance. Physical barriers include skin and the mucous membranes of the eyes, and the digestive, respiratory and reproductive systems. Not only are these surfaces difficult to penetrate, they are packed with white blood cells and lymph capillaries to trap any invaders that might get through. The fluids associated with these physical barriers contain chemicals that act as chemical barriers. These chemicals are contained in tears, saliva, urine, mucous secretions and sweat. One example is the oil secreted by the sebaceous glands of your integumentary system, which can be antibacterial. These barriers, both chemical and physical, prevent some invaders from ever getting inside the body. They are the 'fortress' of the body, part of your innate immunity. After reading this information, can you see why wounds are frequent sources of infection?

Cells

leucocytes (*LOO koh sights*)
leuco = *white*
cyte = *cells*

If an invader has an opportunity to enter the body, white blood cells (**leucocytes**) are responsible for defending the body against invaders. You learned about red blood cells and platelets in Chapter 13. Red blood cells are responsible for carrying oxygen throughout the body, and platelets are responsible for blood's ability to clot. White blood cells, on the other hand, are the mobile units of the immune system. White blood cells, which form in the bone marrow like red blood cells and platelets, move to other parts of the body to grow and mature until they are needed during an invasion. They are generally not released into the bloodstream in large numbers unless an infection is present.

Leucocytes can be divided into two groups. Polymorphonuclear granulocytes are cells with granules or spots in their cytoplasm. Agranulocytes or mononuclear cells have no granules in their cytoplasm. These two groups contain several different types of cells that play a role in the body's defence against infection. Figure 14.4 shows the major white blood cells found in the plasma.

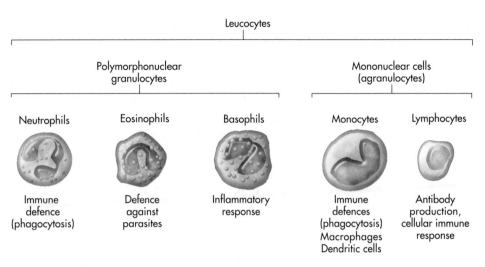

Figure 14.4 Major leucocytes.

Types of white blood cells

To police a large city, several types of jobs are needed in order to function effectively: traffic officers, detectives, scenes of crime officers and more. Each has a specific duty and is called upon as needed, but in essence they all have the same goal – fighting crime. Similarly, the body has various types of white blood cells that are required to protect the body in varying circumstances. The following represents a description of the major white blood cells in the plasma and the additional specialised white blood cells of the lymphatic system.

Neutrophils

neutrophil (*NEW trow fill*)

Neutrophils are granulocytes whose function is phagocytosis. (Remember phagocytosis from Chapter 4?) Phagocytic cells ingest pathogens and cellular debris.

Neutrophils originate in the bone marrow and are the most common leucocyte in the bloodstream. Neutrophils are the first cells to arrive at the site of damage. They immediately begin to clean up the area by ingesting pathogens, and they release chemicals that increase tissue damage and inflammation, stimulating immune response. Neutrophils are part of innate immunity.

macrophages (*MACK row fage ez*)

Macrophages are modified monocytes, a type of agranulocyte, which leave the bloodstream and enter tissues. They are also phagocytic cells, which are active in the later stages of an infection. In addition to phagocytosis, these cells release chemicals to stimulate the immune system and they have an important role to play in the process of wound healing. Macrophages are also part of innate immunity.

basophil (*BAY soh fill*)

Basophils and mast cells release chemicals to promote inflammation. Basophils are granulocytes, which are mobile, entering infected tissues from the bloodstream. Basophil numbers are very low unless an active infection is present. Mast cells are not mobile and are found stationed throughout the body. They are always found in connective tissue. For example, when the mast cells in the connective tissue of the nose are stimulated, chemicals are released that lead to rhinitis, more commonly called 'runny nose'. Mast cell stimulation in the lungs can trigger an asthmatic attack. Both basophils and mast cells are part of innate immunity.

eosinophil (*ee oh SIN oh fill*)

Eosinophils are granulocytes that counteract the activities of basophils and mast cells. The function of eosinophils is to break down the chemicals released by basophils and mast cells, thereby slowing or stopping inflammatory response so it doesn't go too far. Eosinophil numbers are generally low in the bloodstream unless active infection or allergies are present. Eosinophils also have a role in fighting invasion by parasitic worms and are part of innate immunity.

dendritic cells (*den DRIT ick*)

Dendritic cells are another member of a group of cells that are modified monocytes. All of these cells are weakly phagocytic. However, their most important job is as **antigen-displaying cells**, or ADCs. These cells are able to ingest a foreign cell and place the foreign antigens into their own cell membrane. Then the dendritic cell cruises the lymph nodes, displaying the foreign antigens and looking for the lymphocytes that match the antigen. This is an important trigger of adaptive immunity. ADCs are the red flags that alert your adaptive immune system to respond. They are an important bridge between innate and adaptive immunity.

Natural killer cells (NK cells) are a type of lymphocyte. These lymphocytes are part of innate immunity. NK cells are crude weapons, releasing chemicals to kill any cells displaying foreign antigens, whether they are pathogens or the body's own infected cells. NK cells 'take out' the neighbourhood. These cells patrol the body, wiping out any infected cell they encounter. Some early symptoms of a cold or flu are actually due to the action of the NK cells damaging tissues, not from the infection.

T lymphocytes (*T LIMF oh sights*)

T lymphocytes (T cells) are lymphocytes responsible for a portion of adaptive immunity known as cell-mediated immunity. There are several different types of T cells, including **cytotoxic T cells** (literally 'cell poison'), which kill infected cells and release immune stimulating chemicals; **helper T cells**, which help activate other parts of adaptive immunity; **regulatory T cells**, which regulate immune response; and **memory T cells**, which remember pathogens after exposure.

B lymphocytes (*B LIMF oh sights*)

plasma cells (*PLAZ mar*)

B lymphocytes (B cells) are lymphocytes responsible for the part of adaptive immunity known as antibody-mediated immunity. There are two types of B cells: plasma cells, which produce antibodies to non-self-antigens, and **memory B cells**, which remember pathogens.

Chemicals

Not only do blood cells fight invaders, but chemicals found in the body can also assist in neutralising and destroying invaders. Cytokines are proteins produced by damaged tissues and white blood cells that stimulate an immune response in a variety of ways, including increasing inflammation, stimulating

cytokines (*SIGH toe kines*)

lymphocytes and enhancing phagocytosis. Cytokines are involved in both innate and adaptive immunity.

interferon (*in ter FEAR on*)

Interferon is a cytokine produced by cells that have been infected by a virus. Interferon binds to neighbouring, uninfected cells and stimulates them to produce chemicals that may protect these cells from viruses. Interferon has also had some success as an anticancer drug, but it is still considered experimental.

tumour necrosis factor (*neck ROW siss*)

Tumour necrosis factor, or TNF, is a cytokine produced by white blood cells. It stimulates macrophages and also causes cell death in cancer cells. A new class of drugs that inhibit TNF has been very successful in treating rheumatoid arthritis. Many cytokines are types of molecules called **interleukins**.

interleukins (*in ter LOO kins*)

There are at least ten different interleukins. They are involved in nearly every aspect of innate and adaptive immunity. Interleukins have also been used with moderate success in treating some forms of cancer.

Complement cascade is a complex series of reactions that activate 20 proteins that are usually inactive in the blood unless activated by a pathogen invasion. The proteins are C1 to C9, factors B, D and P, and some other regulatory proteins. When these proteins are activated, they have a variety

lysis = *to break down or destroy* (*LYE siss*)

of effects, including **lysis** (destruction) of bacterial cell membranes, stimulation of phagocytosis, attraction of white blood cells to the site of infection (chemotaxis), clumping of cells with foreign antigens, and alteration of the structure of viruses. Complement cascade is part of both innate and adaptive immunity.

The complement cascade can be activated in two ways. The **classical pathway** depends on antibodies binding to invading organisms to form antigen–antibody complexes and then one of the complement proteins (usually C1) attaches itself to the antigen–antibody complex – this is called complement fixation. The **alternative pathway** is caused by an interaction between the surface molecules of certain types of micro-organisms and factors B, D and P.

Both pathways then cause a cascade of complement proteins to be activated in an orderly sequence, leading eventually to enhanced phagocytosis and inflammation plus cell lysis. Cell lysis is achieved via **opsonisation** – a process that involves complement proteins being inserted into the cell membrane of the invading micro-organism, forming an open hole. This process also facilitates enhanced attachment of macrophages and neutrophils to the cell membrane, promoting faster phagocytosis.

Inflammation

Inflammation, or the inflammatory response, is one of the most familiar weapons in the body's arsenal. You have all experienced the swelling, pain, heat and redness associated with inflammation at one time or another. Think back several chapters ago to our example of hitting your thumb with a hammer. What would happen to your thumb within a few minutes of injury? It would swell, turn red, get hot to the touch, and hurt for some time after the injury. What about an infected cut? Or a sore throat when you have a cold or the flu? Again, the symptoms are the same: redness, heat, swelling and pain. There is also often loss of function of the affected area, usually due to the pain and swelling. This reaction is a deliberate action of your body in response to tissue damage, whether a mechanical injury, like hitting your thumb with a

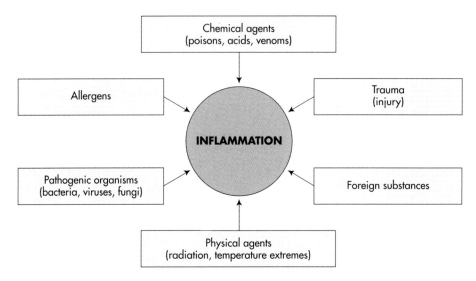

Figure 14.5 Causes of the inflammatory response.

hammer, or damage due to the invasion of a pathogen, as in a wound infection or a sore throat. Part of this response helps to 'wall off' the infected area to prevent further spread and allow the battle to focus at this site. This process is called *margination* and is an attempt to isolate the problem.

When tissue is damaged, the cells send out chemicals such as histamine, a cytokine, which have several effects. These chemicals attract white blood cells to the site of injury, increase the permeability of capillaries and cause local vasodilatation. Extra fluid moves from the capillaries into the damaged tissue, causing swelling. More blood comes to the site, increasing the temperature of the tissue and enhancing local skin colour. White blood cells enter the area, destroying pathogens and clearing away dead and dying cells. The increase in fluid and cells coming to the area increases the pressure and is part of the reason the area remains painful even as the damage is being repaired. Inflammation is an innate immune mechanism, but it is also an important player in adaptive immunity. There are several causes of an inflammatory response. Please see Figure 14.5.

histamine (*HIS tar mean*)

Fever

During an infection, tissues and components of the immune system release a number of cytokines, which promote inflammation and immune responses. These cytokines circulate through the bloodstream and often reach distant targets, including the brain. One of the cytokine targets in the brain is the hypothalamus, which is responsible for setting and maintaining body temperature. Under the stimulation of cytokines, the hypothalamus raises the body's temperature set point. You feel cold and begin to put on more clothes, to huddle under more blankets and even to shiver. Eventually, body temperature rises to the new set point and a fever results. While unpleasant to experience, this rise in body temperature is a deliberate attempt by the immune system to destroy the pathogens that have invaded the body. Like the rest of the innate response system, fever is a crude weapon against invaders. It might help fight off the infection, but it also causes overall discomfort.

Clinical application

Inflammation – a double-edged sword

Inflammation, like many of the body's weapons, has a positive feedback loop. Once inflammation starts, it continues until turned off. This kind of runaway positive feedback can cause problems if localised swelling increases pressure, causing more tissue damage. One of the reasons you apply an ice pack to a sprained ankle is to decrease inflammation to prevent further damage. Inflammation is a particular problem in enclosed or small spaces like the brain, spinal cord, respiratory system and extremities where a small build-up of pressure can cause serious damage. Even more dangerous is inflammation that becomes systemic, spreading throughout the whole body. This type of inflammation, called anaphylaxis, often causes blood pressure to plummet due to widespread vasodilatation. Some people who are allergic to insect stings may experience this kind of inflammation. Anaphylaxis can be fatal unless treated by a medical professional immediately.

Test your knowledge 14.3

Choose the best answer:

1. Which of the following is *not* a function of complement cascade?

    a. lysis of bacterial cell membrane

    b. stimulation of macrophages

    c. chemotaxis

    d. swelling

2. These chemicals may protect the body against viruses and some cancers:

    a. complement

    b. cytokines

    c. interferons

    d. immunoglobulins

3. B lymphocytes are directly responsible for:

    a. cell-mediated immunity

    b. inflammation

    c. complement cascade

    d. none of the above

4. Neutrophils and macrophages aid the innate immune system by doing this:

    a. secreting cytokines

    b. phagocytosis

    c. stimulating immune response

    d. all of the above

5. Redness, heat, swelling, pain and loss of function are all symptoms of:

    a. complement

    b. fever

    c. inflammation

    d. infection

How the immune system works

Think back to when you were in school and all the colds and sore throats you and your friends suffered, not to mention chicken pox. Compare that to the frequency of colds and sore throats you get as an adult. Chances are the number is much lower. What has happened?

Innate immunity (non-specific)

As we have seen, for a pathogen to successfully invade your body, it must first get past your physical and chemical barriers. (Think of your body as a castle and the barriers of innate immunity as the alligator-filled moat protecting the castle from the marauding hoards.) Most of the millions of pathogens you encounter each day are kept out by these barriers. However, some pathogens – influenza or the cold virus, for example – are very good at getting past barriers.

When a pathogen does get past the barriers, the more active portions of innate immunity are activated. The presence of a foreign antigen is detected by neutrophils. Neutrophils ingest the foreign antigen, destroying it, and release chemicals (cytokines, for example) that attract other white blood cells to the site of infection and stimulate inflammation. (Continuing our castle analogy, the neutrophils are the guards who greet any marauders who manage to survive the alligators.)

The release of cytokines and stimulation of inflammation attract macrophages and NK cells to the infection site. Macrophages destroy more infected cells by phagocytosis. NK cells use chemicals to destroy infected cells. Both cells release chemicals that further stimulate inflammation, activate more immune cells and trigger the complement cascade. (The castle guards sound the alarm that the castle has been breached, summoning more troops to fight the invaders.)

At this point, the infected cells, or the pathogens themselves, are under attack on several fronts: phagocytosis, noxious chemicals, membrane rupture, clumping and even alteration of their molecular structure. Chemicals have signalled your hypothalamus to raise your body temperature, and you run a fever. You feel like . . . , well you know how you feel. There is no question in your mind that you are ill. (Even though the guards are protecting the castle, the castle will be damaged.)

You would think this would be enough to fight off most pathogens. But keep in mind that this is crude warfare. Innate immunity simply destroys anything non-self. It does not use surgical strikes or specific weapons. Innate immunity lays waste to the infected area with almost indiscriminate attacks. Defending the castle is warfare in the crudest sense – no specialised weapons, just desperate attempts to defeat the invaders. Uninfected cells can be destroyed in the process.

In some cases, these mechanisms are enough, but often the innate immune system is buying time for adaptive immunity to ready the 'big guns', the B and T cells of adaptive immunity. Indeed, the activities of innate immunity stimulate adaptive immunity. Chemicals released by NK cells, neutrophils and other cells help activate adaptive immunity. When phagocytic cells ingest pathogens, they display the foreign antigen on their cell membranes. This ability to display foreign antigens without being infected is absolutely necessary for activation of B and T cells.

Adaptive immunity (specific)

Fighting specific pathogens is the job of adaptive immunity. This part of the immune system has memory, 'learns' with experience and recognises specific pathogens. It is because of adaptive immunity that people generally only get chicken pox once. (Thank goodness for that!) The cells responsible for the adaptive immune system, B and T lymphocytes, remember pathogens and mount specific responses to those pathogens if they meet again. It is because of adaptive immunity that immunisations are able to prevent illness. When was the last time you heard of somebody in the UK getting polio? Just 60 years ago, polio was an epidemic. Keep in mind that innate and adaptive immunity work hand in hand. One cannot do its job without the other.

Lymphocyte selection

In order to function, lymphocytes must be able to recognise pathogens and to ignore the body's own tissues. Think of these cells as members of the local fire service. They have a very specific job – putting out fires. Like any professional fire-fighter, lymphocytes must be **selected**. During **positive selection**, lymphocytes that actually recognise and bind to antigens are allowed to survive. Lymphocytes that fail to do their job do not survive, much like a fire-fighter who cannot carry a hose or climb a ladder will not become part of the service. Unfortunately, some of these selected lymphocytes actually recognise and bind to your antigens. If not destroyed, they will attack and destroy your own tissues. (By the same token, the fire service takes care not to select arsonists to work for them.) The destruction of self-recognising lymphocytes is known as **negative selection**. Both positive and negative selection must work in order for your immune system to function appropriately. You must have lymphocytes that attack invaders but don't attack you.

Lymphocyte activation

Lymphocytes develop and mature when you are a baby or a very young child. Just like you, they begin as an undifferentiated cell, meaning they have the potential to become anything. They must then undergo a maturation process to become **differentiated** or, in other words, to grow up to be a specialised cell with a specialised function. While undifferentiated lymphocytes are produced in the bone marrow, some migrate to the thymus and are destined to become T cells. Others stay, develop, and mature in the bone marrow to become B cells.

After they are specialised, the lymphocytes hang out in the lymph nodes waiting for a pathogen to come along that they recognise. The lymphocytes can go into suspended animation for all that time. In order for the lymphocytes to fight off the pathogen, they must have a wake-up call. Picture the fire-fighters sleeping in the middle of the night and the alarm going off. This wake-up call is called **lymphocyte activation**. It causes lymphocytes to circulate continuously in the bloodstream and lymph system to combat the pathogen. See Figure 14.6 to see how the immune system activates and differentiates lymphocytes.

Let's go back to innate immune response for a minute. When a pathogen invades your tissues, your innate immunity mounts a response to the pathogen. One part of innate immunity is phagocytosis of infected cells or bits of pathogen. When these cells, macrophages, dendritic cells and others, eat the pathogen, the pathogen's antigens are displayed on the outside of the phagocytic cells. These cells then prowl the lymph nodes, displaying these tiny bits

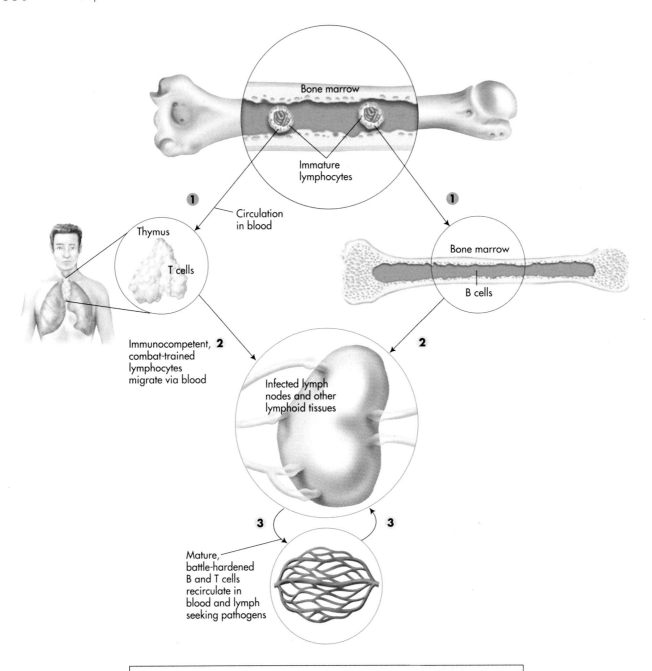

Figure 14.6 Lymphocyte differentiation and activation.

of pathogen, searching for the lymphocytes that can recognise the pathogen bits. When the right lymphocytes meet the right antigen display, the lymphocytes are activated (their wake-up call for battle). This activation is the beginning of adaptive immunity. Remember, water doesn't come out of a fire hose until the hose is turned on.

Lymphocyte proliferation

The body has only a few lymphocytes that recognise each invader to which it has been exposed. But in order to fight off an infection, hundreds of thousands of lymphocytes are needed to attack the infection. Simple activation of lymphocytes is not enough. The activated lymphocytes must make thousands of copies of themselves in order to fight off the thousands of pathogens reproducing in the body. This reproduction of lymphocytes is called **lymphocyte proliferation** (see Figure 14.7). Just like one platoon would not be enough to repel a full-scale invasion of a country, or one fire engine enough to save a street from burning down, a few lymphocytes are not enough to defend the entire body.

When you first met lymphocytes, you discovered that B cells and T cells have very different roles in the defence of the body. While activation and selection is, for our purposes, the same for both B and T cells, proliferation is different depending on the type of lymphocyte that is being reproduced.

There are two types of proliferation: proliferation of helper T cells and proliferation of all other types of lymphocytes. Remember, the job of helper T cells is to help other lymphocytes. There must be lots of helper T cells before any other lymphocytes can be activated. Helper T cells are stimulated to divide by binding to antigen-displaying cells (from innate immunity) and by stimulation by cytokines (some secreted by cells from the innate immune response and inflammation). The helper T cells continue to divide, producing more helper T cells, and then help (hence their name) in the proliferation of B cells and other types of T cells (again, see Figure 14.7).

Clinical application

AIDS

AIDS (acquired immune deficiency syndrome) is caused by infection with the human immunodeficiency virus (HIV), a virus that specifically targets and destroys a type of helper T cell called CD-4. CD-4 cells are necessary for the proliferation of B cells and cytotoxic T cells. As helper T cells are infected with HIV and begin to die, their numbers decrease dramatically, so B cell and T cell proliferation is too slow to respond to an infection. Patients with full-blown AIDS often die from massive infections that they would otherwise be able to fight easily. These same infections are also found in patients with other types of immune deficiency diseases or who are taking immunosuppressant drugs to prevent post transplant organ rejection or to treat autoimmune diseases.

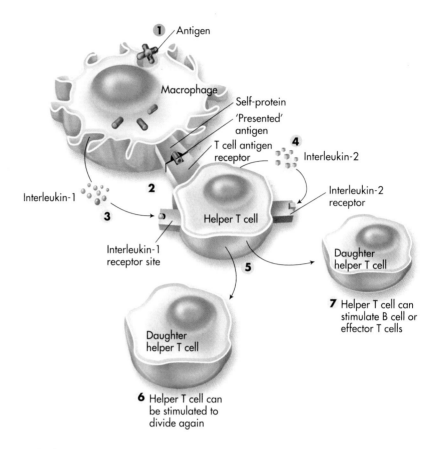

Figure 14.7 Activation and proliferation of helper T cells.

KEY:

1 Antigen processing cell such as a macrophage ingests a pathogen and displays its antigen on the macrophage's cell membrane.

2 Presented antigen is recognised by helper T cell.

3 The macrophage secretes a cytokine called interleukin-1.

4 Interleukin-1 stimulates the helper T cell to secrete interleukin-2 which combines with interleukin receptors.

5 When interleukin-2 binds to the receptor site the helper T cell divides.

6 The daughter cell produced can divide again if exposed to the same antigen. This greatly increases the number of helper T cells which can even further divide proportionate to the antigen exposure.

7 As helper T cells continue to divide they can stimulate B cell activation and produce effector T cells.

Helper T cells are absolutely necessary for the reproduction of B cells and other types of T cells. This is why HIV, the virus that causes AIDS, is so devastating to the immune system. HIV targets helper T cells specifically. To understand the role of the helper T cells, imagine a huge game of tag in which no lymphocytes can be activated or reproduced until they have been tagged by a helper T cell.

Test your knowledge 14.4

Choose the best answer:

1. In order to be activated, B and T cells must bind with:
   a. an antigen-displaying cell
   b. a pathogen
   c. a damaged cell
   d. all of the above

2. Selection for immune-competent cells is called:
   a. negative selection
   b. positive selection
   c. immune selection
   d. make a selection

3. After a lymphocyte is activated, what must it do before it can fight off a pathogen?
   a. die
   b. agglutinate
   c. proliferate
   d. congregate

4. The primary function of these cells is the activation of other lymphocytes:
   a. cytotoxic T cells
   b. helper T cells
   c. memory T cells
   d. regulatory T cells

B and T cell action

So, let's get back to the body's response to invasion. The innate immune system has been attacking the pathogen or cells infected with the pathogen on a number of fronts, using phagocytic cells, NK cells, fever and a variety of noxious chemicals. A number of the body's own cells have been destroyed in the process, but the pathogen has not been defeated. However, while innate immunity has been holding down the fort, antigen-displaying cells (ADCs) have sent out a signal calling on the weapons of adaptive immunity, B cells and T cells.

B cells

B cells are responsible for a type of adaptive immunity known as **antibody-mediated immunity**. B cells fight pathogens by making and releasing antibodies to attack a specific pathogen. As we saw previously, B cells develop into **plasma cells** and **memory B cells**. Antibodies are made by plasma cells and released into the bloodstream. Antibodies bind to the antigens of infected cells or the antigens on the surface of freely floating pathogens. They are missiles programmed to home in on a specific target. Antibodies destroy pathogens using several methods, including inactivating the antigen, causing antigens to clump together, activating the complement cascade, causing the release of chemicals to stimulate the immune system, and enhancing phagocytosis. This response to the pathogen is called **primary response**.

All of these antibody-mediated mechanisms not only destroy pathogens specifically, they also further stimulate both adaptive and innate immune responses, continually increasing the response to the pathogen. Remember that the immune response is a positive feedback loop that must be deliberately turned off. It will not stop on its own. This makes sense if you think in terms of protection. Your protective systems should not give up until the danger is past. Smoke alarms keep wailing until there is no more smoke.

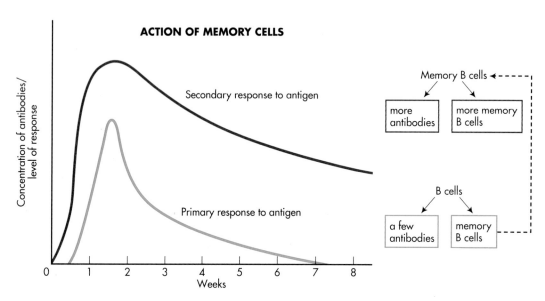

Figure 14.8 The primary response causes B cells to produce *memory* B cells and a few antibodies. The second exposure causes the secondary response to produce more memory B cells and even more antibodies to fight the invaders. Now that the body has antibodies and more memory B cells, the secondary response begins more rapidly after exposure, produces more antibodies, and lasts a longer time.

Other B cells, memory B cells, are stored in lymph nodes until they are needed at some future date. If the body is exposed to the same pathogen in the future, memory cells allow it to mount a much faster response to the invasion. This response is known as **secondary response** and is responsible for the ability of adaptive immunity to improve with experience (see Figure 14.8).

T cells

As previously explained, there are at least four types of T cells: helper T cells, cytotoxic T cells, regulatory T cells (formerly known as suppressor T cells) and memory T cells. We have already seen the action of helper T cells. Helper T cells are responsible for activation of B and T lymphocytes.

Cytotoxic T cells are responsible for a part of adaptive immune response known as **cell-mediated immunity**, so called because the cytotoxic T cells are directly responsible for the death of pathogens or pathogen-infected cells. Cytotoxic T cells release a cytokine called **perforin**, which causes infected cells to develop holes in their membranes and die. Cytotoxic T cells also release other cytokines that stimulate both innate and adaptive immunity, especially attracting macrophages to the site of infection to dispose of cellular debris. The response of cytotoxic T cells is the primary response of cell-mediated immunity. Some T cells, rather than becoming cytotoxic T cells, give rise to memory T cells. Like memory B cells, memory T cells are responsible for secondary response, storing the recognition of the pathogen until the next encounter (see Figure 14.9). This memory of the pathogen is responsible for the secondary response.

Regulatory T cells are the off-switch for the immune system. Immunity is controlled largely by positive feedback. Tissue damage causes the release of stimulatory chemicals that cause inflammation and increased immune response. Increased immune response results in the release of more chemicals, which causes more stimulation, which causes more chemicals to be released,

cytotoxic = *cell death*

cytotoxic T cells (*sigh toe TOX ick*)

perforin = *causes the cell to become perforated and die*

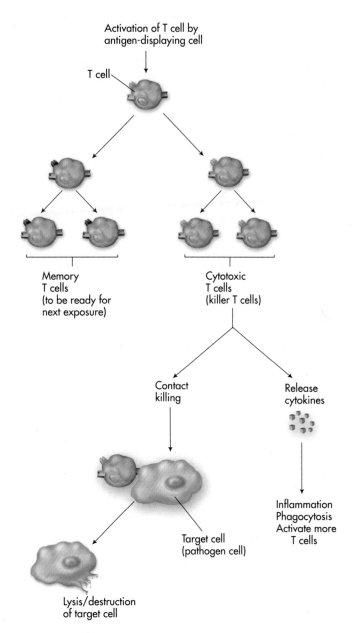

Figure 14.9 Cell-mediated immunity, primary and secondary response.

and so on. Something must actively work to turn off the response when the threat is over or the immune response could become rampant and out of control and thus cause damage. This is the job of the regulatory T cells (along with the eosinophils of innate immunity).

Regulatory T cells are a bit mysterious. For years, scientists have suspected their existence, but these cells have only been found in the last ten years. Exactly how they work is still a mystery, though evidence suggests that they directly inhibit B cells and cytotoxic T cells and that they release cytokines that decrease the immune and inflammatory response. One thing is clear from the research, however. A malfunction of regulatory T cells is implicated in some types of allergy, asthma and autoimmune disease. This makes sense, for example, in the cases of allergy and asthma, where this is an unregulated immune system going too far and actually causing widespread effects.

Clinical application

How you acquire immunity to pathogens

Your adaptive immune system is able to acquire immunity to new pathogens by creating memory cells each time you meet a pathogen. This process, called immunisation, whether natural or done by medical procedures, trains your immune system by creating memory cells for a pathogen. When you are exposed to the pathogen a second time, you do not get ill. Your immune system fights off the pathogen very quickly. In active acquired immunity, *you make* antibodies to fight the pathogens. In passive acquired immunity, antibodies *are introduced* to your body and therefore you don't make your own.

You can acquire immunity in several different ways. **Natural active immunity** is acquired in the course of daily life. When you catch a virus or are infected by bacteria, your immune system fights it off, and memory cells are created for the next meeting. Anybody old enough to have had chicken pox as a child is usually protected from a second round of chicken pox.

Artificial active immunity is acquired during vaccinations. Getting a measles shot exposes you to small amounts of weakened virus, not enough to make you sick, but enough for your immune system to create memory cells. If you meet the virus later in life, you will be able to fight it off. Babies acquire **natural passive immunity** to many pathogens via antibodies passed across the placenta or during breast feeding. These antibodies, which babies can't yet make, protect them from infection for several months after birth. **Artificial passive immunity** is acquired when antibodies from one person are injected into another to help fight infection. See Figure 14.10.

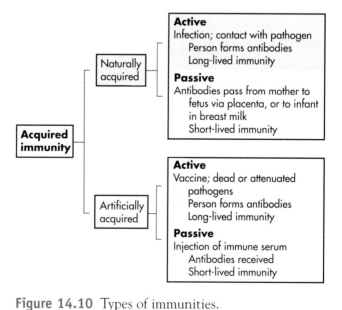

Figure 14.10 Types of immunities.

Test your knowledge 14.5

Choose the best answer:

1. B cells, once activated, can become two different kinds of cells: memory B cells and

   a. cytotoxic T cells

   b. natural killers

   c. plasma cells

   d. macrophages

2. Secondary response is mediated by:

   a. cytotoxic T cells

   b. macrophages

   c. memory cells

   d. all of the above

3. _____ cells are responsible for antibody-mediated immunity while cell-mediated immunity is performed by _____ cells.

   a. B cells, cytotoxic T cells

   b. B cells, plasma cells

   c. plasma cells, memory cells

4. These cells, which were only discovered recently, are part of the 'off-switch' for the immune system.

   a. cytotoxic T cells

   b. memory T cells

   c. regulatory T cells

   d. helper T cells

The big picture

At this point we have talked about the lymphatic system, innate immunity and adaptive immunity as separate means to the same end: ridding the body of invading pathogens. We have mentioned repeatedly that the divisions are *not* separate but are intimately connected. Now that we have inspected them separately, it is time to put the entire defence system back together as a single, integrated fighting force (see Figure 14.11).

A nasty army of pathogens attempts to invade your body. First, they must get past your barriers. Many invaders will be repelled simply by your intact skin or the secretions of your mucous membranes. If the invader gets inside your body, a series of weapons is stimulated by the introduction of a non-self antigen. Cells (neutrophils, macrophages, basophils, etc.) are stimulated. Chemicals (cytokines) are released, which stimulate inflammation and phagocytosis.

Macrophages and other cells, which have ingested some of the invaders and are now wearing the foreign antigens, move to the lymphatic system and search the lymph nodes, looking for the T and B cells that will recognise the intruder. Helper T cells are activated and cause the proliferation of B cells and cytotoxic T cells as well as release chemicals that further stimulate the phagocytic cells and inflammation. B cells produce antibodies that destroy the invaders and further stimulate the immune response. Cytotoxic T cells destroy invaders directly and release chemicals that further stimulate the immune response.

The immune response, both innate and adaptive, will continue to be stimulated until the feedback loop is stopped, at least in part by regulatory T cells. Memory B cells and T cells will be stored in the lymph nodes for later use if another army of those same type of pathogens attempts another invasion.

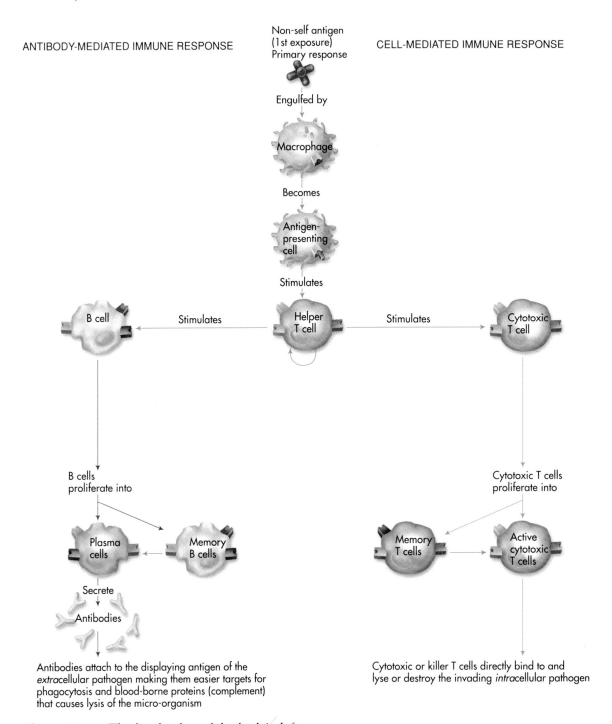

Figure 14.11 The battle plan of the body's defences.

Macrophages and other phagocytic cells will clean up the debris left by the warfare waged by your immune system. Once the danger has passed, your body will return to normal.

Common disorders of the immune system

Immune disorders can be life-threatening because of a weakened or immuno-compromised body defence system. The system can become compromised by factors outside of the body, such as invading viruses, or by a situation in which the body attacks itself.

Immunodeficiency disorders

Patients with any of several types of immunodeficiency disorders have immune systems that are underactive. Their immune systems do not respond adequately to the invasion of pathogens and therefore do not protect them from infection. Patients who are immune-deficient become unwell very easily and do not recover quickly. Minor infections can be fatal. Some immune-deficient patients become infected by pathogens that usually cannot infect humans. Immune deficiency can be caused by viruses, genetics, chemical or radiation exposure, or even medication.

Immune-compromised patients include those with AIDS, SCID (severe combined immune deficiency, a genetic disorder), leukaemia (cancer of the white blood cells), some forms of anaemia, and patients undergoing chemotherapy, taking steroids or immunosuppressant drugs after organ transplant.

Autoimmune disorders

Autoimmune disorders are the opposite of immunodeficiency disorders, exactly what you would expect, given their name. Autoimmune disorders occur when the immune system attacks some part of the body. For some reason, the body fails to recognise 'self' and destroys its own tissue as if it were an invader. There are literally hundreds of autoimmune disorders. Any part of the body can come under attack by mistake. Some of the more common disorders (and what they attack) are as follows:

- rheumatoid arthritis (joint linings)
- multiple sclerosis (myelin sheath in central nervous system)
- lupus erythematosus (every tissue, perhaps DNA)
- Type 1 diabetes mellitus (beta cells in pancreas)
- myasthenia gravis (acetylcholine receptors in skeletal muscle)
- Graves' disease (thyroid gland)
- Addison's disease (adrenal gland)

Just this short list illustrates how devastating an autoimmune disorder can be. Imagine the full power of your immune system turning against your thyroid gland or myelin sheath. The effects are devastating. Most autoimmune disorders can be treated with immunosuppressant drugs, but treatment may not be spectacularly successful and side effects are often severe.

Hypersensitivity reactions

During a hypersensitivity reaction, more commonly known as an allergy, the immune system mounts a hyperactive response to a foreign antigen (referred to in these circumstances as an allergen), often treating a harmless antigen, like grass, mould or an insect bite, as an invading pathogen. It is often on second exposure to an allergen that a harmful reaction is most likely to occur.

On first exposure plasma cells are prompted to produce large quantities of class IgE antibodies in response to the allergen. The IgE antibodies then attach themselves to mast cells in body tissues. On second exposure the same allergen can now attach itself to the IgE antibodies that are attached to the mast cell surface. As the allergen binds to the IgE antibodies so the mast cells begin to release vast quantities of histamine.

Local hypersensitivity reactions, like hay fever, hives, skin rashes and asthma, are generally mild and not life-threatening (asthma is an obvious exception). Systemic hypersensitivity reactions, known as anaphylaxis, are life-threatening. During anaphylaxis, mast cells and basophils release their immune-stimulating chemicals throughout the body. The chemicals cause widespread vasodilatation, which can lead to dangerously low blood pressure and heart failure in certain circumstances. Hives and asthma may also accompany an anaphylactic reaction.

See Figure 14.12, which shows stimulation of the mast cells within the nose due to an allergen, which causes allergic rhinitis (runny nose). Note that mast cells are found throughout the body. If overstimulated in the eyes, mast cells cause red and runny eyes; if overstimulated in the lungs, they cause allergic asthma.

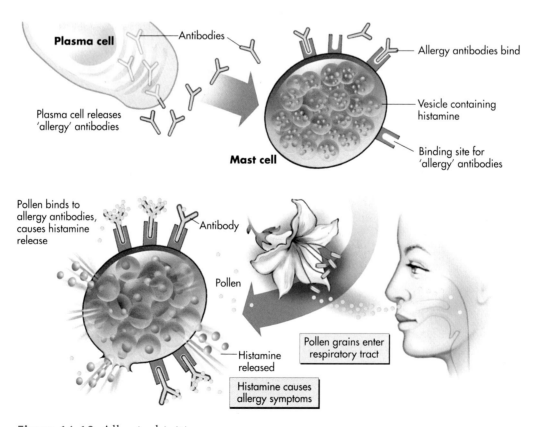

Figure 14.12 Allergic rhinitis.

Summary

- The lymphatic system is the transport system for the immune system and houses the lymphocytes. It consists of lymph capillaries, vessels, trunks and ducts containing lymphatic fluid (lymph) and lymph nodes, which house white blood cells.

- Lymph nodes are concentrated in several regions of the body: cervical, axillary, inguinal, pelvic, abdominal, thoracic and supratrochlear. These patches are in areas where pathogens are most likely to enter. Tonsils, adenoids, spleen, thymus and small intestine all contain lymphatic tissue.

- Fluid leaking from blood capillaries enters the interstitial spaces as tissue fluid and flows into lymph capillaries. The fluid is carried through lymph capillaries to lymph vessels to lymph nodes. In the nodes, any pathogens are destroyed by white blood cells. Fluid then flows from nodes to vessels to lymphatic trunks to collecting ducts and into either the right or left subclavian vein, returning the fluid to the bloodstream.

- The thymus and spleen are lymphatic organs. The spleen contains blood sinuses and removes dead and dying red blood cells as well as pathogens. The thymus is the birthplace of T lymphocytes.

- Immune system function is based on its ability to recognise cell surface molecules called antigens. The immune system must ignore self-antigens (the body's antigens) and respond to non-self antigens (foreign cells).

- The immune system is divided into two separate but extremely interdependent parts, innate immunity and adaptive immunity. Innate immunity is non-specific, has no memory, and cannot improve performance with experience. Adaptive immunity is specific, has memory, and can improve performance with experience.

- Barriers prevent pathogens from getting into the body. Barriers can be either physical or chemical. Skin and tears are examples of barriers.

- The immune system uses a dozen or more different types of cells to combat pathogens. All of these cells are leucocytes or modified leucocytes. Some are part of innate immunity and some are part of adaptive immunity. Their functions range from phagocytosis, chemical stimulation of other cells and antigen display to antibody secretion and direct destruction of pathogens.

- Immune response is stimulated by a variety of chemicals, including the cytokines, histamine and complement.

- Inflammation, the familiar redness, heat, swelling, pain and possible loss of function associated with infection, is a powerful tool in the immune system's arsenal. During inflammation, white blood cells are stimulated and attracted to the site of infection to destroy pathogens and clean up cellular debris. Inflammation is, like much of the immune response, a two-edged sword. Too much inflammation may be more damaging than the infection itself.

- Fever, like inflammation, is a deliberate attempt by the body to destroy a pathogen. Chemicals trick the hypothalamus into raising the temperature set point in an attempt to make body temperature too hot for pathogens.

- Innate immune mechanisms are triggered by the presence of foreign antigens in the body. These mechanisms hold off the infection and stimulate adaptive immune mechanisms.

→

- Adaptive immunity uses T and B lymphocytes to fight specific pathogens. Lymphocytes must be selected during development to recognise foreign antigens but to ignore the body's own antigens. In order to fight a pathogen, lymphocytes must be activated by binding with antigen-displaying cells. Once activated, lymphocytes must proliferate, making thousands of copies of themselves. Helper T cells are required for activation of most types of lymphocytes.

- B cells are mediators of antibody-mediated immunity. B cells are activated by binding to antigen-presenting cells and helper T cells. Once B cells begin to proliferate, they become either plasma cells or memory B cells. Plasma cells secrete antibodies during a primary response. Antibodies help destroy pathogens by binding to antigens on infected cells. Memory B cells are stored for the next time the pathogen is encountered. They mediate the secondary response.

- Cytotoxic T cells mediate cell-mediated immunity by directly killing infected cells. This is the primary T cell response. Like B cells, cytotoxic T cells are activated by helper T cells. Some T cells become memory T cells and mediate the secondary response if the pathogen is encountered again.

- Regulatory T cells are one of the off-switches for the immune system. We don't know very much about how these cells work because they were only discovered in the last ten years.

- Keep in mind that innate and adaptive immunity do not work separately. They work together. One stimulates the other in a huge positive feedback loop. If either innate or adaptive immunity stops working, the whole system breaks down.

Case study

John has made big changes in his life. An IV drug user for more than five years, he recently completed an inpatient rehabilitation programme and is determined to stay off drugs. His cousin, Bill, suggested that John move to a quiet rural area and work with Bill's small business. John takes Bill up on his offer, hoping it will be a true fresh start.

John had lived in the city for his whole life, and had never been to live in the country. He loved the peace and quiet and adjusted to his new role. Things went well for a few weeks, and then John fell ill. His head was stuffy, his nose ran, he had a cough, ran a low-grade fever and his glands (cervical lymph nodes) swelled. No matter how much John slept, he was still tired. The mystery illness would hang around for a few days and then seem to go away, only to come back in a few days. Frustrated and a little scared, John finally went to see his General Practitioner.

The GP requests some blood tests, including a differential white cell count, and, to his delight, finds that John's T cell count is normal and there are no signs of HIV antibodies. However, John's eosinophil, basophil and neutrophil counts are elevated.

1. Given John's history and symptoms, what might the GP suspect?

2. What do the results of John's blood tests reveal?

Review questions

Multiple choice

1. Lymphocytes are selected for their ability to:
   a. recognise antigens
   b. ignore self-antigens
   c. a only
   d. both a and b

2. Mounting an excessive immune response to a harmless antigen is called a(n):
   a. autoimmune disorder
   b. allergy
   c. immunodeficiency
   d. AIDS infection

3. Which of the following is an innate immune cell?
   a. neutrophil
   b. memory cell
   c. helper T cell
   d. plasma cell

4. Lymphocytes are activated by binding with antigens on:
   a. bacteria
   b. antigen-displaying cells
   c. viruses
   d. lymph nodes

5. Innate immunity is not stimulated by foreign antigens:
   a. true
   b. false
   c. it depends
   d. all of the above

6. Fever and inflammation are both part of what kind of immunity?
   a. auto
   b. adaptive
   c. acquired
   d. innate

Short answers

1. List the regions of the body containing many lymph nodes.

   _____
   _____
   _____

2. Trace the circulation of lymph from the cardiovascular system through the lymphatic system and back to the cardiovascular system.

   _____
   _____
   _____

3. List four types of cells and their functions in immunity.

   _____
   _____
   _____

4. Explain the differences between innate and adaptive immunity.

_____

_____

_____

Suggested activities

1. Draw a flow chart of innate and adaptive immunity and their interactions.

2. Form a group and role-play parts of the immune system. Assign some participants to be macrophages, others to be NK cells, and others to be B and T lymphocytes. Describe the process of fighting off a pathogen dependent on your role.

3. Using the Internet, do a quick search of autoimmune disorders. How many can you find? What tissues can be attacked by your immune system?

Answers

Answers to case study

1. The GP might suspect that John has a parasitic infection.
2. The results show elevated oesinophils, that further suggest a parasitic infection.

Answers to multiple choice questions

1. d
2. a
3. a
4. b
5. b
6. d

Answers to short answer questions

1. Tonsils; Thymus; Spleen; Peyer's patches
2. Arterial system – capillary – lymphatic collecting vessels with valves – lymph node – lymph duct – subclavian vein
3. Neutrophils – phagocytosis; Macrophages – phagocytosis and stimulation of immune system, role in wound healing; Basophils – promote inflammation; Eosinophils – counteract the activities of basophils and mast cells; Dendritic cells – antigen displaying cells; Natural killer cells – type of lymphocyte that release chemicals that kill cells displaying foreign antigens; T-lymphocytes – responsible for cell-mediated immunity; B-lymphocytes – role in antibody-mediated immunity
4. Innate immunity provides a 'general' defence; adaptive immunity fights specific pathogens.

The respiratory system

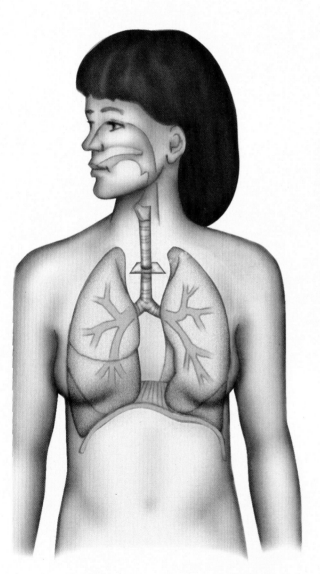

Without fuel, our horizons would be very limited. Our car wouldn't start without petrol, our plane would be grounded without jet fuel, and our bodies would die without the fuel necessary for metabolism. The respiratory system's primary function is to transport the vital fuel oxygen from the atmosphere into the bloodstream to be utilised by cells, tissues and organs for the processes necessary to sustain life. This amazing system is often taken for granted. We don't even consciously realise that the respiratory system is moving 9000 litres of air a day in and out of our lungs. But, just as a car produces waste gases that it eliminates through its exhaust system, so the body also produces a waste gas, carbon dioxide (CO_2), during metabolism that needs to be eliminated via the respiratory system so it does not build up in toxic levels. In this chapter, we look at how oxygen molecules travel from the outside atmosphere to our cells and tissues. In addition, we consider how carbon dioxide leaves the respiratory system and is placed back into the atmosphere.

Learning objectives

At the end of this chapter, you will be able to:

◆ List and state the basic functions of the components of the respiratory system
◆ Differentiate between respiration and ventilation
◆ Explain how the respiratory system warms and humidifies inhaled air
◆ State the purpose and function of the mucociliary escalator
◆ Discuss the process of gas exchange at the alveolar level
◆ Describe the various skeletal structures related to the respiratory system
◆ Explain the actual process of breathing
◆ Discuss several common respiratory system diseases

Pronunciation guide

adenoid (*AD en oid*)
alveoli (*al vee OH lye*)
apnoea (*app NEE yah*)
atelectasis (*a tell ek ta siss*)
bronchi (*BRONG kie*)
bronchioles (*BRONG kee olz*)
bronchitis (*BRONG kie tiss*)
carina (*ca REEN ah*)
cilia (*SIL ee uh*)
conchae (*con CHAY*)

diaphragm (*DIE uh fram*)
emphysema (*emph uh SEE muh*)
empyema (*em pie EE muh*)
epiglottis (*ep ee GLOT iss*)
erythropoiesis (*eh RITH roh poy EE siss*)
hilum (*HIGH lum*)
laryngitis (*lah in JIGH tiss*)
laryngopharynx (*lah RIN goh FAH rinks*)

larynx (*LA rinks*)
lingula (*LIN gue lah*)
mediastinum (*meh dee uh STY num*)
nasopharynx (*NAY zoh FA rinks*)
oropharynx (*OR oh FA rinks*)
parietal pleura (*pa ree tall PLUR uh*)
pharyngitis (*fah rin JIGH tiss*)
pharynx (*FA rinks*)
trachea (*TRACK ee uh*)
tuberculosis (*tue ber cue LOW sis*)

System overview

Have you ever wondered why you feel out of breath when walking or why breathing faster and deeper helps you to recover from strenuous exercise? Our body uses energy from the food we eat, but cells can obtain the energy from foodstuffs only with the help of the vital gas oxygen (O_2), which allows for cellular respiration. Luckily, oxygen is found in relative abundance in the atmosphere and therefore in the air we breathe. However, when the cells use the oxygen, they produce the gaseous waste carbon dioxide (CO_2). If allowed to build up in the body, carbon dioxide would become toxic, so the bloodstream carries the carbon dioxide to the lungs to be exhaled and eliminated from the body. The respiratory system's primary role, therefore, is to bring oxygen from the atmosphere into the bloodstream and to remove the gaseous waste by-product carbon dioxide. As you can see from this discussion, the respiratory system is closely interrelated with the heart and circulatory system. Due to their close relationship, these two systems can be grouped together in medicine to form the **cardiorespiratory system**.

cardio = *heart*
pulmono = *lungs*
thorac/o = *chest*

The respiratory system consists of the following major components:

- Two lungs, the vital organs of the respiratory system.
- Upper and lower airways that *conduct* or move gas in and out of the system.
- Terminal air sacs called alveoli surrounded by a network of capillaries that provide for *gas exchange*.
- A **thoracic** cage that houses, protects and facilitates function for the system.

Amazing body facts

Autocontrol of the cardiorespiratory system

The cardiovascular and respiratory or **pulmonary** systems function without any conscious effort on your part. You probably didn't realise it, but as you read the previous paragraph and these last two sentences, your heart beat approximately 70 times and pumped approximately 5 litres of blood around your body. During that same time, you breathed approximately 12 times, moving over 6000 millilitres (ml) or 6 litres (l) of air.

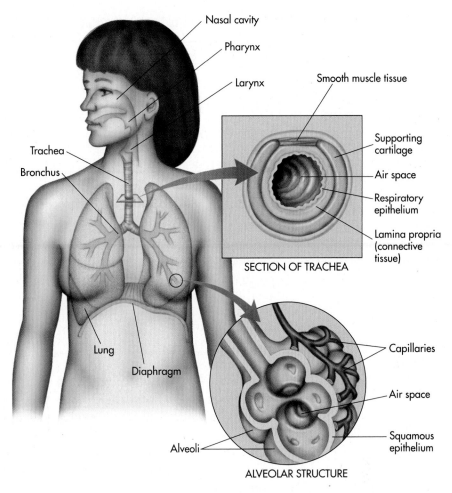

Figure 15.1 The components of the respiratory system.

- Muscles of breathing that include the main muscle, the diaphragm, and accessory muscles.

Take a few minutes to look at Figure 15.1. We will explore each of these components as we travel through the respiratory system.

Ventilation versus respiration

It is important to start with a solid understanding of some commonly confused concepts. The air we breathe is a mixture of several gases, as can be seen from Table 15.1. The predominant gas is nitrogen (N_2), but this is an *inert* gas, which means it does not combine or interact in the body. Even though nitrogen travels into the respiratory system and comes out virtually unchanged, it is vitally important as a support gas that keeps the lungs open with its constant volume. The gas with the next highest concentration is oxygen, and it is very physiologically active within our bodies. You'll notice that carbon dioxide is in low concentration in the air we inhale, but it is in much higher concentration in the air we exhale.

Table 15.1 Gases in the atmosphere

Gas	% of atmosphere
nitrogen (N_2)	78.08
oxygen (O_2)	20.95
carbon dioxide (CO_2)	0.03
argon	0.93

Note: The atmosphere also contains trace gases such as neon and krypton.

ventilation (*vent iu LAY shun*)
respiration (*ress pir AY shun*)
perfusion = *blood flow*

The respiratory system contains a very intricate set of tubes that move, or conduct, gas from the atmosphere deep into the lungs. This movement of gas is accomplished by what we call breathing. However, a more precise look at the process of breathing shows that it is actually two separate processes. The first is **ventilation**, which is the bulk movement of the air down to the terminal end of the lungs where the actual *gas exchange* takes place with the bloodstream. The process of gas exchange, in which oxygen is added to the blood and carbon dioxide is removed, is termed **respiration**. Since the gas exchange in the lungs occurs between the blood and the air in the external atmosphere, it is more precisely called *external respiration*. The oxygenated blood is transported internally via the cardiovascular system to the cells and tissues where gas exchange is now termed *internal respiration*, and oxygen moves into the cells as carbon dioxide is removed. Ventilation, therefore, is not the same thing as respiration. When you watch a television programme that says place the patient on a respirator, it is incorrect and the person should say 'a ventilator' because the machine is only moving the gas mixture (ventilating) into the patient and not causing gas exchange. See Figure 15.2, which contrasts these important processes.

Applied science

Gas exchange in plants

Fortunately for the Earth's ecosystem, the plant physiology of gas exchange is exactly opposite that of humans. Plants take in the CO_2 in the atmosphere and utilise it for energy, then release oxygen into our atmosphere. The Earth's largest source of oxygen released is the Amazon rainforest, which is unfortunately being destroyed at a high rate every day. We truly need a green Earth to survive, so thank the next plant you see.

Applied science

Patient's ventilation

In high dependency or intensive care units, nurses will be responsible for the patient's ventilation through a variety of treatment modalities, for example, CPAP (continuous positive airway pressure), mechanical ventilation or Intermittent Positive Pressure Ventilation (IPPV) delivers air (with normal or increased levels of oxygen) through the use of ventilators.

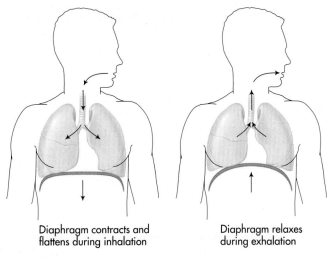

Diaphragm contracts and
flattens during inhalation

Diaphragm relaxes
during exhalation

VENTILATION

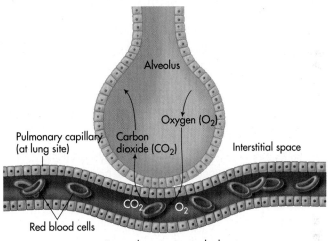

Alveolus

Oxygen (O$_2$)

Pulmonary capillary
(at lung site)

Carbon
dioxide (CO$_2$)

Interstitial space

CO$_2$ O$_2$

Red blood cells

External respiration in the lungs

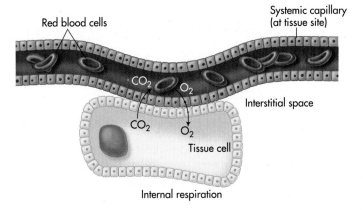

Red blood cells

Systemic capillary
(at tissue site)

CO$_2$ O$_2$

Interstitial space

CO$_2$ O$_2$

Tissue cell

Internal respiration

RESPIRATION

Figure 15.2 Contrast of ventilation and external and
internal respiration.

The respiratory system

The respiratory system is responsible for providing all of the body's oxygen needs and removing carbon dioxide. Unlike cars that have large petrol tanks for fuel storage, the human body has a very small reserve of oxygen. In fact, its oxygen reserve lasts only about four to six minutes (or less in some organs, for example, the heart has oxygen reserves of eight seconds). If that reserve is used up and additional oxygen is unavailable, then tissue damage and even death is the obvious outcome. Therefore, the body must continually replenish its oxygen by bringing in the oxygen molecules from the atmosphere. As you will recall, this process is called ventilation. We will now discuss the internal structures of the respiratory system by following the path oxygen molecules take.

The airways and the lungs

bronchi (*BRONG kye*)
bronchioles (*BRONG kee ohlz*)
alveoli (*al VEE oh lye*)

In a general sense, the respiratory system is a series of branching tubes called **bronchi** and **bronchioles** that transport the atmospheric gas deep within our lungs to the small air sacs called **alveoli**, which represent the terminal end of the respiratory system. To better visualise this system, look at a stalk of broccoli held upside down. The stalk and its branchings represent the airways, and the green bumpy stuff on the end is like the terminal alveoli. See, not only is broccoli good for you, but it can also be a learning tool.

capillaries (*CAP ill air eez*)

Each alveolus is surrounded by a network of small blood vessels called **capillaries**. This combination of the alveoli and the capillary is called the alveolar–capillary (respiratory) membrane and represents the connection or, for you computer buffs, the interface, between the respiratory and cardio-vascular systems. This is where the vital process of gas exchange takes place. Before going any further into this process, let's trace the journey that oxygen molecules must take in order to arrive at the alveolar–capillary membrane.

The upper airways of the respiratory tract

The upper airways, which start at the nose, perform several important functions.

Upper airway functions

nares (*NAIR eez*)

The upper airways begin at the two openings of the nose, called **nostrils** or **nares**, and end at the **vocal cords** (see Figure 15.3). The functions of the upper airway include:

- Heating or cooling (inhaled) inspired gases to body temperature (37°C).
- Filtering (inhaled) particles from the inspired gases.
- Humidifying inspired gases to a relative humidity of 100 per cent.
- Providing for the sense of smell, or *olfaction*.
- Producing sounds, or *phonation*.
- Ventilating, or conducting, the gas down to the lower airways.

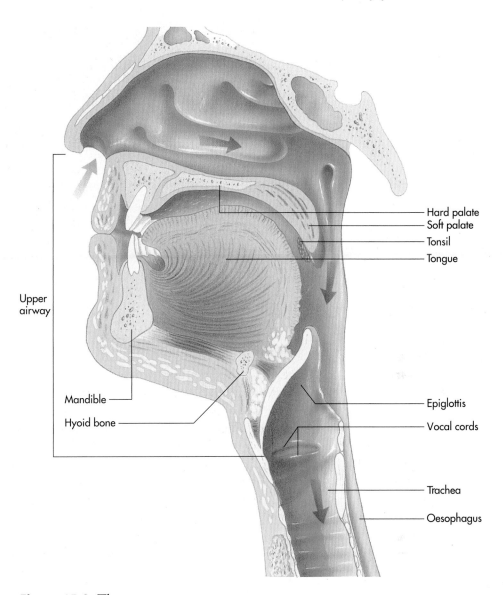

Upper airway

Hard palate
Soft palate
Tonsil
Tongue

Mandible

Hyoid bone

Epiglottis

Vocal cords

Trachea

Oesophagus

Figure 15.3 The upper airway.

The nose

While some people do breathe in through their mouths, under normal circumstances we were meant to breathe in through our nose for reasons that will
become clear as this discussion progresses. The nose is a rigid structure made
of cartilage and bone. There are three main regions contained within the space
behind the nose, called the **nasal cavity**. The two nasal cavities are separated
by a wall called the *septal cartilage*. The regions contained within each nasal
cavity are the vestibular, olfactory and respiratory regions of the nose (see
Figure 15.4).

The vestibular region is located inside the nostrils and contains the coarse
nasal hairs that act as the first line of defence for the respiratory system. These
hairs, called *vibrissae*, are covered with sebum, a greasy substance secreted by
the sebaceous glands of the nose. Sebum helps to trap large particles and also
keeps the nose hairs soft and pliable – those unwanted nose hairs really do

nasal cavity (*NAYZ all CAV it ee*)

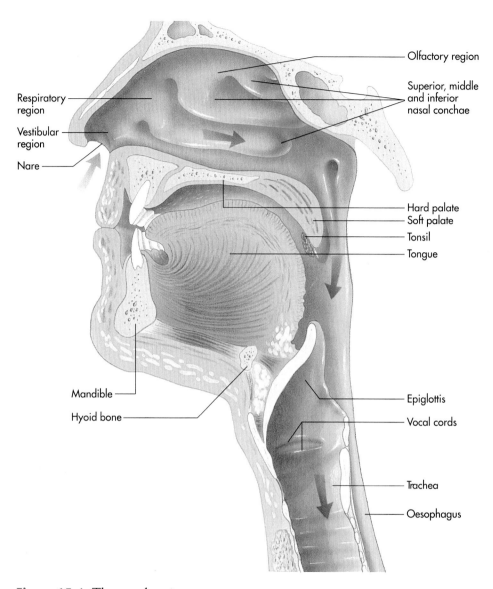

Figure 15.4 The nasal regions.

have a role as a large particle filter. The secretions from the nose (snot) is commonly green because when bacteria become trapped in the nose, parts of the immune system attack the invaders, releasing chemicals that break down the bacterial cell wall. Part of this process turns the dead cells green! The vestibular region helps to filter out large particles so they do not enter the lungs, where they could irritate and clog the airways.

The olfactory region is strategically placed on the roof of the nasal cavity. The advantage to this is that sniffing inspired gas into this region keeps it there and does not allow the gas to reach deeper into your lungs. This is a safe way to sample a potentially noxious or dangerous gas without taking a deep breath into the lungs, where it could cause severe damage. Only four molecules of a substance are required for the olfactory senses to detect a gas. Not everyone though has a sense of smell; those without this sense are said to be anosmic (without functioning olfaction). Some can be completely anosmic, whereas others may have hyposmia (reduced sense of smell) or hyperosmia (heightened sense of smell). It is interesting to note that your ability to taste is related to your sense of smell. If you ever had a nasty head cold and could not taste your food, you now know why.

conchae (con CHAY)

The air from the atmosphere must be warmed to body temperature and must also be moistened so that the airways and lungs do not dry out. This is a job for the respiratory region. Keep in mind that the respiratory region resides in the nasal cavity, which is lined with mucous membranes that are richly supplied with blood. The respiratory region possesses three scroll-like bones known as **turbinates**, or conchae (again see Figure 15.4). These split up the gas into three channels, thereby providing more surface area for incoming air to make contact with the nasal mucosa. While the respiratory region has a small volume of 20 millilitres, if you could unfold the turbinates, you would have a surface area of *106 square centimetres*.

In addition, because the nasal cavity is no longer a straight passageway, the air current becomes turbulent so that more air makes contact with these richly vascularised mucous membranes that transmit heat and moisture to the inspired gas. Incredibly, these moist mucous membranes add 650 to 1000 millilitres of water *each day* to moisten inspired air to 80 per cent relative humidity within the respiratory region of the nose. In such a short distance, this is a pretty impressive humidification process. When the central heating boiler in your house turns on in cold weather, this may dry the inspired air significantly and make it harder for your respiratory region to work. Therefore, humidifying this dry gas with water, such as from a room humidifier, may keep added stress from your body's natural humidification system.

Amazing body facts

Why do we breathe through our mouth?

You may have seen your favourite football player wearing an odd-looking strip of plastic across the nose. This is to help make breathing easier by increasing the diameter of the nostrils. The nose is responsible for one half to two thirds of the total airway resistance in breathing. Airway resistance represents the work required to move the gas down the tube. The larger the tube, the less resistance and therefore less work involved in breathing. Therefore, mouth breathing predominates during stress and exercise because it is easier for the gas to travel through the larger oral opening (less resistance). Of course, when you get a head cold and your nasal passages become blocked by secretions, it becomes necessary to breathe through the easier or open route of the mouth.

pseudo = *false*
stratified = *layers*

cilia (*SIL ee ah*)

Going to ride the mucociliary escalator

The epithelial lining of the respiratory region of the nose plays a very important role in keeping this region of the nose clean and free of debris build-up. Cells in the epithelial layer (or respiratory mucosa) are called **pseudostratified** *ciliated columnar cells* and are found not only in the respiratory region of the nose but throughout most of the airways (see Figure 15.5). The epithelium is a single layer of tall, column-like cells with nuclei located at different heights, giving the false appearance of two layers of cells when in fact there is only one – hence the term *pseudostratified columnar*. Now all we need to add is the cilia. Each columnar cell has 200 to 250 cilia on its surface. Cilia are hair-like projections that can beat at a fantastic rate. Think of them as Olympic standard rowers in a boat.

Test your knowledge 15.1

Choose the best answer:

1. Which gas is found in the atmosphere?
   a. oxygen
   b. nitrogen
   c. carbon dioxide
   d. all the above

2. The process of moving gas into and out of the respiratory system is:
   a. ventilation
   b. external respiration
   c. internal respiration
   d. diffusion

3. The process of gas exchange at the tissue sites is called:
   a. ventilation
   b. external respiration
   c. internal respiration
   d. osmosis

4. Gas exchange takes place across the:
   a. bronchi
   b. bronchioles
   c. alveolar–capillary membrane
   d. heart

5. The bones in the respiratory region of the nose are called:
   a. sinuses
   b. turbinates
   c. tectonic plates
   d. bronchioles

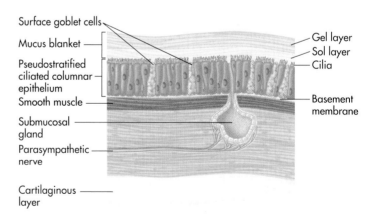

Figure 15.5 The mucociliary escalator.

Goblet cells and submucosal glands are interspersed in the respiratory mucosa and produce about 100 millilitres of mucus per day. The mucus actually resides as two layers. The cilia reside in the sol layer, which contains thin, watery fluid that allows them to beat freely. The gel layer is on top of the sol layer; as its name suggests, it is more viscous or gelatinous in nature. This sticky gel layer traps small particles, such as dust or pathogens, on the mucus blanket, much like flypaper. Once debris is trapped on the mucus blanket, it must be removed from the lung.

So how does this mucous layer actually work? The microscopic, hair-like cilia act as tiny 'oars', and in Figure 15.5 you can see that these oars rest in the watery sol layer. They beat at an incredible rate of 1000 to 1500 times per minute and propel the gel layer and its trapped debris onward and upward at a rate of about 2.5 centimetres per minute to be expelled from the body. When this process occurs in the nose, the debris-laden secretions are pushed toward the front of the nasal cavity to be expelled through the nose. The pseudo-stratified ciliated columnar epithelium, located in the airways of the lungs, propels the gel layer toward the oral cavity to be either expectorated with a cough or swallowed into the stomach. Some texts refer to this epithelial layer as the *mucociliary escalator*, which gives a better picture of what it does. This escalator never stops, that is unless something paralyses it, such as smoking. In some diseases, for example chronic bronchitis, excess mucus is produced which will need expectorating or swallowing. The colour of the expectorant is clinically useful; green sputum indicates an infection but also indicates that the immune system is helping to fight an infection because the mucus goes green when enzymes (from the defence cells) that fight bacteria are activated. Mucus that is red or brown would indicate the presence of fresh or old blood, and tends to indicate that bleeding has occurred, which is seen in lung cancer or tuberculosis, for example.

The sinuses

Have you ever heard of someone being called an airhead? Technically, we are all airheads because the skull contains air-filled cavities (commonly called **sinuses**) that connect with the nasal cavity via small passageways. Because these are located around the nose, they are called **paranasal** sinuses. They are lined with respiratory mucosa that continually drain their secretions into the nose. The sinuses are named after the specific facial bones in which they are located (see Figure 15.6).

These cavities of air in the bones of the cranium are believed to help in the prolongation and intensification of sound produced by the voice. If you have ever shouted inside a cave, you will have noticed a more resonant quality to the sound. In addition, it is theorised the air-filled sinuses help to lighten the heavy head that sits atop the neck.

sinus = *curve, fold or hollow*
para = *around*
nasal = *referring to nose*

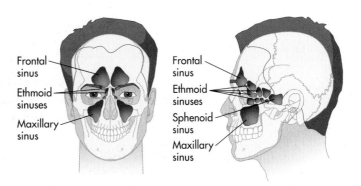

Figure 15.6 The paranasal sinuses.

Sinuses do not exist at birth but develop as you grow and influence facial changes as you mature. Sinuses also provide further warming and moisturising of inhaled air.

The pharynx

The pharynx is a hollow, muscular structure lined with epithelial tissues. The throat (or **pharynx**) begins behind the nasal cavities and is divided into the following three sections, as shown in Figure 15.7:

- nasopharynx
- oropharynx
- laryngopharynx

The **nasopharynx** is the uppermost section of the pharynx and begins right behind the two nasal cavities. Air that is breathed through the nose passes through the nasopharynx. This section also contains lymphatic tissue of the immune system, called the adenoids, and passageways to the middle ear called

adenoid (*AD eh noid*)

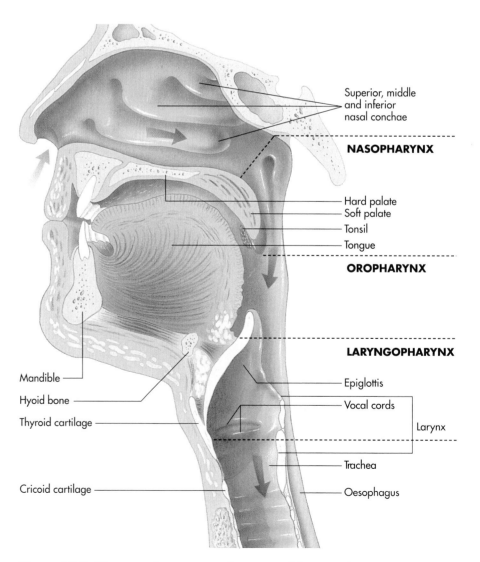

Figure 15.7 The nasopharynx, oropharynx and laryngopharynx.

Eustachian tubes (*you STAY shun*)

Eustachian tubes. You can understand how an infection located in the nasal cavities can lead to an ear infection, and vice versa.

The **oropharynx** is the next structure and is located right behind the oral or buccal cavity (mouth). Air that is breathed through the mouth as well as air that is breathed through the nose passes through here. It is important to note that anything that is swallowed also passes through this section. Therefore, the oropharynx conducts not only atmospheric gas but also food and liquid.

palatine tonsils (*PAL a teen TON sills*)

The oral entrance is a strategic area to place 'guardians' for the immune system because this is where pathogens can easily enter the body. Lymphoid tissue, or tonsils, such as the **palatine tonsils**, are located in this area. Another set of tonsils, the lingual tonsils, are found at the back of the tongue. During the process of swallowing, the uvula and soft palate move in a posterior and superior position to protect the nasopharynx and nasal cavity from allowing food or liquid to enter. Actually swallow and feel this happening within your oral cavity. This protection can be overcome by forceful laughter, and that is why when you laugh with liquid or food in your mouth, the food or liquid can sometimes come up through your nose.

The larynx

hypo = *below*
otomy = *cutting into*

The larynx houses the important structure needed for speech. The **laryngopharynx** is the lowermost portion of the pharynx; an older term for it was the **hypopharynx** because of its position. Commonly known as the voice box, the larynx is a semi-rigid structure composed of several types of cartilage connected

---

Clinical application

Keeping the vital airway open

Just like any vital motorway, the airway needs to remain open to a flow of traffic, or the oxygen molecules will cease to flow into the alveoli and therefore not get into the bloodstream to supply the tissues. While a traffic jam can last for hours, oxygen flow can be disrupted for only a few minutes without tragic results. For example, if the upper airway swells shut from a severe allergic reaction to a bee sting, an emergency airway must be established. Referring back to Figure 15.7, you will see space between the thyroid and cricoid cartilages and this is where an emergency **cricothyroidotomy** is performed. This space has few blood vessels or nerves, which makes it ideal in an emergency situation to place a temporary breathing tube.

Sometimes a longer-term breathing tube must be inserted into the lungs via a technique called intubation. This tube passes through the vocal cords and sits above the juncture (carina) between the right and left lung. A machine called a ventilator can move air into and out of the damaged lungs at this juncture. The tube has an inflatable cuff to seal the airway once it is in place. Knowledge that the vocal cords open and close during breathing becomes clinically significant. Adduction (moving towards the midline) seals the vocal cords and occurs during expiration, while abduction (moving away from the midline) opens the cords, increasing the size of the glottic opening during inspiration. If the patient is breathing, it is better to pass the tube with the deflated cuff through the narrow opening of the vocal cords during a breath in when the cords are open. Conversely, when removing the tube (extubation), the healthcare professional must always remember to deflate the cuff so it doesn't damage the cords as the tube is pulled back through them. It should go without saying (no pun intended) that a patient cannot talk when intubated because the vocal cords cannot function properly. If you ever see a TV soap opera or film where someone is talking with a tube going down his or her mouth into the lungs, you will know it is impossible. More permanent airways, called tracheostomy tubes, can be placed in the neck and are made so the patient is able to talk.

by muscles and ligaments that provide for movement of the vocal cords to control speech. The 'Adam's Apple' is the largest of the cartilages found in the larynx, and is more prominent in men. This cartilage is also anatomically known as the thyroid cartilage, beneath which is the cricoid cartilage. Both cartilages in the exposed areas of airways found in the neck are necessary to provide structure and support for airways so they do not collapse and block the flow of air in and out of the lungs.

Air that is breathed and anything that is swallowed passes through the laryngopharynx. Swallowed materials *should* pass through the **oesophagus** to get to the stomach, and air *should* travel through the larynx and then the trachea on its way to the lungs. What directs the flow of 'traffic' (air to the lungs and food and liquid to the stomach)? It is directed by a mechanism termed the *glottic mechanism* or swallowing reflex. The space between the vocal cords is called the *rima glottis*, or simply the glottis. The **glottis** is the opening that leads into the larynx and eventually the lungs. Fortunately, there is a leaf-shaped, fibrocartilage, flap-like structure located above the opening, or glottis, called the **epiglottis**. The epiglottis closes over the opening to the larynx when you swallow and opens up when you breathe. This selective closure is called the glottic or sphincter mechanism and facilitates the closing of the epiglottis over the glottic opening, thus sealing it so food does not enter the lungs.

Therefore, the lungs are 'closed to traffic' when swallowing and the food and liquid travels down the only open tube or route, which is now the oesophagus, leading into the stomach. When we breathe in, the gas preferentially travels into the lungs through a process that actually draws it into the lungs because of pressure differences. More on this soon.

The vocal cords are the area of division between the upper and lower airways, representing the point of transition to the lower airways. The lower airways start below the vocal cords, and we will soon continue to the lower airways all the way to the end point, the alveoli.

oesophagus (*eh SOFF ah guss*)

glottis (*GLOT iss*)

epiglottis (*ep ee GLOT iss*)
    epi = *above*

Test your knowledge 15.2

Choose the best answer:

1. The hair-like structures that propel mucus in the airways are:

    a. sol layer

    b. gel layer

    c. pathogens

    d. cilia

2. Which of the following is not true about the sinuses?

    a. air-filled cavities

    b. located in the skull and around the nose

    c. help to lighten the head

    d. gas exchange occurs there

3. Food is prevented from entering the _____ when eating by the closure of the _____.

    a. oesophagus, glottis

    b. oesophagus, epiglottis

    c. trachea, epiglottis

    d. epiglottis, glottis

4. The vocal cords are found in the:

    a. laryngopharynx

    b. nasopharynx

    c. oropharynx

    d. alveoli

The lower respiratory tract

The airway that leads to the lungs and then branches out into the various lung segments resembles an upside-down tree and is sometimes called the tracheobronchial tree. See Figure 15.8. Upon leaving the vocal cords in the larynx, the inspired air enters the **trachea**, the lay term for which is windpipe. The trachea extends from the cricoid cartilage of the larynx to the sixth thoracic vertebrae (approximately to the midpoint of the chest). The cartilage found in the trachea is in the form of C-shaped structures in the anterior portion of the trachea to provide rigidity and protection for the exposed airway in the neck. This C shape also serves another important purpose: the oesophagus lies in the area where the C opens up posteriorly. Without the cartilage, there is some 'give' in the posterior aspect of the larynx and trachea, so the oesophagus can expand when you swallow larger chunks of food and they won't get stuck against the rigid cartilage of the trachea.

The trachea is the largest bronchus and can be thought of as the trunk of the tracheobronchial tree. Often it is represented as generation 0 (zero) because it is the trunk or start of the tree that has not yet begun to branch. Once the trachea reaches the centre of the chest, it begins its first branching, or bifurcates, into two bronchi (bronchus is the singular form), the right mainstem and the left mainstem, which can be termed generation 1 because they represent the first branching of the tree. Each new branching is considered

trachea (TRACK ee uh)

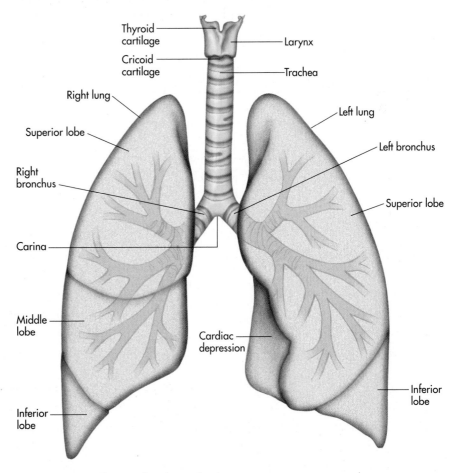

Figure 15.8 The tracheobronchial tree.

another generation. The site of bifurcation is called the **carina** (again, see Figure 15.8). One bronchus goes to the right lung and the other bronchus goes to the left lung. The mainstem bronchi are sometimes also referred to as the primary bronchi. Now the bronchi must branch into the five lobar bronchi (generation 2) that correspond to the five lobes of the lungs.

Each lung lobe is further divided into specific segments, and the next branching of bronchi is called the segmental bronchi (generation 3). At the point from the trachea down to the segmental bronchi, the tissue layers of the bronchi are all the same, only smaller, as they branch downward. The layers are the epithelial layer, which is the mucociliary escalator – or pseudostratified ciliated columnar cells – that keeps the area clean of debris. A middle lamina propria layer contains smooth muscle, lymph and nerve tracts. The third layer is the protective and supportive cartilaginous layer (see Figure 15.9).

Clinical application

The angle makes a difference

The angle of branching is not the same for both sides of the tracheobronchial tree. The right mainstem branches off at a 20 to 30 degree angle from the midline of the chest. The left mainstem branches off at a more pronounced 40 to 60 degree angle. This is important because the lesser angle of the right mainstem branching allows foreign bodies that are accidentally breathed in to more often lodge in the right lung. This is useful to know if a child has aspirated (taken into the lung) an object and the physician must enter the lung with a bronchoscope to remove the object. Time may be critical, and it may make a difference if the search is begun immediately in the right lung because its anatomic structure makes it a high probability that the object has lodged there. In addition, a breathing tube or endotracheal tube may be placed too far into the lung and instead of sitting above the carina so both lungs are ventilated most likely will pass into and ventilate only the right lung. This is why an X-ray for proper tube placement is so important.

The branching becomes more numerous with tiny subsegmental bronchi (generations 4 to 9) branching deep within each lung segment. The diameter of subsegmental bronchi ranges from 1 to 6 millimetres. Cartilaginous rings are now irregular pieces of cartilage and will soon fade away completely. Notice as we move toward the gas exchange regions that the airways simplify to make it easier for gas molecules to pass through. Now we reach the very tiny airways called bronchioles (generations 10 to 15) that average only 1 millimetre in diameter. They have no cartilage layer, and the epithelial lining becomes ciliated cuboidal (short squat cells as opposed to large columns). The cilia, goblet cells and submucosal gland are almost all gone by generation 15. There is no gas exchange yet, just simple conduction of the gas mixture containing the oxygen molecules down the tree. Now we reach the terminal bronchioles (generation 16), which have an average diameter of 0.5 millimetres, no goblet cells, no cartilage, no cilia and no submucosal glands. This marks the end of the conducting areas, and we now journey into the gas exchange or respiratory zone of the lung.

The next airway beyond the terminal bronchiole is called the respiratory bronchiole (generations 17 to 19) because a small portion of gas exchange takes place here. The epithelial lining is simple cuboidal cells interspersed with actual alveoli-type cells, which are flat, pancake-like cells called *simple squamous pneumocytes*. Alveolar ducts (generations 20 to 27) originate from the respiratory bronchioles wherein the walls of the alveolar ducts are completely made up of simple squamous cells arranged in a tubular configuration.

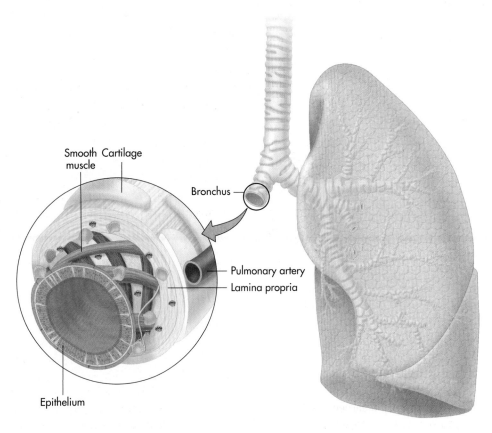

Smooth Cartilage
muscle

Bronchus

Pulmonary artery

Lamina propria

Epithelium

Figure 15.9 Tissue layers in the bronchi.

The alveolar ducts give way to the grape-like structures of several connected alveoli, better known as the alveolar sacs (generation 28). See Figure 15.10.

Clinical application

Therapeutic oxygen

Often, a distressed respiratory system and sometimes the cardiac system needs supplemental oxygen to assist its function and meet its needs. There are many ways to deliver an enriched oxygen supply to the lungs, including an oxygen mask, nasal cannula (tube) and specialised devices to deliver both oxygen and extra humidity to the lungs to assist their function. Oxygen therapy can dry out the respiratory tract, and therefore if supplemental oxygen is required for more than 24 hours, it should be humidified.

Oxygen carrying and the respiratory system

Every breath of air is around 500 ml of gas. This gas contains, amongst other things, oxygen. The percentage of oxygen is very high compared to the level of oxygen found inside the body, so when that breath gets to the site of gaseous exchange (through the alveoli membrane), there is a relatively high proportion of oxygen compared to all of the tissues surrounding it. Because the concentration is so high, the gas moves from an area of relative high concentration, to an area of relatively low concentration (through the alveoli). Haemoglobin, which is found in red blood cells, can carry four oxygen molecules at any one time. As

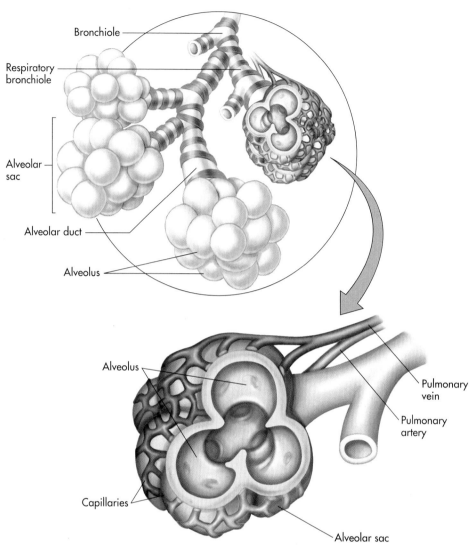

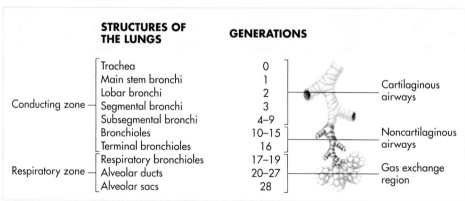

STRUCTURES OF THE LUNGS		GENERATIONS		
Conducting zone	Trachea	0		Cartilaginous airways
	Main stem bronchi	1		
	Lobar bronchi	2		
	Segmental bronchi	3		
	Subsegmental bronchi	4–9		
	Bronchioles	10–15		Noncartilaginous airways
	Terminal bronchioles	16		
Respiratory zone	Respiratory bronchioles	17–19		Gas exchange region
	Alveolar ducts	20–27		
	Alveolar sacs	28		

Figure 15.10 Conduction and gas exchange structures and functions.

the red cells pass along the blood flow to the lungs, they pick up oxygen and lose carbon dioxide. This oxygen-rich blood then flows around the body until it reaches an area where the oxygen levels in the surrounding tissues are low. When it reaches such a point, oxygen is released from the haemoglobin in the red blood cells to the surrounding tissue. Any carbon dioxide is then picked up and begins its journey to the lungs.

The alveolar capillary membrane: where the action is

The alveoli are the terminal air sacs that are surrounded by numerous pulmonary capillaries and together make up the functional unit of the lung known as the *alveolar capillary membrane*. The average number of alveoli in an adult lung ranges from 300 million to 600 million. This gives a total 80 square metres (m^2) surface area for the oxygen molecule to diffuse across (about the size of a tennis court) into the surrounding pulmonary capillaries, which have about the same cross-sectional area (70 m^2). The blood entering the pulmonary capillaries comes from the right side of the heart and is low in oxygen and high in carbon dioxide because it just came from the body tissues. Gas exchange or external respiration takes place, and the blood leaving the pulmonary capillary is high in oxygen and travels to the left side of the heart via the pulmonary veins to be pumped around to the tissues. Conversely, carbon dioxide molecules are in high concentration in the pulmonary capillary blood and very low in the lung (remember, there is little CO_2 in the atmosphere), so CO_2 leaves the blood and enters the lung to be exhaled.

Upon closer inspection of the alveolar capillary membrane, you will see four distinct components. The first layer is the liquid **surfactant** layer that lines the alveoli. This phospholipid helps lower the surface tension in these very tiny spheres (alveoli) that would otherwise collapse due to the high surface tension.

surfactant (*sir FACT ant*)

erythrocytes (*eh RITH row sights*)

haemoglobin (*HEE mow GLOW bin*)

erythropoiesis (*eh RITH roh poy EE siss*)
    erythr/o = *red*
    poiesis = *to make*

erythropoietin (*eh RITH roh poy EE tin*)

Clinical application

What can go wrong with gas exchange?

The membrane between the alveoli and the capillaries is quite thin. In fact, it is only 0.004 millimetres thick! The thinness of this membrane aids in the diffusion of the gases between the lungs and the blood. Anything that would act as a barrier to oxygen molecules getting to or through this barrier would decrease the amount of oxygen that gets into the blood. For example, excessive secretions and fluid such as in pneumonia act as a barrier and reduce the oxygen levels in the blood, which can be measured by sampling arterial blood and analysing the amount of each gas dissolved in it. This is called an arterial blood gas, or ABG. In the case of severe pneumonia, the level of oxygen known as the PaO_2 (the partial pressure of oxygen) goes down because less oxygen can get into the blood, and the $PaCO_2$ (the partial pressure of carbon dioxide) in the arterial blood goes up because less CO_2 crosses into the lungs to be exhaled.

The blood cells can also be affected. Red blood cells, or **erythrocytes**, are responsible for the bulk of the transportation of oxygen and carbon dioxide in the blood via a protein- and iron-containing molecule called **haemoglobin**, which performs the actual transportation. It is estimated that there are about *280 million* haemoglobin molecules found in *each* erythrocyte!

In general, if the haemoglobin is carrying large amounts of oxygen, the blood will be bright red. If there is less oxygen and more carbon dioxide being carried, then the blood will be darker in colour. An obvious example of this can be seen when you look at the veins in your arm. Venous blood has lower levels of oxygen and higher levels of carbon dioxide. As a result, venous blood has a dark red tint. Low levels of red blood cells or anaemia would limit the number of haemoglobin molecules that could transport oxygen and thus greatly reduce the amount in the blood available for the tissues. Therefore, the number of red blood cells and the amount of haemoglobin in your blood (both of which can be measured) are important in oxygen delivery to your tissues.

Your body can attempt to respond to low haemoglobin levels by producing more red blood cells by a process called **erythropoiesis**. This process begins once the kidneys detect low levels of oxygen coming to them from the blood. The kidneys release into the bloodstream a hormone called **erythropoietin**. This substance travels through the blood and eventually reaches specialised cells found in the red bone marrow. Once stimulated, these specialised cells begin to increase their production of erythrocytes until demand is met. Having too little iron in the body can also affect oxygen delivery because the iron in the haemoglobin is what holds onto the oxygen molecules. The terms *iron-poor* blood and *tired* blood come from the fact that a patient with low levels of iron tires easily due to low oxygen levels.

The second component is the actual tissue layer, or alveolar epithelium, comprised of simple squamous cells of two types. The majority type (95 per cent) of alveolar epithelia are flat, thin, pancake-like cells called squamous pneumocytes or Type I cells. These are where the gas molecules can easily pass through in the process of gas exchange. The alveoli also need to produce the valuable surfactant, and this is where the Type II, or granular, pnuemocytes come in. These highly metabolic cells not only produce surfactant but aid in cellular repair responsibilities. Finally, this area needs to be free of debris that would act as barriers to the vital process of gas exchange. This is where the 'clean-up' cells, called Type III cells or wandering macrophages, ingest foreign particles as the macrophages wander throughout the alveoli. There are even small holes between the alveoli called *pores of Kohn* that allow the macrophages to move from one alveolus to another.

Applied science

The amazing surfactant

Not only does surfactant lower the surface tension when the alveoli are small (end-expiration), thereby preventing alveoli collapse, but when you take a deep breath (end-inspiration), your alveoli get larger and the surfactant layer thins and becomes less effective, its surface tension increasing because of its thinning. This prevents over-expansion or rupture of the alveoli. Lack of surfactant can cause 'stiff' lungs that resist expansion. Surfactant develops late in fetal development (at around 32 weeks' gestation), and premature babies therefore may not have sufficient levels. Without immediate intervention, their tiny lungs would collapse (*atelectasis*) and thus prevent vital gas exchange. If they are given too much volume to re-expand their stiff lungs, the alveoli may rupture, again because surfactant is not there to prevent over-expansion. Surfactant also has an antibacterial property that helps to fight harmful pathogens. Fortunately, medical science has developed surfactant replacement therapy that can instil surfactant into the lungs to maintain their function until babies have matured and can produce it on their own.

The third component of the alveolar capillary membrane is the *interstitial space*. This is the area that separates the basement membrane of alveolar epithelium from the basement membrane of the capillary endothelium and contains interstitial fluid. This space is so small that the membranes of the alveoli and capillary appear fused. However, if too much fluid gets into this space (interstitial oedema), the membranes separate, which makes it harder for gas exchange to occur because the gas has to travel a greater distance and through a congested, fluid-filled space.

The fourth component is the *capillary endothelium* that forms the wall of the capillary. The capillary contains the blood with the red blood cells that carry the precious gas cargo to its destination.

The housing of the lungs and related structures

The lungs reside in the thoracic cavity and are separated by a region called the **mediastinum**, which contains the oesophagus, heart, great vessels (superior and inferior vena cavae and aorta), and trachea (see Figure 15.11).

Breathing in and out causes the lungs to move within the thoracic cavity. Over time, an irritation could occur as the lungs rub the inside of the thoracic cage. To prevent such damage, each lung is wrapped in a sac, or serous membrane, called the **visceral pleura**. The thoracic cavity and the upper side of the diaphragm are lined with a continuation of this membrane called the **parietal pleura**. Between these two pleural layers is an intrapleural space (pleural cavity) that contains a slippery liquid called *pleural fluid*. This fluid greatly reduces the friction as an individual breathes.

mediastinum (*meh dee uh STY num*)

visceral pleura (*VISS er all PLOO rah*)
parietal pleura (*pah RYE et all PLOO rah*)

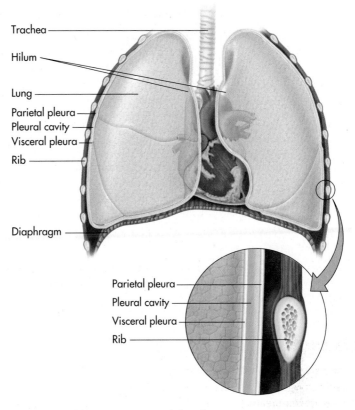

Trachea

Hilum

Lung

Parietal pleura
Pleural cavity
Visceral pleura
Rib

Diaphragm

Parietal pleura
Pleural cavity
Visceral pleura
Rib

Figure 15.11 Structures of the thoracic cavity.

Test your knowledge 15.3

Choose the best answer:

1. The largest bronchus or trunk of the tracheobronchial tree is the:
   a. right mainstem bronchus
   b. left mainstem bronchus
   c. bronchiole
   d. trachea

2. The site of bifurcation of the right and left lungs is called the:
   a. alveoli
   b. carina
   c. trachea
   d. capillary

3. If an object is aspirated into the airways, it is most likely to go to:
   a. the right lung
   b. the left lung
   c. the stomach
   d. the oropharynx

4. The first portion of the airway where gas exchange begins is the:
   a. terminal bronchiole
   b. trachea
   c. mainstem bronchus
   d. respiratory bronchiole

5. The alveolar layer that lowers surface tension to keep the alveoli expanded is the:
   a. surfactant layer
   b. capillary layer
   c. epithelium layer
   d. macrophage layer

6. The alveolar cell that allows for gas exchange is the:
   a. squamous cell
   b. granular cell
   c. macrophage cell
   d. Kohn cell

The actual lungs

The right and left lungs are conical-shaped organs; the rounded peak is called the apex of the lung. The apices of the lung extend up to 5 centimetres above the clavicle. The bases of the lungs rest on the right and left hemidiaphragm. The right lung base is a little higher than the left to accommodate the large liver lying underneath. The medial surface of the lung has a deep, concave cavity that contains the heart and therefore is called the cardiac impression and is deeper on the left side. The **hilum** is the area where the root of each lung is attached. Each root contains the mainstem bronchus, pulmonary artery and vein, nerve tracts and lymph vessels.

The right lung has three lobes – the upper, middle and lower lobes – that are divided by the horizontal and oblique fissures. The left lung has only one fissure, the oblique fissure, and therefore only two lobes, called the left upper and left lower lobes. You may hear the term **lingula**. This is an area of the left lung that corresponds with the right middle lobe. Why only two lobes in the left, you may ask? Remember that the heart is located in a space (cardiac impression) in the left anterior area of the chest and therefore takes up some space of the left lung. In fact, the right lung is larger and about 60 per cent of gas exchange occurs there. The lobes are even further divided into specific segments related to their anatomical position. For example, the apical segment of the right upper lobe is the top portion or tip of the right upper lobe.

The protective bony thorax

The lungs, heart and great vessels are all protected by the *bony thorax*. This bony and cartilaginous frame provides protection and also movement of the thoracic cage to accommodate breathing. The bony thorax includes the rib cage, the sternum or breastbone, and the corresponding thoracic vertebrae to which the ribs attach (see Figure 15.12).

hilum (*HIGH lum*)

lingula (*LING ue lar*)

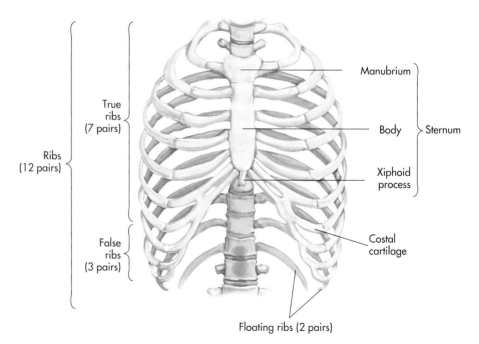

Figure 15.12 The thoracic cage.

sternum (*STER num*)

The **sternum**, or breastbone, is centrally located on the anterior portion of the thoracic cage and is comprised of the manubrium, body and xiphoid process. This anatomical landmark is very important for proper hand placement in cardiopulmonary resuscitation (CPR). The hand is placed over the body of the sternum where compressions squeeze the heart between the body of the sternum and the thoracic vertebrae. If the hand placement is too low on the xiphoid process, it can break off and lacerate the internal organs.

thoracic cage (*thor AS sick*)

The **thoracic cage** consists of 12 pairs of elastic arches of bone called ribs. The ribs are attached by cartilage to allow for their movement while breathing. The true ribs are pairs 1 to 7 and are called *vertebrosternal* because they connect anteriorly to the sternum and posteriorly to the thoracic vertebrae of the spinal column. Pairs 8, 9 and 10 are called the false ribs, or *vertebrocostal*, because they connect to the costal cartilage of the superior rib and again posterior to the thoracic vertebrae. Rib pairs 11 and 12 are called the floating ribs because they have no anterior attachment.

How we breathe

Pulmonary ventilation is the movement of air from an area of relative high pressure to an area of low pressure. To achieve respiratory ventilation the size of the chest must change and so alter the volume and pressure in the lungs. There is what is called an inverse relationship between pressure and volume, which means that if there is a decrease in the volume of a gas, the pressure will rise; if you increase the volume of a gas, its pressure will fall. This was first recognised by Robert Boyle in the 1600s, and is now called Boyle's law.

To achieve changes in volume and pressure, the diaphragm flattens, pulling the lungs downwards. At the same time, the ribs move upwards and outwards. Because the lungs adhere to the pleural membranes, they are moved upwards and outwards. This increases the volume of the lungs and decreases the pressure. Air will then flow from the relative high pressure of the atmosphere, down the trachea and into the low pressure of the alveoli. Because air is a gas, it will fill all the available alveoli, and therefore facilitate gaseous exchange.

The ease by which ventilation occurs is referred to as compliance. Low compliance means that it is more difficult to expand the lungs, whereas high compliance means that less effort is required to expand the lungs. This becomes clinically significant when looking after people who have asthma, emphysema, chronic bronchitis, or any other lung disease. In asthma, the airways narrow, making it more difficult to breathe, and therefore decreasing compliance. In emphysema, the bronchioles become damaged and wider, allowing air into the lungs, i.e. increasing compliance, but making it much more difficult to breathe out.

What makes the brain tell the lungs how fast or how slowly to breathe? Although we can consciously speed up or slow down our breathing, *our breathing rate is normally controlled by the level of carbon dioxide in our blood.* If carbon dioxide levels rise, it means that not enough CO_2 is being ventilated, so the medulla sends signals to the respiratory muscles to increase the rate and depth of breathing.

medulla oblongata (*meh DULL lar ob long GAR ta*)

diaphragm (*DIE ah fram*)

The control centre that tells us to breathe is located in the brain in an area known as the **medulla oblongata**. Inspiration is an active process of ventilation in which the main breathing muscle, the **diaphragm** (a dome-shaped

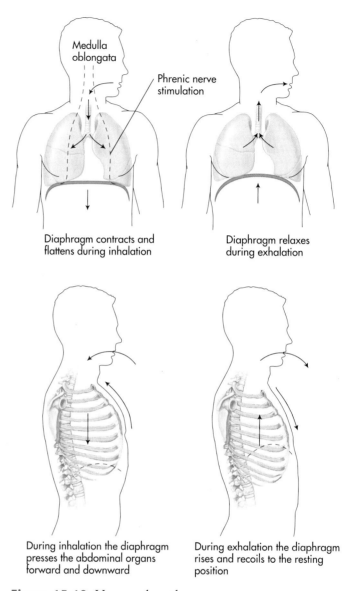

Figure 15.13 How we breathe.

muscle when at rest), is sent a signal via the phrenic nerve and contracts and flattens, thereby increasing the space in the thoracic cavity (see Figure 15.13). Normal respiratory rate is 12 breaths per minute (ranging between 12 and 20 breaths per minute).

Sometimes the body needs help to breathe beyond resting or normal breathing. For example, during increased physical activity or in disease states in which more oxygen is required, **accessory muscles** are used to help pull up your rib cage to make an even larger space in the thoracic cavity. The accessory muscles used are the scalene muscles in the neck, the sternocleido-mastoid, and the pectoralis major and pectoralis minor muscles of the chest.

Although exhalation is a passive process, there are times, especially with certain disease states, when exhalation may need to be assisted. Again, the body has accessory muscles of exhalation that assist in a more forceful and active exhalation by increasing abdominal pressure. The main accessory

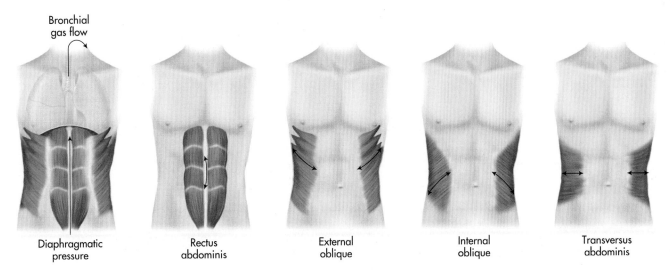

Figure 15.14 The accessory muscles of exhalation.

muscles of exhalation are the various abdominal muscles that push up the diaphragm or the back muscles that pull down and thus compress the thoracic cage. See Figure 15.14 for the specific accessory muscles of exhalation.

When we talk about breathing, we often mean *quiet* breathing. Quiet breathing requires respiratory effort from the diaphragm and external intercostal muscles for inspiration, but expiration is largely passive, i.e. does not involve muscular activity. In forced breathing, both inhalation and exhalation become active. The accessory muscular activity can be seen on clinical examination; for example, abdominal muscles may be used during exhalation in individuals with cystic fibrosis, emphysema or chronic bronchitis.

Lung volumes and capacities

The tidal volume is the amount of air moved into or out of the lungs during a single respiratory cycle. The normal tidal volume is 500 ml, although there is considerable variation depending on general fitness; for example, swimmers tend to develop large tidal volumes. The total volume of the lungs can be divided into volumes and capacities (see Figure 15.15):

- Functional residual capacity (FRC) – the volume of air remaining in the lungs at the end of a normal expiration.
- Inspiratory reserve volume (IRV) – the amount of air that can be forcefully inhaled after a normal inspiration.
- Expiratory reserve volume (ERV) – the amount of air that can be forcefully exhaled after a normal expiration.
- Residual volume – the volume of air remaining in the lungs after a maximum expiration.
- Vital capacity (VC) – the maximum amount of air that can be moved into and out of the respiratory system in a single respiratory cycle.

With each breath, not all of the air contributes to gaseous exchange because some air remains in the conduction passageways, i.e. in the trachea, bronchus, etc. This volume is referred to as the anatomical dead space or dead space

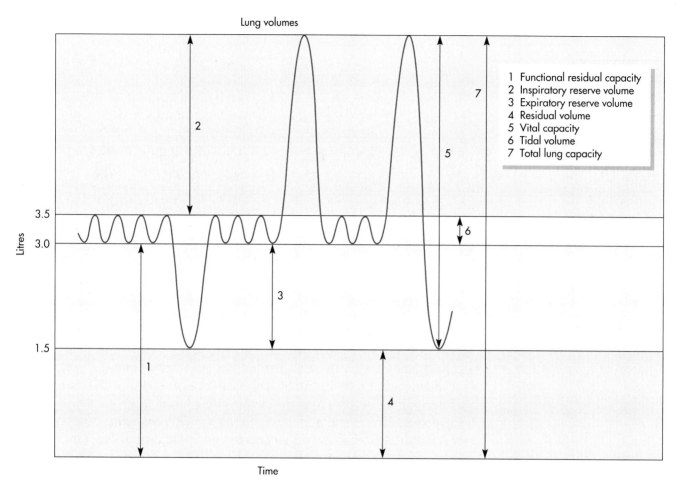

Figure 15.15 Respiratory volumes and capacities – the relationships between the respiratory volumes and capacities of an average male.

(DS), and is simply defined as that part of tidal volume that does not contribute to gaseous exchange. Dead space is usually 150 ml, so for every tidal volume taken in, only 350 ml of air is actually involved in gas exchange. This can be mathematically shown as: $V_T - DS$, or $500 - 150 = 350$ ml, where V_T is the tidal volume.

To calculate the amount of air that is moved into or out of the respiratory system per minute, we can multiply tidal volume by respiratory rate ($V_T \times$ RR), which, if we take V_T to be 500 ml and respiratory rate to be 12, gives us $500 \times 12 = 6000$ ml per minute. By convention, this figure is recorded in litres per minute, so we need to divide by 1000 (the number of millilitres in a litre) to give us 6.0 l/minute. This is called minute volume (MV) or pulmonary ventilation rate (PVR). Since we know that not all of the air contributes to gaseous exchange, we can calculate how much air is being used by making the equation slightly more sophisticated and calculating alveolar ventilation rate. The equation for alveolar ventilation rate (AVR) is $(V_T - DS) \times$ RR. The 'rule' here is that you calculate the information in the brackets first, so using the same figures as before, we get (V_T 500 ml – DS 150 ml) × RR 12, which gives us $(500 - 150) = 350 \times 12$, which is 4200 ml (4.2 litres) per minute.

These calculations may seem daunting, but there is a clinical significance here. If we increase respiratory rate, we tend to reduce tidal volume, so although we can move a lot of air, less air contributes to gaseous exchange. To illustrate, if we increase RR to 30 we can expect a reduction in V_T to 300 ml. PVR would be $30 \times 300 = 9.0$ l/min, but AVR becomes $(300 - 150) \times 30 = 4.5$ l/min. The clinical significance here is that if a patient has a high respiratory rate, their gaseous exchange will fall.

Many people will need to have assessment of their lung function; the reason for this is that asthma, chronic bronchitis and emphysema are common diseases. To assess 'lung' function, patients may undergo assessment of vital capacity, forced vital capacity, forced expiratory volume in one second and peak expiratory flow rate. Vital capacity and forced vital capacity have already been defined, but the forced expiratory volume in one second, or FEV_1 is calculated by dividing FEV_1 into FVC. To illustrate, we can work through an example (see Figure 15.16).

From the graph the forced expiratory volume in one second is 3.6 litres and the forced vital capacity is 4.9 litres, but 3.6 divided by 4.9 looks quite daunting as a calculation. What you can do is to make both numbers 'whole' so 3.6 becomes 36 and 4.9 becomes 49, so you have:

$$\frac{36}{49} \times 100$$

This gives us 0.734, which is multiplied by 100 to give you a percentage, so the final result is 73 per cent. This means that 73 per cent of the total amount of air that this patient got out of his or her lungs was expelled within the first

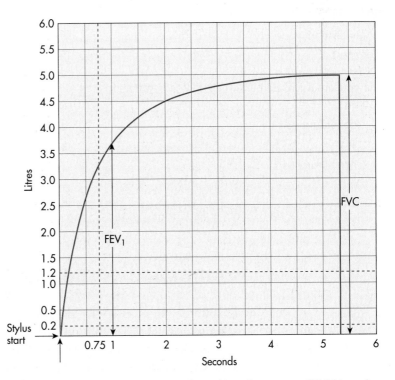

Figure 15.16 Graph showing forced vital capacity (FVC) and forced expiratory volume in 1 second (FEV_1).

Amazing body facts

Exhaled CO₂ and mosquitoes

Since mosquitoes are too small to carry flashlights, how do they find you in the dark? They do it by using carbon dioxide sensors to locate increased concentrations of CO_2 emitted by, you guessed it, your exhaled breath. Once they detect your exhaled CO_2, they use their heat sensors to find an area of skin on which to begin their banquet!

Another amazing carbon dioxide fact is that many people believe swimmers hyperventilate to get more oxygen in their systems before a sustained underwater dive. In reality, they are 'blowing off' CO_2 to get their levels low. Since higher levels of CO_2 cause us to want to breathe, the body is fooled into believing it doesn't need to for a while longer, so swimmers can remain underwater longer. Unfortunately, the body still uses up its oxygen at a regular rate and sometimes uses enough of the reserve that swimmers risk losing consciousness.

second. This test and accompanying calculation is used in helping to diagnose emphysema and chronic bronchitis, for example, as well as other lung diseases.

Another test that helps to establish whether the airways have become 'narrower' than normal (as would be seen in an asthma episode for example) is *peak expiratory flow rate*, or PEFR, which is the maximum flow rate or speed of air a person can rapidly expel after taking the deepest possible breath. PEFR is measured in litres per minute, and should be within a predicted range (see www.brit-thoracic.org.uk for further information on the significance of these tests).

Test your knowledge 15.4

Choose the best answer:

1. The _____ pleural lines the thoracic cavity.
   a. visceral
   b. parietal
   c. mediastinum
   d. chest

2. The portion of the sternum where CPR is performed is the:
   a. xiphoid process
   b. ribs
   c. manubrium
   d. body

3. The _____ nerve innervates the main breathing muscle, called the _____.
   a. thoracic, internal intercostals
   b. thoracic, diaphragm
   c. phrenic, external intercostals
   d. phrenic, diaphragm

4. Rib pairs 1 to 7 are:
   a. attached to the vertebral column only
   b. attached to the vertebral column and the sternum
   c. attached to the sternum only
   d. free floating

Common disorders of the respiratory system

atelectasis *(at eh LEK ta sis)*

Respiratory disease is one of the most common diseases seen in healthcare settings. **Atelectasis**, commonly found in the hospital setting, is a condition in which the air sacs of the lungs are either partially or totally collapsed. Atelectasis can occur in patients who cannot or will not take deep breaths to fully expand the lungs and keep the passageways open. Surgery or an injury of the thoracic cage (such as broken ribs) often makes deep breathing painful.

Taking periodic deep breaths is important not only to expand the lungs but also to stimulate the production of surfactant, which helps to keep the small alveolar sacs open between breaths.

Patients with large amounts of secretions who cannot cough them up are also at risk for atelectasis because the secretions block airways and lead to areas of collapse. Quite often, if atelectasis is not corrected and secretions are retained, **pneumonia** can develop within 72 hours. **Pneumonia** is a lung infection that can be caused by a virus, fungus or bacterium. Inflammation occurs in the infected areas, with an accumulation of cell debris and fluid. In certain pneumonias, lung tissue is destroyed. Pneumonias, if severe enough, can lead to death.

COPD, or chronic obstructive pulmonary disease, is a group of diseases in which patients have difficulty getting all the air out of their lungs and often large amounts of secretions and lung damage is involved. COPD refers to one of or a combination of **asthma**, **emphysema** and **chronic bronchitis**, although there remains some debate about the inclusion of asthma within COPD.

Asthma is a potentially life-threatening lung condition in which the body reacts to an allergy by causing constriction of the airways of the lungs, known as *bronchospasm* (see Figure 15.17). It is difficult to get air in and even more

pneumo = *air or lung*

asthma (*ASS ma*)
emphysema (*emph ah SEE mah*)
chronic bronchitis (*BRONG kye tiss*)

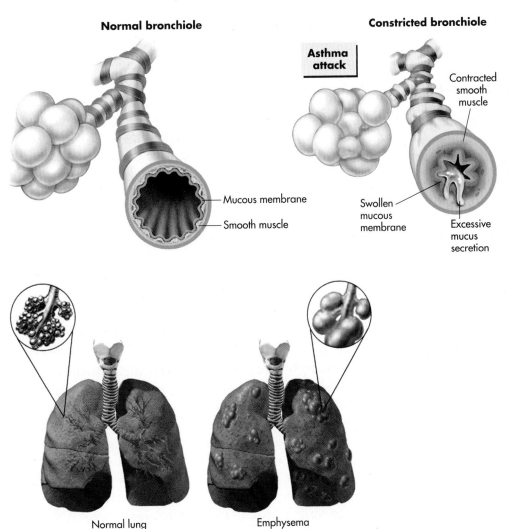

Figure 15.17 Asthma and emphysema.

difficult to get air out of the lungs. The inability to get air out of the lungs is known as *gas trapping*. As a result of gas trapping, fresh air cannot get into the lungs, so the victim breathes the same air over and over. This lowers the amount of oxygen in the blood and increases the blood levels of carbon dioxide. Because this is an inflammatory process of the airways, there is also an increase in the amount of mucus secretions that the airways produce. These increased secretions can block the airways (a phenomenon known as *mucus plugging*) and further reduce the flow of fresh air to the lungs. Although asthma can be a life-threatening disease, it can be controlled with the use of medication.

Emphysema is a non-reversible lung condition in which the alveolar air sacs are destroyed and the lung itself becomes 'floppy' (see Figure 15.17), much like a worn area or bubble in a tyre. As the alveoli are destroyed, it becomes more difficult for gases to diffuse between the lungs and the blood. The lung tissue becomes fragile and can easily rupture (again, much like a worn tyre), causing air to escape into the thoracic cavity and further inhibit gas exchange. This causes a pneumothorax, which is explained shortly.

Chronic bronchitis is a potentially reversible lung disease in which there are inflamed airways and large amounts of sputum being produced. As inflammation occurs, the airways swell and the inner diameter of the airways get smaller. As they get smaller, it becomes difficult to move air in and out, which increases the work of breathing. Because of this increased work level, more oxygen is used and more carbon dioxide is produced.

pneumothorax (*NEW mow THOR aks*)

A **pneumothorax** is a condition in which there is air inside the thoracic cavity and outside of the lungs. Air can enter the thoracic cavity from two directions. A stab wound or gunshot wound to the chest would allow air to rush into the thoracic cavity from the outside. The lung might develop a leak as a result of either a structural deformity or a disease process (such as in emphysema). In this situation, air would enter the thoracic cavity from the lung as air is breathed in. In either case, if the gas cannot escape, it will continue to fill a space in the thoracic cavity and provide less space for the lung or lungs to expand when breathing. If the lungs are too greatly restricted to expand, or if the pneumothorax presses on the heart, a life-threatening situation may occur. A haemopneumothorax is a condition where blood (rather than air) is inside the thoracic cavity. As the blood fills the space, it compresses the lung, shifting the lung to one side, which is also a life-threatening situation.

pleural effusion (*PLEW rall eff YOU shun*)
empyema (*em pie EE muh*)
hydrothorax (*HIGH dro THOR ax*)
haemothorax (*HEEM oh THOR ax*)

A **pleural effusion** is a condition in which there is an excessive build-up of fluid in the pleural space between the parietal and the visceral pleura. This fluid may be pus (in which case it is known as **empyema**), serum from the blood (called a **hydrothorax**) or blood (called a **haemothorax**). Because fluids are affected by gravity, pleural effusions tend to move to the lowest point in the pleural space. If a pleural effusion is large enough, it can have the same effect as a large pneumothorax. It can restrict the amount of expansion of a lung or lungs. Since less air can flow in and out of the lungs, the patient has to work harder by breathing in and out more rapidly to meet the body's demands for more oxygen and the removal of carbon dioxide. This additional work of breathing may exhaust an individual to the point that he or she can no longer breathe without intervention.

tuberculosis (*tue ber cue LOW sis*)

Tuberculosis (TB) is an infectious disease that has seen a recent rise in occurrence. Tuberculosis thrives in areas of the body that have high oxygen content such as in the lung. Tuberculosis can lay dormant in the body for years before beginning to multiply. If it continues unchecked, vast lung damage can occur. There has been recent concern about a multi drug-resistant form that is very resistant to the drugs we normally use to treat tuberculosis, and has a high mortality rate.

Patients with lung disease may exhibit signs and symptoms such as dyspnoea, tachypnoea, cyanosis due to low oxygen levels, and use of accessory muscles of ventilation to assist normal breathing. In addition, the cardiac system may exhibit tachycardia to speed up oxygen delivery and may increase the number of red blood cells that carry oxygen.

The major preventable cause of many of the respiratory diseases is smoking. It has been estimated that smoking kills more people than road accidents, suicides and AIDS combined. This is the equivalent of one person dying every 15 minutes from a smoking-related illness.

Summary

- Moving approximately 9000 litres of air each day, the respiratory system is responsible for providing oxygen for the blood to take to the body's tissues and removing carbon dioxide, one of the waste products of cellular metabolism.
- Ventilation is the movement of gases in and out of the lungs; during respiration, oxygen is added to the blood and carbon dioxide is removed.
- The lungs contain continually branching airways called bronchi and bronchioles.
- At the end of bronchioles are alveolar sacs.
- Each alveolar sac is surrounded by a capillary network where gas exchange occurs with the blood.
- The purpose of the upper airways is to filter, warm and moisten inhaled air for its journey to the lungs.
- In addition, the upper airways provides for *olfaction* (sense of smell) and *phonation* (speech).
- The mucociliary escalator captures foreign particles, and the hair-like cilia constantly move a layer of mucus up to the upper airways to be swallowed or expelled.
- Adenoids and tonsils aid in preventing pathogens from entering the body.
- Since activities of breathing and swallowing share a common pathway, the epiglottis protects the airway to the lungs from accidental aspiration of food and liquids.
- Vocal cords are the gateway between the upper and lower airways.
- The tracheobronchial tree is like an upside-down tree with ever-branching airways, where the trunk of the tree is represented by the trachea and the leaves by the alveoli.
- The alveolar capillary membrane is where external respiration or gas exchange occurs.
- The bony thorax provides support and protection for the respiratory system.
- The main muscle of breathing is the diaphragm, and accessory muscles assist in times of need such as exercise and disease.
- The medulla oblongata in the brain is the control centre for breathing and sends impulses via the phrenic nerve to the diaphragm.

Case study

John is a 20-year-old student who works in a bar. He has been short of breath and wheezy at night time for a couple of months. He goes to his GP for advice.

1. What do you think is wrong with John's respiratory system?

2. What disease do you think he might have?

3. What do you think his peak expiratory flow rate will be: high, normal or low?

4. What factors may exacerbate his breathing trouble?

Review questions

Multiple choice

1. The process of gas exchange between the alveolar area and capillary is:
    a. external ventilation
    b. internal ventilation
    c. internal respiration
    d. external respiration

2. The bulk movement of gas within the lung is called:
    a. internal respiration
    b. ventilation
    c. diffusion
    d. gas exchange

3. Which of the following is *not* a function of the upper airway?
    a. humidification
    b. gas exchange
    c. filtration
    d. heating or cooling gases

4. The largest cartilage in the upper airway is the:
    a. cricoid
    b. Eustachian
    c. mega cartilage
    d. thyroid

5. Which structure controls the opening to the trachea?
    a. oesophagus
    b. hypoglottis
    c. epiglottis
    d. hyperglottis

Fill in the blanks

1. Small bronchi are called _____.

2. The sense of smell is termed _____, and the act of speech is called _____.

3. The hair-like projections called _____ beat within the _____ layer and propel the _____ layer toward the oral cavity to be expectorated.

4. The _____ are thought to lighten the head and provide resonance for the voice.

5. The _____ and the _____ are part of the immune system and are found in the nasopharynx and oral pharynx.

Short answers

1. Describe the tissue layers in the bronchi.

    _____
    _____
    _____
    _____

2. Explain how gas exchange takes place in the lungs.

    _____
    _____
    _____
    _____

3. Discuss the importance of surfactant.

    _____
    _____
    _____
    _____

4. Describe the process of normal breathing beginning with the brain.

    _____
    _____
    _____
    _____

Suggested activities

1. Research the effects of second-hand smoking and chewing tobacco-containing products, and share results with your group or develop posters on these subjects.

2. Working as a group, research occupational causes of lung diseases and see how long a list you can develop.

Answers

Answers to case study

1. John's respiratory system is being adversely affected by the smoky atmosphere at his work.
2. Asthma.
3. His peak expiratory flow rate will be low, due to the narrowing of his airways caused by the inflammation.
4. As well as a smoky atmosphere, factors such as cold weather, animal dander and dust mites will exacerbate his breathing trouble.

Answers to multiple choice questions

1. c
2. b
3. b
4. a
5. c

Answers to fill in the blanks

1. Subsegmental bronchi
2. Olfaction; phonation
3. Cilia; Sol; Gel
4. Sinuses
5. Adenoids; Tonsils (palatine tonsils)

Answers to short answer questions

1. From the lumen out – epithelium; smooth muscle; lamina propria; cartilage

2. The alveoli are the site of gaseous exchange. Blood with high levels of carbon dioxide (deoxygenated blood) pass by the alveoli. Gas moves from an area of high concentration to a relative area of low concentration. The gas in the alveoli is oxygen rich and carbon dioxide poor, therefore the CO_2 passes from the haemoglobin to the alveoli and O_2 passes from the alveoli to blood, binding to haemoglobin

3. Surfactant is a fluid that stops the alveoli from sticking together when you breathe out. It is produced from specialised cells (Type 2 cells) in the alveoli

4. The medulla oblongata detects the level of carbon dioxide (CO_2). As CO_2 rises, a breath is triggered, thus the CO_2 is expelled from the body via the lungs. The diaphragm flattens or contracts and the rib cage comes upwards and outwards, thus increasing the size of the thoracic cavity. This increase in size decreases the pressure inside the lungs. Air, at atmospheric pressure, is now at higher pressure, so air flows down the trachea into the lungs, flooding the airways with gas and equalising pressure. Expiration is largely passive. The diaphragm relaxes, reducing the space in the thorax and increasing intrathoracic pressure. The pressure is now higher than the atmospheric pressure, so air flows from the area of high pressure to the relatively lower atmospheric pressure.

The gastrointestinal system

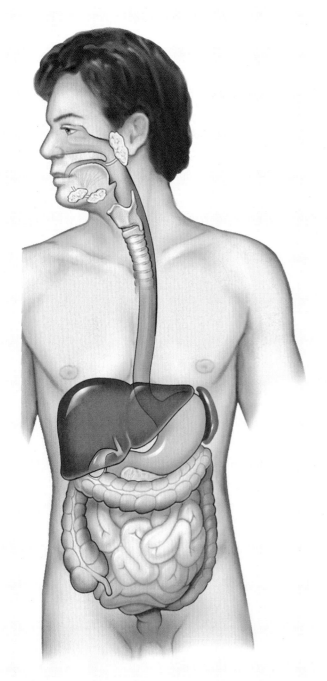

So far, we have discussed a variety of body systems and learned how they are put together and how they function. All of these systems need to function together to create a smooth-running machine. But, just like a Rolls-Royce, no matter how well it is designed nor how precisely it is put together, the body cannot function without fuel. This chapter focuses on the **gastrointestinal** (GI) system and how it:

- *takes in* (*ingests*) raw materials
- *breaks them down* (*digests*) both physically and chemically to usable elements
- *absorbs* those elements
- *eliminates* what isn't usable.

These processes are accomplished through an amazing array of main and accessory organs and substances. In one sense, the food that enters your mouth, travels through your digestive system and is eventually eliminated is never *inside* your body but basically remains in an exterior tube, with materials contained within the food exiting in specific areas!

Learning objectives

At the end of this chapter, you will be able to:

◆ Locate and describe the functions of the main organs of the digestive system
◆ Locate and describe the function of the accessory organs for digestion
◆ Differentiate between ingestion and digestion and between chemical and mechanical processing of food
◆ Trace the journey of material from the mouth to the anus
◆ Discuss the structure of the tooth
◆ Describe the various enzymes and chemicals needed for digestion
◆ Describe common disorders of the gastrointestinal system

Pronunciation guide

adventitia (*add ven TISH ah*)
alimentary tract (*al eh MEN tar ee*)
appendicitis (*ah PEN dee SIGH tiss*)
appendix (*ah PEN dicks*)
caecum (*SEE kum*)
cementum (*si MEN tum*)
cholecystitis (*KOH lee siss TYE tiss*)
cholelithiasis (*KOH lee lith EYE ah siss*)
chyle (*KILE*)
defecation (*deh fi CAY shun*)
duodenum (*due OH dee num*)

emulsification (*ee mull sih fih KAY shun*)
epiglottis (*ep ih GLOT iss*)
frenulum (*FREN you lum*)
fundus (*FUN duss*)
gingivae (*JIN gie vay*)
hepatic duct (*hep PAT ic*)
ileum (*ILL ee um*)
jejunum (*ju JEE num*)
labia (*LAB ee ah*)
mastication (*MASS tih CAY shun*)
mesentery (*ME sen tare ee*)

oesophagus (*eh SOFF ah guss*)
pancreatitis (*PAN kree ah TYE tiss*)
peristalsis (*pair ih STAL siss*)
pharynx (*FAR inks*)
plicae circulares (*PLY kay sir cue LAR es*)
pyloric sphincter (*pye LOR ik SFINK ter*)
pylorus (*pye LOR uss*)
rugae (*ROO gay*)
serosa (*seh ROSE ah*)
villi (*VILL eye*)

System overview

alimentary tract (*al eh MEN tar ee*)

anus (*AY nuss*)

The digestive tract (also called the **gut**, **alimentary tract** or **alimentary canal**) is a muscular tube or tunnel-like structure that contains the organs of the digestive system. This tube begins at the mouth and ends at the **anus**. Between these two points are the pharynx, oesophagus, stomach, and small and large intestines. In addition, *accessory organs* (such as teeth, salivary glands, liver, pancreas and gallbladder) are necessary for processing materials into usable substances. Refer to Figure 16.1.

The components of the digestive system work together to perform the following general steps:

1. ingestion
2. mastication
3. digestion
4. secretion
5. absorption
6. excretion (defecation)

mastication (*MASS tih CAY shun*)

Food first enters the mouth, an activity called **ingestion**. Once food is ingested, the tongue and teeth work together to *mechanically process* the food by *physically* breaking it down. The chewing action is called **mastication**. This mechanical mixing process also continues in the muscular motions of the digestive tract, as you will soon see. **Digestion** is the *chemical process* of breaking down food into small molecules. This is necessary so nutrients can be absorbed by the lining of the digestive tract.

The **secretion** of acids, buffers, enzymes and water aid in the breakdown of food. Once the food is broken down both physically and chemically, it is ready for **absorption** through the lining of the digestive tract for use by the body. Finally, waste products and unusable materials are prepared for **excretion** and are eliminated by the body through **defecation**.

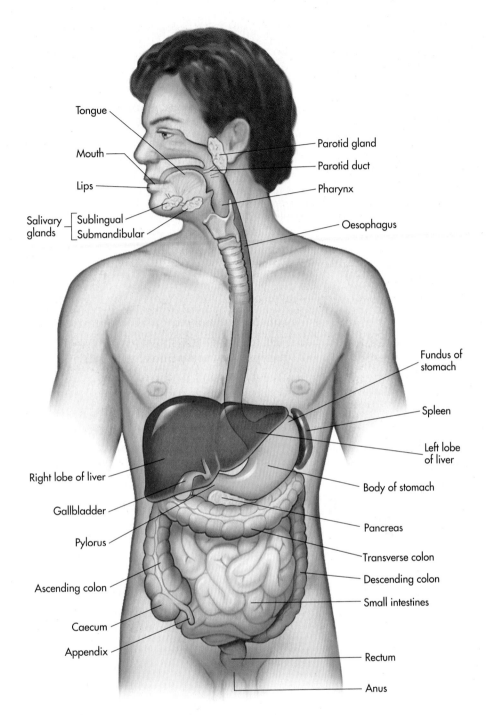

Tongue
Mouth
Lips
Salivary glands — Sublingual
                    Submandibular

Parotid gland
Parotid duct
Pharynx
Oesophagus

Fundus of stomach
Spleen
Left lobe of liver
Body of stomach
Pancreas
Transverse colon
Descending colon
Small intestines
Rectum
Anus

Right lobe of liver
Gallbladder
Pylorus
Ascending colon
Caecum
Appendix

Figure 16.1 The digestive system.

Now that you have a general idea of what's going on, let's begin the journey through the digestive system from beginning to end as a slice of double cheese, pepperoni pizza would.

The mouth and oral cavity

Your mouth is the opening that leads to the oral cavity, which is also called the *buccal cavity*. Your lips, or labia, act as a door to this chamber. The *hard*

labia (*LAY bee yah*)

palates *(PAL ates)*
uvula = *little grape*

pharynx *(FAR inks)*

and *soft* **palates** create the roof of the chamber, while the **tongue** acts as the floor. The tongue's base (area of attachment) and the **uvula**, that punch bag-shaped object dangling down from the soft palate, act as the boundary between the oral cavity and the next part of the digestive system, the **pharynx**. As we discovered when discussing the respiratory system, the uvula aids in swallowing because it helps direct food toward the pharynx and helps block food from coming out of your nose. There is a pair of lingual tonsils back there too. Although they aren't important for digestion, the tonsils help in fighting infection as part of the lymphatic system. The sides of the cavity are created by your cheeks. Your mouth and oral cavity region receives, or *ingests*, food. The food is tasted, *mechanically* broken down into smaller pieces, and *chemically* broken down to some degree. Liquid is added to make it easier to swallow. See Figure 16.2 to view the oral cavity.

Tongue

Your tongue is a muscle that performs many duties. It provides taste stimuli to your brain, senses temperature and texture (as does the rest of your mouth), manipulates food while chewing and aids in swallowing. As the tongue moves the food around in the oral cavity, saliva is added to moisten and soften the food, while teeth continue to crush the food until it reaches the right consistency. The tongue pushes the food into a ball-like mass called a **bolus** so it can be passed on to the pharynx. If you can push that bolus into the pharynx with your tongue, why don't you swallow your tongue too? A membrane under your tongue, called the lingual **frenulum**, which you can see when you lift up your tongue, prevents this from happening. Not only is the frenulum important for swallowing, it also aids in proper speaking. An abnormally short frenulum prevents clear speech, hence the term 'tongue-tied'. The tongue helps prepare the bolus of food, then the tip of the tongue lifts up, so that the bolus moves to the back of the mouth, and comes into contact with the soft palate. So far all the processes involved with the ingestion have been under voluntary control – you can choose how to chew your food, and for how long. As soon as the food comes into contact with the soft palate, however, the process comes under autonomic control, and you cannot voluntarily control the remainder of the digestive process.

bolus *(BOW luss)*

frenulum *(FREN you lum)*

Taste

For most people, this is an important contributor to the enjoyment of life. It is also a safety device, as poisons often taste bitter, and deteriorating food tastes sour.

In order to taste food, it must be in solution; dry foods have no taste. Taste occurs because of the presence of taste papillae, comprising of several taste buds. These occur throughout the mucous membranes of the mouth, tongue, pharynx and upper oesophagus. There are five different types of taste buds: bitter, sweet, sour, salty and umami. Additionally, and perhaps controversially, water is thought to be a recognisable taste. Umami is the most recently identified, and is a savoury, meaty taste. Taste buds are specific for one of these flavours, but traditional maps of the tongue, suggesting that different tastes are perceived at specific locations, are over-simplifications at best. Once a taste

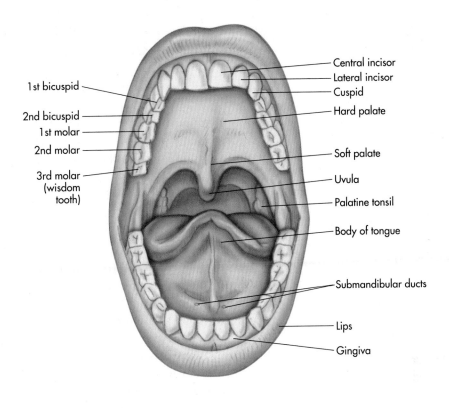

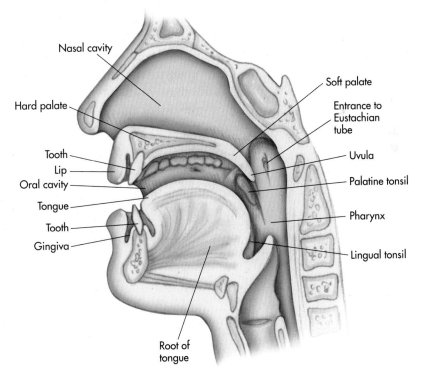

Figure 16.2 The mouth and oral cavity.

bud has been stimulated, the gustatory cell membrane of the taste bud depolarises, and this triggers an action potential (nerve impulse) in the facial nerve (VII), glossopharyngeal nerve (IX) or the vagus nerve (X), depending on the location of the taste bud. This is passed to the medulla in the base of the brain, which, as well as enhancing enjoyment and safety, also triggers a number of reflexes that facilitate digestion.

Salivary glands

sub = *under*

lingual = *pertaining to tongue*

parotid salivary gland (*pah ROT id SAL ih vair ee*)

sublingual salivary glands (*sub LING you all SAL ih var ee*)

submandibular salivary glands (*sub MAN dib you lar SAL ih var ee*)

As you can see in Figure 16.3, there are three pairs of salivary glands, which are controlled by the autonomic nervous system. A large **parotid salivary gland** is found slightly inferior and anterior to each ear. These are the ones that swell up and make you look like a hamster when you get the mumps. The ducts from these glands empty into the upper portion of the oral cavity. The smallest of the salivary glands, the **sublingual salivary glands**, are located under the tongue. The **submandibular salivary glands** are located on both sides along the inner surfaces of the mandible, or lower jaw.

On average, these glands collectively produce *1 to 1.5 litres of saliva daily*. Small amounts of saliva are continuously produced to keep your mouth moist, but once you start eating or even think about eating, look out! The flood gates open! Although saliva is almost totally water (99.4 per cent), it also contains some antibodies, buffers, ions, waste products and **enzymes**. Enzymes are formed by cells. They are organic catalysts whose job is to speed up chemical reactions. Salivary amylase is one of the digestive enzymes that speed up the chemical activity that breaks down carbohydrates, such as starches, into smaller molecules, such as glucose, that are more easily absorbed by the digestive tract once they get there. So, before you even swallow a bite of pizza, amylase is breaking down the starch in the crust for digestion. Saliva plays an important role even *after* you eat. As the saliva is continuously secreted in small amounts, it cleans the oral surfaces and also aids in reducing the amount of bacteria that grows in your mouth. Just remember, that action is *not* a substitute for brushing your teeth.

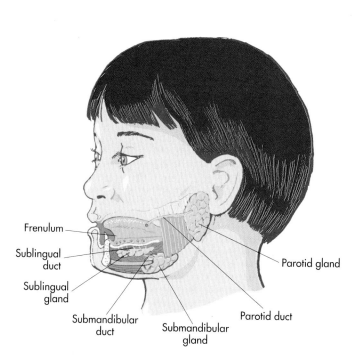

Figure 16.3 The salivary glands.

Clinical application

Sublingual medication

Go to a mirror and open your mouth. Lift your tongue and you will notice that there are blood vessels everywhere just barely underneath the surface of the skin. This sublingual blood vessel network can readily absorb substances such as certain drugs. Rapid absorption into the blood system is vital in some cases, such as for those who need the drug glyceryl trinitrate to treat angina quickly. Angina is caused by insufficient blood flow to the heart and consequently not enough oxygen to coronary tissues. Glyceryl trinitrate (GTN) improves coronary blood flow and in most cases relieves angina.

Teeth

The final important components of the mechanical aspect of digestion in the oral cavity are the teeth. It is unfortunate that we get only two sets of them in a lifetime.

The first set of teeth are called *baby teeth* or, more properly, **deciduous teeth**. Like the leaves on a deciduous tree, they fall away in time. Beginning at around 6 months of age, they begin to appear, the first being the lower central incisors, with all 20 usually in place by 2 years of age. Between the ages of 6 and 12 years the Tooth Fairy is kept busy as these teeth are pushed out and replaced by the 32 larger *permanent teeth*. The exception to this are the *wisdom teeth*, which may not appear until an individual is as old as 21 years.

Now, not all teeth are the same, as you can see by their shapes and locations. Each has special responsibilities. Look at Figure 16.4 as we discuss the various types of teeth. The first tooth type is the **incisor** which is located at the front of your mouth, unless you are an aggressive hockey player. Incisors are blade-shaped teeth used to tear and cut food. **Canine teeth** are for holding, tearing or slashing food. Canine teeth are also known as eyeteeth, or **cuspids**, and are located next to the incisors. Next in line are the **bicuspids**, or pre-molars, which are transitional teeth. **Molars** are the final type of teeth and have flattened tops. Both the bicuspids and molars are responsible for crushing and grinding. The jaws move up and down to crush food, and also sideways, in the opposite direction to each other, so as to grind food.

Regardless of its type, the structure of each tooth is pretty much the same. As you can see in Figure 16.4, each tooth has a **crown, neck** and **root**. The *crown* is the visible part of the tooth. It is covered by the hardest biologically manufactured substance in the body, **enamel**. The *neck* is a transitional section that leads to the *root*.

Internally, most teeth are made up of a mineralised, bone-like substance called **dentine**. The next internal layer is a connective tissue called **pulp**, which is located in the *pulp cavity*. The pulp cavity also contains blood vessels and nerves that provide nutrients and sensations. The nerves and blood vessels get to the pulp cavity via the infamous root canal.

The root is nestled in a bony socket and is held in place by fibres of the periodontal ligament. In addition, **cementum** covers the dentine of the root, aiding in securing the **periodontal ligament**. Cementum is a soft version of

deciduous (*deh SID you uss*)

cementum (*si MEN tum*)
periodontal (*perr ee oh DON til*)

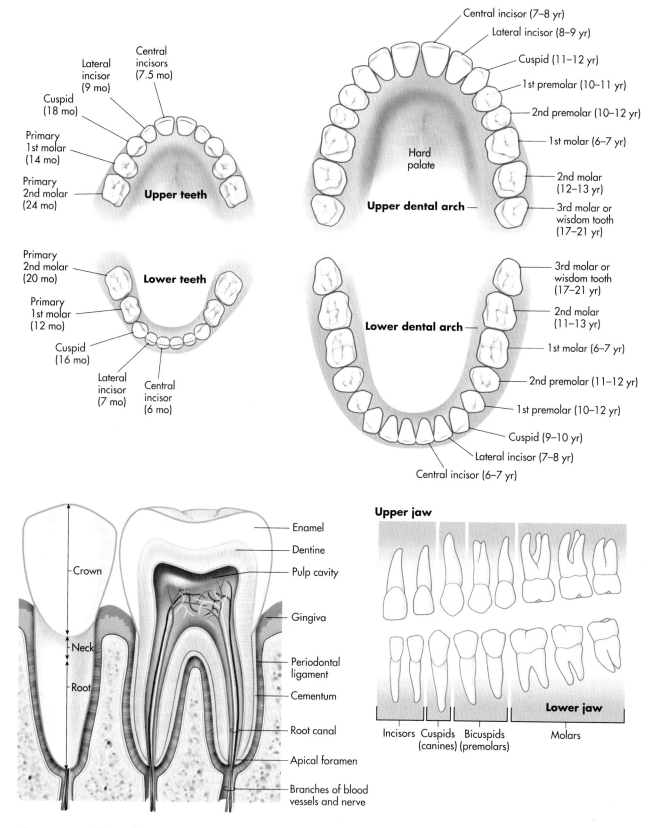

Figure 16.4 Types, location and structures of teeth.

gingiva (*jin gie VA*)

epiglottis (*ep ih GLOT iss*)

oesophagus (*eh SOFF ah guss*)

pharyngo-oesophageal sphincter (*far IN go ee SOFF ih g eel*)

peristalsis (*pair ih STAL siss*)

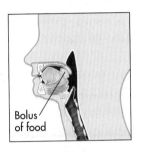

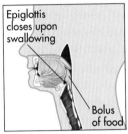

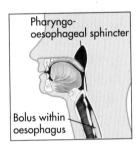

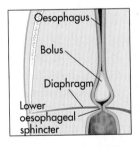

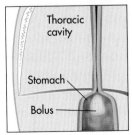

Figure 16.5 The movement of a bolus of food from the mouth to the stomach via the oesophagus.

bone. Healthy gums, or gingiva, also help to hold the teeth in place. Epithelial cells form a tight seal around the tooth to prevent bacteria from coming into contact with the tooth's cementum.

Pharynx

The pharynx is a common passageway not only for food and water but also for air. It has three parts: the *nasopharynx*, the *oropharynx* and the *laryngopharynx*. The nasopharynx is primarily part of the respiratory system. The oropharynx and the laryngopharynx serves two functions: a passageway for food and water and for air. During swallowing, it is imperative that the passageways to the nasal and respiratory regions are protected from the accidental introduction of food and liquids. The nasopharynx is blocked by the soft palate, and a flap of tissue, called the epiglottis, covers the airway to the lungs as the trachea rises during swallowing. These actions force the food to enter the only possible route, the oesophagus.

Oesophagus

The oesophagus is approximately 25 centimetres long and is responsible for transporting food from the pharynx to the stomach. It extends from the pharynx, through the thoracic cavity and diaphragm, to the stomach, which is located in the peritoneal cavity (see Figure 16.5).

The oesophagus is normally a collapsed tube, much like a deflated balloon, until a bolus of food is swallowed. As this bolus moves to the oesophagus, a muscular ring at the beginning of this structure, known as the pharyngo-oesophageal sphincter, relaxes. This is like opening the door to the oesophagus so food can enter. We can't rely on gravity to move food through the oesophagus, so the muscles of the oesophagus begin rhythmic contractions that work the food down to the stomach. This rhythmic muscular contraction is known as peristalsis. Once the bolus reaches the end of the oesophagus, a second door must be opened to allow entry to the stomach. This is the **lower oesophageal sphincter** (which used to be called the *cardiac sphincter*) that relaxes to let food into the stomach and then closes to prevent acidic gastric juices from squirting up into the oesophagus. Heartburn (pyrosis) is what happens if that door opens inappropriately.

The oesophagus also helps move the bolus by excreting mucus so its walls are slippery. The oesophageal walls are lined with stratified squamous epithelium, which makes the oesophagus resistant to abrasion, temperature extremes and irritation by chemicals.

The whole process of swallowing food takes about nine seconds on average. Dry or 'sticky' food may take longer with repeated attempts to work it down (ever tried to swallow too large a bite of a cheese sandwich and have it stick, hammering on the LES door?). Fluids take only seconds to get to the stomach.

The walls of the alimentary canal

It is interesting to note that the same four basic tissue types form the wall of the entire alimentary canal from the oesophagus onward (see Figure 16.6). The innermost layer that lines the lumen of the canal is the **mucosa**. This layer is composed mostly of surface epithelium with some connective tissue and a thin smooth muscle layer surrounding it. The mucosa also possesses cells that secrete *digestive enzymes* to break down foodstuffs, and *goblet cells* that secrete *mucus* for lubrication.

The **submucosa** is the next layer and is composed of soft connective tissue. This layer contains blood and lymph vessels, lymph nodes (called Peyer's patches which are similar to your tonsils), and nerve endings. The next layer is the **muscularis externa** and is composed of two layers of smooth muscle. The innermost layer encircles the canal, while the outer layer is longitudinal in nature so it lies in the direction of the canal. There is an additional third layer of oblique smooth muscle, but it is only found surrounding the stomach.

The outermost layer is the **serosa**, composed of a single thin layer of flat, serous, fluid-producing cells supported by connective tissue. For most of the

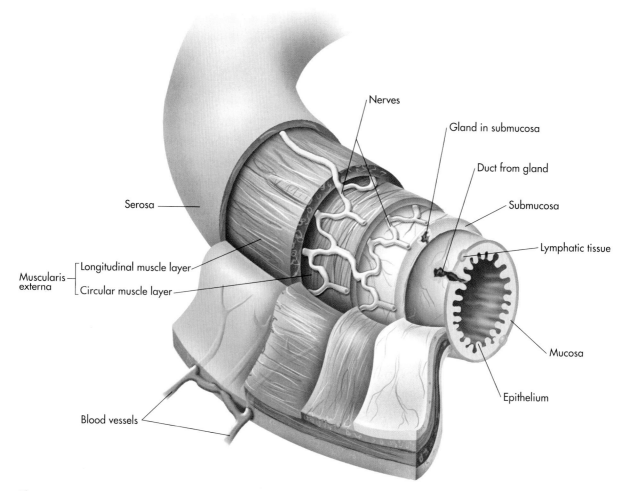

Figure 16.6 Basic tissue types of the alimentary canal.

canal, the serosa is the *visceral peritoneum*. The peritoneum is a serous membrane in the abdominopelvic cavity. Like all serous membranes it has two layers. The visceral peritoneum covers the organs and the parietal peritoneum lines the wall of the abdominopelvic cavity. Between the layers is a fluid-filled potential space called the peritoneal cavity. This fluid is important for both keeping the outer surface of the intestines moist and allowing friction-free movement of the digestive organs against the abdominopelvic cavity. Some abdominal organs such as the urinary bladder and the duodenum are not surrounded by peritoneum and are called **retro**peritoneal organs. The oesophagus differs in that it possesses only a loose layer of connective tissue called the **adventitia**.

retro (*behind*)

adventitia (*add ven TISH ah*)

Test your knowledge 16.1

Choose the best answer:

1. The digestive tract is also called the:
   a. elementary canal
   b. integumentary canal
   c. alimentary canal
   d. panama canal

2. A ball-like mass of food is called a:
   a. bolus
   b. wad
   c. chyme
   d. mastion

3. What is the punch bag-shaped object dangling from the soft palate?
   a. tonsil
   b. uvula
   c. labia
   d. incisor

4. Which of the following is *not* a salivary gland?
   a. parotid
   b. sublingual
   c. submandibular
   d. substernal

Complete the following:

5. Externally, the three main structural parts of a tooth are the _____, _____ and _____.

Stomach

The stomach is located in the left side of the abdominal cavity under the diaphragm and is covered almost completely by the liver. This organ is approximately 25 centimetres long with a diameter that varies, depending on how much you eat at any given time. Although the stomach can hold up to 4 litres when totally filled, it can expand or decrease in diameter thanks to deep folds, called **rugae** (singular **rugus**), in the stomach wall that allow for these size changes. As the stomach receives food from the oesophagus, it performs several functions:

rugae (*ROO gay*)

- acts as a temporary holding area for the received food
- secretes gastric acid and enzymes, which it mixes with the food, causing chemical digestion

- regulates the rate at which the now partially digested food (a thick, heavy, cream-like liquid called **chyme**) enters the small intestine
- absorbs small amounts of water and substances on a very limited basis (although the stomach does absorb alcohol).

On average it takes between four and six hours for the stomach to empty after a meal. Liquids and carbohydrates pass through in about four hours. Protein takes a little more time, and fats take longer, usually about six hours.

The stomach is divided into four regions. Located near the heart, the *cardiac region* surrounds the lower oesophageal sphincter (see Figure 16.7). The **fundus**, which is actually lateral and slightly superior to the cardiac region, temporarily holds the food as it first enters the stomach. The *body* is the mid-portion and largest region of the stomach. The funnel-shaped, terminal end of the stomach is called the **pylorus**. Most of the digestive work of the stomach is performed in the pyloric region. This is also the region where chyme must pass through another door, the **pyloric sphincter**, in order to travel on to the small intestine.

The muscular action of the stomach works much like a cement mixer and is achieved by the three layers of muscle found in its walls. One layer is *longitudinal*, one is *circular* and the third is *oblique* in orientation. This arrangement of muscles enables the stomach to churn food as it mixes with gastric juices excreted by *gastric glands* from *gastric pits* in the columnar epithelial lining of the stomach as well as to work the food toward the pyloric sphincter through the muscle activity, peristalsis. With the combined efforts of muscle and gastric juices, both physical and chemical digestion occur in the stomach.

Gastric juice is a general term for a combination of *hydrochloric acid* (HCl), *pepsinogen* and *mucus*. About 1500 millilitres of gastric juice is produced each day by gastric glands. **Pepsinogen** is secreted by the *chief cells*; it is the inactive precursor to **pepsin**, the stomach's main digestive enzyme. HCl is secreted by *parietal cells*. These two cell types have a special relationship: it is HCl that promotes the conversion of pepsinogen to pepsin. Pepsin is needed to break down proteins (like the ones found in meat). The stomachs of children also release rennin, which facilitates protein digestion, particularly milk protein. Even though it doesn't actually digest food itself, HCl does break down connective tissue in meat. HCl must be very strong to work. Normally, it has a very acidic pH of 1.5 to 2. This highly acidic environment plays another important role: killing off most pathogens that enter the stomach. Why doesn't the stomach digest itself? A healthy stomach is protected by *mucous cells*, which generate a thick layer of mucus to shield the stomach lining from the effects of HCl! Also, pepsinogen is not converted to pepsin until it comes into contact with HCl, which does not occur until it is safely within the lumen of the stomach, away from the mucosal walls. Other specialised cells secrete what is known as *intrinsic factor*, which is needed for absorption of vitamin B_{12}. See Table 16.1, which describes gastric glands and their functions.

The stomach's activity is controlled by the parasympathetic nervous system, particularly the *vagus nerve*. Once the vagus nerve is stimulated, the stomach's *motility* (churning action) increases, as do the secretory rates of the gastric glands.

fundus (FUN duss)

pylorus (pye LOR uss)

pyloric sphincter (pye LOR ik SFINK ter)

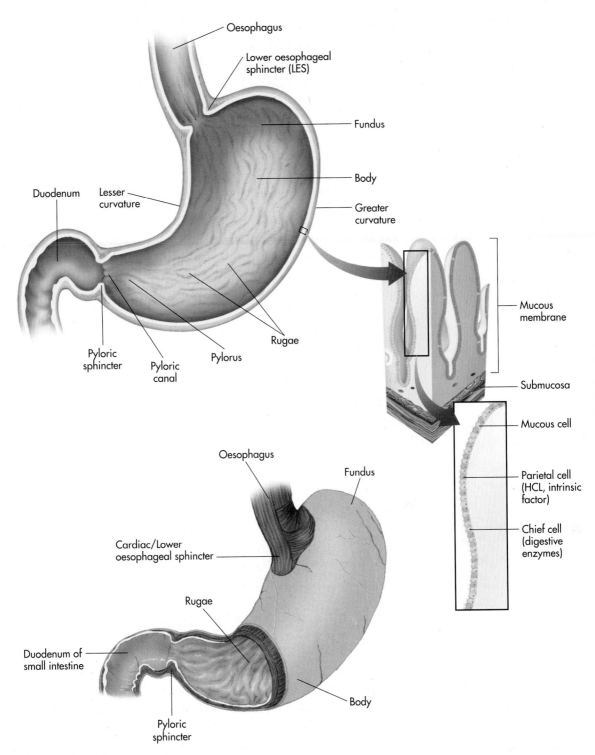

Figure 16.7 The stomach.

The three distinct phases of gastric juice production are illustrated in Figure 16.8. The *cephalic phase* occurs as a result of sensory stimulation such as the sight or smell of food. This sensory input stimulates the parasympathetic nervous system (via the medulla oblongata) and the release of the hormone **gastrin** is increased. Once gastrin travels through the blood and reaches the stomach, gastric gland activity is increased. So, the sight or smell of that pizza literally does get your gastric juices flowing.

gastrin (*GAS trin*)

Table 16.1 Gastric glands and their functions

Digestive cells	Secretion type	Function
chief cells	pepsinogen	begins digestion of protein
parietal cells	HCl	kills pathogens; activates pepsinogen, which is converted to pepsin; breaks down connective tissue in meat
mucous cells	alkaline mucus	protects stomach lining
endocrine cells	the hormone gastrin	stimulates gastric gland secretion

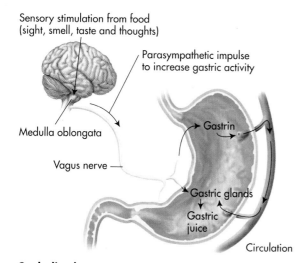

Cephalic phase

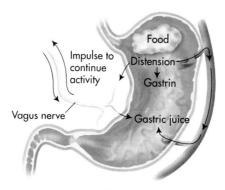

Gastric phase

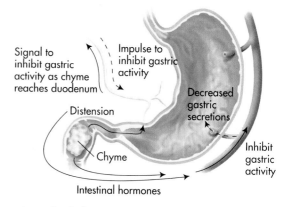

Intestinal phase

Figure 16.8 Phases of gastric secretions.

This leads us to the *gastric phase* in which over two thirds of the gastric juices are secreted as the food moves into the stomach. As the food moves in, the stomach begins to distend. As the stomach distends, it sends signals back to the brain, which fires a reply to the gastric glands to step up their work. As chyme is formed, it is passed through the pyloric sphincter to the first part of the small intestine, the **duodenum**.

duodenum (*deu OH dee num*)

This begins the *intestinal phase* of gastric juice regulation. As the duodenum distends and senses the acidity of chyme, intestinal hormones are released that cause the gastric glands in the stomach to decrease gastric juice production. The brain is also signalled and sends a message to inhibit gastric juice secretion because it is no longer needed now that the food bolus (now called chyme) has left the stomach. Once the chyme begins its movement through the duodenum and on to the rest of the small intestine, those inhibitory responses are halted so gastric juice production can continue once again when a new bolus of food enters the stomach.

Interestingly, the rate of the movement of chyme is very important. If it moves too slowly and slowly empties from the stomach, the rate of nutrient digestion and absorption is decreased and may allow the acidity of the chyme to cause erosion of the stomach lining. If chyme passes too quickly through the stomach, the food particles may not be sufficiently mixed with gastric juices, leading to insufficient digestion. Chyme that is not given time to be neutralised may lead to acidic erosion of the intestinal lining.

Small intestine

Located in the central and lower abdominal cavity, the small intestine is, surprisingly, *the* major organ of digestion. It is where most food is digested (see Figure 16.9). The small intestine is small in diameter, not in length. Beginning at the pyloric sphincter, the small intestine is also the longest section of the alimentary canal, with a length up to 6 metres and a diameter ranging from 4 centimetres where it connects with the stomach to 2.5 centimetres where it meets the large intestine.

The pancreas empties into the small intestine at the first curvature of the duodenum. Pancreatic juice includes most of the enzymes required for the initial stages of digestion, and so digestion will occur throughout the small intestine, although as the material progresses, digestion is completed and absorption becomes more important.

The walls of the small intestine secrete several digestive enzymes important for the final stages of chemical digestion and two hormones that stimulate the pancreas and gallbladder to act and that control stomach activity.

In the small intestine, almost 80 per cent of the absorption of usable nutrients takes place when chyme comes into contact with the mucosal walls. Amino acids, fatty acids, ions, simple sugars, vitamins and water are all absorbed here. Some of the remaining 20 per cent was already absorbed by the stomach, with the rest being absorbed by the large intestine. Any residue that cannot be utilised is sent on to the large intestine for removal from the body.

There are three regions of the small intestine. The duodenum is approximately 25 centimetres long and is located near the head of the **pancreas**. The duodenum gets it name from *duo* (two), and *denum* (ten), which equal 12, the number of finger-widths long that this organ is (25 centimetres)!

pancreas (*PAN kree yass*)

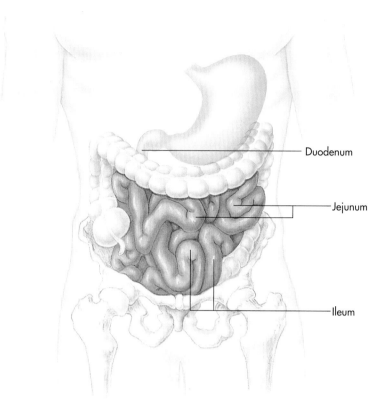

Figure 16.9 The small intestine.

jejunum (*ju GEE num*)
ileum (*ILL ee um*)

The **jejunum** is the middle section and is approximately 2.5 metres long.

The terminal end of the small intestine is the **ileum**. This two-metre section attaches to the large intestine at the *ileocaecal valve*.

The previously discussed pyloric valve is important in allowing small portions of chyme to enter the first part of the small intestine (duodenum) because the small intestine can process only small amounts of food at a time. At the duodenum, additional secretions are added from the pancreas and gall-bladder. The pancreas provides pancreatic juices and the gallbladder provides bile. Bile *emulsifies* fat; that is, it makes fat able to disperse in water, making the fat found in the cheese of the pizza easier to break down. Pancreatic juice contains enzymes and sodium bicarbonate, which neutralises the acidic chyme. The pancreas is stimulated to secrete as a result of the hormone *secretin* that is produced by the small intestine. Gallbladder activity is caused by the hormone **cholecystokinin**, also known as **CCK**, which is also produced by the small intestine. Two types of muscular action occur in the small intestine. **Segmentation** is the muscle action that mixes chyme and digestive juices, working much like a cement mixer. Peristalsis also occurs, moving undigested food remains toward the large intestine. See Table 16.2 for the hormones active in the digestive process.

As previously stated, the small intestine also produces digestive enzymes that are needed to complete chemical digestion. These enzymes (and mucus) are produced by exocrine cells. *Lactase, maltase* and *sucrase* are needed for the digestion of double sugars called *disaccharides* that are contained in starches (such as in the pizza crust). *Peptidase* is needed to digest small proteins called peptides. Internal *lipase* is needed for digestion of certain fats. It is interesting

Table 16.2 Hormones in the digestive process

Hormone	Secreting organ	Action
gastrin	stomach	stimulates release of gastric juice
secretin	duodenum	stimulates release of bicarbonate and water from pancreas and bile from liver; slows stomach activity
cholecystokinin (CCK)	duodenum	stimulates digestive enzyme release from pancreas and bile release from gallbladder; slows stomach activity

to note that the secretion of these substances is mainly due to the presence of chyme in the small intestine! Because of the acidity of chyme, both chemical and mechanical irritation occurs. This irritation plus the distension of the intestinal wall causes the localised reflex action that results in the release of the enzymes and the two hormones.

The structure of the wall of the small intestine is rather interesting. The wall possesses circular folds called *plicae circulares* and finger-like protrusions into the lumen called **villi** (see Figure 16.10). The villi also have outer layers of columnar epithelial cells, which possess microscopic extensions known as *microvilli*. These villi are tightly packed, giving a velvety texture and appearance. The purpose of the microvilli, villi and circular folds is to provide an incredible increase in the surface area of the small intestine. This area, almost the surface area of a tennis court, increases the efficiency of the absorption of nutrients.

Each villus (singular form of villi) contains a network of capillaries and a lymphatic capillary called a **lacteal**. Intestinal glands are located between villi. The capillaries absorb and transport sugars (the result of carbohydrate digestion) and amino acids (the result of protein digestion) to the liver for further processing before they are sent throughout the body. Glycerol and fatty acids (obtained from the digestion of fat) are absorbed by the villi and converted into a lipoprotein that travels on to the lacteal where it is now a white, milky substance called **chyle**. Chyle goes directly into the lymphatic system for distribution throughout the body.

villi (*VILL eye*)

lacteal (*LACK teal*)

chyle (*KILE*)

Clinical application

Lipoproteins

There are different classifications of lipoprotein, depending on the relative proportions of protein and lipid present. High density lipoproteins (HDLs) tend to be more easily metabolised by the liver, whereas low density lipoproteins (LDLs) tend not to be carried well in the blood, and may encourage the promotion of atheroma, a fatty layer inside the arterial walls. It is beneficial then, to encourage the formation of high density lipoproteins, and this can be achieved by eating lipids containing unsaturated fatty acids. Foods high in unsaturated fatty acids tend to be those of plant origin or fish, rather than animal fats. If dietary changes fail, then **statins** are a group of drugs that can help lower high density lipoproteins, and are popularly known as cholesterol-lowering drugs. For people with high blood LDL, statins can help protect them from cardiovascular events such as myocardial infarction and cerebrovascular accident. However, the age at which these drugs should be introduced is quite controversial.

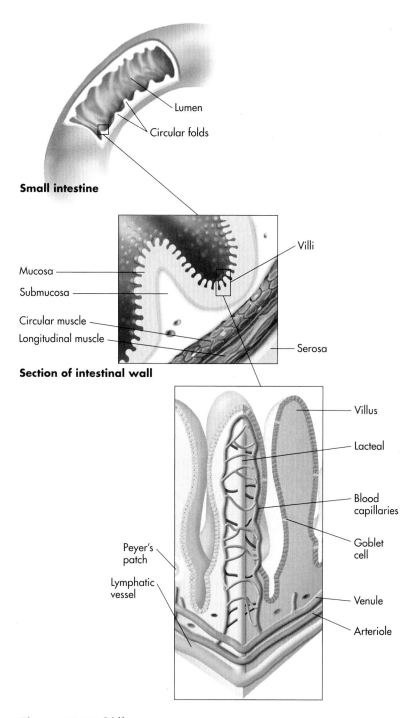

Small intestine

Lumen

Circular folds

Mucosa

Submucosa

Circular muscle

Longitudinal muscle

Section of intestinal wall

Villi

Serosa

Villus

Lacteal

Blood capillaries

Goblet cell

Peyer's patch

Lymphatic vessel

Venule

Arteriole

Figure 16.10 Villi.

Learning hint

Emulsifiers

Think about what Italian salad dressing is like before you shake it. Emulsifiers allow the oil and vinegar to blend together.

Learning hint

The ending ase

Enzymes end in *ase* and break substances down.

Clinical application

Lactose intolerance

This unfortunate condition is the inability to digest the sugar (lactose) found in milk and dairy products such as cheese and ice cream. A person with lactose intolerance has a deficiency of lactase, an intestinal enzyme. As a result, lactose is not sufficiently digested. Normal bacteria found in the intestine utilise those undigested sugars with gas production as a by-product. This is what causes that 'bloated feeling' you see on TV advertisements. In addition, the undigested lactose prevents normal water absorption by the small intestine, so diarrhoea is formed. So a milk pudding would not be chosen by these individuals who are lactose-intolerant.

Interestingly, there is a genetic basis for this condition. In some populations, lactase production continues throughout their entire lives. Approximately 15 per cent of the Caucasian population develop lactose intolerance, while 80 to 90 per cent of the Afro-Caribbean and Asian populations develop this condition to some degree.

To avoid this situation, individuals must either avoid milk or take an oral form of the enzyme lactase before consuming such products. Cheese can often be tolerated because the lactose has already been broken down.

Nutrition

The gastrointestinal tract exists primarily to enable us to extract nutrients from our food, and to absorb them. The foods we eat are broken down to simpler molecules as already described, and it is these molecules that are the nutrients we require. The nutrients can be classified into groups:

- carbohydrates
- proteins
- lipids (fats and oils)
- vitamins
- minerals
- water

Carbohydrates, often abbreviated to CHO, are composed of carbon, hydrogen and oxygen. Rather unusually, they have just one function in human health, and that is to yield energy. However, energy is essential to life, and if there is insufficient carbohydrate in the diet, amino acids or lipid will be utilised to generate energy. Fortunately, carbohydrates are abundant in the diet. They can be further classified into simple carbohydrates or complex carbohydrates. Simple carbohydrates, sometimes called sugars, are small molecules that are readily broken down and quickly yield energy. Whilst this is often useful, it can also lead to rapid changes in blood sugar. They almost always taste sweet, and are abundant in cakes and sweets, and also most fruits. Complex carbohydrates are much bigger, more complex molecules that take time to be broken down in the gastrointestinal tract. Foods containing complex carbohydrates tend to be bulky and, on their own, rather bland. These foods include bread, potatoes, rice and pasta. Some complex carbohydrates, such as some starches, cannot be broken down by humans, and this provides fibre, which is essential to the efficient functioning of the gastrointestinal tract.

Proteins are formed of hundreds of amino acids that contain nitrogen. The proteins that we eat are broken down by the gastrointestinal tract to amino acids. There are approximately 20 different amino acids, but only eight of these (or nine in the case of children) need to be taken in the diet, and are called essential amino acids. The remainder can be made in our liver from other amino acids and are called non-essential amino acids, a process called transamination. Amino acids are crucial for forming and repairing almost all new structures. Proteins are present in most foods: those from meat and dairy products can be particularly useful as they tend to include most of the essential amino acids in approximately the proportions that we require.

Test your knowledge 16.2

Choose the best answer:

1. The deep folds of the stomach wall that allow for size changes of the stomach are called:

   a. rugby

   b. sphincter

   c. rugae

   d. glottal folds

2. The final 'door' of the stomach that needs to open for chyme to travel to the small intestine is located at the end of the:

   a. fundus

   b. pylorus

   c. epiglottis

   d. adventitia

3. This chief digestive enzyme is needed to break down protein:

   a. guafinesin

   b. pepsin

   c. pylorin

   d. rugelin

4. A full two thirds of the gastric juices secreted in the stomach happen as food passes through this gastric juice production phase:

   a. cephalic phase

   b. intestinal phase

   c. gastric phase

   d. pharyngeal phase

5. Proteins are composed of:

   a. amino acids

   b. fatty acids

   c. lipoproteins

   d. glycerol

Complete the following:

6. The stomach's activity is controlled by the _____ nervous system.

Lipids are mostly in the form of triglycerides, which are a glycerol molecule with three fatty acids attached. Fatty acids are either saturated or unsaturated, depending on how many carbon atoms the molecule contains. As a general rule, unsaturated fatty acids come from plant sources, such as nuts, and are liquid at room temperature, so are termed oils. Saturated fatty acids are mostly derived from meat and dairy products, and are solids at room temperature. One class of unsaturated fatty acids are monounsaturated fatty acids, often deriving from fish oils, and may be particularly beneficial to human health. Lipids are essential to life – they are required for making all cell membranes, and also biological messengers such as steroid hormones. They can also yield very high amounts of energy, and that is where they have derived rather a poor reputation. In fact we should aim to have about one quarter of our energy from lipid sources, mainly those containing mostly unsaturated fatty acids.

Vitamins and minerals are sometimes termed micronutrients because they do not need to be taken in large amounts, yet they are vital for the maintenance of health, often acting as enzymes for biological processes. Vitamins are from animal or plant sources, whereas minerals derive from rocks and soils, although they often come to us via water, or through plants and animals that have incorporated them into their structure.

Large intestine

Beginning at the junction with the end of the small intestine (*ileocaecal orifice*) and extending to the **anus**, the **large intestine** almost totally borders the small intestine (see Figure 16.11). The large intestine is responsible for:

- water reabsorption
- absorption of vitamins produced by normal bacteria in the large intestine
- packaging and compacting waste products for elimination from the body.

Since there are no villi in the walls of the large intestine, little nutrient absorption occurs here.

Approximately 1.5 metres long and 6 centimetres in diameter, the large intestine is divided into three main regions: the **caecum**, **colon** and **rectum**. The large intestine is large in diameter, not in length.

caecum (*SEE kum*)

A pouch-shaped structure, the caecum receives any undigested food (such as cellulose) and water from the ileum of the small intestine. The infamous appendix is attached to the caecum. About 9 centimetres long, the appendix is a slender, hollow, dead-end tube lined with lymphatic tissue. Since it is worm-like in appearance, it is often called the **vermiform** appendix. There is no current reason why we have an appendix. It is considered a *vestigial organ* – an organ whose size and function seem to have been reduced as humans evolved. Researchers feel that because it possesses lymphatic tissue, it somehow fights infection. Ironically, if the appendix becomes blocked, inflammation can occur, causing **appendicitis**. Treatment for this is either antibiotics or the surgical removal of the appendix (*appendectomy*).

vermiform = *worm-like*

appendicitis (*ah PEN dee SIGH tiss*)

Some of the water (used in digestion) and electrolytes are reabsorbed by the caecum and the ascending colon, which we discuss momentarily. Although this is a relatively small amount of water reabsorption, it is crucial in maintaining the proper fluid balance in the body.

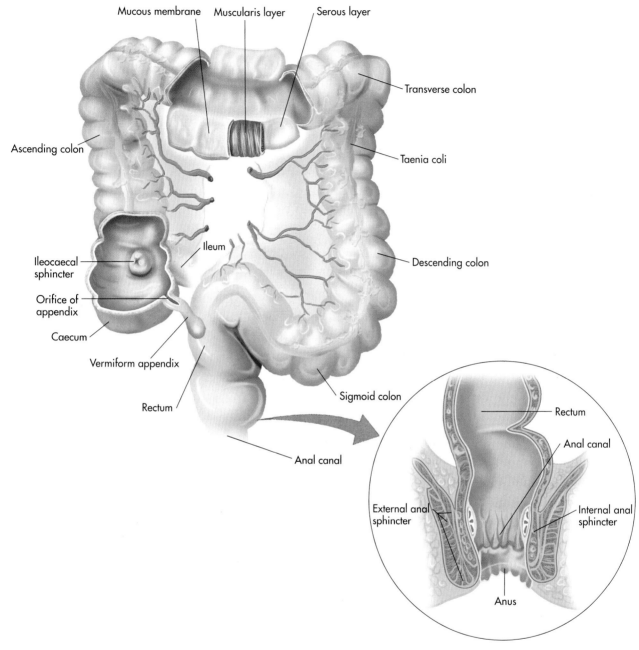

Figure 16.11 The large intestine.

sigmoid = S-shaped

The colon can be further divided into four sections: ascending, transverse, descending and sigmoid. The *ascending* colon travels up the right side of the body to the level of the liver. The *transverse* colon travels across the abdomen just below the liver and the stomach. Bending downward near the spleen, the *descending* colon goes down the left side, where it becomes the **sigmoid** colon. The sigmoid (S-shaped) colon extends to the rectum. The rectum opens to the anal canal, which leads to the anus that relaxes and opens to allow the passage of solid waste (faeces).

Clinical application

Colostomy

Sometimes a portion of the colon must be by-passed because of disease to allow for healing and/or surgical repair. A new opening needs to be made, and this procedure is called a colostomy. A colostomy can be temporary or permanent depending on the condition. The sites where this procedure is formed are shown in Figure 16.12.

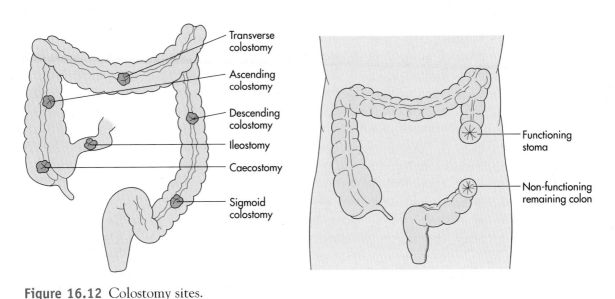

Figure 16.12 Colostomy sites.

Peristalsis continues in the large intestine but at a slower rate. As these slower, intermittent waves move faecal matter toward the distal parts of the colon, water is removed, turning faeces from a watery soup to a semi-solid mass. The material builds up in the sigmoid colon. The material is initially prevented from entering the rectum by the shape of the large bowel at this point – there is a sharp 'corner' between the colon and the rectum. Eventually a more powerful peristaltic wave occurs, and a significant quantity of material moves into the rectum. The distension this causes to the rectal walls leads to the **defecation reflex**. This is a spinal reflex that causes the walls of the sigmoid colon and rectum to relax and those of the anal canal to open. At this point the individual becomes aware of the urge to defecate, and can choose whether to relax the external (voluntary) sphincter and defecate, or to tense it and avoid defecation. It is worth noting that this ability to override a spinal reflex is unusual in physiology, and is quite late developing in children.

If faecal material moves through the large intestine too rapidly, not enough water is removed and diarrhoea occurs. Conversely, if faecal matter remains too long in the large intestine, too much water is removed and constipation occurs.

Bacteria found in the large intestine play two important roles: the bacteria help to (1) further break down indigestible materials and (2) produce B complex vitamins as well as most of the vitamin K that we need for proper blood clotting. Their presence also prevents other, potentially harmful, microorganisms from colonising the large bowel.

Here is a case where bacteria in the right place keep us healthy. If these same bacteria left the intestinal wall and entered the bloodstream, it could be fatal.

Accessory organs

In addition to the salivary glands of the mouth, other accessory organs are necessary for digestion: the liver, gallbladder and pancreas.

Liver

Weighing in at approximately 1.5 kilograms and located inferior to the diaphragm, the liver is the largest glandular organ in the body *and* the largest organ in the abdominopelvic cavity. This organ performs many functions that are vital for survival. As you can see in Figure 16.13, the liver is divided into a larger right lobe and a smaller left lobe (remember, this is from the patient's perspective). The right lobe also has two smaller, inferior lobes.

The liver receives about 1.5 litres of blood *every minute* from the hepatic portal vein (carrying blood full of the end products of digestion) and hepatic (referring to liver) artery (providing oxygen-rich blood).

Although this chapter is on the digestive system and the liver plays a central role in regulating the metabolism of the body, it is important to understand *all* of the functions that this amazing organ performs. Here is a list of what the liver does:

- Detoxifies (removes poisons) the body of harmful substances such as certain drugs and alcohol.
- Creates body heat.
- Destroys old blood cells and recycles their usable parts while eliminating unneeded parts such as the pigment *bilirubin*. Bilirubin is eliminated in bile and gives faeces its distinctive colour.
- Transamination of amino acids.
- Forms blood plasma proteins, such as *albumin* and *globulins*.

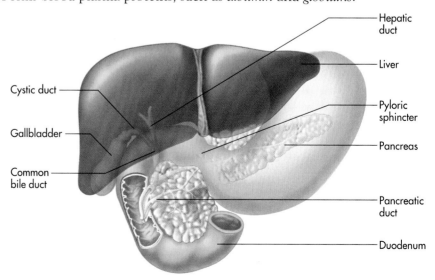

Figure 16.13 The liver.

- Produces the clotting factors *fibrinogen* and *prothrombin*.
- Creates the anticoagulant *heparin*.
- Manufactures *bile*, which is needed for the digestion of fats.
- Stores and modifies fats for more efficient usage by the body's cells.
- Synthesises *urea*, a by-product of protein metabolism, so it can be eliminated by the body.
- Stores the simple sugar *glucose* as *glycogen*. When the blood sugar level falls below normal, the liver reconverts glycogen to glucose and releases enough of it into the bloodstream to bring blood sugar back to an acceptable concentration.
- Stores iron and vitamins A, B_{12}, D, E and K.
- Produces cholesterol.

Stimulated into action by the duodenum's secretion of the hormone *secretin*, bile production is a critical liver digestive function. The salts found in bile act like a detergent to break up fat into tiny droplets that make the work of digestive enzymes easier. This mechanical action of breaking up fat into smaller particles is called **emulsification** and provides more surface area for the enzymes to do their job of chemically digesting fat.

emulsification (*eh mull sih fih KAY shun*)

In addition, bile helps in the absorption of fat from the small intestine and transports bilirubin and excess cholesterol to the intestines for elimination. Once produced by the liver's cells, bile leaves the liver via the **hepatic duct** and travels through the cystic duct to the gallbladder where it is stored until needed by the small intestine.

Gallbladder

The gallbladder is a sac-shaped organ approximately 7.5 to 10 centimetres long and is located under the liver's right lobe. Again, please refer to Figure 16.13. While it is storing the bile, your gallbladder also concentrates it by reabsorbing much of its water content. This makes the bile six to ten times more concentrated than it was in the liver. This is a bit of a balancing act: if too much water is reabsorbed and the bile is constantly too concentrated, bile salts may solidify into gallstones.

When fatty foods enter the duodenum, the duodenum releases the hormone CCK (cholecystokinin). This release causes the smooth muscle walls of the gallbladder to contract and squeeze bile into the cystic duct and on through the common bile duct and then into the duodenum.

Pancreas

Although discussed in the endocrine chapter, your pancreas also plays an extremely important role in digestion. This 15-centimetre-long organ is located posterior to the stomach and extends laterally from the duodenum to the spleen (see Figure 16.14). The exocrine portion of this organ secretes buffers and digestive enzymes through the *pancreatic duct* to the duodenum. The buffers are needed to neutralise the acidity of the chyme in the small intestine. With a pH ranging from 7.5 to 8.8, the chyme is neutralised, saving the intestinal wall from damage. This secretory action is activated by the release of hormones by the duodenum.

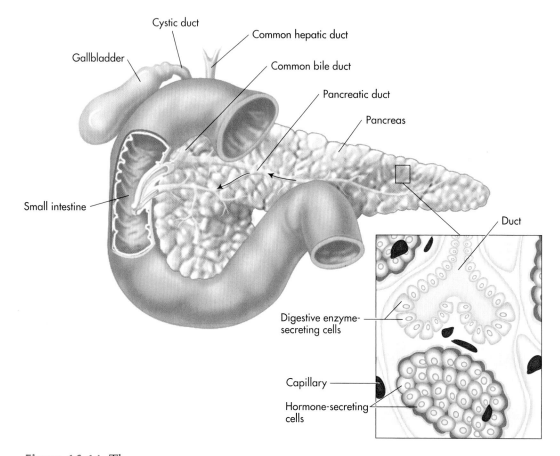

Figure 16.14 The pancreas.

chole = *gall*
cyst = *bladder*
lith = *stone*
cholelithiasis (*KOH lee lith EYE ah siss*)
itis = *inflammation*
cholecystitis (*KOH lee siss TYE tiss*)
tripsy = *to rub*

Clinical application

Cholecystitis and pancreatitis

As we previously discussed, stones can form from substances in the bile while it is stored in the gallbladder. This condition is called **cholelithiasis**. These stones, which are most often formed from cholesterol, can range in size from grains of sand to marble-size and larger. This condition can worsen if the stones lodge in the bile ducts, causing extreme pain, which, surprisingly, often radiates to the right shoulder. If inflammation develops, the condition is called **cholecystitis**.

If the bile backs up into the liver, the disease *obstructive jaundice* can occur. In this scenario, bilirubin is reabsorbed back into the blood, giving the victim a yellowish tint to the skin and eyes.

The problem must be resolved by unblocking the bile ducts. This can done by dissolving the stone through medication, using shock waves to smash the stones (**lithotripsy**), or surgically removing them.

Problems can also occur with the pancreas when the bile duct becomes blocked. In some cases, the pancreatic enzymes back up into the pancreas. As a result, those enzymes begin to inflame and destroy the pancreas. This condition is known as **pancreatitis** and can be caused by excessive alcohol consumption, gallbladder disease or some irritation that causes an abnormally high rate of pancreatic enzyme activation. If this situation is not stopped, death can eventually occur.

The general digestive enzymes excreted by the pancreas are *carbohydrases* that work on sugars and starches, *lipases* that work on lipids (fats), *proteinases* that break down proteins, and *nucleases* that break down nucleic acids.

Clinical application

Cancer of the large bowel

This is the third most common cause of cancer in the UK with more than 40,000 individuals diagnosed in each year. Ninety five per cent of all diagnoses are in the over-50 age group. If diagnosed it usually requires treatment with extensive surgery and often chemotherapy or radiotherapy. One reason for its fearsome reputation is that it is often diagnosed late, or too late for active intervention. Many of the cancers often start as polyps, a non-cancerous lump on the epithelial lining of the colon. These changes occur several years before any symptoms appear (for example blood in faeces or altered bowel habit), and, if they are identified, often could be easily treated. The causes of colorectal cancer are unknown, and almost certainly very varied, but it has been linked to a diet that is high in animal fats, low in fruit and vegetables, and is also associated with inactivity.

Common disorders of the digestive system

Symptoms of digestive disorders generally include one or more of the following:

- vomiting
- **diarrhoea**
- constipation
- abdominal pain

dia = *through*

rrhoea = *flow; literally diarrhoea means 'flow through', which is exactly what happens*

Vomiting is a protective means of ridding the digestive tract of an irritant or overload of food. Sensory fibres are stimulated by the irritant or over-distension and send signals to the vomiting centre (yes, you have a vomiting centre) in the brain. Motor impulses are then sent to the diaphragm and abdominal muscles to contract, which squeezes the sphincter at the oesophageal opening, and the contents are *regurgitated*.

Diarrhoea results when the fluid contents in the small intestines are rushed through the large intestines before they can adequately reabsorb the water. Proper absorption of electrolytes and nutrients is also prevented, which can cause serious problems.

Constipation is the opposite of diarrhoea; the faeces travel so slowly through the colon that too much water is reabsorbed and the stool becomes hard and dry and difficult to push through the system.

One of the more common diseases sometimes blamed on our fast-paced society is **peptic ulcer disease** (PUD), which can affect the lining of the oesophagus, stomach or duodenum. The most commonly affected region is the

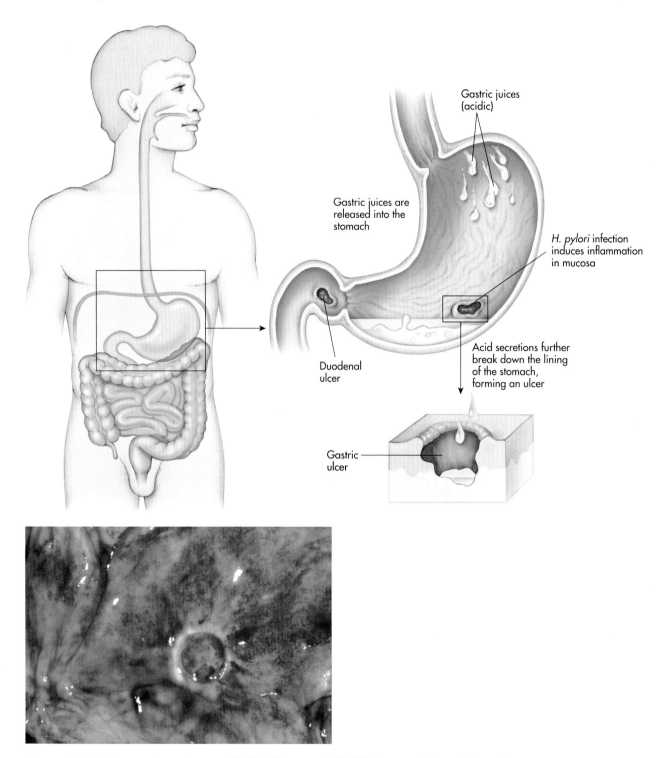

Gastric juices (acidic)

Gastric juices are released into the stomach

H. pylori infection induces inflammation in mucosa

Duodenal ulcer

Acid secretions further break down the lining of the stomach, forming an ulcer

Gastric ulcer

Figure 16.15 Peptic ulcer disease (PUD). (*Source:* © CNRI/Science Photo Library)

upper part of the small intestine, or duodenum. It is caused by an imbalance in the juices of the stomach that produce excess acid and erode the mucosal lining of the digestive tract. It is believed the majority of ulcers are caused by the bacteria *Helicobacter pylori*, which opens a wound in the lining that is made worse by exposure to digestive juices and stomach acids. See Figure 16.15.

There are a host of diseases associated with the digestive system. Please see Table 16.3 for some of the more common ones.

Table 16.3 Pathology of the digestive system

abscess (*AB sess*)	Swelling of soft tissues and the release of pus as a result of infection.
anorexia (*an or REK see ah*)	Loss of appetite that can accompany other conditions such as a gastrointestinal (GI) upset.
bulimia (*bull ee mee ah*)	Eating disorder that is characterised by recurrent binge eating and then purging of the food with laxatives and vomiting.
caries (*KAIR eez*)	Also known as a dental cavity, this is the gradual decay of teeth that can result in inflamed tissue.
cholecystitis (*KOH lee sis TYE tiss*)	Inflammation of the gallbladder.
cholelithiasis (*koh lee lih THIGH al sis*)	Formation or presence of stones or calculi in the gallbladder or common bile duct.
cirrhosis (*sih ROH sis*)	Chronic disease of the liver.
cleft lip	Congenital anomaly in which the upper lip fails to come together. Often seen along with a cleft palate. Corrected with surgery.
cleft palate	Congenital anomaly in which the roof of the mouth has a split or fissure. Corrected with surgery.
Crohn's disease (*CRONES diz EEZ*)	Form of chronic inflammatory bowel disease affecting the ileum and/or colon. Also called *regional ileitis*. Named for Burrill Crohn, an American gastroenterologist.
diverticulitis (*dye ver tik yoo LYE tis*)	Inflammation of a diverticulum or sac in the intestinal tract, especially in the colon.
enteritis (*en ter EYE tis*)	Inflammation of only the small intestine.
fissure (*FISH ure*)	Crack-like split in the rectum or anal canal.
gastritis (*gas TRY tis*)	Inflammation of the stomach that can result in pain, tenderness, nausea and vomiting.
gastroenteritis (*gas troh en ter EYE tis*)	Inflammation of the stomach and small intestines.
gingivitis (*jin gie VIGH tis*)	Inflammation of the gums that is characterised by swelling, redness and a tendency to bleed.
gum disease	Inflammation of the gums, leading to tooth loss, which is generally due to poor dental hygiene.
haemorrhoids (*HEM or roydz*)	Varicose veins in the rectum.
hepatitis (*hep ah TYE tis*)	Inflammation of the liver.
hiatal hernia (*high AY tal HER nee ah*)	Protrusion of the stomach through the diaphragm and extending into the thoracic cavity; reflux oesophagitis is a common symptom.
ileitis (*ill ee EYE tis*)	Inflammation of the ileum.
impacted (*im PAK ted*) wisdom tooth	Wisdom tooth that is tightly wedged into the jaw bone so that it is unable to erupt.
inflammatory bowel disease (IBD) (*in FLAM ah tor ee BOW el dee ZEEZ*)	Ulceration of the mucous membranes of the colon of unknown origin. Also known as *ulcerative colitis*.

Table 16.3 (*continued*)

inguinal hernia (*ING gwi nal HER nee ah*)	Hernia or out-pouching of intestines into the inguinal region of the body.
intussusception (*in tuh suh SEP shun*)	Result of the intestine slipping or telescoping into another section of intestine just below it. More common in children.
irritable bowel syndrome (IBS) (*IR itah bul BOW el SIN drohm*)	Disturbance in the functions of the intestine from unknown causes. Symptoms generally include abdominal discomfort and an alteration in bowel activity.
malabsorption syndrome (*mal ab SORP shun SIN drohm*)	Inadequate absorption of nutrients from the intestinal tract. May be caused by a variety of diseases and disorders, such as infections and pancreatic deficiency.
oesophageal stricture (*eh soff ah JEE al STRIK chur*)	Narrowing of the oesophagus that makes the flow of fluids and food difficult.
oesophageal varices (*eh soff ah JEE al VAR ih seez*)	Enlarged and swollen veins in the lower end of the oesophagus; they can rupture and result in serious haemorrhage.
peptic ulcer (*PEP tik ULL sir*)	Ulcer occurring in the lower portion of the oesophagus, stomach and duodenum thought to be caused by the acid of gastric juices.
periodontal disease (*pair ee oh DON tal dee ZEEZ*)	Disease of the supporting structures of the teeth, including the gums and bones.
polyposis (*poll ee POH siss*)	Small tumours that contain a pedicle or foot-like attachment in the mucous membranes of the large intestine (colon).
pyorrhea (*pye oh REE ah*)	Discharge of purulent material from dental tissue.
reflux oesophagitis (*REE fluks ee soff ah JIGH tis*)	Acid from the stomach backs up into the oesophagus causing inflammation and pain.
ulcerative colitis (*ULL sir ah tiv koh LYE tiss*)	Ulceration of the mucous membranes of the colon of unknown origin. Also known as *inflammatory bowel disease* (*IBD*).
volvulus (*VOL vyoo lus*)	Condition in which the bowel twists upon itself and causes an obstruction. Painful and requires immediate surgery.

Summary

- The digestive tract is a hollow tube extending from the mouth to the anus. It contains a variety of structures that allow the digestion of food and the absorption of nutrients necessary for life.
- Food is processed mechanically and chemically to efficiently break it down to usable substances.
- Following are the main components of the digestive system and their functions:

Organ	Digestive activity	Substance digested	Required digestive secretions
mouth, or oral cavity	chews food and mixes it with saliva; forms food into a *bolus*, and swallows	starch	salivary amylase
oesophagus	moves bolus to the stomach through *peristalsis*	not applicable	not applicable
stomach	stores food, also churns food while mixing in digestive juices	proteins	hydrochloric acid, pepsin
small intestine	secretes enzymes, receives secretions from the pancreas and liver, neutralises the acidity in chyme, absorbs nutrients into the bloodstream and lymphatic system	carbohydrates, fats, nucleic acid, proteins	intestinal and pancreatic enzymes, bile from the liver
large intestine	creates and absorbs fat-soluble vitamins, reabsorbs water, forms and eliminates faeces	vitamins B$_{12}$ and K	not applicable

- Enzymes are formed by special cells and act as a catalyst that speeds up chemical reactions.
- The bulk of the digestive process and the absorption of most nutrients occur in the small intestine.
- Although not *directly* a part of the gastrointestinal system, the accessory organs (liver, gallbladder and pancreas) are needed for proper and efficient functioning of the digestive process.
- The rate of speed that food travels through the gastrointestinal system affects the acidity of the digesting food, the absorption of nutrients, and the quality of the faeces.
- Diseases of the gastrointestinal system can be a result of heredity, the type and amount of food consumed, substance abuse, or emotions.

Case study

Paul presents to Accident and Emergency around 3 am with a severe burning sensation in his chest. He is anxious and thinks he is having a heart attack. All vital signs are within normal limits, and no other pain or discomfort is noted in other regions of the body. He states he ate a large bowl of spicy spaghetti around 11:30 pm. Before falling asleep, he was uncomfortable and felt his large volume of food hadn't digested. He also states this burning sensation has happened in the past after eating, especially at night and when laying on his right side.

1. Do you think this is a heart attack?

2. What do you think the problem is?

3. What physiologic process is malfunctioning?

4. Why is the position of the patient important?

5. What suggestions would you make to the patient to prevent future episodes?

Review questions

Multiple choice

1. Which of the following is not a responsibility of the large intestine?
   a. production of vitamin K
   b. absorption of water
   c. digestion of carbohydrates
   d. elimination of faeces

2. Starches begin to be digested in the:
   a. oral cavity
   b. oesophagus
   c. stomach
   d. large intestine

3. This structure prevents food and liquid from entering the lungs:
   a. larynx
   b. pharynx
   c. epiglottis
   d. glottis

4. The liver receives approximately this much blood every minute:
   a. 500 millilitres
   b. 10 litres
   c. 2000 millilitres
   d. 1.5 litres

5. Which of the following is *not* a colon segment?
   a. transverse
   b. ascending
   c. descending
   d. absorbing

Fill in the blanks

1. _____ is the muscle action that mixes chyme with digestive juices, while _____ is the muscular action that moves food through the digestive system.

2. This vermiform structure is attached to the large intestine and is considered a vestigial organ: _____.

3. The exocrine portion of this important organ secretes buffers needed to neutralise the acidity of chyme and also secretes several digestive enzymes: _____.

4. The end result of faecal matter remaining in the large intestine too long, with too much water being removed from it, is _____.

Short answers

1. Explain the difference between *chyme* and *chyle*.
   _____
   _____
   _____

2. Explain the importance of bacteria in the large intestine.
   _____
   _____
   _____

3. Discuss the importance of the liver in the digestive process.

_____

_____

_____

4. Could you live without a gallbladder? Defend your answer.

_____

_____

_____

Suggested activities

1. Nutrition is a very important consideration when planning a healthy lifestyle. Research the different food groups and the daily recommended amounts that you should consume. List your daily consumption of food for a week and compare what you eat to what you should eat. List suggestions for changes that you should make in your diet, if any.

2. Although it is felt that most stomach ulcers are a result of bacteria, research ways that you can help to prevent their formation.

Answers

Answers to case study

1. Given Paul's history, a heart attack is unlikely to be the cause.
2. Paul's history suggests that he is suffering from indigestion or reflux.
3. Acid from the stomach is backing up into the oesophagus, resulting in the pain and inflammation.
4. Lying down will exacerbate the problem of acid reflux whilst a more upright position should bring some relief.
5. To prevent further episodes, Paul should reduce his intake of spicy food and avoid eating large meals late at night.

Answers to multiple choice questions

1. c
2. a
3. c
4. d
5. d

Answers to fill in the blanks

1. Chewing; peristalsis
2. Appendix
3. Pancreas
4. Constipation

Answers to short answer questions

1. Chyme is mixed food and digestive juices in the stomach and small intestines; chyle is a milk-like substance formed from digested and absorbed fats.
2. Bacteria found in the large intestine play two important roles: the bacteria help to (1) further break down indigestible materials and (2) produce B complex vitamins as well as most of the vitamin K that we need for proper blood clotting. Their presence also prevents other, potentially harmful, micro-organisms from colonising the large bowel.
3. The liver detoxifies harmful substances, manufactures bile, stores and modifies fats, stores iron and vitamins (A, B_{12}, D, E and K), produces cholesterol.
4. Yes; cholecystectomy is removal of the gallbladder due to gallstones.

The urinary system

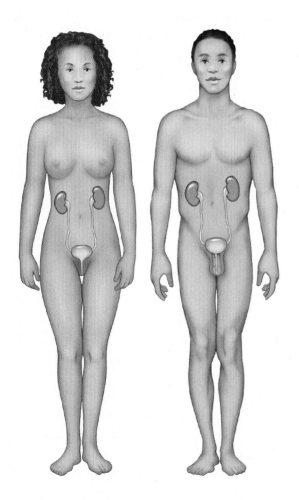

A city must have a safe, reliable water supply, usually obtained by pumping water from a reservoir or diverting a river. Sometimes cities have a series of wells to supply drinkable water. In desert areas, there are even special water treatment plants to remove salt from seawater to make fresh water for human consumption. Either way there must be a means to clean the water. Water purification plants remove chemicals and debris and disinfect the water before it ever gets to your tap. You are completely unaware of the activities of your city's water purification plant, but if it stopped working, you would be extremely unhappy. Imagine a glass full of muddy, bacteria-laden water!

Your body also must have a purification plant for its fluid – blood. Your liver does some of the purification, but your urinary system is responsible for controlling the electrolyte (ion) and fluid balance of your body. The kidneys filter blood, *reabsorb* and secrete ions, and produce urine. Without them, fluid and ion imbalance, blood pressure irregularities, and nitrogen waste build-up would cause death in a matter of days.

Learning objectives

At the end of this chapter, you will be able to:

◆ Present an overview of the organs and functions of the urinary system
◆ Describe the internal and external anatomy and physiology of the kidneys
◆ Discuss the importance of renal blood flow
◆ Describe the process of urine formation
◆ Trace the pathway of reabsorption or secretion of vital substances
◆ List and discuss the importance of hormones for proper kidney function
◆ Describe the anatomy and physiology of the bladder and urine removal from the body
◆ Discuss several common disorders of the urinary system

Pronunciation guide

afferent arterioles (*AFF er ent are TIER ee ohlz*)

aldosterone (*al DOSS ter ohn*)

antidiuretic hormone (ADH) (*AN tea dye you RET ik*)

atrial natriuretic peptide (*AY tree all nat ree your RET ick PEP tide*)

calyx, calyces (*KAY licks, KAY leh seez*)

cortical nephron (*CORE tih cull NEFF ron*)

efferent arterioles (*EFF er ent are TIER ee ohlz*)

external urethral sphincter (*EKS ter nal yal REE thral SFINK ter*)

glomerular capsule (*glom er ROO lar CAPS yule*)

glomerulus (*glom er ROO luss*)

juxtaglomerular cells (*JUX ta glom AR YOU lar*)

juxtamedullary nephron (*JUX ta glom AR YOU lar*)

renal hilum (*REE nal HIGH lum*)

renal medulla (*REE nal meh DULL lah*)

renin–angiotensin–aldosterone (*REE nen an gee oh TEN sen al DOSS ter ohn*)

ureter (*you REE ter*)

urethra (*you REE thrah*)

System overview

ureter (*you REE ter*)

urethra (*you REE thrah*)

The urinary system (see Figure 17.1) consists of two **kidneys**, bean-shaped organs located in the superior dorsal abdominal cavity, and accessory structures that filter blood and make urine. A **ureter** is a tube that carries urine from each kidney to the single **urinary bladder**, located in the inferior ventral pelvic cavity. The urinary bladder is basically an expandable muscle (called the detrusor) that holds the urine. The **urethra** is the tubing that transports urine from the bladder to the outside of the body.

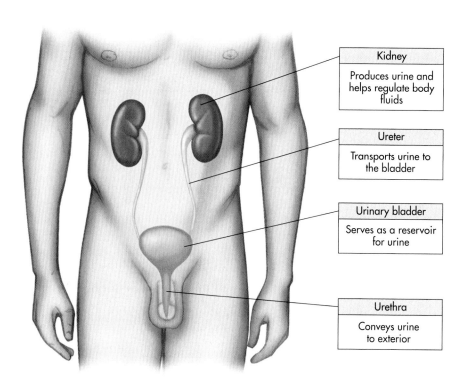

Kidney
Produces urine and helps regulate body fluids

Ureter
Transports urine to the bladder

Urinary bladder
Serves as a reservoir for urine

Urethra
Conveys urine to exterior

Figure 17.1 Anatomy of the urinary system.

The purpose of the urinary system is to make urine, thereby controlling the body's fluid and electrolyte (ion) balance, and eliminating waste products. However, there are other critical processes undertaken by the urinary system that contribute to health. These include:

- regulation of blood volume
- concentration of blood chemicals (for example, sodium, chloride, potassium, calcium, etc.)
- control of acid/base balance (the pH of blood is 7.35–7.45) in conjunction with the respiratory system
- blood cell synthesis (releasing erythropoietin)
- synthesis of vitamin D.

To make urine, three processes are necessary: **filtration, reabsorption** and **secretion**. The kidneys receive about 1250 ml of arterial blood each minute (25 per cent of cardiac output). Filtration involves filtering the blood to a filtrate (that which is allowed to pass through the filter). This filtrate, which contains various substances such as water, glucose, acid, etc., can then be either reabsorbed back into the bloodstream or secreted from the body as urine. Keep in mind the part of the filtrate that is reabsorbed is conserved and brought back into the bloodstream, while the part that is secreted is eliminated from the body. Erythropoietin is released from the peritubular capillaries when the oxygen levels fall in the blood. Erythropoietin results in erythropoiesis, which is production of more red blood cells (that carry oxygen). Vitamin D (vitamin D_3) is produced in the skin following exposure to sunlight. It is converted to its active form (calcidiol) in the kidney, which helps to control the levels of calcium and phosphate in the body. Dark-skinned people who live in temperate climates have been found to suffer from low levels of vitamin D (it is thought that the melanin in their skin hinders vitamin D production), so may benefit from vitamin supplementation.

The anatomy of the kidney

The kidney is a very intricate filtration system. Though you have two kidneys, you can actually function well with only one healthy kidney, which is why someone can donate a kidney while he or she is alive. Furthermore, the function of the kidneys can deteriorate to around 60 per cent of normal before the person notices changes in their body. Urinalysis (a test of the urine) may seem like a simple test, but it can reveal changes to the urinary system before the person feels unwell.

External anatomy

renal capsule (*REE nall CAPS yule*)
renal hilum (*REE nall HIGH lum*)
hilum = *root*

The external anatomy of the kidney is relatively simple. The kidney is covered by a fibrous layer of connective tissue called the **renal capsule**. The indentation that gives the kidney its bean-shaped appearance is called the **renal hilum**. At the hilum, renal arteries bring blood into the kidneys to be filtered. Once filtered, the blood leaves the kidney via the renal vein. The ureter is also attached at the hilum to transport the urine away from the kidney to the bladder (see Figure 17.2).

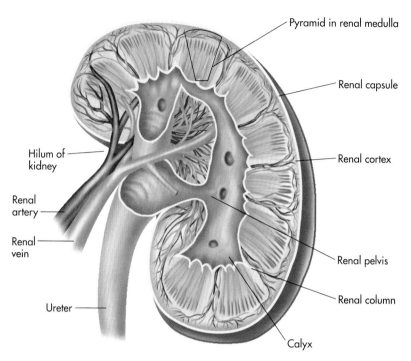

Figure 17.2 The internal and external anatomy of the kidney.

Internal anatomy

The internal anatomy of the kidney (Figure 17.2) is considerably more complicated than its external anatomy. The kidney can be divided into three layers. The outer layer is the **renal cortex**, the middle layer is the **renal medulla**, and the innermost layer is the **renal pelvis**. Adding the word *renal* is important here. Remember that the brain and the adrenal gland both have a **medulla** and a **cortex**. Also keep in mind that the body also has more than one pelvis, as in the bony pelvis. The renal cortex is grainy in appearance and has very little in obvious structure to the naked eye. The renal cortex is where the blood is actually filtered.

The renal medulla contains a number of triangle-shaped striped areas called **renal pyramids**. The renal pyramids are composed of collecting tubules for the urine that is formed in the kidney. Adjacent pyramids are separated by narrow **renal columns**, which are extensions of the cortical tissue.

The renal pelvis is a funnel. The funnel is divided into two or three large collecting cups, called **major calyces**. Each major calyx is divided into several **minor calyces**. The calyces form cup-shaped areas around the tips of the pyramids to collect the urine that continually drains through the pyramids. The kidney is essentially a combination of a filtration and collection system. The blood is filtered by millions of tiny filters in the cortex, the filtered material flows through tiny tubes in the medulla, and the resulting urine is collected in the renal pelvis. The renal pelvis, which is simply the enlarged proximal portion of the ureter, empties into the ureter tube. The ureter then carries the urine to the urinary bladder where it is stored and eventually eliminated from the body (again, see Figure 17.2).

renal cortex (*REE nall CORE tex*)

renal medulla (*REE nall meh DULL lah*)

renal pelvis (*REE nall PELL vis*)

medulla = *inner portion*

cortex = *Latin for 'rind,' or outer layer, like the rind of a watermelon*

calyces = *cup or cup-shaped (KAY leh seez)*

major calyx (*KAY licks*)

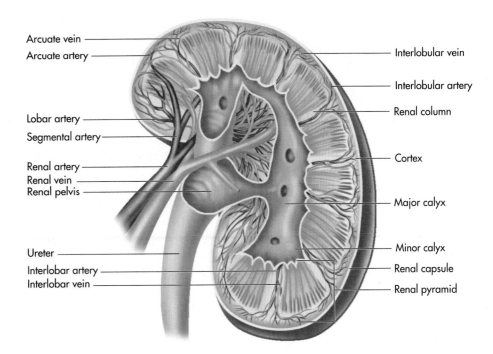

Renal artery → Segmental arteries → Lobar arteries → Interlobar arteries → Arcuate arteries → Interlobular arteries → Afferent arterioles → Glomerulus → Efferent arterioles → Peritubular capillaries → Interlobular veins → Arcuate veins → Interlobar veins → Lobar veins → Renal vein

Figure 17.3 Renal blood vessels and the pathway of blood through the renal system.

Blood vessels

Because the kidney's role is to filter blood, the blood must reach every part of the kidney. To accomplish this filtration process there is a network of blood vessels throughout kidney tissue. A single **renal artery** enters each kidney at the hilum (see Figure 17.3). The renal artery then branches into five segmental arteries. The segmental arteries branch into lobar arteries. The lobar arteries branch into interlobar arteries, which pass through the renal columns. Arcuate arteries originate from the interlobar arteries. The *arcuate* arteries are so named because they arch around the base of the pyramids in the renal medulla. Many tiny interlobular arteries branch from the arcuate arteries, supplying blood to the renal cortex. These interlobular arteries give rise to numerous **afferent arterioles**.

Each afferent arteriole *leads to* a ball of capillaries called a **glomerulus**. **Efferent arterioles** then *leave from* the glomerulus and travel to a specialised series of capillaries called the **peritubular** **capillaries** and *vasa recta* that are associated with the **renal nephron**, the functional unit of the kidney. The peritubular capillaries wrap around the tubules of the nephron. You have seen a situation like this in the lungs, where the pulmonary capillaries surround the alveoli. Having blood vessels close to the nephron allows efficient movement of ions between blood and the fluid in the nephron, just as having pulmonary capillaries near the alveoli allows efficient diffusion of respiratory gases between the alveoli and the bloodstream.

afferent arterioles (*AFF er ent are TIER ee ohlz*)

glomerulus (*glom er ROO luss*)

efferent arterioles (*EFF er ent are TIER ee ohlz*)

peritubular (*peri TUBE you ler*)

nephron (*NEFF ron*)

Learning hint

Visualising the peritubular system

Think of the peritubular capillaries as red wool wrapped around plastic pipe, which represents the actual tubular system. You will soon learn that the filtrate that stays in the pipe eventually becomes urine and the filtrate that is reabsorbed into the red wool (peritubular capillaries) is brought back into the body.

From each set of peritubular capillaries, blood flows out into the interlobular veins. From there, the blood flows out a series of veins that are the direct reverse of the arteries, with one exception. There are no segmental veins. The blood finally leaves the kidney via the **renal vein**, and flows into the inferior vena cava, and thus back to the heart. Please see Figure 17.3 for a diagram of the renal blood vessels.

Microscopic anatomy of the kidney: the nephron

So far we have looked at an overview of the kidney's structure and function. Now let's take a closer look at what actually happens within the kidneys. The business end of the kidney, the part that performs the real functions of the kidney, consists of millions of microscopic funnels and tubules. These fundamental functional units of the kidney are called **nephrons** (see Figure 17.4) and each kidney has a little over a million nephrons. The nephron is divided into two distinct parts: the renal corpuscle and the **renal tubule**. The renal corpuscle is a filter, much like a coffee filter.

Blood enters the renal corpuscle via the **glomerulus**, a capillary ball. Surrounding the glomerulus is a double-layered membrane called the glomerular capsule

renal corpuscle (*REE nall KOR puss all*)

glomerular capsule (*glom er ROO lar CAPS yule*)

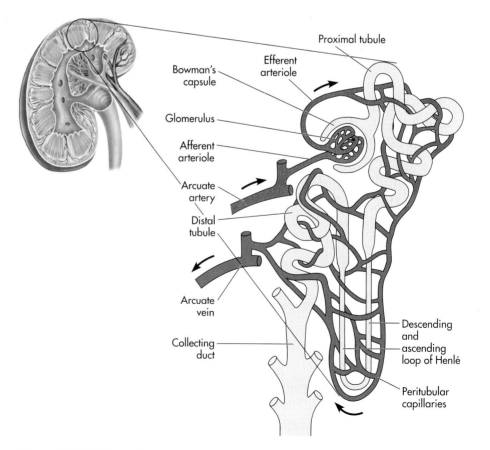

Figure 17.4 The nephron.

(Bowman's capsule). The layers of the glomerular capsule are similar to the layers of a serous membrane like the pleura or pericardium. The inner layer of the glomerular capsule, the visceral layer, surrounds the glomerular capillaries. The visceral layer is made of specialised squamous epithelial cells called **podocytes**. The combination of podocytes and the simple squamous epithelium making up the walls of the glomerular capillaries make a very effective filter. The glomerular capillaries have holes in, called fenestrations or fenestrae, which are 3–7 nanometres in diameter, which means that only small substances, for example, water, glucose, ions, acids, etc. can pass through the fenestrations, but large substances, for example, red blood cells, white blood cells and proteins, are too big to move through the filter. The material filtered from the blood into the glomerular capsule is called glomerular filtrate. Can you see why blood cells or excessive protein found in urine may indicate a kidney filtration problem?

glomerular filtrate (*glom er ROO lar FILL trate*)

The rest of the nephron is a series of tubes known as the *renal tubules* (see Figure 17.5). Just like the water filtration system in our towns and cities, the

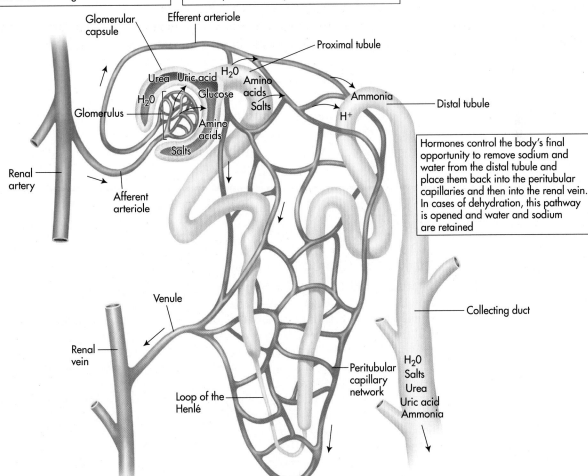

Glomerular filtration

Blood from the renal artery is filtered in the glomerulus. The filtered product which contains water, salts, nutrients and waste products is called the glomerular filtrate

Tubular reabsorption

Nutrients and salts are actively reabsorbed and transported to the peritubular capillary network and some water is passively reabsorbed into the peritubular capillaries

Tubular secretion

Some substances are actively secreted from the peritubular capillaries into the distal tubule for removal from the body

Hormones control the body's final opportunity to remove sodium and water from the distal tubule and place them back into the peritubular capillaries and then into the renal vein. In cases of dehydration, this pathway is opened and water and sodium are retained

Figure 17.5 The renal tubule.

water (glomerular filtrate) travels through a network of pipes (tubules) where the impurities remain in the pipes to be discharged while the filtered water is collected (peritubular capillaries) and recycled back into the city's water supply.

Now back to the kidney. Glomerular filtrate flows from the glomerular capsule into the first part of the renal tubule, the **proximal convoluted tubule**. The wall of the proximal convoluted tubule is made of cuboidal epithelium with microvilli, which dramatically increases the available surface area available for reabsorption. From the proximal convoluted tubule, glomerular filtrate flows into the loop of Henlé (also called the **nephron loop**). The loop of Henlé consists of five parts: the thick descending limb, the thin descending limb, the 'bend', the thin ascending limb and the thick ascending limb of the loop of Henlé. There are two types of nephrons. There are cortical nephrons (which have short loops of Henlé) and juxtamedullary nephrons (with long loops of Henlé). 85 per cent of nephrons are cortical; 15 per cent are juxtamedullary. The longer the loop of Henlé, the more water than can be reabsorbed from it! We call the measure of concentration **specific gravity (SG)**, and this is used clinically to determine how hydrated (or dehydrated) someone is. From the loop of Henlé, the glomerular filtrate flows into the **distal convoluted tubule**. The wall of the distal tubule is like that of the ascending branch of the loop of Henlé.

From the distal convoluted tubule, glomerular filtrate flows into one of several **collecting ducts** also made of cuboidal epithelium. The collecting ducts lead to the minor calyces, then to the major calyces, renal pelvis and ureter. At this point, the glomerular filtrate is urine. Again, if you refer to our previous learning hint, what stays in the tubular system (plastic pipe) eventually gets eliminated as urine.

As you might expect, blood vessels are in close proximity to the nephrons because certain substances within the filtrate must be brought back into the bloodstream. Blood approaches the nephron via the afferent arteriole. Blood flows from the afferent arteriole into the glomerulus, a capillary ball surrounded by

---

Clinical application

Trauma, ischaemia and kidney damage

The kidney is obviously very well vascularised. Each nephron is literally surrounded by blood vessels, and the flow of blood around the nephron is controlled by flow through the afferent arteriole. Ischaemia is a condition of tissue injury resulting from too little oxygen delivery to tissues, usually caused by decreased blood flow. When blood flow to the nephrons decreases for a period of time, oxygen delivery to the nephron decreases and ischaemia can result. Decreased blood flow to the kidney results from anything that causes prolonged constriction of the afferent arterioles. Probably the most common cause of prolonged vasoconstriction in the kidney is any number of hormonal mechanisms used to increase blood pressure, such as after severe blood loss.

For example, a man runs into a door, puncturing his femoral artery, which begins to bleed profusely. As his blood volume falls, so does his blood pressure. His body fights desperately to bring his blood pressure back to normal, causing widespread vasoconstriction. The afferent arterioles, under the influence of sympathetic hormones and other vasoconstrictors, get smaller and smaller, greatly decreasing blood supply to the nephrons. If the situation continues long enough, the tissues will become ischaemic and eventually begin to die. Even if the man survives the initial blood loss from the wound, his kidneys may be damaged, resulting in temporary or permanent renal failure. It is not uncommon for trauma patients to survive the initial trauma only to become the victim of organ damage due to ischaemia.

the glomerular capsule. Blood flows from the glomerulus via the **efferent arteriole** into the peritubular capillaries and vasa recta, a series of blood vessels surrounding the renal tubules (or the red wool from the learning hint earlier). These surrounding blood vessels allow for reabsorption back into the bloodstream from the filtrate that is within the tubular system. Blood then leaves the area of the nephron via the interlobular veins.

Test your knowledge 17.1

Choose the best answer:

1. What carries urine from the kidneys to the bladder?
   a. urethra
   b. ureter
   c. vagina
   d. all of the above

2. The renal _____ is the outer layer of the kidney.
   a. medulla
   b. pelvis
   c. hilum
   d. cortex

3. These vessels carry blood into the glomerulus:
   a. peritubular capillaries
   b. afferent arterioles
   c. segmental arteries
   d. none of the above

4. The fundamental functional unit of the kidney is the
   a. renal corpuscle
   b. renal pelvis
   c. nephron
   d. pyramid

5. Glomerular filtrate flows from the _____ into the _____.
   a. proximal tubule, distal tubule
   b. ascending loop, descending loop
   c. glomerular capsule, proximal tubule
   d. proximal tubule, collecting duct

Urine formation

As the body's water purification system, the job of the kidneys is to control fluid and electrolyte balance by carefully controlling urine volume and composition. The kidneys also filter nitrogen-containing waste and other impurities from blood. In order to form urine, the nephron must perform three processes: **glomerular filtration**, **tubular reabsorption** and **tubular secretion**. Please see Figure 17.6 for a diagram of these three processes.

During glomerular filtration, fluid and molecules pass from the glomerular capillaries into the glomerular capsule, across a filter composed of the wall of the capillaries and the podocytes of the glomerular capsule. The filtrate flows into the renal tubule where the composition of the filtrate is controlled by reabsorption and secretion. Substances that are reabsorbed pass from the renal tubule into the peritubular capillaries and will not end up in urine but stay within the body. Substances that are secreted pass from the peritubular capillaries into the renal tubule and eventually leave the body via the urine. The

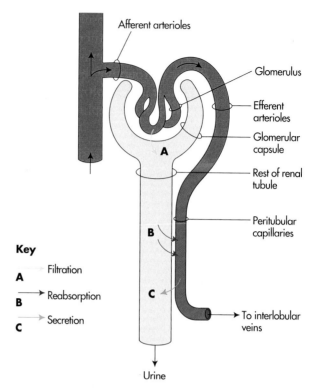

Afferent arterioles

Glomerulus

Efferent arterioles

Glomerular capsule

Rest of renal tubule

Peritubular capillaries

A

B

C

To interlobular veins

Urine

Key

A ──→ Filtration

B ──→ Reabsorption

C ──→ Secretion

Schematic view of the three stages of urine production:
(A) filtration; (B) reabsorption; and (C) secretion

Figure 17.6 The processes involved in urine formation.

combination of all three processes is necessary for the formation of urine. Filtration moves fluid and chemicals into the nephron from blood, and re-absorption and secretion control the concentration of chemicals and volume of urine. Glomerular filtrate is chemically similar to blood, while urine is chemically very different. Some substances, like glucose, are completely re-absorbed (in health), and other substances, like the metabolic waste products urea and creatinine, are secreted such that urea and creatinine are much more concentrated in urine than in blood. See Table 17.1 for a comparison of plasma, glomerular filtrate and urine chemistry and Table 17.2 for details of what might be found when urine is analysed (urinalysis).

Table 17.1 Kidney fluid chemistry

Substance	Plasma	Glomerular filtrate	Urine
protein	3900–5000	none	none
glucose	100	100	none
sodium	142	142	128
potassium	5	5	60
urea	26	26	1820
creatinine	1.1	1.1	140

All concentrations in mg/100 ml

Table 17.2 Urinalysis

Substance	If found in urine is called	Reason it should not be found in urine	Examples of potential diseases
protein	proteinuria	proteins are too big to pass through the fenestrations in Bowman's capsule	urinary tract infection; renal disease
glucose	glycosuria	glucose should be completely reabsorbed in the proximal convoluted tubule	diabetes mellitus; use of steroids
blood	haematuria	blood cells are too big to pass through the fenestrations in Bowman's capsule	trauma; infection; tumour; renal disease
ketones	ketonuria	ketones are the breakdown products of fat metabolism; they are only found in fasting states	starvation; diabetes mellitus

Control of filtration

Think for a moment about filters you are familiar with. What controls these filters? What force drives filtration? What determines whether a substance passes through the filter or stays on one side? One example is filtration of coffee. The purpose of the filter is to prevent coffee grains from getting into your drink and spoiling the taste. So what determines whether the coffee grounds get through? The size of the filter holes. The glomerulus acts like a filter. All filters are *selective*: only some substances pass through, mainly depending on the size of the openings in the filter (see Figure 17.7).

Pressure pushes material through the holes. The higher the pressure on one side of the filter compared to the other side, the faster substances are filtered. An example of this is making coffee in a cafetière. The filters have holes too small to let any but the tiniest coffee grounds pass through. The pressure of the water on top of the filter pushes the water, and the chemicals that make coffee grounds into drinkable coffee, through the filter, but the grounds stay on the other side.

The glomerulus and glomerular capsule work in much the same way. Like most filters, the podocytes and capillary walls of the renal corpuscle create a filter with fixed openings. Plasma and many of the substances dissolved in plasma pass through the filter, but red and white blood cells, platelets and some large molecules, like proteins, do not pass through the filter, in a healthy kidney, but remain in the bloodstream. What passes through the filter is predetermined by the size of the openings. This explains why protein in urine is a sign of kidney

Amazing body facts

Glomerular filtration

The millions of tiny filters in your kidneys perform a truly Herculean task. The average glomerular filtration rate is 110 millilitres per minute, or 160 litres per day.

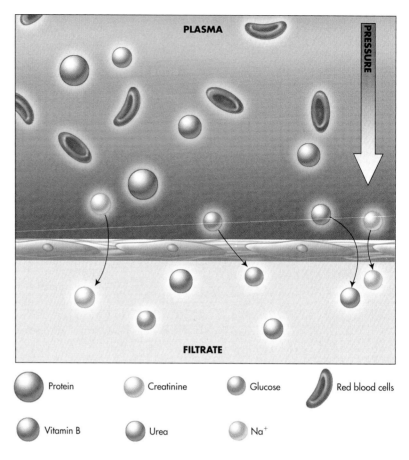

Figure 17.7 Filter selectivity.

nephropathy *(neff ROPP ah thee)*
  nephro = *kidneys*
  pathy = *disease*

damage. Under normal circumstances, protein molecules are too large to fit through the glomerular filter, so proteins do not get into the renal tubule or into the urine. Only when the filter is damaged can protein pass into the filtrate and then into the urine (see Figure 17.8).

Clinical application

Diabetic nephropathy

Diabetes mellitus is a condition characterised by abnormally high blood glucose. This hyperglycaemia, caused by production of too little insulin or by insensitivity to insulin, wreaks havoc with blood chemistry, including osmotic balance. One of the functions of your kidney is to remove all that extra glucose from your blood. When the blood glucose is high, the kidneys must work that much harder to remove it. Patients with elevated glucose eliminate a much greater volume of urine than do patients with healthy levels of blood glucose because their kidneys try to excrete the excess blood glucose.

Prolonged high blood glucose causes a type of kidney damage known as diabetic **nephropathy**. It begins with a thickening of the filter surface of the glomerular capsule and eventually leads to a complete breakdown of that tissue. Once the tissue is damaged, the selectivity of the filter is destroyed. Substances that usually would not pass through the filter and into glomerular filtrate, like proteins, albumin and blood cells, begin to appear in urine. The efficiency of filtration is compromised and kidney function begins to deteriorate. Protein, albumin and blood cells in the urine are early indicators of renal failure. Diabetics can help prevent the onset of kidney damage by keeping blood sugar levels as tightly controlled as possible, preventing high blood pressure, and reducing blood cholesterol to safe levels.

damaged
kidney

healthy
kidney

Figure 17.8 Comparison of damaged and healthy kidneys.
(*Source:* © Dr E. Walker/Science Photo Library)

Clinical application

Kidney stones (nephrolithiasis)

Kidney stones are exactly what their name implies: hard bodies (stones) in the kidney. Kidney stones result when substances in the urine crystallise in the renal tubule, often because the concentration of the molecule is higher than normal in the renal tubule. However, sometimes the cause of kidney stones is a mystery. Stones can be made of many different chemicals, including calcium or uric acid, or can be caused by kidney infection. Some individuals appear to be more susceptible to stones than others, and once you have had a kidney stone, you are more susceptible to kidney stones in the future. Many kidney stones pass through the kidney unnoticed. However, larger or irregularly shaped stones may lodge in the kidney tubules, obstructing flow and irritating nearby tissues. Most patients diagnosed with kidney stones are driven to seek treatment because of blood in their urine (haematuria) or excruciating lower back pain. Patients with kidney stones often describe the pain as the worst they have ever felt. Even painful stones will often pass on their own without medical intervention (particularly if they are less than 5mm in diameter). Patients are treated for pain and sent home with instructions to drink lots of water and wait for the stone to pass. (Often, they are asked to save the stone as it passes so it can be sent for chemical analysis.) Twenty years ago, the only treatment for a stone too large to pass on its own was surgery to remove it. Today, however, a non-invasive technique called lithotripsy, which uses shock waves applied to the outside of the body, can often break up the stone so it is small enough to pass through the kidney, allowing the patient to avoid surgery.

Filtration rate

The filtration rate can be controlled by changing the pressure difference across the filter. The most obvious way to control the pressure is to change the pressure of the blood in the glomerular capillaries. Higher pressure in the glomerulus will increase filtration, lower pressure will decrease filtration. You might think that every minor change in systemic blood pressure would affect glomerular filtration rate. That would be the logical conclusion. However, the glomerulus is protected from minor changes in blood pressure by a mechanism called **autoregulation**. Specialised cells called the macula densa and the juxtaglomerular cells detect the volume and pressure of the blood entering the glomerulus from the afferent arteriole. As systemic blood pressure increases, the afferent arterioles leading into the glomerulus constrict, decreasing the amount of blood getting into the glomerulus. If less blood gets into the glomerulus, the pressure doesn't rise. Autoregulation protects the delicate filter from repeated rapid changes in blood pressure caused, for example, by walking up steps.

Autoregulation can be overridden in situations when blood pressure must be regulated. Because the kidney controls fluid volume, it is often part of mechanisms, along with the cardiovascular system, that regulate systemic blood pressure. For example, if there is a decrease in systemic blood pressure or volume, such as during severe blood loss, glomerular filtration decreases dramatically in an attempt to conserve fluid volume by producing less urine.

Remember that flight-or-fight response includes decreased urine production. The sympathetic nervous system and the hormones of the adrenal medulla, adrenaline (epinephrine) and noradrenaline (norepinephrine), decrease glomerular filtration by causing dramatic vasoconstriction of the afferent arterioles. This prevents blood from flowing to the glomerulus, decreasing the glomerular filtration rate. Small changes in systemic blood pressure do not affect glomerular filtration, because autoregulation keeps glomerular pressure relatively constant, but during shock, for example, glomerular filtration decreases significantly. This is one of the reasons that urine output is monitored in trauma and surgery patients. Remember, in the discussion on blood vessels, we noted that one serious complication of severe blood loss is permanent kidney damage due to decreased blood flow and subsequent death of kidney tissue.

The reverse is also true: if blood volume is elevated, sympathetic output to the afferent arterioles decreases, the arterioles dilate and the glomerular filtration rate increases, allowing the kidneys to get rid of excess fluid.

Control of tubular reabsorption and secretion

Glomerular filtration controls the speed of filtration and ultimately the amount of urine formed. Tubular reabsorption and secretion control the chemistry and volume of the urine. Substances that are reabsorbed move from the tubule back to the bloodstream via the peritubular capillaries and stay in the body. Substances that are secreted stay in the tubule and eventually leave the body via the urine. Thus, anything that affects reabsorption and secretion affects urine chemistry.

The first thing that affects tubular reabsorption and secretion is tubule permeability. Each portion of the tubule can reabsorb and secrete different substances. Remember that molecules can move across membranes via several different methods. *Diffusion* is the movement of molecules from high to low concentration. *Osmosis* is the movement of water across a semi-permeable membrane. Some molecules can only move across membranes by being carried across by proteins. This type of movement is called *active transport*. Differences in tubule permeability result in dramatic differences in which molecules are reabsorbed or secreted in each part of the tubule. See Table 17.3 for a list of substances reabsorbed or secreted in each part of the tubule.

The second factor that affects tubular reabsorption and secretion is a special type of circulation around the loop of Henlé called **counter current circulation**. When ions move across cell membranes, they move from areas of high to low concentration. They are said to move 'down their concentration gradient'. In a solution, if there is a lot of solvent (water), there is less solute

Table 17.3 Individual tubule functions

Tubule	Substances reabsorbed or secreted
proximal convoluted tubule	potassium, chloride, sodium, magnesium, bicarbonate, phosphate, amino acids, glucose, fructose, galactose, lactate, citric acid, water, hydrogen (H^+), neurotransmitters, bile, uric acid, drugs, toxins, ammonia
descending loop	water
ascending loop	sodium, potassium, chloride
distal convoluted tubule	sodium, potassium, chloride, hydrogen (H^+), water
collecting duct	sodium, potassium, chloride, water

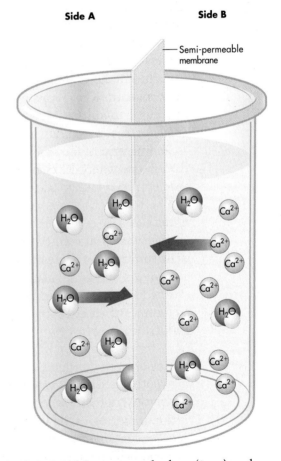

Figure 17.9 Movement of solute (ions) and solvent (water) down their concentration gradient.

(dissolved substances), and vice versa, so you would expect that solute and solvent would move in opposite directions down their concentration gradients. Figure 17.9 shows two solutions separated by a diffusible membrane. In solution A, the solvent water is more concentrated and therefore it will move down its concentration gradient into B. The dissolved solutes, in this

Amazing body facts

Tubular reabsorption

You already know about the amazing ability of your kidneys to filter blood. However, the ability of your renal tubule to reabsorb the fluid filtered by the glomerulus is just as incredible. Your kidneys filter 160 litres (40 gallons) per day, yet you only produce between 1 and 3 litres of urine per day!

case ions, are more concentrated in solution B, so they move down their concentration gradient into A. As always in the body, this is an attempt to maintain the homeostatic balance.

Because of the tendency for water and ions to move in opposite directions, as shown in Figure 17.9, the kidney could not reabsorb both ions and water without the special selective environment around the loop of Henlé. One would be reabsorbed and one would be secreted if it weren't for selected areas of the nephron being impermeable to one or the other. The characteristics of the nephron that make the **counter current circulation** work include the concentration gradient in the fluid surrounding the nephron, with low ion concentration at the beginning of the descending loop of Henlé and high concentration at the tip of the loop of Henlé, and the differences in permeability between the descending (water) and ascending (ions) loops of Henlé.

As filtrate flows into the *descending loop of Henlé, water* is reabsorbed, and the concentration of ions in the loop of Henlé increases as water leaves the tubule. As the filtrate turns the corner and enters the *ascending loop of Henlé,* much of the water has left and the fluid is extremely concentrated. The ascending loop of Henlé is permeable only to ions, so *ions* are reabsorbed from the ascending loop of Henlé. Water and ions that leave the renal tubule enter the capillaries and go back to the bloodstream (see Figure 17.10).

The third set of factors that affect reabsorption and secretion are several hormones that regulate blood pressure. You were introduced to these mechanisms when you learned about regulation of blood pressure. It should come as no surprise that these mechanisms affect kidney function, since the kidneys control ion and fluid balance.

The hormones secreted by the kidneys perform a variety of functions. You met some of these hormones during your visit to the cardiovascular and endocrine systems.

antidiuretic hormone (ADH) (*AN tea dye you RET ik*)

- Antidiuretic hormone (ADH) (also known as vasopressin) is made by the hypothalamus and secreted from the posterior pituitary (also called the neurohypophysis) when blood pressure decreases or blood ionic concentration increases. ADH increases the permeability of the distal tubule and the collecting duct so that more water is reabsorbed, thereby increasing the blood volume and blood pressure and diluting the ionic concentration. Less urine is produced (and it is more concentrated or darker) as more water is reabsorbed, hence the name antidiuretic hormone. All fluids effectively inhibit ADH, but alcohol and caffeine are thought to have a greater inhibitory effect for ADH working than some other fluids and thus prevent the distal tubules and collecting duct from becoming permeable to water. The water then stays in the collecting tubules and is sent to the bladder, thereby increasing urination. If someone is dehydrated, alcohol and caffeine-containing drinks (tea has more caffeine than coffee) should be avoided as it will result in further dehydration.

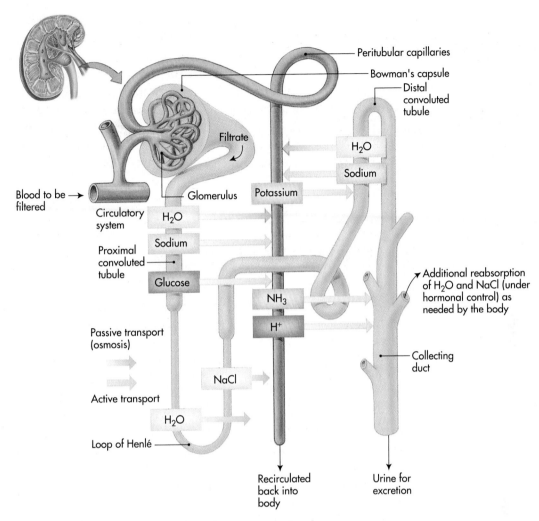

Figure 17.10 Sites of tubular reabsorption and secretion.

aldosterone (*al DOSS ter own*)

- **Aldosterone** is an adrenocorticosteroid, a steroid secreted by the adrenal cortex. Aldosterone is secreted when plasma sodium decreases or plasma potassium increases. It increases the reabsorption of sodium ions (thus bringing more back into the blood) and secretion of potassium ions (thus decreasing the plasma levels) by the distal tubule and ascending limb of the nephron loop. Because sodium is reabsorbed back into the bloodstream, so is more water, and urine volume therefore decreases under the influence of aldosterone.

atrial natriuretic peptide (*AY tree all nat tree your RET ick PEP tide*)

- **Atrial natriuretic peptide** (ANP) is secreted by the atria of the heart when blood volume increases. ANP decreases sodium reabsorption and therefore increases urination because the sodium 'traps' the water in the tubule and it cannot be reabsorbed so is lost in the urine.

renin–angiotensin–aldosterone (*REE nen-an ghee oh TEN sen-al DOSS ter ohn*)

juxtaglomerular (*JUX ta glom AR YOU lar*)

- **Renin–angiotensin–aldosterone** system is a series of chemical reactions that regulate blood pressure in several different ways. When there is a decrease in blood flow to the kidney, a special group of cells near the glomerulus called the **juxtaglomerular** apparatus secretes renin into the bloodstream. Renin converts angiotensinogen, a protein made by the liver, into angiotensin I. Angiotensin I has (as yet) no known biological activity. Another enzyme made by the lungs, angiotensin-converting enzyme (ACE), converts angiotensin I to angiotensin II. Angiotensin II (active angiotensin) is a potent vasoconstrictor. In addition, it increases ADH secretion, increases aldosterone secretion and thirst. Blood pressure is therefore increased either by increased fluid volume due to higher water intake or decreased urination caused by increased levels of ADH or aldosterone. Notice how the kidneys, lungs and liver work together to regulate blood pressure. Many patients with high blood pressure may be given an ACE inhibitor, which inhibits all the previous effects and therefore lowers blood pressure.

Applied science

Electrolytes and acid–base balance

The kidney maintains electrolyte balance by selectively excreting or reabsorbing the electrolytes within the tubular system. One very important interaction is the relationship of hydrogen ions (H^+) and bicarbonate ions (HCO_3^-). The relationship between these ions determines the blood's pH (level of acidity or alkalinity). This is referred to as the acid–base relationship. If too much acid is present in the blood, H^+, which causes acidity and the pH to drop, will be excreted to a greater level in the urine. At the same time, more bicarbonate ions (base that neutralises acids) will be reabsorbed back into the acidic blood to bring the pH value back toward normal. The respiratory system also plays a role in maintaining the acid–base balance by increasing ventilation to 'blow off' more acid in the form of exhaled carbon dioxide (carbonic acid) if the blood is too acidic.

Clinical application

Specific gravity

Specific gravity is a measure of the density of a liquid. Water has a density of 1.000. It is physiologically impossible for a human being to pass 'water'. Urine always contains products that have been filtered by the blood and therefore the urine will always have a specific gravity (SG) of greater than 1.000. In someone who is well hydrated, there is no need for the body to release ADH and conserve water, therefore in someone well hydrated the SG will be around 1.001. In someone who is dehydrated, ADH and aldosterone will be released and so water will be reabsorbed from the nephron. In these individuals, SG will be near to the maximum concentration of 1.035. So, imagine that someone has had surgery, would you expect a concentrated urine or a dilute urine? If the patient had not been given lots of intravenous fluid during surgery, you would expect the urine to be relatively concentrated, thus the SG will be nearer to 1.035. If the SG is low (towards 1.001) and the patient has not had lots of intravenous or oral fluid, it could be a sign of acute renal failure. As the kidneys lose pressure, their ability to concentrate urine is lost, thus dilute urine is often produced even though the individual is dehydrated.

Test your knowledge 17.2

Choose the best answer:

1. Glomerular filtrate is most similar in composition to:
   a. blood
   b. lymph
   c. urine
   d. none of the above

2. When substances move from the tubule into the bloodstream, this is known as:
   a. secretion
   b. reabsorption
   c. filtration
   d. all of the above

3. Which of the following is usually completely reabsorbed in the proximal tubule?
   a. sodium
   b. urea
   c. glucose
   d. water

4. The descending nephron loop is permeable to _____, while the ascending is permeable to _____.
   a. water, ions
   b. ions, water
   c. water, water
   d. ions, ions

5. If blood pressure increases beyond normal range, what happens to the glomerular filtration rate?
   a. it increases
   b. it decreases
   c. it stays the same
   d. none of the above

6. Which of the following is not typically found in urine?
   a. glucose
   b. protein
   c. blood cells
   d. all of the above

The urinary bladder and urination reflex

Glomerular filtrate flows out of the collecting ducts into the minor calyces and then into the major calyces that form the renal pelvis. Once the glomerular filtrate leaves the collecting ducts, its concentration cannot be changed. At this point, the filtrate is urine. Urine collects in the renal pelvis and flows down the ureters to the urinary bladder where it is stored.

The urinary bladder is a small, hollow organ posterior to the symphysis pubis and behind the peritoneum. It is lined with transitional epithelium, the only epithelium stretchy enough to expand as the bladder fills. The ability of the bladder to stretch is enhanced by a series of pleats called *rugae*. The bladder has a muscular wall consisting of several layers of circular and longitudinal smooth muscle and is covered by connective tissue and parietal peritoneum (see Figure 17.11).

As urine accumulates, the bladder fills and stretches. At some point, this stretch triggers urination, or voiding – the emptying of the bladder. For years, urination was thought to be a spinal cord reflex, influenced but not controlled by the brain. New research, however, suggests that the brain actually controls urination. When the bladder is full, signals are sent from the bladder to the

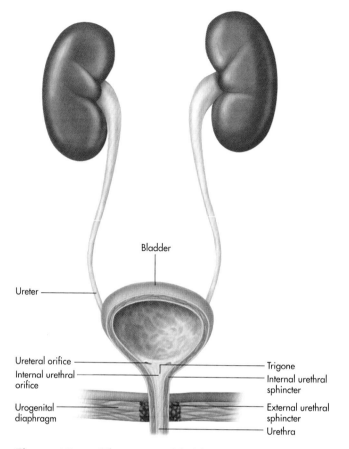

Figure 17.11 The urinary bladder.

spinal cord and then to the pons. The pons sends parasympathetic signals down the spinal cord that cause contraction of the muscular walls of the bladder, and the bladder empties.

Clinical application

Urinary tract infection (UTI)

Urinary tract infection can be caused by the movement of faecal bacteria into the urinary tract. Symptoms include frequent, painful urination. Sometimes the urine is bloody or cloudy or has an unusual odour. Urinary tract infections must be treated promptly because they can cause damage to the kidneys if the infection moves from the bladder up the ureter and into the kidneys. Urinary tract infections are more common in women than in men, because the female urethra is short and closer to the anus, which means gut flora can tract up the urethra into the bladder and form a colony. Urinary tract infections may be prevented in susceptible individuals by drinking plenty of water.

internal urethral sphincter (*in TER nall yoo REE thral SFINK ter*)

external urethral sphincter (*ek STERN all yoo REE thral SFINK ter*)

Urine leaves the bladder via the urethra, a thin muscular tube lined with several different types of epithelium along its length. (For details on the anatomy of the urethra, including sexual differences, see Chapter 18.) Parts of the brain can inhibit urination by controlling the internal urethral sphincter, a valve at the junction of the bladder and the urethra, and the external urethral

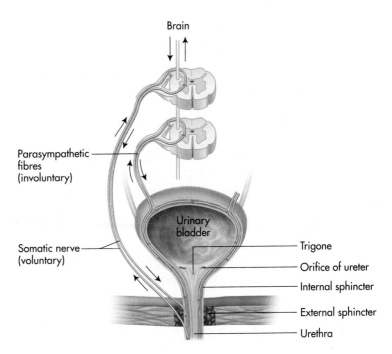

Figure 17.12 Control of urination.

sphincter, a valve that is part of the muscles of the pelvic floor. Sympathetic stimulation of these sphincters prevents urine from leaving the body (see Figure 17.12). Fortunately, although you have little control over contractions of the bladder wall, you have very good control over the sphincters.

Common disorders of the urinary system

The most common disorders of the urinary system are **urinary tract infection** (abbreviated to UTI); **renal failure** (either acute or chronic; if chronic this is called **chronic kidney disease**), **prostate enlargement** and **prostate cancer** (in men). The causes of urinary tract infection can be a structural anomaly in the urinary system, which is more commonly seen in children but resolves as their urinary system matures.

Chronic kidney disease (also known as chronic renal failure) is commonly caused by hypertension (high blood pressure) over a number of years. The condition is made worse if the individual has diabetes because diabetes accelerates deposition of fat into the arteries, which causes hypertension, and changes the basement membrane in the glomerulus, thus increasing pressure in the glomeruli which leads to damage of the delicate filter system. Once the damage has occurred, proteins (particularly albumin) can pass from the blood-stream into the filtrate, and are lost in the urine. For this reason, checking albumin levels in urine can identify the presence of renal disease.

Overuse or abuse of drugs can severely affect renal function. Analgesic nephropathy is caused by long-term use of pain relievers, particularly non-steroidal anti-inflammatory drugs (NSAIDs) like ibuprofen and naproxen, particularly when combined with caffeine or codeine. Even over-the-counter dosages can cause chronic kidney damage leading to kidney failure.

Diabetes insipidus is an endocrine disorder characterised by too little ADH or insensitivity of the kidney to ADH. Without normal ADH activity, copious amounts of urine are produced. Even as the patient gets dehydrated, the production of urine cannot be slowed. A rarer form of the disorder is caused by abnormalities of the thirst mechanism, in which patients drink uncontrollably. Patients with uncontrollable thirst actually drink so much water that they have water toxicity (dangerously low blood sodium because of large volume of water dilution), which can cause brain damage or death.

Glomerulonephritis is inflammation of the glomerulus. Glomerulosclerosis is scarring of the glomerulus. Both cause damage to the delicate filter apparatus. When the filter is damaged, blood cells and blood proteins enter the filtrate and eventually appear in the urine. Removal of waste products is decreased and electrolyte balance is generally abnormal because of the change in urine chemistry. There are many causes of glomerulonephritis and glomerulosclerosis, including bacterial infection, diabetic nephropathy, systemic lupus erythematosus, and genetic disorders such as Alport syndrome and Goodpasture's syndrome.

Haemolytic uraemic syndrome is a disorder caused by an infection with the bacteria *E. coli*, typically from eating undercooked meat. The bacteria infect the digestive tract and release toxins that destroy red blood cells. The damaged red blood cells lodge in the blood vessels in the kidney, blocking them and preventing blood flow to the nephrons. Without treatment, permanent kidney damage may result.

Amazing body facts

Urine

In a healthy renal system, the urine produced is sterile. The urea found in urine is the same substance that is used to melt ice in the winter. Urea is also a component in plant fertiliser: although nitrogen is a waste product for humans, it is an essential product for plant growth. Finally, pain from a jellyfish sting can be alleviated by the application of urine, although we wouldn't recommend this as an alternative to medical management. We'll leave you to decide whether you want to risk your own urine or someone else's!

Summary

- The urinary system consists of paired kidneys and paired ureters, which carry urine to the single urinary bladder. The urethra transports urine from the bladder to outside the body. The function of the urinary system is control of fluid and electrolyte balance and elimination of nitrogen-containing waste.

- The kidney is bean-shaped and covered in a capsule. It has an indentation known as the renal hilum and an interior cavity known as the renal sinus. The kidney can be divided into three layers: renal cortex, renal medulla and renal pelvis. The renal pelvis is a funnel that is divided into large pipes, the major calyces. Each major calyx is divided into several minor calyces. The renal pelvis empties into the ureter.

- The kidney is very well vascularised. Blood is supplied to each kidney by a renal artery. The blood vessels split into smaller and smaller branches until there are millions of tiny arterioles, the afferent arterioles. The afferent arterioles supply millions of nephrons, the functional unit of the kidney, with blood. Blood leaves the kidney by a series of veins and ultimately returns to the circulation via the renal vein.

- The nephron is the functional unit of the kidney. There are millions of nephrons in each kidney. The nephron is divided into two parts. The renal corpuscle, consisting of the glomerulus (capillaries) and the glomerular capsule, filters blood and produces glomerular filtrate. The renal tubule, consisting of the proximal tubule, nephron loop, distal tubule and collecting ducts, controls the concentration and volume of urine by reabsorbing and secreting water, electrolytes and other molecules. The walls of the nephron are made of epithelium. The type of epithelium changes depending on the specific function of each part of the nephron.

- Urine is formed by a combination of three processes: glomerular filtration, tubular reabsorption and tubular secretion. The selectivity of the glomerular filter is determined by the size of the openings in the filter and the difference between the blood pressure of the glomerulus and the pressure in the glomerular capsule. The size of the filter does not change unless the glomerulus is damaged. Protein, for example, cannot pass through the filter. However, the filtration rate will change if the pressure in the glomerulus changes. Most of the time, autoregulation, control of the diameter of the afferent arteriole, keeps glomerular pressure and the glomerular filtration rate constant. But sympathetic stimulation can regulate (in this case decrease) glomerular filtration and urine output due to constriction of afferent arterioles.

- Tubular reabsorption and secretion is controlled by differences in tubular permeability. The proximal tubule is the most versatile, reabsorbing dozens of different molecules. The nephron loop is part of an elaborate counter current mechanism, with the descending loop permeable to water and ascending loop permeable to ions. The distal tubule and collecting ducts reabsorb water. The permeability of the renal tubule can be regulated by a number of hormones that control blood pressure. These hormones, aldosterone, ADH, atrial natriuretic peptide and others, regulate blood pressure by regulating urine volume and ion secretion. Changes in urine volume change total body fluid volume and thereby change blood pressure.

- The urinary bladder is a collecting and storage device for urine and is located in the pelvic cavity. It has a muscular wall. Contractions of the muscle result in voiding (urination), emptying the bladder. Urination is a reflex controlled by parasympathetic neurons in the pons. Signals from a full bladder reach the pons. The neurons in the pons then send signals for the bladder to contract. Sympathetic neurons control two valves, the internal and external urethral sphincters, which allow significant conscious control of the urination reflex.

Case study

Jane has recently developed very annoying symptoms. She has to go to the toilet several times a day. Sometimes it seems that she spends every waking moment in there. She hasn't slept through the night in more than a week. She goes to see her doctor, who requests a series of tests to differentiate between several disorders that cause urinary frequency, for example: diabetes mellitus, overactive bladder and urinary tract infection.

Jane's test results:

Urine bacteria	yes
Blood	no
Leucocytes	yes
Glucose	normal
Proteins	yes

1. What is your diagnosis and why?

2. What would be the usual treatment?

3. Given the normal level of glucose found in the urine, which disease is unlikely to be present?

Review questions

Multiple choice

1. The function of this part of the renal tubule is filtration of blood.
   a. renal calyx
   b. renal corpuscle
   c. renal cortex
   d. renal columns

2. The collecting ducts are found in this part of the kidney:
   a. renal cortex
   b. renal capsule
   c. renal pelvis
   d. renal pyramids

3. This tube leads from the urinary bladder to the outside:
   a. collecting ducts
   b. distal tubule
   c. ureter
   d. none of the above

4. The ion responsible for causing acidic blood is:
   a. Na^+
   b. H^+
   c. K^+
   d. HCO_3^-

5. The renal hormone secreted by the hypothalamus when blood pressure decreases to promote the reabsorption of water is:
   a. aldosterone
   b. atrial natriuretic peptide
   c. antidiuretic hormone
   d. adrenaline

Fill in the blanks

1. Most substances are reabsorbed or secreted in this part of the renal tubule: _____.

2. This part of the renal tubule has an elaborate counter current mechanism for reabsorption of sodium and water: _____.

3. This hormone is released by the heart when fluid volume increases: _____.

Short answers

1. List and explain the activity of three regulators of kidney function.

   _____
   _____
   _____
   _____

2. Describe the structure of the wall of the urinary bladder.

   _____
   _____
   _____
   _____

3. Explain the control of urination reflex.

_____

_____

_____

_____

Suggested activities

1. Draw and label a nephron! Some patients will benefit from a visual explanation of what is happening in their body and a diagram can replace a thousand words, as the saying goes. Try to label the key parts of the nephron, e.g. proximal convoluted tubule, and try explaining to a friend what each section does.

2. Quiz yourself. Do you know how the afferent arteriole constricts? What does ADH do to the distal tubule and collecting duct?

3. Can you explain why haematuria, glycosuria, proteinuria and ketonuria are abnormal? What diseases can cause these? How common are they? What clinical signs might there be?

4. Why should blood pressure be less than 135 mmHg systolic and less than 90 mmHg diastolic? What condition will result from high blood pressure?

Answers

Answers to case study

1. A urinary tract infection, evidenced by the presence of leucocyes, bacteria and proteins. [An overactive bladder is another possible additional cause.]
2. A course of antibiotics.
3. Diabetes mellitus.

Answers to multiple choice questions

1. b
2. d
3. d
4. b
5. c

Answers to fill in the blanks

1. Proximal convoluted tubule
2. Loop of Henlé
3. Atrial natriuretic peptide

Answers to short answer questions

1. Filtration – blood is filtered in Bowman's capsule; Secretion – some substances are actively secreted into the distal tubule for removal from the body; Reabsorption – the proximal convoluted tubule reabsorbs 80 per cent of material that enters the PCT.
2. The lining of the urinary bladder is with transitional cells. The bladder is actually a muscle (the detrusor) that compresses when one passes urine (voids)
3. To pass urine (micturate) the detrusor contracts at the same time as the internal and external urinary sphincters relax. Once the bladder is empty, the detrusor stops contracting and the internal and external sphincters contract, maintaining continence.

The reproductive system

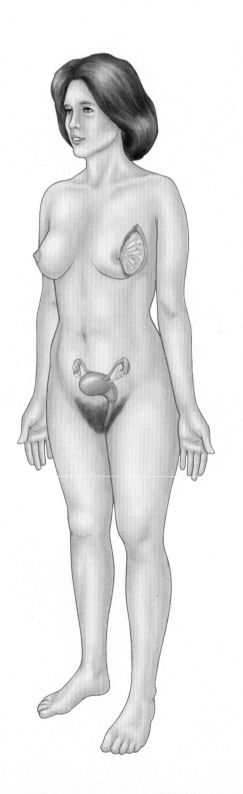

In order for a city to run smoothly, it must have an infrastructure: buildings, roads, playgrounds, schools, hospitals etc. All of these structures must be repaired on a regular basis, or if damaged beyond repair, they must be replaced. As a city grows along with its population, it needs the resources and means to expand. The same is true of your body. Cells and tissues get damaged or simply wear out. Damaged or worn-out cells and tissues must be repaired or replaced. Asexual reproduction, or mitosis, the process by which cells make exact copies of themselves, is absolutely necessary to maintain a healthy body. Mitosis was discussed in Chapter 4. Ultimately, cellular reproduction leads to the complicated process by which humans produce new humans: sexual reproduction. Without this ability, the human species would die out and the journey would end for the human race. Thankfully, all the splendid diversity of the human race is passed on for generations to come, who will also get to enjoy this amazing journey called 'life'.

Learning objectives

At the end of this chapter, you will be able to:

- ◆ Differentiate mitosis from meiosis
- ◆ Locate and describe the male and female reproductive organs
- ◆ Describe the function of the male and female reproductive organs
- ◆ Discuss the phases of the menstrual cycle
- ◆ Discuss the effects of hormonal control on the male and female reproductive systems
- ◆ Describe the stages of labour and delivery
- ◆ Relate common disorders of the male and female reproductive systems

Pronunciation guide

bulbourethral gland (*BUHL boh you REE thral*)
clitoris (*KLIT oh riss*)
corpus luteum (*KOR pus LOO tee um*)
cytokinesis (*SIGH toe kin EE siss*)
endometrium (*EN doh MEE tree um*)
epididymis (*ep ih DID ih miss*)
eukaryotic cell (*YOO care ee OH tic sell*)
fimbria (*FIM bree yah*)
follicle-stimulating hormone (*FOLL I cul stim yoo LAY ting HOR mone*)
gametes (*gay MEETS*)
genitalia (*jen ih TAY lee ahh*)
gonadotrophin-releasing hormone (*GO nad oh TROW pin*)
human chorionic gonadotrophin (*HUE man core ee OH nik GO nad oh TROW pin*)

labia majora (*LAY bee ah maj OR ah*)
labia minora (*LAY bee yah my NOR ah*)
luteal phase (*LOO teal faze*)
luteinising hormone (*LOO ten IZE ing*)
meiosis (*me OH sis*)
menses (*MEN seez*)
menstruation (*MEN stroo AY shun*)
myometrium (*MY oh MEE tree yum*)
oocyte (*OO oh site*)
oogenesis (*oo JENN eh siss*)
perimetrium (*pair ee MEE tree um*)
primordial follicles (*pry MORE dee all FALL ik alls*)
progesterone (*proh JESS ter ohn*)
pudendal cleft (*pew DEN dall*)
seminal vesicles (*SEM ih nal VEE ik alls*)

seminiferous tubules (*SEM ih NIFF er uss TUBE yules*)
Sertoli cells (*sir TOW lee*)
spermatids (*sper MAT ids*)
spermatocytes (*sper MAT oh sites*)
spermatogonia (*sper MAT oh GO nee ah*)
spermatozoa (*sper MAT oh ZOE ah*)
testicles (*TESS tick alls*)
testis, testes (*TESS tiss, TESS teez*)
testosterone (*tess TOSS ter own*)
urethra (*yoo REE thrah*)
uterus (*YOO ter uss*)
vagina (*vah JYE nah*)
vas deferens (*VAS DEFF er enz*)
vulva (*VULL vah*)
zygote (*ZIGH goat*)

Tissue growth and replacement

Although mitosis was discussed in Chapter 4, we will briefly review that form of cellular reproduction in this chapter. Cellular reproduction is the basis for *all* more complex reproduction, so this is a good starting point.

Mitosis

a = *without*

Cellular reproduction is the process of making a new cell. It is also known as **cell division**, because one cell divides into two cells when it reproduces. Cells can only come from other cells. When cells make *identical* copies of themselves *without the involvement* of another cell, the process is called **asexual reproduction**. Most cells can reproduce themselves asexually, whether they are animal cells, plant cells or bacteria.

eukaryotic cell (*you care ee OH tic sell*)

The cells that make up the human body are a type of cell known as a eukaryotic cell. Eukaryotic cells have a nucleus, cellular organelles, and usually several chromosomes in the nucleus. (Reminder: the genetic material of the cell, DNA, is bundled into 'packages' of chromatin known as chromosomes.) Since chromosomes carry all the instructions for the cells, all cells must have a complete set after reproduction. These instructions include how the cell is to function within the body and blueprints for reproduction. No matter whether a cell has one chromosome, like bacteria, or 46 chromosomes, like humans, all the chromosomes must be copied before the cell can divide.

Now let's go to the more complex form of cellular reproduction. Eukaryotic cells, like yours, must go through a more complicated set of manoeuvres in order to reproduce. Not only do your cells have to duplicate all 46 of their chromosomes, they have to make sure that each cell gets all of the chromosomes and all of the right organelles. The process of sorting the chromosomes so that each new cell gets the right number of copies of all of the genetic material is called **mitosis**. Mitosis is the only way that eukaryotic cells can reproduce asexually. For a quick review on mitosis, go back to Chapter 4.

mitosis (*MY toe sis*)

Mitosis in the body

Mitosis, asexual cellular reproduction, serves many purposes in your body. Any time your cells must be replaced, mitosis is the method used to replace them. Many of your tissues, such as bone, epithelium, skin and blood cells, are replaced on a regular basis. Repair and regeneration of damaged tissue is accomplished by mitosis as well. If you cut your hand, the skin is replaced, first by collagen, but eventually by the original tissue. Mitosis increases in cells near the injury so that the damaged or destroyed cells can be replaced. A broken bone is replaced in much the same way.

Growth is also accomplished by mitosis. Lengthening of bones as you grow, increases in muscle mass due to exercise, indeed, most ways that tissue gets bigger, are due to mitosis of cells in the tissues or organs. Without mitosis, your body would not be able to grow or replace old or damaged cells.

Sexual reproduction

Thus far we have discussed cell division for growth and repair, but we must also be able to perpetuate the species. This requires sexual reproduction.

Reduction division: meiosis

Many animals reproduce sexually; that is, they produce, with the aid of another individual, offspring that are not identical to themselves. Sexual reproduction involves the union of a cell from one organism with a cell from another organism of the same species. In animals, females produce **eggs** and males produce **sperm** for the purpose of reproducing sexually. These special cells are known as **gametes**.

gametes (*gay MEETS*)
meiosis (*me OH sis*)

Gametes are produced by a specialised type of cell division known as **meiosis**, or **reduction division**. Meiosis is called reduction division because the daughter cells produced at the end of meiosis have half as many chromosomes as the original mother cell. (Note: the cells are called mother cells and daughter cells even when we are talking about the process in males.) These daughter cells must have half as many chromosomes because they will fuse together during sexual reproduction. In humans, the total number of chromosomes in a cell is 46. If the gametes did not lose half of their chromosomes somewhere along the way, then the cell that resulted from sexual reproduction would have twice as many

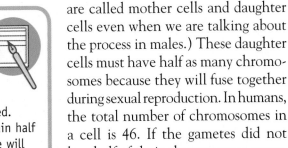

Learning hint

Mitosis vs. meiosis

These words sound and look alike and are therefore often confused. Remember, *meiosis* produces *gametes* or sexual cells, which contain half of the chromosomes because the sexual union of male and female will contribute the other half. *Mitosis* (I reproduce myself) is asexual and produces exact copies of the cell and the full complement of chromosomes because no union is needed.

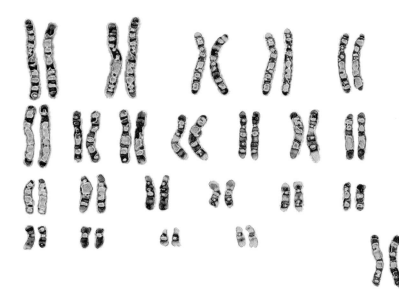

Figure 18.1 Photo of human chromosome profile: both an X and a Y chromosome appear in the image; therefore, this is a male sample. (*Source*: Phototake Inc./Alamy)

chromosomes (92 chromosomes) as necessary. It is absolutely necessary to control the number of chromosomes in a cell. Cells with too few or too many chromosomes often die.

The fact that you were produced by the fusion of an egg from your mother and a sperm from your father means that your 46 chromosomes can be thought of as being 23 *pairs* of chromosomes. Each pair of chromosomes consists of one from your father and one from your mother. They can be thought of as pairs because they can be matched based on size, shape, and which genes they carry. For example, you get a chromosome 1 from your father and a chromosome 1 from your mother, the same for chromosome 2, 3, 4 and up to 22.

The 23rd 'pair' of chromosomes is a set of sex chromosomes. These are called sex chromosomes because their identity determines the sex of a baby. XX is female, XY is male (see Figure 18.1). The female always contributes an X, but the male can contribute either an X or Y chromosome. The male actually determines the sex of the baby by either contributing an X chromosome for a girl or a Y chromosome for a boy.

Clinical application

Down's syndrome

Down's syndrome is a relatively common birth defect that causes short stature, heart defects, increased risk of leukaemia, Alzheimer's disease and mental retardation. Down's syndrome is caused by the presence of an extra chromosome 21 in a person's cells. Some time during meiosis, usually in the mother, chromosomes fail to separate, leaving some daughter cells without a chromosome 21 and others with two copies of chromosome 21. If the egg with the extra chromosome is fertilised, the resulting fetus will have three copies of chromosome 21 instead of two.

The possibility of having a baby with Down's syndrome increases in women over 35 years old. Down's syndrome is one of very few disorders resulting from abnormal chromosome numbers, probably because most fertilised eggs with abnormal numbers of chromosomes do not develop normally enough to survive.

The human life cycle

Both mitosis and meiosis are absolutely necessary parts of human life. Without them, cells could not be replaced, injuries could not be repaired, new humans could not be produced. The relationship between mitosis and meiosis and their importance can be easily explained by looking at human life as a cycle (see Figure 18.2). Eggs and sperm, with only half as many chromosomes as other cells, are produced by meiosis in specialised organs known as the gonads. The male gonads are called testes. The female gonads are the ovaries. During sexual reproduction, the gametes (egg and sperm) unite and combine their genetic material. This union and combination of genetic material is called **fertilisation**. The fertilised egg is known as a zygote. Unlike gametes, which have only 23 chromosomes, the zygote has the typical number of chromosomes for a human cell, 46. The zygote undergoes millions of rounds of mitosis and development within the female to change from an *embryo* to a *fetus* (the infant that is not born yet). The rest of this chapter is devoted to describing sexual reproduction in humans.

gonad (*GO nad*)

testis, testes (*TESS tiss, TESS teez*)

ovary, ovaries (*OH vah ree, OH vah reez*)

zygote (*ZIGH goat*)

Test your knowledge 18.1

Choose the best answer

1. Cells with half the number of chromosomes used for sexual reproduction are known as:
   a. zygotes
   b. daughter cells
   c. gametes
   d. half cells

2. Organs that produce sperm and egg are called:
   a. gametes
   b. gonads
   c. zygotes
   d. chromosomes

3. After _____, the egg is called a _____ and has 46 chromosomes.
   a. reproduction, gamete
   b. meiosis, daughter cell
   c. mitosis, zygote
   d. fertilisation, zygote

The human reproductive system

genitalia (*jen ih TAY LEE ah*)

Reproductive organs are called genitalia. The primary genitalia are the gonads, which produce the gametes. The secondary genitalia are all the other structures that aid in the reproductive process. In this section, we will first discuss female anatomy and physiology and then male anatomy and physiology.

Female anatomy

uterus (*YOU ter uss*)

vagina (*vah JYE nah*)

vulva (*VULL vah*)

In females, the primary genitalia are the ovaries. The secondary genitalia are the **uterine tubes** (also known as **oviducts** or **Fallopian tubes**), the uterus, the vagina, and the external genitalia, called the vulva.

The primary genitalia, or ovaries, are paired structures, about 3 centimetres long, in the peritoneal cavity. There is one ovary on each side of the uterus.

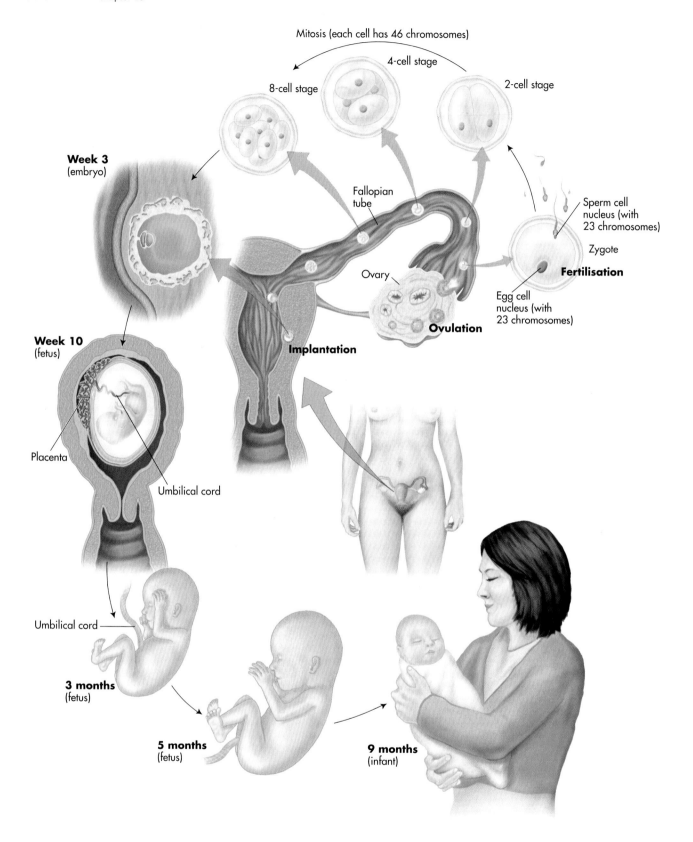

Figure 18.2 The early stages of the human life cycle.

Several ligaments suspend or anchor each ovary. The *mesovarium* suspends the ovary, the *suspensory ligament* attaches the ovary to the lateral pelvic wall, and the *ovarian ligament* anchors the ovary to the uterine wall. Blood vessels, the ovarian artery and the ovarian branch of the uterine artery, travel through the mesovarium and suspensory ligament, supplying the ovary with oxygenated blood. Please see Figure 18.3 for a diagram of internal reproductive anatomy.

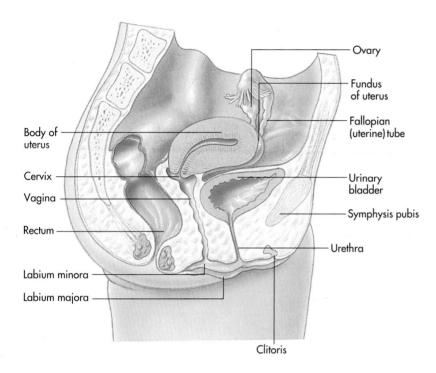

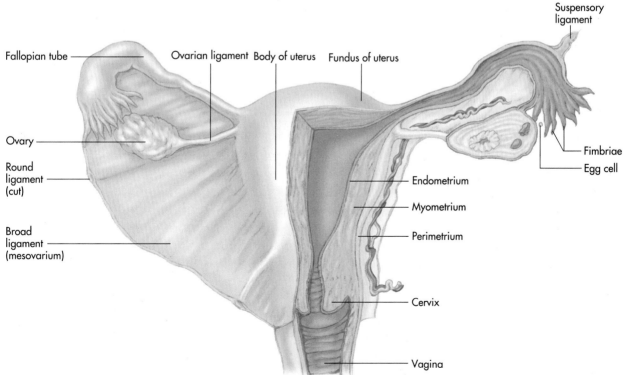

Figure 18.3 Internal female reproductive organs.

The ovary

The ovary is covered by a fibrous capsule called the *tunica albuginea* made of cuboidal epithelium. The interior of the ovary is divided into the cortex, which contains the eggs, and the medulla, which contains blood vessels, nerves and lymphatic tissue surrounded by loose connective tissue. The anatomy of the cortex is relatively complicated and will be described during our discussion of physiology.

The uterine tubes

fimbria (*FIM bree ah*)

The uterine tubes, also known as oviducts or Fallopian tubes, are the passageway for the egg to get to the uterus. The uterine tubes begin as a large funnel, the *infundibulum*, surrounded by ciliated projections, the **fimbria**. The infundibulum leads to a widened area, the *ampulla*, followed by a longer, narrower portion known as the *isthmus*. The uterine tubes are connected to the superior portion of the uterus. The tube is constructed of sheets of smooth muscle lined with highly folded, ciliated, simple columnar epithelium. The outside of the tube is covered by the visceral peritoneum and suspended by a mesentery known as the *mesosalpinx*.

The uterus

cervix (*SER viks*)

The uterus is in the pelvic cavity, posterior and superior to the urinary bladder and anterior to the rectum. The major portion of the uterus is called the body. The rounded superior portion between the uterine tubes is the fundus, and the narrow inferior part is the isthmus. The **cervix** is a valve-like portion of the uterus that protrudes into the vagina while the cervical canal connects with the vagina. Like the ovaries, the uterus is suspended and anchored by a series of ligaments. The *mesometrium* attaches the uterus to the lateral pelvic walls. (The combination of the mesometrium and mesovarium is called the *broad ligament*.) The lateral cervical ligaments attach the cervix and vagina to the lateral pelvic walls. The uterus is anchored to the anterior wall of the pelvic cavity by the round ligaments.

Clinical application

Endometriosis

Each month, women of child-bearing age shed and replace the endometrium, the lining of the uterus. In many women, this endometrial tissue escapes the uterus and implants in the abdominal and pelvic cavities. There, the tissue responds to the woman's hormonal cycles, continuing to build up and decay each month. Unfortunately, the continued proliferation and decay and bleeding of endometrial tissue in the abdominal and pelvic cavities can cause scarring and damage to the organs. Many women who have endometriosis have no noticeable symptoms, while other women experience severe abdominal and back pain around the time of their period. Untreated, endometriosis can cause adhesions on the intestines and urinary bladder, which can be extremely painful and must be removed surgically. Endometriosis is the most common cause of infertility, blocking the uterine tubes and therefore blocking the journey of the egg and scarring the ovaries and uterus.

Like most of the hollow organs or tubes we have visited, the walls of the uterus consist of three layers; the **perimetrium**, the outermost layer, is also the visceral peritoneum. (Just like the outermost layer of the heart is the visceral pericardium.) The **myometrium** consists of smooth muscle, and the **endometrium**, or inner lining, is a mucosa layer of columnar epithelium and secretory cells. The mucosa has two divisions. The **basal layer** is responsible for regenerating the uterine lining each month. The **functional layer** sheds about every 28 days when a woman has her period.

The endometrium is highly vascular. (No big surprise to any woman of child-bearing age.) Blood is supplied by the uterine arteries that branch from the internal iliac arteries on each side. The uterine arteries split into arcuate arteries, which supply the myometrium, and radial arteries, which supply blood to the endometrium. As you might expect, since the endometrium has two separate divisions, there are two different types of radial arteries. Straight radial arteries supply the basal layer. Spiral radial arteries supply the functional layer. The spiral arteries actually decay and regenerate every month, as part of the menstrual cycle, and undergo spasms that contribute to the shedding of the endometrium each month. (For more on control of the menstrual cycle see the section 'Hormonal control' later in this chapter.) Blood returns to circulation via a network of venous sinuses.

The vagina

The vagina is a tube, approximately 10 centimetres long, which runs from the uterus to the outside of the body. Its purpose is to receive the penis during intercourse and to allow for the passage of menstrual fluid out of the uterus. The vagina is also known as the birth canal, since its primary function is to allow the movement of a baby out of the uterus during childbirth. The external opening of the vagina may be covered by a perforated membrane called the hymen. A torn hymen was once thought to 'prove' that a woman had had intercourse. However, many hymens are highly perforated and easily ruptured by day-to-day activities such as riding a bicycle or jogging. An intact hymen is no longer considered a litmus test for virginity; however, some cultures still hold this erroneous belief to be true.

The external genitalia

The external genitalia (Figure 18.4), collectively known as the **vulva**, while not perhaps as obvious as the male external genitalia, are a complex and important part of reproduction. The vulva is surrounded on each side by two prominences called the **labia majora**. The labia majora are rounded fat deposits that meet and protect the rest of the external genitalia. The labia majora meet anteriorly to form the mons pubis. Both the mons pubis and the labia majora are covered by pubic hair.

Between the two halves of the labia majora is an opening known as the **pudendal cleft**. Within the pudendal cleft lies the **vestibule**, a space into which the urethra (anterior) and vagina (posterior) empty. (Unlike the male reproductive system, the female urinary system is completely separate from the reproductive system.) The lateral border of the vestibule is formed by the thin **labia minora**, which meet anteriorly to form the **prepuce**. Several glands surround the vestibule, helping to keep it moist.

---

perimetrium (*pair ee MEE tree yum*)

myometrium (*MY oh MEE tree yum*)
endometrium (*EN doh MEE tree yum*)

labia majora (*LAY bee ah maj OR ah*)

pudendal cleft (*PEW den dall cleft*)

labia minora (*LAY bee yah my NOR ah*)
prepuce (*PREE puce*)

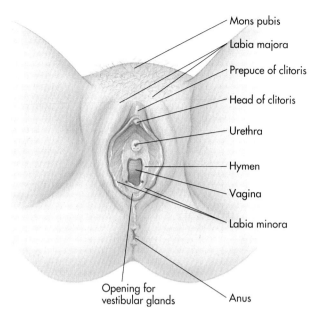

Figure 18.4 The external female genitalia (vulva).

clitoris (KLIT oh riss)

Just posterior to the prepuce, and anterior to the vestibule, is the **clitoris**. The clitoris is a small erectile structure, 2 centimetres in diameter. Like the penis (see 'Male anatomy' later in this chapter), the clitoris has a shaft, or body, and a glans (tip), and it becomes engorged with blood during sexual arousal. However, the clitoris increases in diameter, but not in length.

Mammary glands

mammary glands (MAM ah ree)

There is another set of external accessory sexual organs in the female, far removed from the pubic area, the **mammary glands**. The mammary glands are milk production glands housed in the breasts (see Figure 18.5). In young children, mammary tissue is virtually identical in boys and girls. At puberty, oestrogen and progesterone stimulate breast development in girls. In adult

Test your knowledge 18.2

Choose the best answer

1. The _____ is also known as the birth canal.
    a. ovary
    b. uterus
    c. uterine tubes
    d. vagina

2. The inner lining of the uterus that is shed each month is the:
    a. perimetrium
    b. endometrium

    c. myometrium
    d. all of the above

3. This part of the female reproductive system is erectile.
    a. ovary
    b. labia majora
    c. clitoris
    d. penis

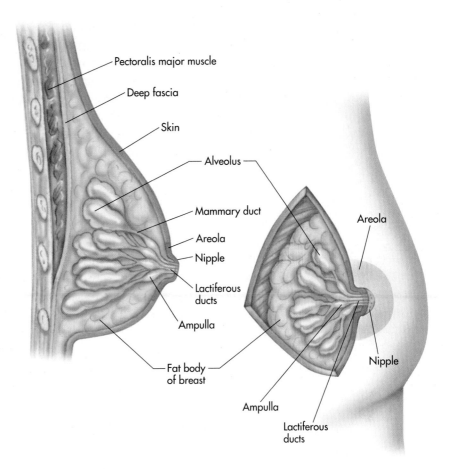

Figure 18.5 The mammary glands.

females, the breasts consist of 15 to 20 lobes, which are glandular, and lots of adipose tissue. Each lobe is divided into smaller lobules, which house milk-secreting sacs called alveoli, when a woman is lactating (producing milk). Milk, made in the alveoli, travels through a series of ducts and sinuses, eventually reaching the areola or nipple. Milk production is controlled by the hormone prolactin.

Reproductive physiology: female

The female reproductive physiology is closely tied to a regulated cycle. This cycle is normally regulated by hormonal control.

The menstrual cycle

menstrual cycle (*MEN stroo all*)
ovarian cycle (*oh VAIR ee an*)
uterine cycle (*YOU ter ihe*)

Female reproduction is organised around an approximately 28-day cycle involving both the ovaries and the uterus, collectively known as the **menstrual cycle**. The **ovarian cycle** involves the monthly maturation and release of eggs from the ovary. The **uterine cycle** consists of the monthly build-up, decay and shedding of the uterine lining. The cycles begin during a woman's teen years at the beginning of puberty and end during her 40s or 50s, with the end of menopause. The ultimate goal of the cycle is to release an egg that might be fertilised and to prepare the uterus to receive and nourish the fertilised egg should pregnancy result. If pregnancy does not occur, the uterine lining sheds and the cycle begins again.

The menstrual cycle, occurring approximately every 28 days in sexually mature women who are not pregnant, begins with the first day of menses. Menses is the time period when the uterine lining is shed. You are probably more familiar with the term menstruation. Menstruation is the actual shedding of the endometrium, while menses is the time during which a woman is menstruating. In more common terms, menses is the time during which a woman is having her 'period', and menstruation is the 'period' itself. Menses typically lasts four to five days but can be longer or shorter in different women and can even vary month to month in the same woman.

Once menses is over, the endometrium begins to proliferate, or build up, readying itself for the egg that is about to be released from the ovary, during ovulation. From day 1 to day 14, the ovary is also busy. In the ovary, an egg cell, or oocyte, is undergoing a number of developmental changes getting ready for ovulation on day 14. Ovulation is the release of a mature egg from the ovary. The egg travels from the ovary to the uterus, which has been getting ready to receive the egg. If the oocyte has been fertilised by a sperm, it will implant in the thickened endometrium. If the egg does not implant within a few days, the endometrium will begin to decay and menstruation will occur about two weeks after ovulation. The time between the end of menses and ovulation is known as the follicular or proliferative phase, because the endometrium is proliferating and the follicles are maturing in the ovary. Follicle is the term used to refer to an egg and associated helper cells. The time between ovulation and menses is known as the luteal phase or secretory phase, because of the development of a structure called the corpus luteum in the ovary and the beginning of secretion in the uterus.

Sounds simple enough, right? The cycle is deceptively simple when described without details or information about control of the cycle. Now that you understand the cycle itself, let's get down to the nitty gritty. Just how does an egg mature, and how is the cycle controlled?

Oogenesis, follicle development and ovulation

The process by which eggs are produced is called oogenesis. Oogenesis begins with the birth of **oogonia**, or egg stem cells, in the ovary. The oogonia undergo mitosis, producing millions of **primary oocytes**. This happens very early in a woman's life. There are millions of primary oocytes produced in a fetus. That's right: women have all the eggs they will ever have five months before they are born!

Primary oocytes, since they are born via mitosis, still have all 46 chromosomes. In order to become gametes, they must undergo meiosis to cut their chromosome number to 23. Remember, gametes must have only 23 chromosomes in order for fertilisation (successful combining of sperm and egg) to produce a cell with the normal number of 46 chromosomes. So, primary oocytes begin meiosis. However, they do not fully complete their development. The primary oocytes stay in a kind of suspended animation until puberty when they finish developing. (That's at least ten years!)

These primary oocytes eventually are surrounded by helper cells, called *granulosa cells*. Once surrounded by granulosa cells, the primary oocyte and

menses (*MEN seez*)

menstruation (*MEN stroo AY shun*)

ovulation (*OV yoo LAY shun*)
oocyte (*OO oh site*)
follicular (*foll lik you lar*)
proliferative (*pro LIFF er ah TIV*)
luteal (*LOO teal*)

oogenesis (*OO oh JENN eh siss*)

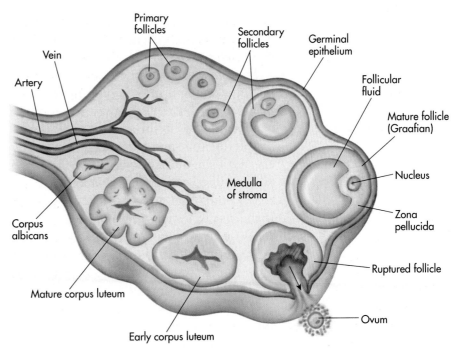

Figure 18.6 Maturation of follicle.

primordial (*pry MORE dee all*)

surrounding cells are known as **primordial follicles**. These primordial follicles stay dormant until puberty. Hormonal signals during puberty cause some primordial follicles to enlarge and increase the number of granulosa cells. These enlarged cells are then called **primary follicles** (see Figure 18.6).

Once a girl reaches puberty, one primary follicle will become a **secondary follicle**. The secondary follicle will not complete its development unless it is ovulated and fertilised. Just before ovulation, the secondary follicle fills with

Test your knowledge 18.3

Choose the best answer:

1. The time from the end of menses to ovulation is what phase?

   a. luteal phase

   b. follicular phase

   c. mitotic phase

   d. ovarian phase

2. The union of an egg and a sperm is called:

   a. sex

   b. fertilisation

   c. oogenesis

   d. ovulation

3. The act of the egg being expelled from the ovary is called:

   a. oogenesis

   b. ovulation

   c. puberty

   d. pregnancy

4. If _____ does not occur, the endometrium will decay and a woman will menstruate.

   a. sex

   b. ovulation

   c. fertilisation

   d. oogenesis

fluid and moves toward the surface of the ovary, where it becomes a visible lump. The fimbria of the uterine tubes brush the surface of the ovary, causing the follicle to rupture. When the follicle ruptures, the egg (now an oocyte) is released into the peritoneal cavity. The fimbria then pulls the egg toward the funnel, drawing it into the uterine tube.

As the egg travels down the uterine tube (also called the oviduct or Fallopian tube), it either will or will not be fertilised. If sperm are present in the uterine tube and all the conditions are right, the sperm will penetrate the egg, fertilising it and triggering the rest of the egg development. The successfully fertilised egg has 46 chromosomes and is now called a **zygote**. When the zygote enters the uterus, the newly proliferated endometrium is ready for it. If the zygote successfully implants in the uterus, pregnancy will result and the woman will not menstruate. The ruptured follicle left behind in the ovary during ovulation will become the corpus luteum (remember the luteal phase?) and secrete hormones to help maintain the thickened endometrium, which will serve to nourish the growing fetus.

If there is no sperm in the uterine tube, conditions are not right, or something goes wrong after fertilisation, the zygote will not implant in the uterus. If there is no implantation within a few days, the uterine lining will begin to degenerate and the woman will have her period. The corpus luteum will become a corpus albicans and eventually disappear (see Figure 18.7).

Hormonal control

It seems obvious that something must control the complex cyclic changes in the female reproductive system. The control system that keeps the uterus and ovaries in synch is the endocrine system, specifically hypothalamic, pituitary and ovarian hormones.

Remember that hormones are chemical signals released by one organ into the bloodstream to control another organ or tissue some distance away. Hormone levels are generally controlled by negative feedback. As hormone levels rise, the organ releasing the hormone decreases the amount of hormone released. Hormones are often released as part of a hierarchy, with the hypothalamus releasing a hormone that controls the pituitary, which then releases a hormone that controls another organ.

The hormones controlling female reproduction are no exception. The menstrual cycle is controlled by a combination of four hormones: **oestrogen** and **progesterone** from the ovary, and **luteinising hormone** (LH) and **follicle-stimulating hormone** (FSH) from the pituitary gland. At the beginning of puberty, oestrogen and progesterone secretion from the ovaries increases greatly, increasing the secretion of LH and FSH.

The release of gonadotrophin-releasing hormone (GnRH) from the hypothalamus causes an increase in the secretion of LH and FSH from the pituitary. FSH initiates the development of primary follicles each month, and LH triggers ovulation. During this part of the cycle (the follicular or proliferative phase), oestrogen levels continue to rise as more and more is secreted by the developing follicle. Oestrogen exerts a positive influence on the hypothalamus, increasing secretion of GnRH and thus increasing LH and FSH secretion. This positive feedback loop increases the levels of LH and FSH, stimulating follicle development and triggering ovulation. Rising oestrogen levels also stimulate proliferation of the uterine lining (endometrium).

---

corpus luteum (*KOR puss LOO tee um*)

progesterone (*proh JESS ter ohn*)
luteinising hormone (*LOO ten IZE ing*)
follicle-stimulating hormone (*FOLL I cul STIM yoo LAY ting HOR mohn*)
gonadotrophin-releasing hormone (*GO nad oh TROW pin*)

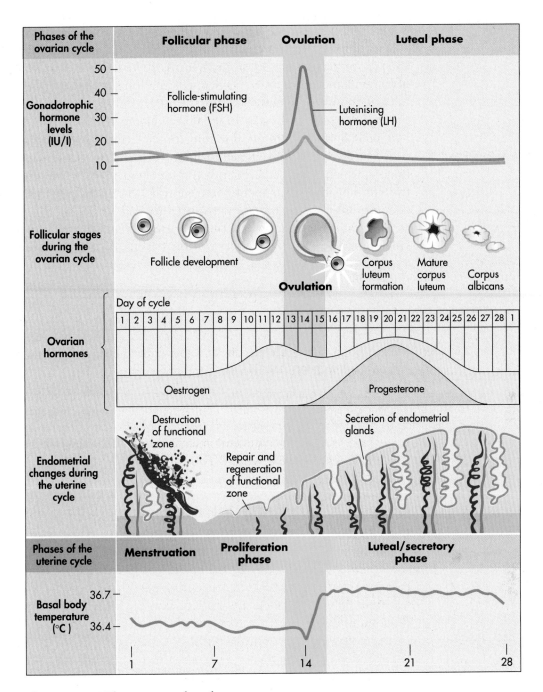

Figure 18.7 The menstrual cycle.

Once ovulation occurs, the feedback loop reverses itself. The leftover ruptured follicle, now the corpus luteum, begins to secrete progesterone and secretes a little oestrogen. Under the influence of progesterone, oestrogen exerts negative feedback on the hypothalamus and pituitary, decreasing GnRH, LH and FSH secretion. Progesterone also exerts negative feedback on the hypothalamus and pituitary. Thus, during the luteal or secretory phase, LH, FSH and oestrogen levels drop while progesterone levels rise. These hormonal changes prevent another egg from maturing.

For about ten days after ovulation, progesterone levels remain high as the corpus luteum continues to secrete the hormone. Progesterone's effect on the uterus is to *maintain* the build-up of the endometrium and to decrease uterine contractions. If no pregnancy results, then the corpus luteum will degenerate and stop producing progesterone. Decreased progesterone causes degeneration of the endometrium, followed by menstruation. Decreased progesterone also releases the hypothalamus and pituitary from inhibition. FSH and LH levels rise and the cycle begins again (again see Figure 18.7).

If pregnancy does result, the implanted fertilised egg secretes a hormone called **human chorionic gonadotrophin (HCG)**. HCG stimulates the corpus luteum, which keeps secreting progesterone and a little oestrogen to maintain the uterine lining. At about three months' gestation (pregnancy), the placenta begins to secrete its own progesterone and oestrogen, thereby becoming an endocrine organ. For a list of hormones and functions related to pregnancy, see Table 18.1.

human chorionic gonadotrophin (HUE man core ee YOH nik GO nad oh TROW pin)

Table 18.1 Hormones controlling pregnancy

Hormone	Where secreted	Causes
human chorionic gonadotrophin	implanted fertilised egg	maintains the function of corpus luteum; this is what gives a positive pregnancy test
oestrogen and progesterone	corpus luteum for first two months of pregnancy; then by the placenta	both stimulate development of uterine lining and mammary glands; progesterone prohibits uterine contractions during pregnancy; oestrogen relaxes the pelvic joints; once labour begins, oestrogen negates the effects of progesterone on uterine contractions and makes the myometrium sensitive to oxytocin
prolactin	anterior pituitary gland	stimulates milk production by breasts
oxytocin	posterior pituitary gland	uterine contractions to begin labour; stimulates the release of milk from the breasts (see Figure 18.8)

Test your knowledge 18.4

Choose the best answer:

1. At puberty, the ovaries begin to secrete:
   a. FSH
   b. LH
   c. oestrogen
   d. all of the above

2. These hormones stimulate follicle maturation and ovulation:
   a. oestrogen and progesterone
   b. oestrogen and GnRH
   c. LH and FSH
   d. FSH and progesterone

3. Before ovulation, oestrogen _____ the secretion of FSH and LH.
   a. decreases
   b. increases
   c. not enough information
   d. there is no relationship between oestrogen, FSH and LH

4. The role of progesterone is
   a. to promote ovulation
   b. to keep the endometrium from shedding
   c. to enhance fertilisation
   d. to increase FSH and LH levels

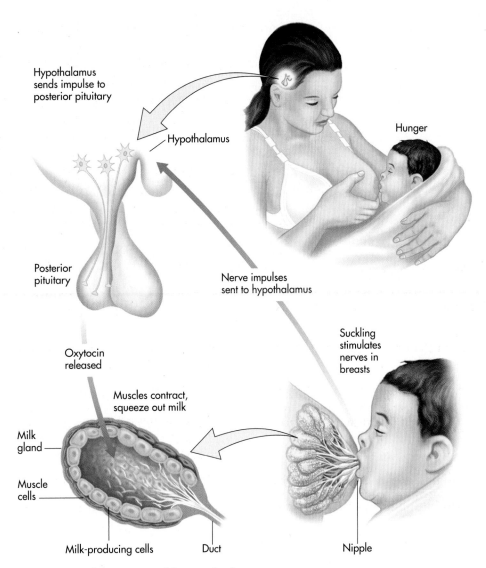

Figure 18.8 Oxytocin and breast feeding.

Male anatomy

Like the female reproductive system, the male reproductive system has a pair of primary genitalia, or gonads, called the *testes*. The testes produce the male gamete or sperm that must travel along its journey to find the egg. Unlike the female, the male primary genitalia are external. In addition, the male has many secondary genitalia – the **penis**, an external sperm-delivery organ; several sperm ducts; the **epididymis**; the **vas deferens**; and the **urethra** – and several accessory glands, the **prostate gland**, the **seminal vesicles** and the **bulbourethral glands**. See Figure 18.9 for the anatomy of the male reproductive system.

The testes (testicles)

The primary genitalia, the testes (or testicles), are paired glands suspended in a sac called the **scrotum**. The testicles are external genitalia, one hanging on

epididymis (*ep ih DID ih miss*)

vas deferens (*VAS DEFF er enz*)

urethra (*yoo REE thrah*)

seminal vesicles (*SEM ih nall VESS ik alls*)

bulbourethral gland (*BUHL boh you REE thral*)

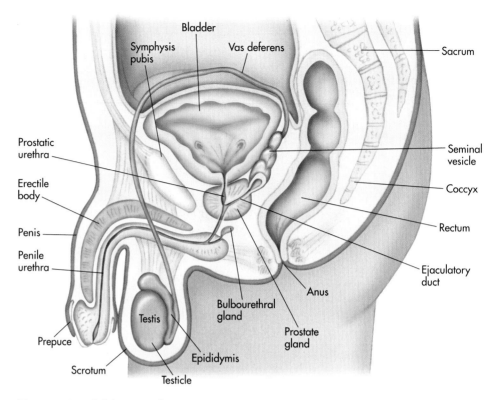

Figure 18.9 Male reproductive anatomy.

either side of the penis. (Viable sperm cannot be made at normal human body temperature.) Each testis is surrounded by a serous membrane, called the tunica vaginalis, which originates from the peritoneum. The inside of the testes are divided into 250 to 300 wedges called lobules, each of which contain one to four seminiferous tubules. The seminiferous tubules, which are made of epithelium and areolar tissue, contain sperm stem cells and sperm helper cells (Sertoli cells or nurse cells).

The epididymis

There are several ducts in the male reproductive system. The epididymis is a comma-shaped duct on the posterior and lateral part of the testes. (It looks something like a stocking cap for each gonad.) The tube is highly coiled. If unravelled, it would be 6 metres long. It is made of pseudostratified ciliated epithelium and smooth muscle. Sperm mature here.

The vas deferens

The vas deferens is a short tube, only 45 centimetres long. It is lined with ciliated pseudostratified epithelium, like the epididymis, but has a thick smooth muscle layer and is surrounded by a connective tissue layer called the *adventitia*. The vas deferens runs from the scrotum to the penis via a relatively complicated pathway. The vas deferens runs from the anterior part of the scrotum as a pair of tubes, one on each side, into the abdominal wall (through the inguinal canal) and pelvic cavity, medially over the urethra and along the

seminiferous tubules (*SEM ih NIF er uss TUBE yules*)

Sertoli cells (*sir TOW lee sells*)

posterior bladder wall. Posterior to the bladder, the vas deferens joins the seminal vesicle to form the **ejaculatory duct**. The ejaculatory duct then passes through the prostate gland and empties into the urethra. (Remember, the urethra carries both sperm and urine in males.) Between the scrotum and the inguinal canal, the vas deferens runs through a tube, with blood vessels and nerves, collectively called the **spermatic cord**.

The urethra

The male urethra differs from the female urethra in several important respects. The female urethra is short (approximately 4 centimetres), straight, and has no role in reproduction. By comparison, the male urethra is long, curved, has three distinct sections (the prostatic urethra, membranous urethra and caveronosal urethra). In addition to urine, the urethra carries semen from the prostate to the outside world and therefore has a role in reproduction.

Amazing body facts

Boxers or briefs?

Normal human body temperature is too warm for healthy sperm development. The testes are on the outside of the body because sperm development is extremely temperature-sensitive and sperm could not develop normally inside the pelvic cavity. Indeed, there is some speculation that very tight clothing might contribute to male infertility by holding the gonads too close to the body and raising their temperature. In addition, there is a testes–blood barrier to prevent the immune system from destroying sperm, which the white blood cells would recognise as invaders.

The penis

The other part of the external genitalia in males is the penis. The penis, from the Latin word for 'tail', is a sperm-delivery organ that transfers sperm from male to female. The attached portion of the penis is called the root, while the freely moving portion is the shaft or body. The tip of the penis, the glans penis, is covered by a loose section of skin called the foreskin, unless a man has been circumcised. The penis has three erectile bodies, tubes with a sponge-like network of blood sinuses.

Accessory glands

There are three accessory glands in the male reproductive system. The seminal vesicles are highly coiled glands posterior to the bladder, made of pseudostratified epithelium, smooth muscle and connective tissue. The prostate gland is a walnut-sized gland surrounding the urethra just inferior to the bladder. There are several parts or zones to the prostate: the transitional zone, which is wrapped around the urethra; the central zone that sits underneath the transitional zone; the peripheral zone that covers both the transitional and central zones; and the anterior fibromuscular stroma, that sits on top of the prostate. It is a dense mass of connective tissue and smooth muscle with embedded glands that has an extensive blood supply. Clinically the prostate can be examined via the rectum (this is called digital rectal examination) to see if the prostate is getting bigger (enlargement), is infected (prostatitis) or has a tumour present (prostate cancer). The bulbourethral glands are pea-sized glands inferior to the prostate that secrete a clear mucus into the urethra during sexual excitement.

Test your knowledge 18.5

Choose the best answer:

1. The _____ are the gonads, where sperm are made.
   a. vas deferens
   b. testes
   c. scrotum
   d. penis

2. After passing the seminal vesicles, the vas deferens becomes the:
   a. prostate gland
   b. urethra
   c. penis
   d. none of the above

3. Sperm mature in the
   a. epididymis
   b. testes
   c. scrotum
   d. penis

Reproductive physiology: male

Unlike female reproduction, male reproductive physiology is not organised around a tightly controlled monthly cycle. Let's explore the physiologic processes of the male reproductive system.

Spermatogenesis

Sperm production, in the testes, is a continuous process beginning when a boy reaches puberty and usually continuing until death. As such, the control of spermatogenesis, sperm production, is much less complicated than control of oogenesis.

spermatogonia (*sper MAT oh GO nee ah*)

spermatocytes (*sper MAT oh sites*)

spermatids (*sper MAT ids*)

spermatozoa (*sper MAT oh ZOE ah*)

The **spermatogonia**, sperm stem cells, undergo mitosis to form **primary spermatocytes**. Unlike primary oocytes, the primary spermatocytes do not wait to go through meiosis. Primary spermatocytes form two **secondary spermatocytes**. Secondary spermatocytes complete meiosis to form **spermatids**, and spermatids go through a period of development to form immature **spermatozoa** (**sperm**). All of this takes place in the testes inside the seminiferous tubules. The distribution of the different stages of sperm development is predictable. Spermatogonia line up against the walls of the tubules, and mature sperm cluster near the lumen of the tubules (see Figure 18.10). Sperm then travel from the seminiferous tubules to the epididymis, where the sperm spend about two weeks maturing and gaining the ability to swim.

Hormonal control of male reproduction

testosterone (*tess TOSS ter own*)

Testosterone is arguably the most important male sex hormone. Before birth, HCG secreted by the placenta causes the embryonic (still developing) testes to secrete testosterone, masculinising the fetus. (Fetuses not exposed to testosterone or insensitive to testosterone look female.) After birth, there is little testosterone secreted until puberty. At puberty, two hormonal changes occur that signal the beginning of sexual maturity. First, testosterone secretion by the testes increases dramatically. Second, there is a change in the relationship

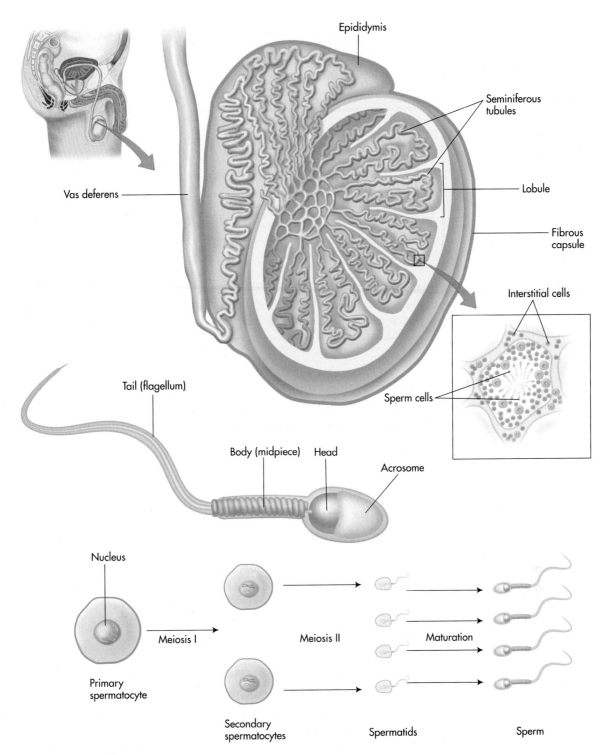

Figure 18.10 Spermatogenesis.

between testosterone and GnRH, LH and FSH. (Yes, the same hormones that control female reproduction control male reproduction!) Before puberty, the small amount of testosterone that is secreted by the testes inhibits GnRH secretion. After puberty, testosterone does not inhibit GnRH secretion. So after puberty, GnRH and therefore FSH and LH secretion increase. In addition, FSH and LH also enhance testosterone secretion in a major positive feedback loop, but only after the onset of puberty. Testosterone secretion at puberty is also responsible for the obvious physical changes of the male secondary sexual characteristics, including body, facial and pubic hair growth, deepening of the voice, and increased muscle and bone mass.

Clinical application

Hypogonadism

This is reduced or absent production of testosterone in men. The normal range of testosterone is 9 to 27 nmol/L (with low-normal levels between 9 and 12 nmol/L) depending on how the laboratory calculates the results. The testosterone level peaks in the morning, so tests for this hormone must be taken between 09.00 and 11.00 am for an accurate result. Testosterone levels can fall by as much as 13 per cent throughout the day. Low levels of testosterone can result in sexual problems, infertility and tiredness, but some researchers believe that low or low-normal testosterone can contribute to developing some cancers and metabolic disorders.

LH and FSH affect males exactly as they do females. They stimulate gamete development and are controlled in the same way in males as in females. GnRH is released from the hypothalamus, which stimulates LH and FSH secretion from the pituitary.

Erection and ejaculation

When a man is sexually aroused, the erectile bodies in the penis (remember the sponge-like tissue with blood sinuses?) become engorged with blood, stiffening and expanding the penis. The change in shape of the penis is called **erection**.

In order for sperm to leave the male reproductive system, ejaculation must occur. **Ejaculation** is the expulsion of **semen** (sperm and assorted chemicals) from the urethra. Smooth muscle contracts throughout the ducts and glands of the male reproductive system and propels the sperm from the epididymis into the vas deferens, which then carries the sperm into the pelvic cavity. As the sperm passes the seminal vesicles, sugar and chemicals are added to the sperm. The sperm and chemicals then enter the ejaculatory duct. As the ejaculatory duct passes through the prostate gland, prostatic fluid is added, liquefying the semen and protecting the sperm from the acid environment of the vagina by the secretion of an alkaline substance. The semen then passes by the bulbourethral glands, which add mucus to the semen. The semen then enters the urethra and is carried outside the body. The time between penetration and ejaculation is called the Ejaculatory Latency Time or ELT and the average time is between three and seven minutes, although various factors influence ELT.

If a man is engaged in sexual intercourse with a woman and ejaculates, the sperm enter the vagina and make their way to the uterus and into the uterine tubes. The female reproductive system is not particularly hospitable to sperm, and many sperm do not survive the journey to the uterine tubes. If there is an egg waiting to be fertilised in the uterine tubes, sperm will find the egg and

semen (*SEE men*)

attempt to penetrate it and fertilise it. New research suggests that the egg is not a passive participant in fertilisation, but may actually engulf the sperm and even choose which sperm to allow inside. One sperm and only one sperm will fertilise the egg. If the egg is fertilised, and all other conditions are met, a pregnancy will result.

Amazing body facts

Hormones

Did you know that females have some natural testosterone and males have natural oestrogen? It's true. The balance between the hormones – not the presence or absence of the hormones – is what is important.

Test your knowledge 18.6

Choose the best answer:

1. The stiffening and expanding of the penis is known as:
   a. ejaculation
   b. erection
   c. emulsification
   d. erectile dysfunction

2. The combination of sperm, sugars, chemicals and mucus is:
   a. semen
   b. urine
   c. spermatozoa
   d. all of the above

3. Male secondary sexual characteristics, masculinisation of the body, and enhancement of FSH and LH levels are caused by:
   a. LH
   b. FSH
   c. oestrogen
   d. testosterone

Pregnancy

Pregnancy occurs when an egg is fertilised by the sperm and implants in the female reproductive system. The period of time during which the developing baby grows within the uterus is called the gestational period, and lasts approximately 40 weeks. A baby born up to the 37th week of gestation is called a premature infant.

From the time the egg is fertilised by a sperm and implants in the uterine wall up until the eighth week, the developing infant is referred to as an embryo. During the embryonic period, the organs and systems are fundamentally formed. Beyond the eight-week period until the birth, the developing infant is called a fetus.

The growing fetus is nourished by a spongy structure called the placenta, which is attached to the fetus and the mother via the umbilical cord. The fetus is encased in a membranous sac called the *amnion*, which contains the amniotic fluid in which the fetus floats. Labour is the actual process whereby the fetus is delivered from the uterus through the vagina and into the outside world.

Labour consists of three stages. In the dilation stage, the uterine smooth muscle begins to contract, thereby moving the fetus down the uterus and causing the cervix to begin to dilate. When the cervix is completely dilated (10 centimetres), the second stage (expulsion) begins during which the baby is actually delivered. Generally, the head presents first, which is called crowning, and this is when the baby's mouth should be suctioned before the baby takes its first breath to prevent aspiration of fluid into the baby's lung. Sometimes, the baby is turned around and the buttocks appear first in a breech presentation, which makes the delivery difficult. The last stage of labour is the placental stage in which the placenta or afterbirth is delivered due to the final uterine contractions (see Figure 18.11).

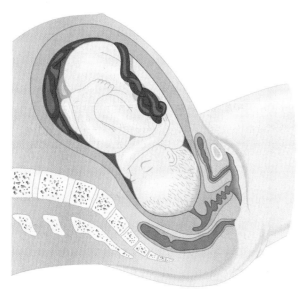

DILATION STAGE:
First uterine contraction to dilation of cervix

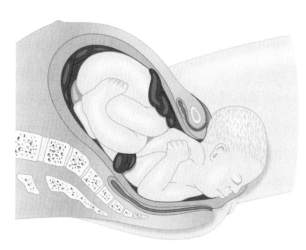

EXPULSION STAGE:
Birth of baby or expulsion

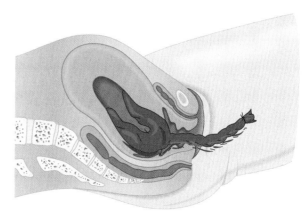

PLACENTAL STAGE:
Delivery of placenta

Figure 18.11 Stages of labour.

Clinical application

Contraception

The prevention of pregnancy is termed contraception ('against conception') and can be accomplished by a number of means, such as intrauterine devices (IUDs), spermicidal agents, birth control pills, or shields such as a condom.

Sterilisation of the male can be accomplished through a vasectomy. The vas deferens are the tubes that carry the sperm into the urethra, and tying off these tubes prevents the sperm from travelling out of the penis during sexual intercourse. Females may be sterilised via tubal ligation. The Fallopian tubes are cut or tied shut, preventing the egg from travelling from the ovary to the uterus.

Common disorders of the reproductive system

Both male and female reproductive systems are susceptible to many disorders. Common to both systems are sexually transmitted diseases (STDs). However, each system has its own specific disorders tied to anatomical and physiologic differences of the sexes.

Diseases of the female reproductive system

The first menstrual period is referred to as *menarche*, and *menopause* represents the ending of menstrual activity, both of which are normal cycles in the female reproductive system. However, this normal cycle can be disrupted. **Amenorrhoea** is the absence of menstruation and can be a result of pregnancy, menopause, or other factors such as emotional distress, extreme dieting or poor health. **Dysmenorrhoea** is difficult menstruation usually resulting in painful cramping.

Pre-menstrual syndrome (PMS) occurs several days prior to the onset of menstruation. This is characterised by effects on many systems of the body, including the emotional realm. These symptoms vary among individuals and usually subside near the onset of menstruation. For a list of the multi-system effects, please see Figure 18.12.

Like any system, the reproductive system can become infected. **Vaginitis** is the inflammation of the vagina that is usually caused by a micro-organism such as bacteria or yeast. Vaginitis is often not a result of sexually transmitted diseases. Sexually transmitted diseases represent a large number of infections found in both male and female reproductive systems and can be bacterial, viral or fungal.

The cervix is the area where a Pap smear (named after George Papanicolaou) is taken that examines scrapings from the cervical cells to detect the presence of cancer. Regular Pap smears allow for early detection, which increases the likelihood of successful treatment.

amenorrhoea (*ay MEN oh REE ah*)
   a = *absence*
dysmenorrhoea (*DISS men oh REE ah*)
   dys = *difficult*

vaginitis (*vaj in NYE tiss*)

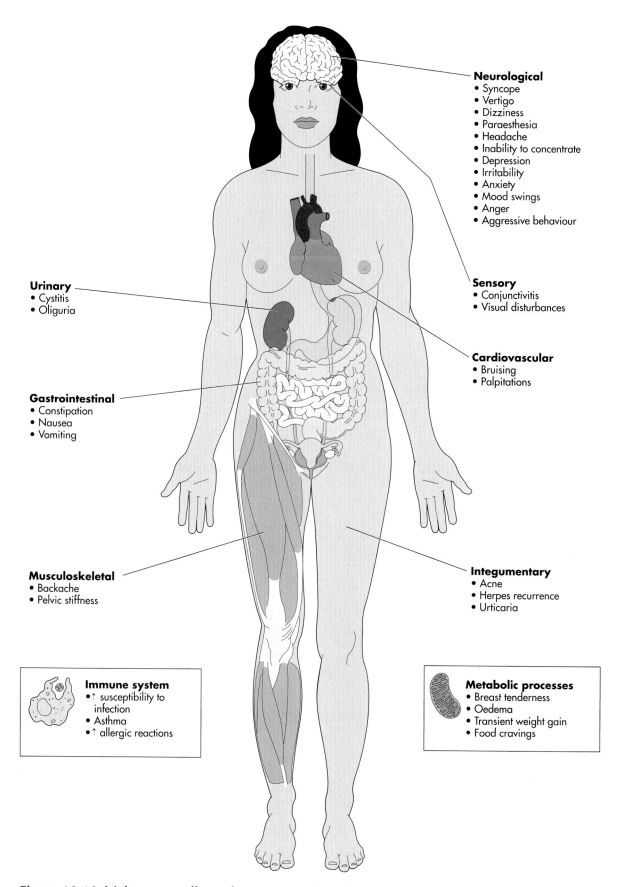

Neurological
• Syncope
• Vertigo
• Dizziness
• Paraesthesia
• Headache
• Inability to concentrate
• Depression
• Irritability
• Anxiety
• Mood swings
• Anger
• Aggressive behaviour

Sensory
• Conjunctivitis
• Visual disturbances

Cardiovascular
• Bruising
• Palpitations

Urinary
• Cystitis
• Oliguria

Gastrointestinal
• Constipation
• Nausea
• Vomiting

Musculoskeletal
• Backache
• Pelvic stiffness

Integumentary
• Acne
• Herpes recurrence
• Urticaria

Immune system
•↑ susceptibility to infection
• Asthma
•↑ allergic reactions

Metabolic processes
• Breast tenderness
• Oedema
• Transient weight gain
• Food cravings

Figure 18.12 Multi-system effects of pre-menstrual syndrome.

ectopic pregnancy (*ek TOP ik PREG nan see*)

abruptio placentia (*ab RUP tee oh plah SEN shah*)

In some cases of pregnancy, the fertilised egg implants in the Fallopian tubes and does not make the full trip down to the uterus. This is an **ectopic** or **tubal** pregnancy, which requires surgical intervention. **Abruptio placentia** is an emergency situation in which the placenta tears away from the uterine wall before the 20th week of pregnancy.

Postpartum depression is a serious psychological state that can occur after childbirth, and mothers should be monitored for signs and symptoms because it can lead to harm to the mother or baby or both.

mastitis (*mass TYE tiss*)

mast = *breast*

ectomy = *removal of*

Mastitis is inflammation of the breast, which can occur at any age and in both males and females, but is usually associated with lactating females. **Breast cancer** is one of the leading causes of death in woman between the ages of 32 and 52 and kills about 46,000 women a year. Men can also get breast cancer but at a lower rate, and it was estimated that there were about 1000 cases of male breast cancer in 2003 compared to 182,000 female cases. Cancer of the breast may require a full or partial *mastectomy*, but the earlier it is detected, the better chance for a positive outcome. Figure 18.13 shows the steps for a breast self-examination.

Diseases of the male reproductive system

In **erectile dysfunction**, or **ED**, the penis cannot attain full erection. In 80 per cent of cases, the causes of ED are related to atheroma or the deposition of fat in the arteries. The pudendal artery (the artery to the penis) is only 1 mm in diameter, so can easily become blocked. Recently ED has been identified as one of the first clinical signs of vascular disease and the average time between onset of ED and a heart attack is 3 ½ years. Various medicines can help to increase blood flow to the penis and thus treat some forms of erectile dysfunction.

cryptorchidism (*kript OR kid izm*)

The failure of the testes to descend into the scrotal sac is termed **cryptorchidism** and may require surgical intervention. If the testes remain undescended, the male may become sterile because of the body heat destroying the sperm. A **hydrocele** is an abnormal collection of fluid within the testes. To determine male fertility, a semen analysis must be performed. Semen is collected three to five days after abstaining from sexual activity. The semen is analysed for the shape, number and swimming strength of the sperm. This is also used to determine the effectiveness of a vasectomy where the blocking of the vas deferens should result in no sperm in the semen sample six weeks post-operatively.

hydrocele (*HIGH droh seel*)

Benign prostatic enlargement, or **BPE**, is the enlargement of the prostate gland and is commonly seen in males over 40 years old (see Figure 18.14). **Prostate cancer** is a slow-growing cancer (doubling times of the tumour can be as slow as one to three years) that also affects males and can be treated effectively if detected early. A PSA blood test (prostate-specific antigen) is used in combination with physical examination and later biopsy, to diagnose this cancer although a PSA test alone is not accurate enough to diagnose prostate cancer; indeed some tumours are independent of changes in PSA. Furthermore, some drugs, for example Finasteride (used to manage prostate enlargement) can suppress PSA levels. Changes to urinary function, for example getting up at night or passing urine frequently or with difficulty, can all point to prostate trouble, although without accurate investigations it is impossible to determine whether the growth is benign or a cancer.

WHY DO THE BREAST SELF-EXAM?	There are many good reasons for doing a breast self-exam each month. One reason is that it is easy to do and the more you do it, the better you will get at it. When you get to know how your breasts normally feel, you will quickly be able to feel any change, and early detection is the key to successful treatment and cure.

REMEMBER: A breast self-exam could save your breast – and save your life. Most breast lumps are found by women themselves, but in fact, most lumps in the breast are not cancer. Be safe, be sure. |
| **WHEN TO DO BREAST SELF-EXAM** | The best time to do breast self-exam is right after your period, when breasts are not tender or swollen. If you do not have regular periods or sometimes skip a month, do it on the same day every month. |
| **NOW, HOW TO DO BREAST SELF-EXAM** | 1. Lie down and put a pillow under your right shoulder. Place your right arm behind your head.

2. Use the finger pads of your three middle fingers on your left hand to feel for lumps or thickening. Your finger pads are the top third of each finger.

3. Press firmly enough to know how your breast feels. If you're not sure how hard to press, ask your health-care provider. Or try to copy the way your healthcare provider uses the finger pads during a breast exam. Learn what your breast feels like most of the time. A firm ridge in the lower curve of each breast is normal.

4. Move around the breast in a set way. You can choose either the circle (A), the up and down line (B), or the wedge (C). Do it the same way every time. It will help you to make sure that you've gone over the entire breast area, and to remember how your breast feels.

5. Now examine your left breast using your right hand finger pads.

6. If you find any changes, see your doctor immediately. |
| **FOR ADDED SAFETY:** | You should also check your breasts while standing in front of a mirror right after you do your breast self-exam each month. See if there are any changes in the way your breasts look: dimpling of the skin, changes in the nipple, or redness or swelling.

You might also want to do a breast self-exam while you're in the shower. Your soapy hands will glide over the wet skin, making it easy to check how your breasts feel. |

Figure 18.13 Breast self-examination.

Amazing body facts

A million to one shot

The adult male produces 200 million sperm daily but it takes only one sperm to fertilise the egg. The usual volume of ejaculate is around 2 ml or more and semen has a pH of 7.2 or more.

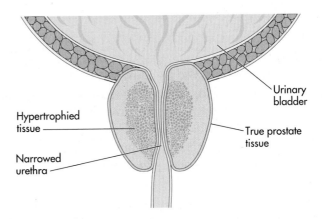

Figure 18.14 Benign prostatic hyperplasia.

Summary

- Tissues grow, are replaced and are repaired by asexual reproduction. Cells make identical copies of themselves. Asexual reproduction takes place all over the body as tissues grow or are repaired. Some tissues, like epidermis, blood and bone, replace themselves continually, always by asexual reproduction.

- Asexual reproduction in eukaryotic cells is accomplished by a relatively complex process called mitosis.

- If an organism is going to reproduce sexually, it must use specialised cells called gametes with only half the typical number of chromosomes for that organism. (It must have half as many chromosomes because, during fertilisation, the gametes will fuse and combine their genetic material.)

- In humans, the gametes are eggs and sperm. Eggs and sperm are produced by a special type of cell division called meiosis, or reduction division. Reduction division produces four daughter cells, each with only 23 chromosomes. These cells are not identical to each other or to the mother cell.

- Cells undergoing meiosis go through two divisions. The chief difference between mitosis and meiosis is the pairing of homologous chromosomes (alike in size, shape and genetic content) during meiosis.

- Human reproduction can be described as a cycle, known as the human life cycle. Adult humans have specialised organs called gonads (ovaries and testes), which produce gametes via meiosis. Gametes get together during sexual reproduction in a process known as fertilisation. The fertilised egg is called a zygote. The zygote undergoes many rounds of mitosis, eventually becoming an embryo, a fetus, a baby, a child and, finally, an adult.

- The female reproductive system consists of several internal genitalia. The gonads (ovaries) produce eggs (gametes). The uterine tubes provide a passageway for the egg to get to the uterus. The uterus is an incubator for the fertilised eggs. The vagina is the birth canal, connecting the uterus with the outside world. The external genitalia include the external opening of the vagina and several protective structures.

→

- Female reproductive physiology is relatively complicated, organised around a monthly cycle of changes in both the ovaries and the uterus, collectively known as the menstrual cycle. The cycle begins with menstruation, the shedding of the uterine lining. After the lining, the endometrium, has finished shedding, it begins to build up again in a process called proliferation. As the endometrium is proliferating, an egg (follicle) is maturing in the ovary. Eventually, the follicle will be released from the ovary (ovulation) and travel to the uterus. If fertilisation occurs, then the fertilised egg will implant in the thickened endometrium and pregnancy will result. If the egg is not fertilised, the endometrium will degenerate and menstruation will occur.

- Control of the menstrual cycle is accomplished by four hormones: oestrogen, progesterone, follicle-stimulating hormone (FSH) and luteinising hormone (LH). During puberty, the ovaries begin to secrete larger amounts of oestrogen and progesterone, which cause the development of female secondary sexual characteristics and set the cycle in motion.

- The cycle works this way: the hypothalamus releases gonadotrophin-releasing hormone (GnRH) that causes the pituitary to increase secretion of FSH and LH. FSH and LH cause follicles to develop and eventually trigger ovulation. Oestrogen levels continue to rise as the developing follicles secrete more and more oestrogen. Oestrogen causes positive feedback to the hypothalamus and pituitary, which increases the levels of FSH and LH. This positive feedback loop continues right up until ovulation.

- At ovulation, the ruptured follicle (left behind after ovulation) begins to secrete progesterone and reduces oestrogen secretion. Under the influence of progesterone, the feedback loop between oestrogen and the hypothalamus and pituitary reverses itself, becoming a negative feedback loop. Thus, LH and FSH levels drop. Progesterone levels continue to rise, maintaining the thickened endometrium in case fertilisation occurs. If fertilisation does not occur, progesterone decreases, the endometrium decays and the cycle begins all over again.

- The male reproductive system has somewhat more obvious external genitalia, the penis and the testes. Internal genitalia include a series of ducts, the vas deferens, ejaculatory duct and urethra, and a series of glands, the seminal vesicles, prostate and bulbourethral glands.

- Male reproductive physiology is not cyclic and is therefore a bit less complicated than female reproduction. Sperm, like eggs, develop via meiosis under the control of LH and FSH in the testes. Sperm mature in the epididymis. When a man is sexually aroused, the penis becomes engorged with blood and an erection occurs. If arousal continues, ejaculation, the movement of sperm from the testes to the penis and out of the man's body, may occur. During ejaculation, sperm move from the epididymis through the vas deferens, ejaculatory duct and out of the urethra. Along the journey, the seminal vesicles, prostate and bulbourethral glands add sugar, chemicals and mucus to the sperm to form semen.

- Testosterone is the chief male hormone. It is secreted by the testes and is responsible for masculinising male fetuses, triggering LH and FSH production, and the development of male secondary sexual characteristics.

Case study

Linda and her husband John have been planning a family for many years. Now that the time seems right, she cannot seem to get pregnant. After two years of trying, they have decided to consult their General Practitioner for a referral to a fertility specialist.

Here are the results:

John is producing sperm at normal levels, and the sperm are healthy and mobile.

Linda's hormone levels and all other blood tests are normal. She has a more or less regular menstrual cycle, but she has always had cramps during the middle of her cycle. On further questioning, she admits to the doctor that in the last few years, the pain seemed to get worse but she thought that was just normal for her. She also admits to some low back pain and some urinary disturbances, particularly around her period.

1. What do you think the problem is?

2. What aspects of Linda's history point to this diagnosis?

3. What effect might this condition have on her fertility?

Review questions

Multiple choice

1. In each lobule of the testes are several _____, tubes in which the sperm are made and develop.
    a. seminal vesicles
    b. seminiferous tubules
    c. bulbourethral glands
    d. all of the above

2. The female primary genitalia are the:
    a. ovaries
    b. uterus
    c. vagina
    d. testes

3. This division of the endometrium sheds each month:
    a. horny layer
    b. basal layer
    c. menstrual layer
    d. none of the above

4. This hormone stimulates ovulation:
    a. oestrogen
    b. progesterone
    c. LH
    d. FSH

5. Which hormone stimulates milk production?
    a. Oxytocin
    b. Prolactin
    c. Oestrogen
    d. Human Chorionic Gonadotrophin

→

Fill in the blanks

1. The stiffening of the penis is known as _____, and the actual movement of sperm out of the penis is _____.

2. In the male reproductive system the _____ is posterior to the bladder, and the _____ gland is inferior to the bladder.

3. _____ is the type of cell division that produces egg and sperm.

4. After ovulation, oestrogen and progesterone send _____ (positive or negative) feedback to the pituitary and hypothalamus.

5. The union of sperm and egg is called _____.

Short answers

1. Describe the process of spermatogenesis.

   _____
   _____
   _____
   _____

2. Explain the role of hormones in controlling the female menstrual cycle.

   _____
   _____
   _____
   _____

3. Contrast mitosis and meiosis.

   _____
   _____
   _____
   _____

Suggested activities

1. List all the hormones involved in reproduction. Describe what they do.

2. Describe the human life cycle in detail. How much do you know about your own reproductive system?

3. With a partner, list the similarities and differences between male and female reproduction. How many can you list?

Answers

Answers to case study

1. Endometriosis – a condition caused when endometrial tissue escapes from the uterus and becomes implanted in the abdominal and pelvic cavities where still it builds up and decays in response to the woman's menstrual cycle.
2. The worsening of mid-cycle cramps; low back pain and urinary disturbances especially around the time of her period.
3. It is likely to be the cause of her inability to conceive and, indeed, is the commonest cause of infertility.

Answers to multiple choice questions

1. b
2. a
3. d
4. c
5. b

Answers to fill in the blanks

1. Erection; ejaculation
2. Seminal vesicle; bulbourethral
3. Meiosis
4. Negative
5. Fertilisation

Answers to short answer questions

1. Primary spermatocyte – meiosis 1 – secondary spermatocytes – meoisis 2 – spermatids – maturation – sperm.
2. Hormones are oestrogren, progesterone, luteinising hormone and follicle-stimulating hormone. GnRH from the hypothalamus results in LH and FSH release from the pituitary. FSH results in development of primary follicles and LH triggers ovulation. Oestrogen stimulates female secondary sexual characteristics. Progesterone prepares the uterine lining for implantation.
3. Mitosis is division of cells to produce a copy with the correct number of chromosomes; meiosis produces cells with half the number of chromosomes.

Ageing

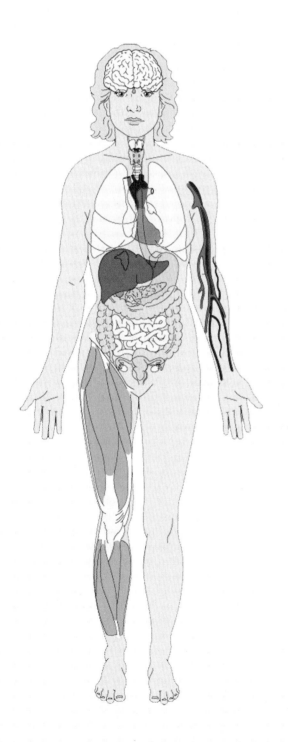

For any of you who have ever taken a memorable journey or have lived through a significant event, you know that you have been changed to some degree forever. We hope that in some positive way this book has also changed you.

The following topics are just a small sampling of how anatomy and physiology plays an important role in many areas. We have added a whole section of amazing body facts so you can impress friends and family. Enjoy the chapter!

Learning objectives

At the end of this chapter, you will be able to:

- Relate anatomy and physiology changes to the process of ageing
- Describe the concept of wellness and personal choices
- List and describe wellness concepts for each body system
- Discuss cancer prevention and treatment
- Dazzle your friends with amazing anatomy and physiology facts

Pronunciation guide

anhidrosis (*an HIGH drow sis*)
bacilli (*ba SILL eye*)
Chlamydia trachomatis (*klah MID ee ah TRAH koh MAH tis*)
herpes simplex virus 2 (*HER peez*)

human papilloma virus (*pap ih LOW ma*)
incontinence (*in CON tin ense*)
Neisseria gonorrhoeae (*nye SEE ree ah gon ah REE ah*)

spina bifida (*SPY nah BIFF ih dah*)
thallium (*thah LEE um*)
Treponema pallidum (*trep ah NEE mah PAL ih dum*)

Older adults

Many regions in the UK are exhibiting a major change in patient demographics. In general, we are seeing an older population due to safer workplaces, healthier lifestyles, effective vaccines and medications, and opportunities to access healthcare. As a result, you probably will deal with a high number of older patients in the healthcare profession that you choose (even the maternity departments may have 50- to 60-year-old mothers in the future, as witnessed by several recent news reports).

Older patients differ in many ways from other patient age groups, so it is important to recognise these differences in order to provide the best healthcare possible for this patient population.

Using the vague term *ageing* as a way to generally describe a patient is interesting in that an individual's body does not age uniformly. For example, an individual may look older due to ageing skin but may have a cardiovascular system of a person 10 to 15 years younger than his or her chronological age. With that said, there is the general 'one per cent rule' in which we see a one per cent decrease in the function of most body systems each year beginning around the age of 30.

So what general characteristics or tendencies do we see in an ageing patient? The hallmark sign of ageing is a decrease in the ability to maintain homeostasis. These individuals may have normal baselines but will begin exhibiting a decrease in the ability to adapt to stressors. Often, disease processes along with ageing will accelerate the loss of body reserves that younger patients take for granted, such as recovery time or complications following an accident or surgery. In addition, you may have increased difficulties in evaluating these patients because of decreases in their vision, hearing, and possibly their mental abilities.

Amazing body facts

You're not getting older, you're getting smarter

There are a lot of misconceptions about the brain and ageing. In the absence of disease, your brain continues to mature up to the age of 50. Interestingly, we have better overall brain function and decision-making skills at age 60 than at age 30. Even though an older patient may lose up to 1000 neurons a day, you have to realise that we all started out with several billion! This is why you may not see a decrease in brain function until the age of 75.

General body changes

Let's take a general look at the ageing process of the body. There is a decrease in the total body water for both males and females. The clinical significance of this is that they have a tendency to dehydrate more rapidly, which can potentially affect the excretion rate of medications they are on.

From ages 20 to 70, we see a loss of lean body mass due to up to a 30 per cent loss in the number of muscle cells, atrophy of remaining muscle cells and a general decrease in muscle strength. Conversely, we see an increase in total body fat, which slowly increases between the ages of 25 and 45, peaking at 40. This trend can continue up to age 70. The fat accumulated is a 'deeper' fat, meaning that it is more abdominal fat and found more in the viscera than subcutaneously.

Bone density usually reaches its greatest peak around age 35. Contrary to what you see on TV, loss of bone mass occurs both in women *and* men. For women, during the first five years postmenopause, 1 to 2 per cent bone loss per year can occur. In general, we see a 1 per cent loss per year between the ages of 55 and 70. After that, it is approximately 0.5 per cent per year.

So in general, as an individual ages, he or she loses muscle mass, gains fat and loses bone density. This varies among individuals and can be affected by lifestyle choices. Later in this chapter, we discuss steps that can be taken to slow (or even reverse) this process.

Gustatory changes

Sensory changes can also occur in this patient group. The senses of taste and smell begin to deteriorate as a natural process of ageing. The number of taste buds decreases by about 50 per cent. Sweet versus bitter tastes become less discernible. The acuity of salt and bitter tastes decline. Orange juice may taste metallic. Patients on oxygen via a nasal cannula may have a decrease in their ability to taste. These changes are clinically significant because they may make it more difficult to insure a properly balanced diet for good health and activity. This is also a concern coupled with the additional functional and physiologic difficulties affecting grocery shopping and food preparation. It is estimated that 33 per cent of this patient group lives alone. These and other sensory impairments affect daily activities, and as a result, we see approximately 5 to 15 per cent of this population exhibiting protein and calorie malnutrition.

Additional barriers to proper nutrition include a loss of teeth, difficulties in swallowing and a decrease in salivary secretion. In addition, there are decreases in digestive juice acidity and secretion, nutrient absorption and bowel function.

The brain and nervous system

The main problem with the ageing process and the nervous system is the slower reaction time that occurs. As a result, there are increased chances for car accidents, falls, burns and so forth.

Pain is something usually associated with the ageing process. Even though the authors of this book are young in spirit, vibrant and good looking, they have occasional creaking joints and stiff, sore muscles! Untreated pain can lead to an overall decrease in the quality of a person's life, including impaired sleep, a decrease in socialisation, confusion, depression, malnutrition, poly-pharmacy (use of multiple medications) and impaired mobility. If any of these conditions were occurring before the pain was introduced, the pain may cause these conditions to worsen.

Often, older patients are under-medicated for pain. This may be because patients who are debilitated, cognitively impaired or have a history of substance abuse are sometimes unable to effectively relate how they feel. Sometimes, it is a result of the ignorance of healthcare professionals who cannot recognise indicators of pain.

Clinically, these are some of the potential behavioural changes related to pain:

- changes in personality such as becoming agitated, quiet, withdrawn, sad, confused, depressed, grumpy
- loss of appetite
- screaming, swearing, name calling, grunting, noisy breathing
- crying, rocking, fidgeting
- rubbing a sore area, wincing
- cold, clammy and pale skin.

The cardiovascular system

Within the cardiovascular system, we see changes such as calcification of the heart valves, which decreases their efficiency. We also see a lessening in the flexibility of the blood vessels, which leads to clogging (arteriosclerosis). This makes them less able to deal with blood pressure changes, so increased blood pressure is common. There is also a decrease in cardiac output and an approximate 25 per cent decrease in the maximum heart rate.

The genitourinary system

Between the ages of 20 and 80, we lose about 50 per cent of our renal function. This is clinically significant when you consider the amount of drugs used by the older population that are excreted by the kidneys. Therefore, a drug taken by someone with impaired renal (and/or liver) function can lead to the drug not being metabolised or excreted at the expected rate. This can lead to an accumulation of the drug in the body at harmful or toxic levels.

The integumentary system

Here we see a loss in skin elasticity, increased skin delicacy, multiple skin lesions and incidences in skin cancer. While many of these problems could have been reduced by limiting sun exposure and not smoking earlier in their lives, the use of medications such as systemic corticosteroids does accelerate many of these conditions.

Polypharmacy

Polypharmacy, the administration of many drugs at the same time, is a major concern for this patient group. Contributing factors to this problem include the fact that these patients typically see several specialists for a variety of diseases, often with one doctor not knowing what the other is prescribing. These multiple diseases have competing therapeutic needs, and the combinations of utilised drugs can cause life-threatening situations. Ageing also can affect the rates of drug absorption, drug distribution throughout the body and metabolism of the drugs, as well as their excretion from the body.

As a result, there are some clinical considerations. Review the drugs that the individual is taking, looking for possible drug interactions or medications that are no longer needed. Look at your patient's liver and kidney functions. They will affect the rate of removal of medications from the body and resultant drug blood levels. When medicating with opioids, dose by age, not by body weight. Remember, opioids slow the gastrointestinal tract. In general, when considering an *equal* dose given to an older patient compared to a middle-aged patient, analgesics are stronger and last longer in older patients. So, start low and go slow.

Test your knowledge 19.1

Choose the best answer:

1. There is a _____ decrease in the function of most body systems each year beginning around the age of 30.
   a. 10 per cent
   b. 20 per cent
   c. 5 per cent
   d. 1 per cent

2. The hallmark sign of ageing is the decrease in the ability to maintain:
   a. cardiac enzymes
   b. exercise potential
   c. homeostasis
   d. sense of humour

3. _____ is a term used for the administration of many drugs at the same time.
   a. addiction
   b. polypharmacy
   c. multimeds
   d. overdose

4. In the absence of disease, your brain continues to mature up to the age of:
   a. 50
   b. 35
   c. 75
   d. 10

Well-being

One of the most important personal choices you will ever make will be the decision on what kind of lifestyle you want to live. Choices you make now may have a *profound* effect on your future health and lifespan. Will you eat properly, exercise within reason, smoke, do drugs and/or alcohol, live or work in a dangerous environment, have risky behaviours?

Let's take a quick review of the body systems that we covered and briefly discuss some ways to improve your health. While there are always some controversial issues in wellness, we have chosen the more commonly held beliefs.

Nervous system

Let's talk a little bit about stress. First of all, stress is a natural part of life. Stress is also good and necessary: it is a motivator; it helps you to protect yourself. The problem occurs when stress becomes chronic and you can no longer effectively deal with it. The high cost of stress is that it can affect some or all of your body's various systems to varying degrees. A poor stress response can lead to an assortment of disorders such as eating and digestive disorders, decreased immune response, decreased memory and work capacities, sleep problems, joint and muscle aches, heart problems and personality changes. Some of these problems can be life-threatening. For an extensive review of a healthy stress management system, remember to consult the MasteringA&P website.

Often, when dealing with a patient, we focus specifically on the presenting illness and fail to realise we should look at the whole person. The 'whole person' includes not only other systems of the body but also the mental and spiritual aspects of the person. Too often in the past, dealing with mental health has had a negative image. We don't hide the fact that we have the flu or a broken arm, so why should we hide the fact that we may be sad for long periods of time (depression) and need help to resolve that condition?

Skeletal system

Diet is extremely important in the growth and protection of your bones. A diet rich in calcium and vitamins helps to maintain good bone growth and development. Weight-bearing exercise has also been shown to be beneficial in maintaining healthy bones over a lifetime.

One common occupational condition related to repetitive motion, such as typing on a keyboard, playing a piano or hammering, is known as carpal tunnel syndrome. While this syndrome is caused by damage to the median nerve, it is the result of the *skeletal* structure of the wrist being too restrictive during extended periods when the wrist is kept in an upward bent position. See Figure 19.1, which illustrates the structures that affect this repetitive syndrome.

Muscular system

Again, proper exercise and diet will help to develop and maintain properly functioning muscles. While there are many types of muscle-training programmes, you need to investigate which type is best for you depending on your needs or desired outcomes. See Figure 19.2, which presents the activity pyramid as a possible guide.

Be careful of muscle-enhancement drugs: many are dangerous and have serious side effects. Again, careful critical investigation is the best rule to follow.

Integumentary system

Proper diet and hydration is important for the functioning of the integumentary system. Smoking also affects this system by causing premature ageing of the skin.

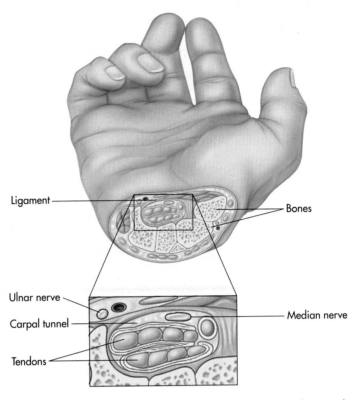

Figure 19.1 Anatomical structures affecting carpal tunnel syndrome.

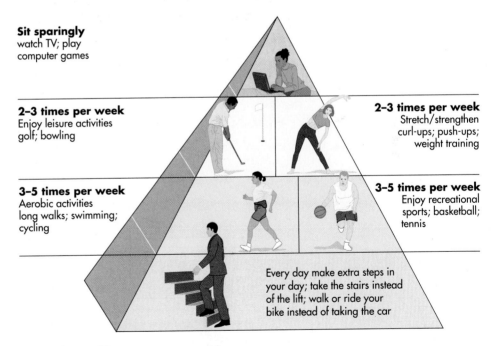

Figure 19.2 The activity pyramid.

While sun exposure is important for the production of vitamin D in your body, limiting the amount of sun exposure to prevent skin cancer is equally important. While some forms of skin cancer are treatable, others can be lethal. The effect of the sun is cumulative, and severe sunburns as a child can have severe consequences in adulthood. There are several ways to prevent excessive sun exposure. First, minimise your time in the sun between 10 am and 4 pm because this is when the sun's rays are most intense. Wear long-sleeved shirts when possible, brimmed hats, sunglasses (preferably wrap-arounds) and, of course, sunblock.

Cardiovascular system

A heart-healthy diet low in saturated fats (remember, we all need some fat for proper function) and high in fibre, rich in fruit and vegetables, will help maintain an optimal cardiovascular system.

Diet alone is not sufficient for a healthy heart and the proper level and regularity of exercise also helps condition your heart for maximum functioning. This can be as simple as brisk walking for 30 minutes a day, three to four times a week. Of course, the level of your exercise programme depends on many individual factors.

Smoking, alcohol and other drugs can adversely affect the cardiovascular and other systems. For an illustration of the effects of chronic alcoholism on the cardiovascular, digestive and nervous systems, please see Figure 19.3.

Respiratory system

Smoking is the primary preventable cause of respiratory diseases. Smoking can lead to damage of lung tissue and chronic diseases such as bronchitis, emphysema and asthma. In addition, smoking increases the occurrence of lung infections and colds as well as sinus infections. Approximately 80 per cent of all lung cancers can be traced to smoking. Smoking also affects the heart by reducing the availability of oxygen to the heart muscle.

Smoking, along with alcohol consumption, leads to an increase in stomach and mouth cancers.

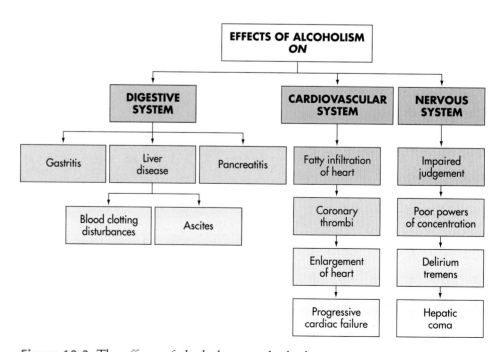

Figure 19.3 The effects of alcoholism on the body systems.

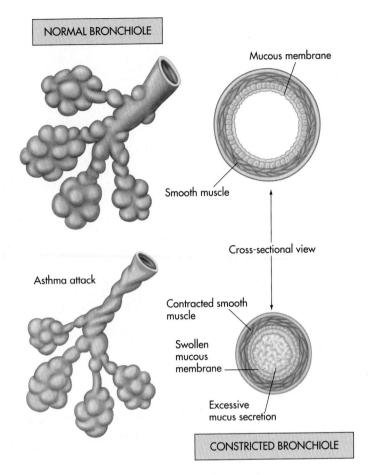

Figure 19.4 The normal and constricted airway.

What we breathe in, depending on where we live or work, can also affect the health of the respiratory system. Both outdoor and indoor pollution can lead to a number of respiratory problems.

Occupational hazards can occur when workers are exposed to dust or vapours. For example, coal miners exposed to coal dust without proper protection can develop black lung. The lung's initial response to an inhaled irritant is to close down or restrict the airway, thereby minimising the inhalation of the substance. This can lead to severe breathing difficulties. See Figure 19.4 which shows the constricted airway in an asthma attack.

Gastrointestinal system

A proper diet is critical for growth, development and good health in general. Lack of a proper diet, which leads to undernourishment, can affect all the systems, as can be seen in Figure 19.5. While most experts agree that the best source of vitamins and minerals comes from natural food sources, the responsible use of supplemental vitamins and minerals can play an important role in health. For example, excessive dosages of fat-soluble vitamins (A, D, E and K) can actually harm the body because they can build up in toxic levels. See Table 19.1 for a list of some of the vitamins and the systems they assist.

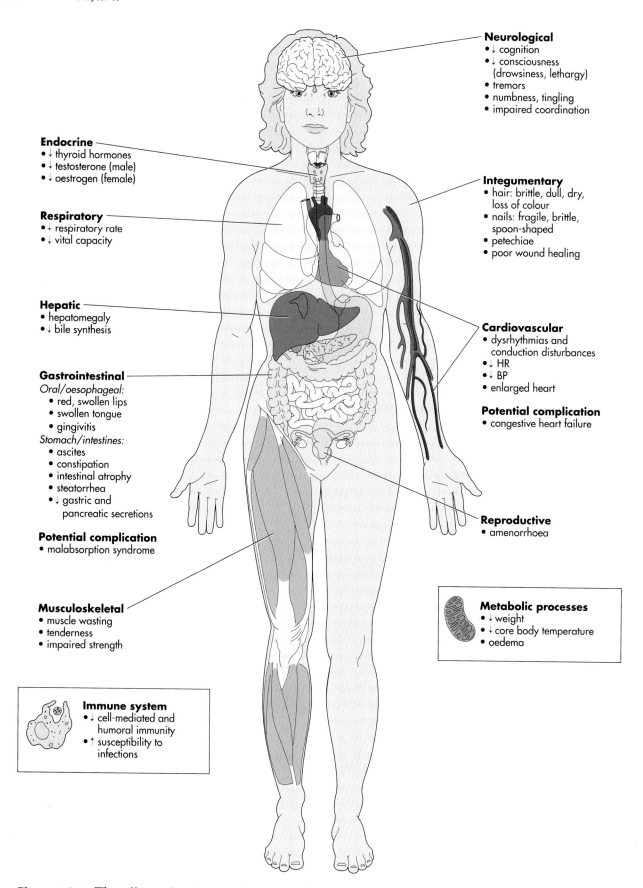

Neurological
- ↓ cognition
- ↓ consciousness (drowsiness, lethargy)
- tremors
- numbness, tingling
- impaired coordination

Endocrine
- ↓ thyroid hormones
- ↓ testosterone (male)
- ↓ oestrogen (female)

Integumentary
- hair: brittle, dull, dry, loss of colour
- nails: fragile, brittle, spoon-shaped
- petechiae
- poor wound healing

Respiratory
- ↓ respiratory rate
- ↓ vital capacity

Hepatic
- hepatomegaly
- ↓ bile synthesis

Cardiovascular
- dysrhythmias and conduction disturbances
- ↓ HR
- ↓ BP
- enlarged heart

Potential complication
- congestive heart failure

Gastrointestinal
Oral/oesophageal:
- red, swollen lips
- swollen tongue
- gingivitis
Stomach/intestines:
- ascites
- constipation
- intestinal atrophy
- steatorrhea
- ↓ gastric and pancreatic secretions

Potential complication
- malabsorption syndrome

Reproductive
- amenorrhoea

Metabolic processes
- ↓ weight
- ↓ core body temperature
- oedema

Musculoskeletal
- muscle wasting
- tenderness
- impaired strength

Immune system
- ↓ cell-mediated and humoral immunity
- ↑ susceptibility to infections

Figure 19.5 The effects of undernourishment on the body systems.

Table 19.1 Vitamins and the body systems

Vitamin	Description
vitamin A	proper night vision; proper development of bones and teeth; mucous membrane and epithelial cell integrity; helps resist infection
vitamin B	B_2 promotes healthy muscular growth; B_{12} is needed for healthy blood cell development and to treat pernicious anaemia; B_1 and B_{12} promote healthy function of nervous tissue; B_1 aids in carbohydrate metabolism, normal digestion and appetite; niacin is necessary for fat synthesis and cellular respiration
vitamin C	aids in absorption of iron; promotes healing of fractures, development of teeth and bone matrix, and wound healing; ensures capillary integrity; bolsters immune system
vitamin D	promotes strong bones and teeth, regulates skeletal calcium reabsorption; aids in absorption of calcium and phosphorus from the intestinal tract; D_3 helps regulate release of parathyroid hormone
vitamin E	promotes muscle growth; E_1 necessary for haemolytic resistance of red blood cell membranes; helps prevent anaemia for proper reproductive system functioning; current research shows that consumption from *natural* sources may reduce the risk of Parkinson's disease; further investigation is needed
vitamin K	needed for proper blood clotting

Endocrine system

Again, proper diet and exercise assist the endocrine system. One of the areas of concern in professional sports is the use of performance-enhancement substances. For example, anabolic steroids are used to increase strength and endurance rapidly and to build muscle mass. Anabolic steroids are closely related to the male hormone testosterone. However, these steroids have *serious* side effects that include kidney damage, liver damage, increased risk of heart disease, irritability and aggression. Women taking steroids can develop facial hair and deeper voices. In men, these substances diminish sperm production. Some of these effects can be permanent even after ceasing the use of the drugs. The use of anabolic steroids is banned and is tested for in sports.

Sensory system

Care of the sensory system includes proper diet, wearing hearing and sight protective devices when necessary, and periodic examination of the eyes and ears. Wearing hearing protection during activities that produce high levels of noise will greatly extend the functional life of your hearing. Damage to the ear is cumulative, so there is no better time to start than right now. In addition, protective eyewear should be worn any time that the risk of eye injury can occur, such as in certain occupations, hobbies and sports. Figure 19.6 shows the Snellen eye chart for determining visual acuity.

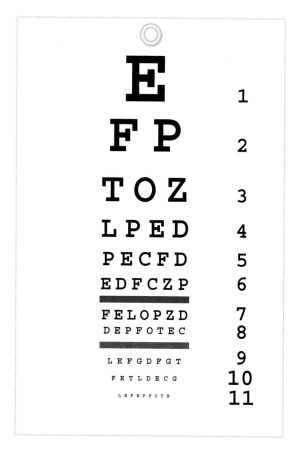

Figure 19.6 The Snellen eye chart.

Immune system

Proper diet and exercise are needed for optimal functioning of the immune system. In addition, other factors can assist your immune system. One of the simplest and most effective ways to protect not only yourself but your patients is hand-washing. Always wash your hands before and after working with each of your patients. Just because you are a staff member in a hospital or healthcare institution, you may still be a carrier of infectious organisms – in addition, you could be susceptible to becoming infected by hospital pathogens. As we previously discussed, correct washing of your hands goes a long way in stopping the spread of infection.

Having current immunisations is important to assist the immune system in being prepared for certain pathogens. Immunisation schedules are recommended by the NHS in the UK. Many individuals mistakenly believe immunisations only occur in childhood. Influenza vaccines are just one example of an immunisation that is particularly important for the older population.

A big issue is when and when not to take antibiotics. Most infections can be handled by the body's immune system in a few days. The overuse of antibiotics has led to several critical health issues. First, many viral infections are mistakenly treated with antibacterial agents, which do nothing to the virus and cause harm to normal bacteria such as in our intestinal system. Second, overuse does not allow the immune system of a child to properly develop and respond to future infections and can cause related disorders, such

Applied science

Antibiotics

The term antibiotics means 'against life' and technically includes medications that inhibit or destroy any micro-organism, including bacteria, viruses and fungi. However, antibiotics in medicine have become associated only with antibacterial agents.

as asthma. Finally, many patients do not properly take their antibiotics. For example, they do not take the full dose for the full length of time but discontinue when they 'feel better'. However, often there are still surviving bacteria that are left that represent a stronger, drug-resistant strain, and they are now free to reproduce stronger drug-resistant offspring. This has led to epidemics of drug-resistant infections, for example MRSA (methicillin-resistant *Staphylococcus aureus*).

Reproductive system

Don't smoke! Smoking mothers tend to have babies of lower birthweights, suffer more premature births, and have a higher rate of SIDS (sudden infant death syndrome). And while we're talking about babies and children, don't forget about the hazards of second-hand smoke in the home. In homes that have at least one smoking parent, children have slower than normal lung development and are predisposed to increased incidences of bronchitis, asthma and ear infections (otitis media).

Diet

As the old saying goes; when you are pregnant you are now eating for two! This doesn't mean that mum should pig out at every chance she gets. What it means is that diets should be followed that provide important vitamins, minerals and nutrients for the developing fetus and to maintain the health of the mother. Think about it. If the diet is lacking in calcium, where does the fetus get calcium for bone development? It has to take it from the bones and teeth of the mother, thus decreasing the integrity of her system. The congenital condition of **spina bifida** can be prevented by a dietary supplement of a member of the vitamin B complex, folic acid. The elimination of alcohol during pregnancy is also important to assure the best chances of normal spinal cord and nervous system development.

spina bifida (*SPY nah BIFF ih dah*)

Sexually transmitted diseases (STDs)

STDs are a growing problem and can have serious effects on the reproductive system and lethal effects on the body. There are various types of diseases and organisms that can be transmitted through unprotected sex (including oral sex). Please see Table 19.2 for a list of sexually transmitted diseases.

Cancer prevention and treatment

All the body systems can be ravaged by cancer. Cancer is the runaway reproduction and spreading of abnormal cells and is a very complicated disorder. Each type of cancer, named for the location or the type of cells that are running amok (for example, colon cancer, prostate cancer, squamous cell

Table 19.2 Sexually transmitted diseases

Disease	Organism	Symptoms
herpes	herpes simplex virus 2	male: fluid-filled vesicles on penis female: blisters in and around vagina
gonorrhoea	*Neisseria gonorrhoeae*	male: purulent discharge from urethra, dysuria, and urinary frequency female: purulent vaginal discharge, dysuria, urinary frequency, abnormal menstrual bleeding, abdominal tenderness; can lead to sterility
chlamydia	*Chlamydia trachomatis*	may be asymptomatic male: mucopurulent discharge from penis, burning and itching in genital area, dysuria, swollen testes; can lead to sterility female: mucopurulent discharge from vagina, inflamed bladder, pelvic pain, inflamed cervix; can lead to sterility
syphilis	*Treponema pallidum*	systemic disease that can lead to lesions, lymph node enlargement, nervous system degradation, chancre sores
genital warts	human papilloma virus (HPV)	cauliflower-like growths on penis and vagina

herpes simplex virus 2 (*HER peez*)

Neisseria gonorrhoeae (*nye SEE ree ah gon ah REE ah*)

Chlamydia trachomatis (*klah MID ee ah TRY koh mah tis*)

Treponema pallidum (*trep ah NEE ma PAL ih dum*)

human papilloma virus (*pap ih LOW ma*)

carcinoma), has its own unique characteristics. However, in the past few years, medical science has learned a number of things about cancer that have made great improvements in cancer prevention and treatment.

Any number of triggers can make a cell cancerous, including genes, radiation, sunlight exposure, smoking, fatty foods, viruses and chemical exposure. Some of these triggers, like genes or some viruses, are difficult to avoid. But others, like smoking, sunlight, radiation exposure and fatty foods, can be pretty easily avoided by eating correctly, avoiding smoking, and wearing sunblock. Many types of cancer can be prevented or managed with a healthy diet and exercise. Even genetic susceptibility to cancer does not make cancer unavoidable. Testing, such as mammograms (for breast cancer), colonoscopy (for colon cancer) and Pap smears (for cervical cancer), can improve survival by catching cancers early, before they have spread, or even allowing the removal of abnormal cells before they become cancerous.

Treatments for cancer typically involve removal of the cancerous cells, if possible, and some form of treatment to kill any cells remaining in the body. Chemotherapy is the treatment of cancer with chemicals that kill rapidly dividing cells. Radiation uses energy waves to shrink tumours. Biological or immunotherapy targets the cancer by manipulating the immune system to hunt down and kill the cancer cells. New treatments are constantly under development to treat cancers that are difficult to fight. Research has made great strides in the treatment of cancer.

Let's use the skin cancer, melanoma, as an example. Melanoma is the most deadly form of skin cancer. It is formed by the runaway reproduction of melanocytes, the pigment-forming cells of the skin. People at the highest risk of melanoma are those with fair skin and light eyes or hair, who have been exposed to lots of sun during their lifetime. However, new evidence indicates that even people who tan easily may develop melanoma if they get enough sun exposure. Melanoma risk is higher for those living near the equator, but people in the more northern parts of the world are not without risk. Exposure to sunlight, particularly sunburns, even as an adult, is the key risk factor for melanoma. Genetic factors are involved in some cases of melanoma.

Melanoma can be easily prevented by decreasing exposure to UV light. Aside from staying indoors all the time, which isn't very practical, sunblock is the best way to protect yourself from melanoma. Individuals at risk should have a skin screening on a regular basis.

Standard treatment for melanoma in early stages (stage I, no spread) has been the 'watch and wait' approach. The melanoma is removed, and the patient is monitored for several years. Patients with more advanced melanomas often have lymph nodes sampled and removed to prevent further spread. This more extensive surgery was deemed unnecessary for patients in very early-stage disease. However, a study in 2005 has shown that even patients with no obvious spread of their cancer benefit from having lymph nodes sampled and removed if they contain cancer cells. Patients who had the procedure, called a sentinel lymph node mapping and biopsy, were 26 per cent less likely to have their cancer return within five years than patients who only had the tumour removed.

Test your knowledge 19.2

Choose the best answer:

1. This vitamin is needed for proper blood clotting:
   a. A
   b. B
   c. C
   d. K

2. The abuse of this substance can have side effects such as facial hair and deeper voices in women, kidney and liver damage, and aggressive behaviour.
   a. antidepressants
   b. aspirin
   c. anabolic steroids
   d. chocolate

3. The _____ eye chart is used to determine normal vision.
   a. Snellen
   b. Seymour
   c. See Clear
   d. Optical

More amazing body facts

Remember those amazing body facts we promised you at the beginning of the chapter? True to our word, here is a list of a few to amaze your friends and family.

- Older adults are more prone to food poisoning not only because of decreases in their senses of smell and taste but because their digestive juices are not as acidic as they used to be and so cannot always efficiently destroy all of the food-borne pathogens they ingest.
- Nerve impulses can travel up to 130 metres per second.
- Approximately 114,000 people die in the UK annually because of smoking-related diseases.
- On average, a healthy kidney filters about *160 litres of fluid* every day. What makes this amazing is that the kidney is only about 10 cm long, 5 cm wide and 2.5 cm thick.
- Hair grows about 0.5 cm each month and grows faster during the day than at night. Hair also grows faster in the summer than in the winter.
- You use a little over 250 ml of oxygen each minute when you are at rest.
- Everyone has one nare (nostril) that is larger than the other. If you don't believe it, take a look at people next time you go out shopping or out to eat.
- Talk about a busy worker! Your heart beats over *36 million* times a year!
- You possess over 25,750 km of capillaries.
- Because viruses are continuously mutating, your immunity to influenza will not last a lifetime.
- More hair facts! You have from 110,000 to 150,000 hairs just on your head. Each strand of hair can support approximately 100 grams of weight, so, at least in theory, a full head of hair could support the weight of two African elephants.
- Vitamins, natural or in pill form, which is best? Research appears to indicate that vitamins and minerals from natural food sources are better utilised than synthetic pills. But the pills are better than nothing!
- The horns of a bull are composed of the same material that makes up your fingernails and toenails.
- You have about a quarter of a million sweat glands on your feet.
- Based on current research of fibroblasts' doubling ability before they can no longer accurately divide, we have the *potential* to live to 120 years of age.
- Your eyes can see approximately *7 million* shades of colour.
- Due to its sterile nature, urine can be used to clean out a wound when no antiseptic is available.
- The ability to roll your tongue into a tube is inherited; not everyone can do it (or cares to).
- Cavities and poor oral hygiene can lead to diabetes and heart attacks. In fact, some health experts believe that daily flossing can add 6.4 years to your life. That's because bacteria that grows in the mouth of an individual with poor oral hygiene can escape into the bloodstream and travel throughout the body, causing problems. As a result, in worst-case scenarios, that individual may be at a four times greater risk for stroke and at a fourteen times greater risk for heart attack. The risk for diabetes is also increased.
- Current research indicates that stomach cancer, which affects about 9100 people in the UK annually, may originate from bone marrow cells that enter the stomach to repair damage to the stomach lining.

- Walking uphill or downhill may make a difference in your desired health outcomes. A recent study showed that individuals who walked uphill cleared fats (especially triglycerides) from their blood faster, while downhill hiking reduced blood sugars more readily and improved glucose tolerance. Hiking either way removed LDL, or bad cholesterol. This information may be applicable for exercise regimens for diabetics who may have trouble with aerobic exercises.

Summary

- The hallmark sign of ageing is the decreased ability to maintain homeostasis.
- Our bodies don't age evenly, and certain systems age more rapidly than others.
- The loss of mental capacities is not directly related to ageing until age 75, when it is still minimal unless disease is present.
- Due to changes in the gastrointestinal, renal and hepatic systems, older patients respond differently to many medications.
- Polypharmacy is the use of many drugs at the same time and often is the result of seeing many specialists at the same time.
- The most important personal choice you will make is a healthy lifestyle.
- To maintain a healthy lifestyle, it is important to eat properly, exercise, manage stress and avoid bad habits.
- Some cancers, such as skin and lung cancer, can be highly preventable. Limiting the amount and intensity of sunlight exposure and not smoking are two ways that can help prevent cancer.

Review questions

Multiple choice

1. From ages 20 to 70, there is up to a
   _____ per cent loss of lean body mass.

   a. 10
   b. 20
   c. 30
   d. 50

2. This vitamin is needed for strong bones and teeth and calcium absorption:

   a. A
   b. B
   c. C
   d. D

3. The overuse of this classification of drugs has caused drug-resistant strains of bacteria:

   a. steroids
   b. antibiotics
   c. diuretics
   d. painkillers

4. The congenital condition of spina bifida can be prevented by the addition of what vitamin during pregnancy?

   a. A
   b. folic acid
   c. niacin
   d. K

→

Fill in the blanks

1. The fat-soluble vitamins are _____, _____, _____ and _____.

2. The test for cervical cancer is called the _____ _____.

Short answers

1. Discuss various ways to prevent skin cancer.

   _____

   _____

2. List and discuss ways to prevent STDs.

   _____

   _____

3. Discuss ways to protect yourself and your patient from the spread of infection.

   _____

   _____

Suggested activities

1. Discuss the various ageing processes you have seen in your parents and grandparents.

2. List the foods you've eaten over the last three days. Share the list with your classmates and determine whether or not it is a healthy diet. Discuss ways to improve it.

3. Make a poster presentation on one of the following topics or choose your own: STDs, passive smoking, hazards of smoking, healthy lifestyles, ageing process, proper diet, healthy pregnancy.

Answers

Answers to multiple choice questions

1. c
2. d
3. b
4. b

Answers to fill in the blanks

1. A, D, E and K
2. Pap smear

Answers to short answer questions

1. Reduce exposure to sunshine; reduce exposure when the sun is at its highest (e.g. 11.00–15.00); cover up – wear loose clothing; use sunscreen and re-apply regularly; avoid burning the skin; prevent children from developing a tan (increases likelihood of developing skin cancer); wear a hat

2. Barrier method of contraception; abstinence; health education; peer education programmes; vaccination (HPV vaccine); education concerning binge drinking – binge drinking can result in sexual dis-inhibition

3. Handwashing; use of tissues when sneezing; aprons/gloves as necessary

Appendices

Appendix A Answers to test your knowledge   502

Appendix B Medical terminology, word parts, and singular and plural endings   508

Appendix C Clinical abbreviations   516

Appendix D Laboratory reference values   523

Appendix A

Answers to test your knowledge

Chapter 1

Page 4, 1.1
1. G
2. M
3. M
4. G
5. M

Page 10, 1.2
1. inflammation of the stomach
2. surgical repair of the nose
3. slow heart rate
4. recording or image of the breast
5. enlarged cell
6. nephritis
7. gastrectomy
8. cardiomegaly
9. osteopathy
10. neurologist

Page 12, 1.3
1. a. vital sign
   b. not a vital sign
   c. vital sign
   d. not a vital sign
   e. not a vital sign
   f. vital sign
   g. vital sign
2. c
3. a

Chapter 2

Page 25, 2.1
1. Person should be standing face forward, palms out, as in Figure 2.1

2. a. prone
   b. Fowler's
   c. Fowler's
   d. supine

Page 29, 2.2
1. a. inferior
   b. anterior
   c. cephalic
   d. dorsal
   e. proximal
   f. internal
   g. deep
   h. central
   i. lateral
2. posterior
3. transverse or horizontal
4. mid-sagittal
5. superficial
6. proximal, distal
7. superior
8. lateral
9. peripheral

Page 31, 2.3
a. thoracic
b. spinal or dorsal
c. abdominal
d. thoracic
e. pelvic
f. cranial or dorsal

Page 35, 2.4
1. oral or buccal
2. axillary

3. umbilical
4. lumbar
5. patellar

Chapter 4

Page 61, 4.1
1. diffusion
2. low, higher
3. filtration
4. diffusion
5. facilitated diffusion

Page 62, 4.2
1. Phagocytosis is 'cell eating' where a solid particle is engulfed. Pinocytosis is a form of endocytosis where liquid is brought into the cell or 'cell drinking'.
2. a. active
   b. passive
   c. passive
   d. active
   e. passive

Page 69, 4.3
1. a. nucleus
   b. endoplasmic reticulum
   c. mitochondria
   d. Golgi bodies
   e. flagella

Page 71, 4.4
1. b
2. a
3. c
4. d

Chapter 5

Page 91, 5.1
1. squamous – flat or scale-like shaped; cuboidal – cubed shaped; columnar – column like; transitional – stretchy or variable shaped.
2. synovial.

3. Because it is a support tissue for many other tissues.
4. Mucous is the adjective that describes the type of membrane, while mucus is the noun or the actual substance produced by the membrane.

Page 106, 5.2
1. respiratory
2. urinary
3. skeletal
4. nervous and sensory system
5. immune
6. cardiovascular
7. digestive
8. integumentary

Chapter 6

Page 116, 6.1
1.

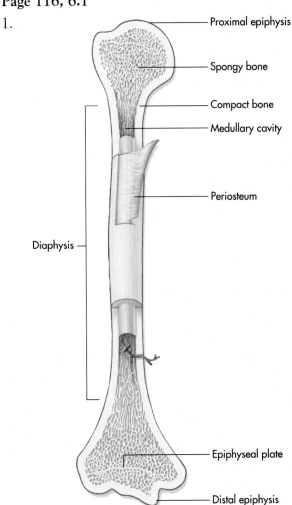

Proximal epiphysis
Spongy bone
Compact bone
Medullary cavity
Periosteum
Diaphysis
Epiphyseal plate
Distal epiphysis

2. Found in the skull, clavicle, vertebrae of the spinal column, sternum, ribs, pelvis and the epiphysis of the long bones, and is needed for the production of red blood cells.

3. Provides support and allows for movement, protects organs, produces red blood cells, and acts as storage for minerals and fats.

4. b

5. a

Page 120, 6.2

1. a
2. d
3. b
4. c
5. b
6. a

Page 130, 6.3

1. The axial skeleton is the central portion that protects and provides support for the organ systems in the dorsal and ventral cavities. The appendicular skeleton is the bones of the limbs.

2. d

3. c

4. Flexion is the movement in the anterior-posterior plane that reduces the angle in articulated elements. Extension occurs in the same plane, but increase the angles, for example when kicking a football.

Chapter 7

Page 142, 7.1

1. c
2. d
3. b

Page 149, 7.2

1. rotation
2. abduction
3. adduction
4. flexion
5. extension

Page 153, 7.3

1. sarcomere
2. myosin
3. actin
4. calcium, ATP

Chapter 8

Page 168, 8.1

1. a. epidermis
   b. dermis
   c. subcutaneous fascia

2. (any four)
   - protects from invasion of pathogens
   - keeps you from drying out
   - storage unit for fatty tissue
   - produces vitamin D
   - helps regulate body temperature
   - helps to provide sensory input

3. d

4. c

Page 174, 8.2

1. b
2. c
3. b
4. d

Chapter 9

Page 191, 9.1

1. b
2. b
3. c
4. b

Page 199, 9.2

1. c
2. b
3. c
4. b
5. c

Page 206, 9.3
1. b
2. b
3. b
4. c

Chapter 10

Page 223, 10.1
1. d
2. c
3. c
4. c
5. d

Page 229, 10.2
1. a
2. b
3. b
4. d

Chapter 11

Page 244, 11.1
1. a
2. c
3. d

Page 248, 11.2
1. c
2. c
3. pinna
4. cerumen
5. hammer
6. incus
7. stapes

Chapter 12

Page 261, 12.1
1. c
2. c
3. c

Page 265, 12.2
1. c
2. a
3. b

Page 270, 12.3
1. c
2. c
3. b
4. a

Chapter 13

Page 291, 13.1
1. d
2. d
3. d
4. left ventricle
5. right

Page 298, 13.2
1. a
2. b
3. b
4. albumin
5. sino-atrial node

Page 307, 13.3
1. a
2. c
3. b
4. tunica interna, tunica media, tunica externa

Chapter 14

Page 325, 14.1
1. d
2. d
3. c
4. b
5. b
6. d

Page 327, 14.2
1. b
2. d
3. c

Page 333, 14.3
1. d
2. c
3. d
4. d
5. c

Page 339, 14.4
1. a
2. b
3. c
4. b

Page 343, 14.5
1. c
2. c
3. a
4. c

Chapter 15

Page 362, 15.1
1. d
2. a
3. c
4. c
5. b

Page 366, 15.2
1. d
2. d
3. c
4. a

Page 373, 15.3
1. d
2. b
3. a
4. d
5. a
6. a

Page 380, 15.4
1. b
2. d
3. d
4. b

Chapter 16

Page 399, 16.1
1. c
2. a
3. b
4. d
5. crown, neck, root

Page 408, 16.2
1. c
2. b
3. b
4. c
5. a
6. parasympathetic

Chapter 17

Page 431, 17.1
1. b
2. d
3. b
4. c
5. c

Page 441, 17.2
1. a
2. b
3. c
4. a
5. a
6. d

Chapter 18

Page 453, 18.1
1. c
2. b
3. d

Page 458, 18.2

1. d
2. b
3. c

Page 461, 18.3

1. b
2. b
3. b
4. c

Page 464, 18.4

1. d
2. c
3. b
4. b

Page 468, 18.5

1. b
2. d
3. a

Page 471, 18.6

1. b
2. a
3. d

Chapter 19

Page 487, 19.1

1. d
2. c
3. b
4. a

Page 497, 19.2

1. d
2. c
3. a

Appendix B

Medical terminology, word parts, and singular and plural endings

Word parts arranged alphabetically and defined

The word parts that have been presented in this textbook are summarised with their definitions for quick reference. Prefixes are listed first, followed by combining forms and suffixes.

Prefix	Definition	Prefix	Definition
a-	without, absence of	macro-	large
ad-	towards	meta-	after, change
an-	without, absence of	micro-	small
ana-	up, towards, apart	mono-	single
ante-	before	multi-	many
anti-	against, opposing	my-	muscle
bi-	two	myo-	muscle
bin-	two	neo-	new
brady-	slow	nulli-	none
di-	two	ortho-	straight, normal
dia-	around, passing through	pan-	all, entire
dys-	bad, abnormal	para-	near, alongside, departure from normal
e-	to remove	per-	through
en-	within, upon, on, over	peri-	around, about, surrounding
endo-	within, inner, absorbing	poly-	many
ep-	upon, on, over	post-	after
epi-	upon	pre-	before
eu-	normal, good	pro-	forwards, preceding
ex-	outside, away from	quad-	four
exo-	outside, away from	re-	back
hemi-	one-half	sub-	beneath
homo-	same	sym-	together, joined
hydro-	water	syn-	together, joined
hyper-	excessive	tachy-	rapid, fast
hypo-	under, below normal	tetra-	four
im-	not	trans-	through, across, beyond
inter-	between	uni-	one
intra-	within		

Combining form	Definition	Combining form	Definition
abdomin/o	abdomen, abdominal cavity	cartil/o	gristle, cartilage
acou/o	hearing	caud/o	tail
acoust/o	hearing	caec/o	blind intestine, cecum
acr/o	extremity, extreme	cel/o	hernia, protrusion
aden/o	gland	celi/o	abdomen, abdominal cavity
adren/o	adrenal gland	cephal/o	head
aer/o	air or gas	cerebell/o	cerebellum (little brain)
aesthesi/o	sensation	cerebr/o	cerebrum (brain)
aeti/o	cause (of disease)	cerumin/o	wax
albumin/o	albumin	cervic/o	cervix, neck
alveol/o	alveolus (air sac)	cheil/o	lip
amni/o	amnion, amniotic fluid	chir/o	hand
amnion/o	amnion, amniotic fluid	chol/e	bile, gall
an/o	anus	choledoch/o	common bile duct
andr/o	male	chondr/o	cartilage
angi/o	blood vessel	chori/o	membrane, chorion
ankyl/o	crooked	chromat/o	colour
anter/o	front	clon/o	spasm
aort/o	aorta	col/o	colon
appendic/o	to hang onto, appendix	colp/o	vagina
aque/o	water	conjunctiv/o	to bind together,
arche/o	first, beginning		conjunctiva
arter/o	artery	cor/o	pupil
arteri/o	artery	core/o	pupil
arthr/o	joint	corne/o	horny, cornea
astheni/o	weakness	coron/o	crown, circle
atel/o	imperfect, incomplete	cortic/o	tree bark, outer covering,
ather/o	fat		cortex
atri/o	atrium	cost/o	rib
aud/o	hearing	cran/o	skull, cranium
audi/o	hearing	crani/o	skull, cranium
aur/i	ear	crin/o	to secrete
aut/o	self	crypt/o	hidden
azot/o	urea, nitrogen	culd/o	cul-de-sac
bacter/o	bacteria	cutane/o	skin
balan/o	glans, penis	cyan/o	blue
bi/o	life	cyes/i	pregnancy
bil/i	bile	cyes/o	pregnancy
blast/o	germ, bud	cyst/o	bladder, sac
blephar/o	eyelid	cyt/o	cell
bronch/i	bronchus (airway)	dacry/o	tear
bronch/o	bronchus (airway)	dent/i	teeth
burs/o	purse or sac, bursa	derm/o	skin
calc/i	calcium	dermat/o	skin
cancer/o	cancer	diaphragmat/o	diaphragm
carcin/o	cancer	dipl/o	double
card/o	heart	dips/o	thirst
cardi/o	heart	dist/o	away
carp/o	wrist	diverticul/o	small blind pouch, diverticulum

Combining form	Definition	Combining form	Definition
dors/o	back	hom/o	sameness, unchanging
duct/o	lead, move	hormon/o	to set in motion
duoden/o	twelve, duodenum	hydr/o	water
dur/o	hard	hymen/o	hymen
ech/o	sound	hyster/o	uterus
electr/o	electricity	iatr/o	to heal
embol/o	a throwing in	idi/o	person, self
embry/o	embryo	ile/o	ileum of small intestine, to roll
encephal/o	brain		
endocrin/o	endocrine	ili/o	flank, groin, ilium of the pelvis
enter/o	small intestine	immun/o	exempt, immunity
epididym/o	epididymis	infect/o	to enter, invade
epiglott/o	epiglottis	infer/o	below
episi/o	vulva	inguin/o	groin
eryth/o	red	ir/o	rainbow, iris
erythr/o	red	irid/o	rainbow, iris
fasci/o	fascia	isch/o	to hold back, deficiency, blockage
femor/o	thigh		
fet/i	fetus	ischi/o	haunch, hip joint, ischium
fet/o	fetus	jejun/o	empty, jejunum
fibr/o	fibre	kal/i	potassium
fibul/o	clasp of buckle, fibula	kerat/o	hard, horny, cornea
fovea/o	small pit	ket/o	ketone bodies
gangli/o	ganglion	keton/o	ketone bodies
gastr/o	stomach	kinesi/o	motion
gen/o	formation, cause, produce	kyph/o	hump
ger/o	old age	labyrinth/o	labyrinth, internal ear
geront/o	old age	lacrim/o	tear
gingiv/o	gums	lact/o	milk
gli/o	glue, neuroglia	lamin/o	thin, lamina
glomerul/o	little ball, glomerulus	lapar/o	abdomen, abdominal cavity
gloss/o	tongue	laryng/o	larynx
gluc/o	glucose, sugar	later/o	side
glut/o	buttock	lei/o	smooth
glyc/o	glycogen, sugar	leuc/o	white
glycos/o	sugar	lingu/o	tongue
gravid/o	pregnancy	lip/o	fat, lipid
gravidar/o	pregnancy	lith/o	stone
gyn/o	woman	lob/o	lobe
gynec/o	woman	lord/o	bent forward
halat/o	breath	lumb/o	loin, lower back
haem/o	blood	lymph/o	clear water or fluid
haemat/o	blood	lys/o	dissolution
hepat/o	liver	mal/o	bad
hern/o	protrusion, hernia	mamm/o	breast
herni/o	protrusion, hernia	mast/o	breast
heter/o	other	meat/o	opening
hidr/o	sweat	medi/o	middle
hist/o	tissue	megal/o	abnormally large

Combining form	Definition	Combining form	Definition
melan/o	dark, black	ovari/o	ovary
men/o	month, menstruation	ox/o	oxygen
mening/o	meninges, membrane	pachy/o	thick
menisc/o	cresent-shaped moon, meniscus membrane	palat/o	roof of mouth, palate
		pancreat/o	sweetbread, pancreas
menstru/o	month, menstruation	par/o	parturition or labour
ment/o	mind	parathyroid/o	parathyroid
metr/o	uterus	pariet/o	wall
mon/o	one	part/o	parturition or labour
muc/o	mucus	patell/o	small pan, patella
my/o	muscle	path/o	disease
myc/o	fungus	pector/o	chest
myel/o	bone marrow, spinal cord, medulla, myelin	ped/o	child
		pediatr/o	child
myelon/o	bone marrow, spinal cord, medulla, myelin	pelv/l	basin, pelvis
		pen/o	penis
myos/o	muscle	peps/o	digestion
myring/o	membrane, eardrum	perine/o	perineum
myx/o	mucus	peritone/o	to stretch over, peritoneum
nas/o	nose	petr/o	stone
nat/o	birth	phac/o	lens
natr/o	sodium	phag/o	eat, swallow
necr/o	death	phak/o	lens
nephr/o	kidney	phalang/o	row of soldiers
neur/o	sinew or cord, nerve, fascia	pharyng/o	pharynx (throat)
noct/i	night	phas/o	speech
nucl/o	kennel, nucleus	phleb/o	vein
nyct/o	night, nocturnal	phot/o	light
nyctal/o	night, nocturnal	physi/o	nature
obstetr/o	midwife, prenatal development	physis/o	growth
ocul/o	eye	plegi/o	paralysis
oesophag/o	gullet, oesophagus	pleur/o	pleura
olig/o	few in number	pneum/o	lung or air
omphal/o	umbilicus (navel)	pneumat/o	lung or air
onc/o	tumour	pod/o	foot
onych/o	nail	poikil/o	irregular
oophor/o	ovary	polyp/o	polyp
opt/o	eye, vision	poster/o	back
opthalm/o	eye	presby/o	old age
or/o	mouth	prim/i	first
orch/o	testis, testicle	proct/o	anus
orchi/o	testis, testicle	prostat/o	prostate gland
orchid/o	testis, testicle	proxim/o	near
organ/o	tool	pseud/o	false
orth/o	straight	psych/o	mind
oste/o	bone	pub/o	grown up
ot/o	ear	puerper/o	childbirth
ov/i	egg	pulmon/o	lung
ov/o	egg	py/o	pus

Combining form	Definition	Combining form	Definition
pyel/o	pelvis (renal)	synovi/o	binding eggs, synovial
pylor/o	pylorus	tars/o	flat surface
quad/o	four	taxi/o	reaction to a stimulus
quadr/i	four	ten/o	to stretch out, tendon
rachi/o	spine	tend/o	to stretch out, tendon
radi/o	spoke of a wheel, radius	tendin/o	to stretch out, tendon
radic/o	nerve root	test/o	testis, testicle
radicul/o	nerve root	testicul/o	small testis, testicle
rect/o	straight, erect, rectum	thalam/o	thalamus
ren/o	kidney	thel/o	nipple
retin/o	net, retina	therm/o	heat
rhabd/o	rod	thorac/o	thorax (chest)
rhin/o	nose	thromb/o	clot
rhytid/o	wrinkles	thym/o	wart-like, thymus gland
sacr/o	sacred, sacrum	thyr/o	shield, thyroid
salping/o	tube: Eustachian tube, fallopian tube	toc/o	birth, labour
		tom/o	cut, section
sarc/o	flesh, muscle	ton/o	tone, tension, pressure
scler/o	thick, hard, sclera	toxic/o	poison
scoli/o	curved	trache/o	trachea
seb/o	sebum, oil	trachel/o	neck, cervix
semin/o	seed	trich/o	hair
sept/o	wall, partition	tubercul/o	little mass of swelling
sial/o	saliva, salivary gland	tympan/o	eardrum
sigm/o	the letter 's', sigmoid colon	umbilic/o	navel
sinus/o	cavity	ur/o	urine
somat/o	body	ureter/o	ureter
somn/o	sleep	urether/o	urethra
son/o	sound	urin/o	urine
sperm/o	seed	uter/o	womb, uterus
spermat/o	seed	uvul/o	grape, uvula
sphygm/o	pulse	vagin/o	sheath, vagina
spin/o	spine or thorn	valvul/o	little valve
spir/o	breathe	varic/o	dilated vein
splen/o	spleen	vas/o	blood vessel, duct
spondyl/o	vertebra	vascul/o	little blood vessel
staped/o	stapes	ven/o	vein
staphyl/o	grape-like clusters (bacterium)	ventr/o	front, belly
stasis/o	standing still	ventricul/o	little belly or cavity, ventricle
steat/o	fat		
sten/o	narrowness, constriction	vers/o	turn
stern/o	chest, sternum	vertebr/o	joint, vertebra
steth/o	chest	vesic/o	bladder, sac
stigmat/o	point	vesicul/o	vesicle (seminal vesicle)
stomat/o	mouth	vitr/o	glassy
strept/o	twisted or gnarled (bacterium)	vitre/o	glassy
super/o	above	vulv/o	vulva
syndesm/o	binding together	xanth/o	yellow
synov/o	binding eggs, synovial	xer/o	dry

Suffix	Definition
-a	singular
-ac	pertaining to
-acusis	hearing condition
-ad	towards
-al	pertaining to
-algesia	pain
-algia	pain
-apheresis	removal
-ar	pertaining to
-ary	pertaining to
-asthenia	weakness
-atresia	closure, absence of a normal body opening
-capnia	carbon dioxide
-cele	hernia, swelling, protrusion
-centesis	surgical puncture
-clasia	break apart
-clasis	break apart
-clast	break apart
-crit	separate
-cusis	hearing condition
-cyte	cell
-desis	surgical fixation, fusion
-drome	run, running
-dynia	pain
-eal	pertaining to
-ectasis	expansion, dilation
-ectomy	excision
-elle	small
-emesis	vomiting
-emetic	vomiting
-emia	blood (conditon of)
-gen	formation, cause, produce
-genesis	origin, cause
-genic	pertaining to formation, causing, producing
-gram	recording
-graph	instrument for recording
-graphy	recording process
-haemia	blood (condition of)
-ia	diseased state (condition of)
-ial	pertaining to
-iasis	condition of
-iatry	treatment, specialty
-ic	pertaining to
-ion	pertaining to
-ior	pertaining to
-is	pertaining to

Suffix	Definition
-ism	condition or disease
-ist	one who practises
-itis	inflammation
-lepsy	seizure
-logist	one who studies
-logy	study of
-lytic, -lysis	to loosen, dissolve
-malacia	softening
-metre	measuring instrument
-metry	measurement
-oid	resemblance to
-oma	abnormal swelling, tumour
-opia	vision
-opsy	view of
-osis	process or condition that is usually abnormal
-otomy	cutting into, excision
-ous	pertaining to
-oxia	oxygen
-paresis	paralysis (minor)
-pathy	disease
-penia	abnormal reduction in number, deficiency
-pepsia	digestion
-pexy	surgical fixation, suspension
-phagia	eating or swallowing
-phasia	speaking
-phil	loving, affinity for
-philia	loving, affinity for
-phobia	fear
-phonia	sound or voice
-phylaxis	protection
-physis	growth
-plasia	shape, formation
-plasm	something shaped
-plasty	surgical repair
-plegia	paralysis (major)
-pnoea	breathing
-poiesis	formation
-practic	one who practises
-ptosis	falling downward (condition of)
-ptysis	spit out a fluid
-rrhagia	bleeding, haemorrhage
-rrhaphy	suturing
-rrhea	excessive discharge
-rrhexis	rupture
-salpinx	trumpet, Fallopian tube
-sarcoma	malignant tumour

Suffix	Definition	Suffix	Definition
-schisis	split, fissure	-stomy	surgical creation of an opening
-sclerosis	hardening	-tic	pertaining to
-scope	viewing instrument	-tocia	birth, labour
-scopy	process of viewing	-tome	cutting instrument
-sis	state of	-tomy	incision
-some	body	-tripsy	surgical crushing
-spasm	sudden, involuntary muscle contraction	-trophy	nourishment, development
		-um	pertaining to
-stasis	standing still	-uria	urine, urination
-stenosis	narrowing, constriction	-y	process of

Endings in medical terminology

1. **Plural endings**. The following list provides a summary of plural endings that are in common use with medical terms. Examples are provided to demonstrate how these endings are applied.

Endings singular	Plural	Examples singular	Plural
-a	-ae	fistula	fistulae
-ax	-aces	haemothorax	haemothoraces
-ex	-ices	cortex	cortices
-is	-es	mastoiditis	mastoidites
-ix	-ices	cicatrix	cicatrices
-ma	-mata	fibroma	fibromata
-on	-a	contusion	contusia
-um	-a	bacterium	bacteria
-us	-i	fungus	fungi
-y	-ies	episiotomy	episiotomies

2. **Adjective endings**. The list below provides a summary of suffixes that mean 'pertaining to' and form an adjective (a description of a noun) when combined with a root.

Ending	Example	Definition
-ac	cardiac	pertaining to the heart
-al	endotracheal	pertaining to within the trachea
-ar	submandibular	pertaining to below the mandible
-ary	pulmonary	pertaining to a lung
-eal	oesophageal	pertaining to the oesophagus
-ic	leukaemic	pertaining to leukaemia
-ous	fibrous	pertaining to fibre
-tic	cyanotic	pertaining to cyanotic (blue)

3. **Diminutive endings**. The endings listed below provide the meaning of 'small' to the word of origin.

Ending	Example	Definition
-icle	ossicle	small bone
-ole	bronchiole	small bronchus (airway)
-ula	macula	small macule (spot)
-ule	pustule	small pimple

4. **Diagnostic endings**. The endings in this list summarise the suffixes that are in common use to indicate measurements, treatments and procedures.

Ending	Meaning	Example	Definition
-gram	record	bronchogram	recording of bronchus image
-graph	recording instrument	sonography	ultrasound instrument
-graphy	process of recording	echocardiography	procedure of heart recording
-iatrics	treatment	paediatrics	treatment of children
-iatry	treatment	psychiatry	treatment of the mind
-ist	one who specialises	optometrist	specialist in eye measurement
-logist	one who studies	audiologist	one who studies hearing
-logy	study of	oncology	study of cancer
-metre	instrument of measure	spirometre	instrument for measuring breathing
-metry	process of measuring	spirometry	process of measuring breathing
-scope	instrument for exam	endoscope	instrument for examination within
-scopy	examination	endoscopy	examination within

Appendix C

Clinical abbreviations

These are the currently acceptable abbreviations; however, different health care facilities may use others as well.

Abbreviation	Meaning	Abbreviation	Meaning
#	fracture	ANS	autonomic nervous system
@	at	ante	before
ā	before	AP	anteroposterior
AB	abortion	aq	aqueous (water)
ABGs	arterial blood gases	ARC	AIDS-related complex
ac	before meals	ARD	acute respiratory disease
ACAT	automated computerised axial tomography	ARDS	adult respiratory distress syndrome
Acc	accommodation	ARF	acute respiratory failure, acute renal failure
ACL	anterior cruciate ligament		
ACTH	adrenocorticotropic hormone	ARMD	age-related macular degeneration
ad lib	as desired	AROM	active range of motion
ADD	attention deficit disorder	AS	aortic stenosis
ADH	antidiuretic hormone	ASCVD	arteriosclerotic cardiovascular disease
ADHD	attention-deficit-hyperactivity disorder		
		ASD	atrial septal defect
ADL	activities of daily living	ASHD	arteriosclerotic heart disease
AE	above elbow	AST	aspartate transaminase
AF	atrial fibrillation	Astigm.	astigmatism
AGN	acute glomerulonephritis	ATN	acute tubular necrosis
AI	artificial insemination	AU	both ears
AIDS	acquired immunodeficiency syndrome	AV, A-V	atrioventricular
		Ba	barium
AK	above knee	BaE	barium enema
ALL	acute lymphocytic leukaemia	BBB	bundle branch block (L for left; R for right)
ALS	amyotropic lateral sclerosis	BC	bone conduction
ALT	alanine transaminase	BCC	basal cell carcinoma
am, AM	morning	bd	twice a day
AMI	acute myocardial infarction	BDT	bone density testing
AML	acute myelogenous leukaemia	BE	barium enema, below elbow
		BK	below knee
amt	amount	BM	bowel movement
Angio	angiography	BMR	basal metabolic rate

Abbreviation	Meaning	Abbreviation	Meaning
BMT	bone marrow transplant	CoA	coarctation of the aorta
BNO	bladder neck obstruction	COLD	chronic obstructive lung disease
BP	blood pressure		
BPD	bipolar disorder	COPD	chronic obstructive pulmonary disease
BPE	benign prostate enlargement		
BPH	benign prostatic hypertrophy	CP	cerebral palsy, chest pain
bpm	beats per minute	CPD	cephalopelvic disproportion
Bronch	bronchoscopy	CPK	creatine phosphokinase
BS	bowel sounds	CPR	cardiopulmonary resuscitation
BSE	breast self-examination		
BUN	blood urea nitrogen	CRF	chronic renal failure
BX, bx	biopsy	C & S	culture and sensitivity test
c̄	with	CSD	congenital septal defect
C	100	C-section	Caesarean section
C1, C2, etc.	first cervical vertebra, second cervical vertebra, etc.	CSF	cerebrospinal fluid
		CT	computerised tomography, clear to auscultation
Ca^{2+}	calcium	CTA	
CA	cancer, chronological age	CTS	carpal tunnel syndrome
CABG	coronary artery bypass graft	CUC	chronic ulcerative colitis
CAD	coronary artery disease	CV	cardiovascular
cap(s)	capsule(s)	CVA	cerebrovascular accident
CAPD	continuous ambulatory peritoneal dialysis	CVD	cerebrovascular disease
		CVS	chorionic villus biopsy
CAT	computerised axial tomography	Cx	cervix
		CXR	chest X-ray
CBD	common bile duct	cysto	cystoscopic exam
CC	clean catch urine specimen, cardiac catheterisation, chief complaint	D	day
		D & C	dilation and curettage
		dB	decibel
CCU	cardiac care unit, coronary care unit	Derm, derm	dermatology
		DI	diabetes insipidus
CD4	protein on T-cell helper lymphocyte	diff	differential
		dil	dilute
CDH	congenital dislocation of the hip	disc	discontinue
		disp	dispense
c.gl.	correction with glasses	DJD	degenerative joint disease
CGL	chronic granulocytic leukaemia	DM	diabetes mellitus
		DOB	date of birth
chemo	chemotherapy	DOE	dyspnoea on exertion
CHF	congestive heart failure	DPT	diphtheria, pertussis, tetanus
chol	cholesterol	dr	dram
CIS	carcinoma in situ	DRE	digital rectal exam
Cl^-	chloride	DSA	digital subtraction angiography
CLL	chronic lymphocytic leukaemia	DSM-IV	*Diagnostic and Statistical Manual for Mental Disorders*, fourth edition
CML	chronic myelogenous leukaemia		
CNS	central nervous system	DTR	deep tendon reflex
CO_2	carbon dioxide	DUB	dysfunctional uterine bleeding

Abbreviation	Meaning
DVA	distance visual acuity
DVT	deep vein thrombosis
Dx	diagnosis
E. coli	*Escherichia coli*
EAU	exam under anaesthesia
EBV	Epstein-Barr virus
ECC	endocervical curettage, extracorporeal circulation
ECCE	extracapsular cataract extraction
ECG	electrocardiogram
Echo	echocardiogram
ECT	electroconvulsive therapy
ED	erectile dysfunction
EDC	estimated date of confinement
EEG	electroencephalogram, electroencephalography
ELISA	enzyme-linked immunosorbent assay
elix	elixir
EM	emmetropia (normal vision)
EMB	endometrial biopsy
EMG	electromyogram
emul	emulsion
Endo	endoscopy
ENT	ear, nose, throat
EOM	extraocular movement
ERCP	endoscopic retrograde cholangiopancreatography
ERV	expiratory reserve volume
ESR	erythrocyte sedimentation rate
ESRD	end-stage renal disease
e-stim	electrical stimulation
ESWL	extracorporeal shock-wave lithotripsy
et	and
ET	endotracheal
FBC	full blood count
Fe	iron
FEF	forced expiratory flow
FEV	forced expiratory volume
FEVR	forced expiratory volume rate
FHR	fetal heart rate
FHT	fetal heart tone
fl	fluid
FOBT	faecal occult blood test
FRC	functional residual capacity
FS	frozen section
FSH	follicle-stimulating hormone
FTND	full-term normal delivery
FVC	forced vital capacity
GA	general anaesthesia
GB	gallbladder
GERD	gastroesophageal reflux disease
GH	growth hormone
GI	gastrointestinal
gm	gram
GOT	glutamic oxaloacetic transaminase
gr	grain
grav I	first pregnancy
gt	drop
gtt	drops
GTT	glucose tolerance test
GU	genitourinary medicine
GVHD	graft vs. host disease
GYN, gyn	gynaecology
H_2O	water
HAV	hepatitis A virus
Hb	haemoglobin
HBOT	hyperbaric oxygen therapy
HBV	hepatitis B virus
HCG, hCG	human chorionic gonadotropin
HCO_3^-	bicarbonate
HCT, Hct	haematocrit
HCV	hepatitis C virus
HD	Hodgkin's disease
HDL	high-density lipoproteins
HDN	haemolytic disease of the newborn
HIV	human immunodeficiency virus
HMD	hyaline membrane disease
HNP	herniated nucleus polposus
HPV	human papilloma virus
HRT	hormone replacement therapy
hs	hour of sleep
HSG	hysterosalpingography
HSV	*Herpes simplex* virus
HZ	Hertz
$\dot{I}$	one
IBD	inflammatory bowel disease

Abbreviation	Meaning	Abbreviation	Meaning
IBS	irritable bowel syndrome	LL	left lateral
IC	inspiratory capacity	LLE	left lower extremity
ICCE	intracapsular cataract cryoextraction	LLL	left lower lobe
ICP	intracranial pressure	LLQ	left lower quadrant
ICU	intensive care unit	LMP	last menstrual period
I & D	incision and drainage	LP	lumbar puncture
ID	intradermal	LUE	left upper extremity
Ig	immunoglobins (IgA, IgD, IgE, IgG, IgM)	LUL	left upper lobe
		LUQ	left upper quadrant
ii	two	LVAD	left ventricular assist device
iii	three	LVH	left ventricular hypertrophy
IM	intramuscular	lymphs	lymphocyte
inj	injection	mA	milliampere
I & O	intake and output	MA	mental age
IOL	intraocular lens	MAO	monoamine oxidase
IOP	intraocular pressure	mcg	microgram
IPD	intermittent peritoneal dialysis	mCi	millicurie
		MCV	mean corpuscular volume
IPPB	intermittent positive pressure breathing	mEq	milliequivalent
		mets	metastases
IRDS	infant respiratory distress syndrome	mg	milligram
		MI	myocardial infarction, mitral insufficiency
IRV	inspiratory reserve volume	ml	millilitre
IU	international unit	mm	millimetre
IUD	intrauterine device	MM	malignant melanoma
IV	intravenous	mmHg	millimetres of mercury
IVC	intravenous cholangiogram	mmol	millimoles
IVF	*in vitro* fertilisation	Mono	mononucleosis
IVP	intravenous pyelogram	monos	monocyte
JVP	jugular venous pulse	MR	mitral regurgitation
K^+	potassium	MRA	magnetic resonance angiography
kg	kilogram		
KS	Kaposi's sarcoma	MRI	magnetic resonance imaging
KUB	kidney, ureter, bladder	MS	mitral stenosis, multiple sclerosis, musculoskeletal
L	left		
l	litre	MSH	melanocyte-stimulating hormone
L1, L2, etc.	first lumbar vertebra, second lumbar vertebra, etc.		
		MUA	manipulation under anaesthesia
LAT, lat	lateral		
LAVH	laparoscopic-assisted vaginal hysterectomy	MV	minute volume
		MVP	mitral valve prolapse
LBW	low birth weight	n & v	nausea and vomiting
LDH	lactate dehydrogenase	Na^+	sodium
LDL	low-density lipoproteins	NG	nasogastric (tube)
LE	lower extremity	NGU	nongonococcal urethritis
LGI	lower gastrointestinal series	NHL	non-Hodgkin's lymphoma
LH	luteinising hormone	NK	natural killer cells
liq	liquid	NMR	nuclear magnetic resonance

Abbreviation	Meaning
no sub	no substitute
noc	night
non rep	do not repeat
NPDL	nodular, poorly differentiated lymphocytes
NPO	nothing by mouth
NSAID	nonsteroidal anti-inflammatory drug
NSR	normal sinus rhythm
O$_2$	oxygen
OA	osteoarthritis
OB	obstetrics
OCD	obsessive-compulsive disorder
OCPs	oral contraceptive pills
OD	overdose
oint.	ointment
OM	otitis media
O & P	ova and parasites
Ophth.	ophthalmology
ORIF	open reduction-internal fixation
Orth, ortho	orthopaedics
OS	left eye
OT	occupational therapy, occupational therapist
OTC	over the counter
Oto	otology
OU	each eye
oz	ounce
p̄	after
P	pulse
PI	first delivery
PAC	premature atrial contraction
PAP	Papanicolaou test, pulmonary arterial pressure
para I	first delivery
PBI	protein-bound iodine
pc	after meals
PCA	patient-controlled administration
PCP	Pneumocystis carinii pneumonia
PCV	packed cell volume
PDA	patent ductus arteriosus
PEG	percutaneous endoscopic gastrostomy
per	with

Abbreviation	Meaning
PERRLA	pupil equal, round, react to light and accommodation
PET	positron emission tomography
PFT	pulmonary function test
pH	acidity or alkalinity of urine
PID	pelvic inflammatory disease
PKU	phenylketonuria
PM, pm	evening
PMNs	polymorphonuclear neutrophil
PMP	previous menstrual period
PMS	premenstrual syndrome
PND	paroxysmal nocturnal dyspnoea, postnasal drip
PNS	peripheral nervous system
PO, po	by mouth
polys	polymorphonuclear neutrophil
pp	postprandial (after meals)
PPD	purified protein derivative (tuberculin test)
preop, pre-op	preoperative
prep	preparation, prepared
PRK	photo refractive keratectomy
PRL	prolactin
prn	as needed
PSA	prostate specific antigen
pt	pint
PT	prothrombin time
PTC	percutaneous transhepatic cholangiography
PTCA	percutaneous transluminal coronary angioplasty
PTH	parathyroid hormone
PUD	peptic ulcer disease
PVC	premature ventricular contraction
q̄	every
qam	every morning
qd	once a day, every day
qds	four times a day
qh	every hour
qhs	every night
qod	every other day
qs	quantity sufficient
Ra	radium
RA	rheumatoid arthritis
rad	radiation absorbed dose

Abbreviation	Meaning	Abbreviation	Meaning
RAI	radioactive iodine	SMD	senile macular degeneration
RAIU	radioactive iodine uptake	SOB	shortness of breath
RBC	red blood cell	sol	solution
RD	respiratory disease	SOM	serous otitis media
RDA	recommended daily allowance	ss	one-half
		st	stage
RDS	respiratory distress syndrome	ST	skin test
REM	rapid eye movement	stat, STAT	at once, immediately
Rh-	Rh-negative	STD	sexually transmitted disease
Rh⁺	Rh-positive	strep.	streptokinase
RIA	radioimmunoassay	STSG	split-thickness skin graft
RL	right lateral	subcut	subcutaneous
RLE	right lower extremity	supp.	suppository
RLL	right lower lobe	susp	suspension
RLQ	right lower quadrant	syr	syrup
RML	right mediolateral, right middle lobe	T1, T2, etc.	first thoracic vertebra, second thoracic vertebra, etc.
RN	registered nurse		
ROM	range of motion	T_3	triiodothyronine
RP	retrograde pyelogram	T_4	thyroxine
RPR	rapid plasma reagin (test for syphilis)	T_7	free thyroxine index
		tab	tablet
RR	respiratory rate	TAH-BSO	total abdominal hysterectomy-bilateral salpingo-oophorectomy
RUE	right upper extremity		
RUL	right upper lobe		
RUQ	right upper quadrant	TB	tuberculosis
RV	reserve volume	tds	three times a day
Rx	take	TENS	transcutaneous electrical nerve stimulation
s̄	without		
S1	first heart sound	TFT	thyroid function test
S2	second heart sound	THA	total hip arthroplasty
SA, S-A	sinoatrial	THR	total hip replacement
SAD	seasonal affective disorder	TIA	transient ischaemic attack
SAH	subarachnoid haemorrhage	TKA	total knee arthroplasty
SBFT	small bowel follow-through	TKR	total knee replacement
SC, sc	subcutaneous	TLC	total lung capacity
SCC	squamous cell carcinoma	TMJ	temporomandibular joint
SCI	spinal cord injury	TNM	tumour, nodes, metastases
SCIDS	severe combined immunodeficiency syndrome	top	apply topically
		tPA	tissue-type plasminogen activator
SG	specific gravity		
SGOT	serum glutamic oxaloacetic transaminase	TPN	total parenteral nutrition
		TPR	temperature, pulse and respiration
SIDS	sudden infant death syndrome		
		TSH	thyroid-stimulating hormone
s/l	sub-lingual (under the tongue)	TSS	toxic shock syndrome
		TUR	transurethral resection
SLE	systemic lupus erythematosus	TURP	transurethral resection of prostate

Abbreviation	Meaning	Abbreviation	Meaning
TV	tidal volume	USS	ultrasound scan
TX, Tx	traction, treatment	UTI	urinary tract infection
UC	uterine contractions, urine culture	UV	ultraviolet
		VA	visual acuity
UE	upper extremity	VC	vital capacity
UGI	upper gastrointestinal, upper gastrointestinal series	VCUG	voiding cystourethrogram
		VD	venereal disease
ung	ointment	VF	ventricular fibrillation
URI	upper respiratory infection	VLDL	very low density lipoproteins
URTI	upper respiratory tract infection	VSD	ventricular septal defect
		VT	ventricular tachycardia
US	ultrasound	WBC	white blood cell

Appendix D

Laboratory reference values

Haematology-tests abbreviations used in reporting laboratory values

cm^3	cubic centimetre
cu μ	cubic microns
dl	decilitre
fl	femtolitre
g	gram
g/dl	grams per decilitre
IU	International Unit
kg	kilogram
l	litre
mol (M)	mole
mEq	milliequivalent
mg	milligram
mg/dl	milligram per decilitre
mm	millimetre
mmol	millimole
mm^3	cubic millimetre
mm Hg	millimetre of mercury
ng	nanogram
ng/dl	nanogram per decilitre
ng/ml	nanogram per millilitre
pg	picogram
U	unit
U/l	units per litre
uIU/ml	units International Unit per millilitre
μg (mcg)	microgram

Haematology tests

Normal ranges (these values may vary between hospital laboratories)

Erythrocytes – red blood cells (RBCs)
   Females 4.2–5.4 million/mm^3
   Males 4.6–6.2 million/mm^3
   Children 4.5–5.1 million/mm^3

Haemoglobin (HGB, Hgb)
   Females 10–12 g/dl
   Males 12–14 g/dl
Haematocrit (HCT) 37.0–54%
   Females 37–47%
   Males 40–54%
Leucocytes – white blood cells (WBCs)
4500–11,000/mm^3
Differential
   Neutrophils 54–62%
   Lymphocytes 20–40%
   Monocytes 2–10%
   Oesinophils 1–2%
   Basophils 0–1%
Thrombocytes (platelets) 200,000–400,000/mm^3

Coagulation tests

Bleeding time 2.75–8.0 min
Prothrombin time (PT) 12–14 sec
Partial thromboplastin time (PTT) 30–45 sec

Chemistries

Normal ranges

Adrenocorticotrophic hormone (ACTH)
(9.00 am) <50 ng/l
Alanine aminotransferase (ALT, SGPT)
10–40 U/l
Albumin 35–50 g/l
Alkaline phosphatase (ALP) 30–150 U/l
Amylase <200 U/L
Aspartate aminotransferase (AST, SGOT)
10–40 U/l
Bicarbonate 22–28 mmol/l
Bilirubin <20 μmol/l
Blood urea nitrogen (BUN) 2.5–6.0 mmol/l
C-reactive protein (CRP) <10 mg/l
Calcium (total) 2.20–2.60 mmol/l

Calcium (ionised) 1.15–1.30 mmol/l
Chloride (Cl) 100–108 mmol/l
Cholesterol <5.0 mmol/l
   High density lipoprotein (HDL) >1.2 mmol/l
   Low density lipoprotein (LDL) <3 mmol/l
PCO_2 4.7–6.7 kPa
Creatine kinase (alternative name is creatine phosphokinase, CPK) 4.0–9.0 ng/ml
Creatinine 55–105 µmol/l
Erythrocyte sedimentation rate (ESR)
Male 1–10 mm/hr
Female 5–20 mm/hr
Ferritin 10–300 µmol/l
Folate 150–700 µg/l
Free T_4 9–26 pmol/l
Glucose (fasting) 3.5–8 mmol/l
Random blood glucose 3.5–8 mmol/l
Haemoglobin: Male – 13–17 g/dl
            Female – 11–15 g/dl
Iron (Fe) 10–30 µmol/l
Magnesium 0.7–1.0 mmol/l
Mean Cell Volume (MCV) 80–95 fl
pH 7.35–7.45
Packed cell volume (PCV)/Haematocrit:
Male 40–52%
Female 36–48%
Parathyroid hormone (PTH) 10–70 ng/l
Phosphate (PO_4) 0.80–1.40 mmol/l
Potassium (K) 3.5–5.0 mmol/l
Prostate specific antigen (PSA) male 0.0–4.0 ng/ml
Sodium (Na) 135–145 mmol/l
Testosterone 9–27 mmol/l
Thyroid-stimulating hormone (TSH) 0.3–4.5 mU/l
Thyroxine (T_4) 9–26 pmol/l
Triiodothyroxine – free (FT 3) 3.0–9.0 pmol/l
Triglycerides (fasting) 0.45–1.80 mmol/l
Urea 2.5–6.5 mmol/l
Vitamin B_{12} 150–1000 ng/l
White Cell Count – Male 3.7–9.5 × 10^9/l
            Female 3.9–6.5 × 10^9/l

Urinalysis

	Normal ranges
Colour	Yellow to amber
Turbidity (appearance)	Clear to slightly hazy
Specific gravity	1.001–1.035
Reaction (pH)	5.0–7.0
Odour	Faintly aromatic
Output	1–2 litres per day
Protein	Negative
Glucose	Negative
Ketones	Negative
Bilirubin	Negative
Blood	Negative
Urobilinogen	0.1–1.0
Nitrite	Negative
Leucocytes	Negative

Normal values of arterial blood gases*

pH 7.35–7.45 Adults; Neonates – 7.31–7.47
PO_2 Adults: 10.6–13.3 kPa (80–100 mmHg)
        Neonates: 4.3–8.1 kPa (32–61 mmHg)
PCO_2 Adults: 4.7–6.0 kPa (35–45 mmHg)
        Neonates: 3.8–6.5 kPa (28–49 mmHg)
HCO_3 Adults 22–28 mmol/l;
        Neonates: 15–25 mmol/l
Base excess −2.0 to +2.0
O_2 saturation 95–100%
Critical values in adults:
pH – <7.2 or >7.6
PCO_2 – <2.7 kPa or >9.3 kPa
PO_2 – <5.3 kPa
Bicarbonate <10 mmol/l or >40 mmol/l

* Some normal values will vary according to the kind of test carried out in the laboratory.

Most common blood chemistries and examples of disorders they indicate

Test	Abbreviation	Normal range	Examples of possible diagnosis	
			Results increased	Results decreased
Alkaline phosphate	ALP	30–150 U/l	Liver disease, bone disease, mononucleosis	Malnutrition, hypothyroidism, chronic nephritis
Blood urea nitrogen	BUN	2.5–6.0 mmol/l	Kidney disease, dehydration, GI bleeding	Liver failure, malnutrition
Calcium	Ca^{2+}	2.20–2.60 mmol/l	Hypercalcaemia, bone metastases, Hodgkin's disease	Hypocalcaemia, renal failure, pancreatitis
Chloride	Cl	100–108 mmol/l	Dehydration, eclampsia, anaemia	Ulcerative colitis, burns, heat exhaustion
Cholesterol	CHOL	<5.0 mmol/l	Atherosclerosis, nephrosis, obstructive jaundice	Malabsorption, liver disease, hyperthyroidism
Creatinine	Creat	55–105 μmol/l	Chronic nephritis, muscle disease, obstruction of urinary tract	Muscular dystrophy
Globulin	Glob	1.0–3.5 g/dl	Brucellosis, rheumatoid arthritis, hepatic carcinoma	Severe burns
Glucose fasting blood sugar	FBS	3.5–8 mmol/l	Diabetes mellitus	Excess insulin
Lactic acid	LDH	100–225 mU/ml	Acute MI, acute leukaemia, hepatic disease	
Potassium	K+	3.5–5.0 mmol/l	Renal failure, acidosis, cell damage	Malabsorption, severe burns, diarrhoea
Serum glutamic oxaloacetic transaminase	SGOT	0–41 mU/ml	MI, liver disease, pancreatitis	Uncontrolled diabetes mellitus with acidosis
Serum glutamic pyruvic transaminase	SGPT	0–45 mU/ml	Active cirrhosis, pancreatitis, obstructive jaundice	
Sodium	Na+	135–145 mmol/l	Diabetes insipitus, coma, Cushing's disease	Severe diarrhoea, severe nephritis, vomiting syndrome
Free thyroxine	T4	9–26 pmol/l	Thyroiditis, hyperthyroidism, Graves disease	Goitre, myledoema, hypothyroidism
Total bilirubin	TB	<17 μmol/l	Liver disease, haemolytic anaemia, lupus erythemia	
Triglycerides	TRIG	0.45–1.80 mmol/l	Liver disease, atherosclerosis, pancreatitis	Malnutrition
Uric acid	UA	2.2–9.0 mg/dl	Renal failure, gout, leukaemia, eclampsia	

Glossary

abdominopelvic cavity (*abb DOM in oh PELL vick CAV ih tee*) Continuous cavity within the abdomen and pelvis that contains the largest organs of the gastrointestinal system.

abduction (*abb DUCK shun*) Moving a body part away from the midline of the body.

absorption (*ab SORP shun*) Process by which digested food nutrients move through villi of the small intestine into the bloodstream or substances move from the kidney tubules into the bloodstream.

accessory muscles Muscles in the neck, chest and abdomen that can be used, if necessary, to help expand the thoracic cavity on inspiration.

acetylcholine (*uh SEAT isle COE leen*) Neurotransmitter between neurons in the brain and spinal cord, also between a neuron and a voluntary skeletal muscle.

acidosis (*a SID oh sis*) pH of arterial blood less than 7.35 or a hydrogen concentration of > 45 nmol/l.

actin (*act IN*) One of the two proteins in muscle fibres needed for contraction.

action potential The change of the electrical charge of a nerve or muscle fibre when stimulated. Action potentials are all or none.

active transport The movement of cellular material that requires energy.

adaptation The adjustment of an organism to changing environments.

adduction (*ad DUCK shun*) Moving a body part toward the midline of the body.

adenoid (*AD eh oid*) Lymphoid tissue in the superior part of the nasopharynx. Also known as the pharyngeal tonsils.

adenosine diphosphate (ADP) (*ah DEN oh seen try FOSS fate*) The compound that can be converted to ATP for energy storage. When ATP is broken down to ADP, energy is released that can be used for cellular energy.

adenosine triphosphate (ATP) (*ah DEN oh seen try FOSS fate*) Energy storage molecule used to power muscle contraction and other cellular reactions.

adrenal (*ad REE nal*) Referring to the adrenal gland.

adrenal cortex (*ad REE nal KOR teks*) Outermost part of the adrenal gland. It produces and secretes three groups of hormones: mineralocorticoids (aldosterone), glucocorticoids (cortisol) and some androgens (male hormones).

adrenal medulla (*ad REE nal meh DULL lah*) Innermost part of the adrenal gland. It produces and secretes the hormones epinephrine and norepinephrine.

adventitia (*add ven TISH ah*) The outmost covering of a structure or organ.

aetiology (*ee tee YOLL oh jee*) The cause or origin of a disease.

afferent arterioles (*AFF er ent are TIER ee ohlz*) Small arteries travelling toward an organ.

agglutinate (*ah GLUE tin ate*) To clump.

agonist (*AGG on ist*) The muscle that contracts while another muscle relaxes at the same time to cause movement.

albumin (*ALB you men*) Most abundant plasma protein.

aldosterone (*al DOSS ter own*) Most abundant and biologically active of the mineralocorticoid hormones secreted by the adrenal cortex. It regulates the balance of electrolytes, keeping sodium (and water) in the blood while excreting potassium in the urine.

alimentary tract (*al eh MEN tar ee*) Alternative name for the gastrointestinal system.

alkalosis A pH > 7.45 or a hydrogen ion concentration less than 35 nmol/l.

alveoli (*al VEE oh lye*) Hollow spheres of cells in the lungs where oxygen and carbon dioxide are exchanged.

amblyopia (*am bly OH pee ah*) To prevent double vision, the brain ignores visual images from a misaligned eye (strabismus). Also known as lazy eye.

anabolism (*an AH bow lizm*) Assembly of new molecules in the body.

anaemia (*ah NEE mee ah*) Any condition in which the number of erythrocytes in the blood is decreased.

anaesthesia (*an ess THEE zee ah*) Condition in which sensation of any type, including touch, pressure, proprioception, or pain, has been completely lost.

anastomosis (*an AST owe mow sees*) The suturing of one blood vessel to another.

anatomical position (*an a TOM ickle*) This is a standard position in which the body is standing erect, the head is up with the eyes looking forward, the arms are by the sides with the palms facing forward, and the legs are straight with the toes pointing forward.

anatomy (*ANNA toe me*) The study of the structures of the human body.

aneurysm (*AN you rizm*) Area of dilation and weakness in the wall of an artery. This can be congenital or where arteriosclerosis has damaged the artery. With each heartbeat, the weakened artery wall balloons outward. Aneurysms can rupture without warning. A dissecting aneurysm is one that is enlarging by tunnelling between the layers of artery wall.

antagonists (*an TAH gon ists*) Something that does the opposite of the agonist.

antebrachial (*an tee BRAY kee all*) The area of the forearm.

antecubital (*an tee KEW bit all*) The area in front of the elbow.

anterior (*an TEE ree or*) The front of the body.

anterior commissure (*an TEE ree ur KOM i sure*) A nerve pathway in the anterior part of the brain that allows communication between the right and left halves of the cerebrum.

antibody (*AN tea bod ee*) Proteins secreted by B lymphocytes (plasma cells) that attack infected cells.

antibody-mediated immunity (*AN tea bod ee*) Immunity that results from a formation of antibodies in response to antigens.

antidiuretic hormone (ADH) (*AN tea dye you RET ik*) Hormone secreted by the posterior pituitary gland. It stimulates the kidneys to move water back into the blood to increase the volume of blood.

antigen-displaying cells (ADCs) (*AN tea jenns*) Cells that engulf antigens and display them as markers.

antigens (*AN tea jenns*) Substance that causes a formation of antibodies. Cell surface markers that help the immune system identify cells.

antiseptic (*AN tea SEP tik*) A substance capable of destroying micro-organisms.

anus (*AYE nuss*) External opening of rectum, located between the buttocks. The external anal sphincter is under voluntary control. The perianal area is around the anus.

anvil One of the three small bones in the ear.

apex (*AY pecks*) The rounded top of each lung and the gently rounded tip of the outer surface of the heart. The ventricles lie beneath the apex.

apocrine glands (*APP oh creen*) Sweat glands found in the pubic and axillary regions which open into the hair follicles.

appendicitis (*ah pen dee SIGH tiss*) Inflammation and infection of the appendix.

appendicular skeleton (*app en DICK you lar SKELL it on*) The bones of the shoulders, arms, hips and legs.

appendix (*ah pen DICKS*) Long, thin pouch on the exterior wall of the caecum. It does not play a role in digestion. It contains lymphatic tissue.

aqueous humour (*AY kwee uss HUW mer*) Clear, watery fluid produced by the ciliary body. It circulates through the posterior and anterior chambers and takes nutrients and oxygen to the cornea and lens.

arachnoid mater (*ah RAK noyd MAY ter*) Web-like connective tissues that attach to the pia mater. The space under the arachnoid mater (subarachnoid space) contains cerebrospinal fluid.

arteries (*ahr TER rees*) Blood vessels that carry blood away from the heart.

arteriole (*ar TEE ree oll*) Smallest branch of an artery.

arteriosclerosis (*ar tee ree oh skler OH sis*) Progressive degenerative changes that produce a narrowed, hardened artery.

arthritis (*arth RYE tiss*) Inflammation of the joints.

articulation (*AR tick yoo lay shun*) A joint where two bones come together and join or articulate.

artificial active immunity Vaccination. Intentional exposure to pathogens so patient makes own antibodies.

artificial passive immunity Injection of antibodies to help patient fight infection.

asthma (*ASS ma*) Sudden onset of hyperreactivity of the bronchi and bronchioles with bronchospasm (contraction of the smooth muscle). Inflammation and swelling severely narrow the lumen of the airways.

astrocyte (*AST row SIGHT*) Star-shaped cell that provides structural support for neurons, connects them to capillaries, and forms the blood–brain barrier.

ataxia (*ah TAX ee uh*) Lack of coordination of the muscles during movement, particularly the gait. Caused by diseases of the brain or spinal cord, cerebral palsy, or an adverse reaction to a drug.

atelectasis (*at eh LEK ta siss*) Incomplete expansion or collapse of part or all of a lung due to mucus, tumour or a foreign body that blocks the bronchus. The lung is said to be atelectatic. Also known as collapsed lung.

atherosclerosis (*ath er oh scler ROW sis*) Hardening of the arteries as a result of plaque build-up in the lumen of the arteries.

atrial natriuretic peptide (*AY tree all NAY tree your RET ick PEP tide*) Hormone released by the atria when blood pressure rises. Cause increased excretion of water by the kidney, thereby decreasing blood pressure.

atrioventricular node (*ay tree oh vehn TRIK yoo lhar*) Small knot of tissue located between the right atrium and right ventricle. The AV node is part of the conduction system of the heart and receives electrical impulses from the SA node.

atrioventricular valve (*ay tree oh vehn TRIK oo lar*) The valve situated between the atrium and the ventricle.

atrium, atria (*AY tree um, AY tree ah*) Two upper chambers of the heart. Intra-atrial structures are located within the atria.

atrophy (*AT row fee*) Loss of muscle bulk in one or more muscles. It can be caused by malnutrition or

can occur in any part of the body that is paralysed and the muscles receive no electrical impulse from the nerves.

auditory canal (*AW dih tor ee kah NAL*) One of two canals that lead to the ear.

auricle (*AW rih kl*) The visible external ear. Also known as the pinna.

autonomic nervous system Division of the peripheral nervous system that carries nerve impulses to the heart, involuntary smooth muscles and glands. It includes the parasympathetic nervous system (active during sleep and light activity) and the sympathetic nervous system (active during increased activity, danger and stress).

autorhythmicity (*aw to rith MISS city*) The heart's ability to generate its own stimulus.

axial skeleton (*AK seal SKELL it on*) The bones of the head, chest and back.

axillary (*ak SILL are ee*) Pertains to the area of the armpit.

axon (*AK son*) Part of the neuron that is a single, elongated branch at the opposite end from the dendrites. It receives an electrical impulse and releases neurotransmitters into the synapse. Axons are covered by an insulating layer of myelin.

B lymphocytes (*B LIMF oh sights*) White blood cells which make antibodies to destroy specific pathogens.

bacteria (*back TEER ree yah*) Asexually reproducing cell capable of creating an infection.

basal nuclei (*BAY sal new KLEE ie*) Clusters of cells (nuclei) deep in the diencephalon, midbrain and cerebrum which help fine-tune voluntary movements.

base The bottom of each lung.

basophil (*BAY so fill*) Least numerous of the leucocytes. It is classified as a granulocyte because granules in its cytoplasm stain dark blue to purple with basic dye. It releases histamine and heparin at the site of tissue damage.

benign (*ben INE*) Not progressive, non malignant.

bicuspid (*bye CUSS pid*) A two-leafed structure.

blood platelets Blood cells responsible for clotting. Also known as thrombocytes.

bolus (*BOW luss*) A mass of masticated food.

brachial artery (*BRAY kee yall ahh TER ee*) Major artery that carries blood to the upper arm.

brainstem Most inferior part of the brain that joins with the spinal cord. It is composed of the midbrain, pons and medulla oblongata.

bronchi (*BRONG kye*) Tubular air passages that branch off the trachea to the right and left and enter each lung. They carry inhaled and exhaled air to and from the lungs.

bronchioles (*BRONG kee ohlz*) Small tubular air passageways that branch off the bronchi. They carry inhaled and exhaled air to and from the alveoli.

buccal (*BUCK al*) Pertaining to cheek or mouth region.

bundle of His (*HISS*) Section of the conduction system of the heart after the AV node. It splits into the right and left bundle branches.

bursa (*BER sah*) Fluid-filled sac that decreases friction where a tendon rubs against a bone near a synovial joint.

caecum (*SEE kum*) First part of the large intestine. A short, pouch-like area. The appendix is attached to its external wall.

calcium ion channels (*cal see um EYE on*) Pathways that allow calcium ions to pass through.

cancellous bone (*CAN sell us*) Spongy bone found in the epiphyses of long bones. Its spaces are filled with red bone marrow that makes blood cells.

canine teeth (*KAY nine*) Long, pointed teeth located between the incisors and the premolars. The canines sink deeply into food to hold it. There are four canines, two in the maxilla and two in the mandible. Also known as cuspids (because they have one large, pointed cusp) or eye-teeth.

capillaries (*CAP ill air eez*) Smallest blood vessels in the body. Connecting blood vessels between arterioles and venules. The exchange of oxygen and carbon dioxide takes place in the capillaries.

capillary bed (*cap ILL air ree*) A network of capillaries.

capsid (*CAP sid*) The protein covering around a virus particle.

cardiac (*CAR dee ack*) Pertaining to the heart.

cardiopulmonary (*CAR dee yoh PULL mun air ee*) Pertaining to the heart and lungs.

cardiovascular system (*CAR dee yoh VASS cue lar*) Body system that includes the heart, arteries, veins and capillaries. It distributes blood throughout the body.

carina (*cah reen uh*) A structure with a projecting central ridge such as occurs at the bifurcation of the mainstem bronchi in the lung.

carotene (*CAR oh teen*) A yellow pigment found in plant and animal tissue. The precursor of vitamin A.

carpal bones The eight small bones of the wrist joint.

cartilage (*KAR tih lij*) Smooth, firm, but flexible connective tissue.

catabolism (*ka TAH bow lizm*) Breaking down of molecules in the body.

cataract (*KAT ah rakt*) Clouding of the lens. Protein molecules in the lens begin to clump together. Caused by ageing, sun exposure, eye trauma, smoking and some medications.

caudal (*CORD aul*) Toward the tailbone, feet or lower part of the body.

cell-mediated immunity Destruction of pathogens by T lymphocytes. T lymphocytes directly attack infected cells, destroying them.

cell membrane Semi-permeable barrier that surrounds a cell and holds in the cytoplasm. It allows water and some nutrients to enter and waste products to leave the cell.

cementum (*sih MEN tum*) Continuous layer of bone-like connective tissue that covers the dentin layer of the tooth and roots below the gum line. It begins at the gum line where the enamel stops. It anchors one side of the periodontal ligaments.

central nervous system (CNS) Division of the nervous system that includes the brain and the spinal cord.

centrioles (*SEN tree ohlz*) An organelle that precedes mitosis.

centrosomes (*SEN troh soams*) Region of cytoplasm usually near the nucleus that contains one or two centrioles.

cephalic (*keff AL ik*) Toward the head of the body.

cerebellum (*ser eh BELL um*) Small rounded section that is the most posterior part of the brain. Monitors muscle tone and position, and coordinates new muscle movements.

cerebrospinal fluid (*ser ree bro SPY nall*) The fluid cushion that protects the brain and spinal cord from shock.

cerebrum (*ser EE brum*) The largest and most visible part of the brain. Its surface contains gyri and sulci and is divided into two hemispheres.

cerumen (*seh ROO men*) Sticky wax that traps dirt in the external auditory canal.

ceruminous glands (*seh ROO men us*) Glands that produce the wax-like substance in the ear.

cervical (*SER vik all*) Pertaining to the region of the neck.

chemical synapse (*KEM ick al SIGH naps*) Site of communication between neurons and other excitable cells. Neurotransmitters are released from the neuron (presynaptic cell) which travel to a muscle or gland cell (the post-synaptic cell), allowing communication between the two cells.

cholecystitis (*koh lee siss TYE tiss*) Acute or chronic inflammation of the gallbladder because of gallstones.

cholecystokinin (CCK) (*koh lee siss toe KYE nin*) Hormone released by the duodenum when it receives food from the stomach. It causes the gallbladder to release bile and the pancreas to release its digestive enzymes.

cholelithiasis (*koh lee lih THIGH ah sis*) One or more gallstones in the gallbladder. Choledocholithiasis occurs when a gallstone becomes lodged in the common bile duct.

cholesterol (*coe LESS ter oll*) Lipid-containing compound that is a component of bile (from the gallbladder), sex hormones, neurotransmitters and cell membranes.

chondrosarcoma (*KON droe sar KOE ma*) Cancer of the cartilage.

choroid (*COR royd*) Spongy membrane of blood vessels that begins at the iris and continues around the eye. In the posterior cavity, it is the middle layer between the sclera and the retina.

chromatin (*CROW mah tin*) Genetic material found in the nucleus of a cell.

chronic bronchitis (*brong KYE tiss*) Chronic inflammation or infection of the bronchi. Inflammation of the bronchi is due to pollution or smoking.

chyle (*KILE*) Milk-like substance formed from digested and absorbed fats.

chyme (*KIME*) Mixed food and digestive juices in the stomach and small intestine.

cilia (*SIL ee uh*) Small hairs that flow in waves to move foreign particles away from the lungs toward the nose and the throat where they can be expelled. Also found inside the Fallopian tube to propel an ovum toward the uterus.

ciliary muscles (*SILL ee air ee*) Smooth muscle that alters the lens of the eye to accommodate for near vision.

clot A thrombus or coagulated blood.

coagulation (*coe GAG you LAY shun*) The formation of a blood clot by platelets and the clotting factors.

cochlea (*COCK lee ah*) Structure of the inner ear that is associated with the sense of hearing. It relays information to the brain via the cochlear branch of the vestibulocochlear nerve.

collecting duct Common passageway that collects fluid from many nephrons. The final step of reabsorption takes place there and the fluid is known as urine.

colon (*COE lon*) The longest part of the large intestine. It has four parts: the ascending colon, the transverse colon, the descending colon and the S-shaped sigmoid colon.

commissures (*COM mi sures*) Transverse bands of nerve fibres; carry information from one side of the nervous system to the other.

compact bone Hard or dense bone forming the superficial layer of all bones.

complement cascade A series of chemical reactions triggered by infection that leads to the destruction of a pathogen.

conchae (*con chay*) Shelf-like structures inside the nasal cavity.

cones Light-sensitive cells in the retina that detect coloured light. There are three types of cones, each of which responds to either red, green or blue light.

congestive heart failure (CHF) (*con JESS tiv*) Inability of the heart to pump sufficient amounts of blood. Caused by chronic coronary artery disease or hypertension.

conjunctiva (*kon JUNK tye vah*) Delicate, transparent mucous membrane that covers the inside of the eyelids and the anterior surface of the eye. It produces clear, watery mucus.

conjunctivitis (*kon JUNK tye VYE tiss*) Inflamed, reddened and swollen conjunctivae with dilated blood vessels on the sclerae. Caused by a foreign substance in the eye or an infection.

cor pulmonale (*KOR pull mun AH lee*) Failure of the pumping ability of the right ventricle.

corium (*CORE ee um*) Layer of skin immediately under the epidermis. Also known as the true skin.

cornea (*COR nee ah*) Transparent layer over the anterior part of the eye. It is a continuation of the white sclera.

coronal plane (*kor ROHN al*) Also called frontal plane; an imaginary vertical plane that divides the entire body into front and back sections. The coronal plane is named for the coronal suture where the anterior and posterior skull bones meet.

corpus callosum (*KOR pus ka LOH sum*) Thick white band of nerve fibres that connects the two hemispheres of the cerebrum and allows them to communicate and coordinate their activities.

cortex (*KOR tex*) Tissue layer of kidney just beneath the renal capsule.

corticobulbar tract (*KOR ti coe BUL bar*) Spinal cord tract that carries impulses to the brainstem from the motor cortex. Carries orders for voluntary movements.

corticospinal tract (*KOR tih coe SPY nal*) Pertains to the tract between the cerebral cortex and spinal cord. Carries orders for voluntary movements.

counter-current circulation Exchange of substances between two streams on either side of a membrane. Helps control concentration of fluids as in the loop of Henlé.

cramp Spasmodic muscle contraction.

cranial (*CRANE nee all*) Pertaining to the skull.

cranial nerves (*CRANE nee all*) Twelve pairs of nerves that originate in the brain. Carry sensory nerve impulses to the brain from the nose, eyes, ears and tongue for the senses of smell, vision, hearing and taste. Also carry sensory nerve impulses to the brain from the skin of the face, and motor nerve impulses from the brain to the muscles of the face, mouth, throat, eye and salivary glands.

cross-sectioning Making slices of a sample for examination purposes.

crown White part of the tooth that is visible above the gum line.

crural (*CRUR all*) Pertains to the leg or thigh.

cuspids (*CUSS pids*) Canine teeth.

cutaneous membranes (*cue TAY nee yuss*) Membranes of the skin.

cuticle (*CUE tickle*) Layer of dead skin that arises from the epidermis around the proximal end of the nail. It keeps micro-organisms from the nail root.

cyanosis (*sign ah NOH siss*) Bluish-grey discoloration of the skin from abnormally low levels of oxygen and abnormally high levels of carbon dioxide in the tissues.

cytokines (*SIGH toe kines*) Chemicals released by injured body tissues that summon leucocytes and cause them to move to the area.

cytoplasm (*SIGH toe plazm*) Gel-like intracellular substance. Organelles are embedded in it.

cytotoxic T cells (*SIGH toe TOX ick*) Type of T lymphocyte that matures in the thymus. Cytotoxic T cells destroy all types of pathogens as well as body cells infected with viruses.

deciduous teeth (*deh SID you uss*) Teeth that erupt during childhood from age six months to two years. Also called the milk teeth, baby teeth or primary teeth.

decubitus (*dee KYOO bih tus*) Lying down position; on the back.

defecation (*deh fi CAY shun*) Process by which undigested food fibre and water are removed from the body in the form of a bowel movement.

dendrite (*DEN dright*) Multiple branches at the end of a neuron that carry information to the cell body.

dendritic cells (*DEN drit ick*) One of several types of antigen-displaying cells that stimulate adaptive immunity.

dentin Hard layer of tooth just beneath the enamel layer.

deoxyribonucleic acid (DNA) (*dee OX ee RYE bow new KLEE ik*) Sequenced pairs of nucleotides that form a double helix. A segment of DNA makes up a gene.

depolarisation (*DEE pow lar is AY shun*) Changing of the permeability of the cell membrane of any excitable cell, for example cardiac muscle or a neuron, that leads to a decrease in charge across the cell membrane.

dermis (*DER miss*) Layer of skin under the epidermis. It is composed of collagen and elastin fibres. It contains arteries, veins, nerves, sebaceous glands, sweat glands and hair follicles.

diagnosis (*dye agg NOH siss*) A determination as to the cause of the patient's symptoms and signs.

diaphragm (*DIE ah fram*) Muscular sheet that divides the thoracic cavity from the abdominal cavity. Most important ventilation muscle.

diaphysis (*dye AFF i siss*) The straight shaft of a long bone.

diastole (*dye ASS toe lee*) Resting period between contractions. It is when the heart fills with blood.

diencephalon (*dye en KEFF ah lon*) Central part of the brain that contains that thalamus and hypothalamus.

differentiation (*DIH feer en she AY shun*) Process by which embryonic cells assume different shapes and functions in different parts of the body.

diffusion (*dih FEW zhun*) The process of movement of a substance from high concentration to low concentration.

digestion (*Dye JESS chun*) Process of mechanically and chemically breaking food down into nutrients that can be used by the body.

digital Pertaining to the fingers or toes.

disease (*dih ZEEZ*) Any change in the normal structure or function of the body.

distal (*DISS tal*) Moving from the body toward the end of a limb (arm or leg).

distal tubule (*DISS tal*) Tubule of the nephron that begins at the loop of Henlé. It empties into the collecting duct. Reabsorption takes place there.

dorsal (*DOOR sall*) Pertaining to the posterior of the body, particularly the back.

dorsal column tract (*DOOR sall*) Spinal cord pathway that carries fine touch sensation from the spinal cord to the brain.

dorsal root ganglion (*DOOR sall root GANG lee yon*) A collection of sensory neurons on the dorsal roots of the spinal cord.

duodenum (*due OH dee num*) First part of the small intestine. It secretes cholecystokinin, a hormone that stimulates the gallbladder and pancreas to release bile and digestive enzymes.

dura mater (*DURE a MAY ter*) Tough, outermost layer of the meninges. The dura mater lies just under the bones of the cranium and vertebrae.

eccrine glands (*EKK reen*) Sweat glands that cover the entire skin surface.

efferent arterioles (*EFF er ent are TIER ee ohlz*) Small blood vessels that carry blood away from the glomeruli of the kidney.

electrolyte (*eh LEK trow light*) Chemical element that carries a positive or negative charge and conducts electricity when dissolved in a solution. Examples include sodium (Na^+), potassium (K^+), chloride (Cl^-), calcium (Ca^{2+}), and bicarbonate (HCO_3^-). Electrolytes are carried in the plasma. Excess amounts in the blood are removed by the kidneys.

electromyography (EMG) (*eh lek troh my OG ruff ee*) Diagnostic procedure to diagnose muscle disease or nerve damage. A needle electrode inserted into a muscle records electrical activity as the muscle contracts and relaxes. The electrical activity is displayed as waveforms on an oscilloscope screen and permanently recorded on paper as an electromyogram.

embolus (*EM boh luss*) Mass of undissolved matter present in the blood or lymphatic vessels that was brought there by the blood or lymph current.

emphysema (*emph uh SEE muh*) Chronic pulmonary disease resulting in destruction of air spaces distal to the terminal bronchiole.

empyema (*em pie EE muh*) Localised collection of purulent material (pus) in the thoracic cavity from an infection in the lungs. Also known as pyothorax.

emulsification (*ee mull sih fih KAY shun*) Process performed by bile of breaking down large fat droplets into smaller droplets with more surface area.

enamel Glossy, thick white layer that covers the crown of the tooth. Enamel is the hardest substance in the body.

endocardium (*ehn doh KAR dee um*) Innermost layer of the heart. It covers the inside of the heart chambers and valves.

endocrine system (*EN doh kreen*) Body system that includes the testes, ovaries, pancreas, adrenal glands, thymus, thyroid gland, parathyroid glands, pituitary gland and pineal gland. It produces and releases hormones into the blood to direct the activities of other body organs.

endocytosis (*EN doe sigh TOE siss*) Ingestion of substances by a cell. Substances are taken into the cells after being surrounded by vesicles.

endolymph (*EN doe limf*) Fluid within the labyrinth of the ear.

endoplasmic reticulum (*EN doh PLAS mik ree TICK you lum*) Organelle that consists of a network of channels that transport materials within the cell. Also the site of protein, fat and glycogen synthesis.

enzyme (*EN sime*) Molecules that speed up the rate of chemical reactions in cells. Enzymes are particularly important in the breakdown and synthesis of biological molecules.

eosinophils (*ee oh SIN oh fills*) Type of leucocyte. It is classified as a granulocyte because it has granules in the cytoplasm. The nucleus has two lobes. Eosinophils are involved in allergic reactions and defence against parasites.

ependymal cells (*ep PEN dye mal*) Specialised cells that line the walls of the ventricles and spinal canal and produce cerebrospinal fluid.

epidermis (*ep ee DER miss*) Thin, outermost layer of skin. The most superficial part of the epidermis consists of dead cells filled with keratin. The deepest part (basal layer) contains constantly dividing cells and melanocytes.

epidural space (*ep ee DURE all*) Area between the dura mater and the vertebral body.

epiglottis (*ep ih GLOT iss*) Lid-like structure that seals off the larynx, so that swallowed food goes into the oesophagus.

epinephrine (*EP ee NEFF rinn*) Hormone secreted by the adrenal medulla in response to stimulation by nerves of the sympathetic nervous system.

epiphyseal plate (*eh PIFF ih SEAL*) The growth plate.

epiphysis (*eh PIFF i siss*) The widened ends of a long bone. Each end contains the epiphyseal plate where bone growth takes place.

epithelial tissue (*ep ee THEE lee all*) Layers of cells that form the epidermis of the skin as well as the surface layer of mucous and serous membranes.

erythrocyte (*eh RITH roh sights*) A red blood cell. Erythrocytes contain haemoglobin and carry oxygen and carbon dioxide to and from the lungs and cells of the body.

erythropoiesis (*eh rith roh poy EE siss*) Formation of red blood cells.

erythropoietin (*eh RITH roh poy EE tin*) In the body, a hormone secreted by the kidneys when the number of red blood cells decreases. It stimulates the bone marrow to make more red blood cells. As a drug, erythropoietin does the same.

Eustachian tube (*yoo STAY shun*) Tube that connects the middle ear to the nasopharynx and equalises the air pressure in the middle ear.

excretion (*ik SCREE shun*) Removal of waste matter from the body.

exocytosis (*EK soh sigh TOE siss*) Secretion. The expulsion of material from a cell using vesicles.

extension (*eks TEN shun*) Straightening a joint to increase the angle between two bones or two body parts.

extensor muscle (*ek STEN sor MUSS all*) Muscle that produces extension when it contracts.

external Near or on the outside surface of the body or an organ.

external auditory meatus (*AW dih tor ee mee AY tuss*) Opening at the entrance to the external auditory canal where sound waves enter.

external urethral sphincter (*EKS ter nal you REE thral SFINK ter*) One of two valves, made of circular muscle, which allows voluntary control of urination.

facilitated diffusion (*fah SILL ih TATE ed dih FEW zhun*) Also known as carrier-mediated passive transport; the movement of substances into cells via carrier proteins.

femoral artery (*FEE mor all AH ter ee*) Major artery that carries blood to the upper leg.

fibrin (*FYE brin*) Fibre strands that are formed by the activation of clotting factors. Fibrin traps erythrocytes and this forms a blood clot.

fibrinogen (*fye BRINN oh jenn*) Blood clotting factor.

fibroblasts (*FYE bro blasts*) Any cell from which connective tissue is created.

fibromyalgia (*fye bro my AL jar*) Pain located at specific, small trigger points along the neck, back and hips. The trigger points are very tender to the touch and feel firm. Caused by injury or trauma.

filtration (*fil TRAY shun*) Process in which water and substances in the blood are pushed through the pores of the glomerulus. The resulting fluid is known as filtrate.

fissure (*FISH ure*) Deep division on the surface of the brain and spinal cord.

flaccid (*FLAS sid*) Limp or without muscle tone.

flagella (*flah JELL ah*) Hair-like processes on bacteria or protozoon that cause movement.

flexion (*FLEK shun*) Bending of a joint to decrease the angle between two bones or two body parts. Opposite of extension.

flexor muscle (*FLEK sor*) Muscle that produces flexion when it contracts.

flora (*FLOOR ah*) Plant life occurring in a specific environment.

follicle (*FOLL i cul*) 1. Mass of cells with a hollow centre. It holds an oocyte before puberty and a maturing ovum after puberty. The follicle ruptures at the time of ovulation and becomes the corpus luteum. 2. Also a site where a hair is formed. The follicle is located in the dermis.

fornix (*FOR niks*) 1. Tract of nerves that joins all the parts of the limbic system. 2. Area of the superior part of the vagina that lies behind and around the cervix.

Fowler's position A semi-sitting position with the torso at a 45–60 degree angle.

frenulum (*FREN you lum*) Structure that attaches the lower side of the tongue to the gum.

frontal lobe Lobe of the cerebrum that predicts future events and consequences. Exerts conscious control over the skeletal muscles.

frontal plane An imaginary plane parallel with the long axis of the body that divides the body into an anterior and posterior section.

fundus (*FUN duss*) A larger part, base, or body of a hollow organ such as the dome-shaped top of the bladder, uterus above the Fallopian tubes, or rounded, most superior part of the stomach.

fungi (*FUN gee*) A plant-like organism that includes mould and yeasts.

furuncle (*FERN ickle*) A boil.

ganglion (*GANG lee on*) Mass of nervous tissue composed mostly of nerve cell bodies and lying outside the brain and spinal cord.

gastrin (*GAS trin*) Hormone produced by the stomach that stimulates the release of hydrocholoric acid and pepsinogen in the stomach.

gastrointestinal (GI) system (*gas troh in TESS tie nall*) Body system that includes the oral cavity, pharynx, stomach, oesophagus, small and large intestines, and the accessory organs of the liver, gallbladder and pancreas. Its function is to digest food and remove undigested food from the body. Also known as the gastrointestinal tract and digestive system or tract.

gene (*jean*) An area on a chromosome that contains all the DNA information needed to produce one type of protein molecule.

genitourinary system (*genn i toe YOU rin air ee*) Combination of two closely related body systems: the male genitalia and the urinary system. Also known as the urogenital system.

gingival (*jin gie VAL*) Referring to the gum.

glaucoma (*gl OW KOH mah*) Increased intraocular pressure (IOP) because aqueous humour cannot circulate freely. In open-angle glaucoma, the angle where the edges of the iris and cornea touch is normal and open, but the trabecular meshwork is blocked. Open-angle glaucoma is painless but destroys peripheral vision, leaving the patient with tunnel vision. In closed-angle glaucoma, the angle is too small and blocks the aqueous humour. Closed-angle glaucoma causes severe pain, blurred vision and photophobia. Glaucoma can progress to blindness.

glia (*GLY ah*) Non-nervous or supporting tissue found in the brain and spinal cord; made of glial cells.

glial cells (*GLY all sells*) Cells that include astrocytes, oligodendrocytes, ependymal cells, microglia cells, Schwann cells and satellite cells.

glomerular filtrate (*gla MARE you ler FILL trate*) The filtered fluid within the glomerulus.

glomerulus (*glom er ROO luss*) Network of intertwining capillaries within Bowman's capsule in the nephron. Filtration takes place in the glomerulus.

glottis (*GLOT is*) V-shaped structure of mucous membranes and vocal cords within the larynx.

glucose (*GLOO kons*) 1. A simple sugar found in foods and also the sugar in the blood. 2. Glucose is not normally found in the urine. Its presence (glycosuria) indicates uncontrolled diabetes mellitus with excess glucose in the blood 'spilling' over into the urine.

gluteal (*GLOO tee all*) Pertains to buttocks.

glycogen (*GLY co jin*) The form that glucose (sugar) takes when it is stored in the liver and skeletal muscles.

Golgi apparatus (*GOAL ghee app ah RAH tuss*) Organelle of the cell that packages cellular material for transport.

Guillian-Barre syndrome (*gil ann barr ee SIN drome*) A neuromusclular disease that ususally leads to ascending flaccid paralysis.

gustatory sense (*JUSS ta TOR ee*) Sense of taste.

gyri (*JIE rie*) Convolutions of the cerebral hemispheres of the brain.

haemoglobin (*HEE moh GLOW binn*) Substance in an erythrocyte that binds to oxygen and carbon dioxide. Its globin chains give it a round shape. When it is bound to oxygen it forms the compound oxyhaemoglobin.

haemophilia (*HEE moh FILL ee ah*) Inherited genetic abnormality of a gene on the X chromosome that causes an absence or deficiency of a specific clotting factor. When injured, haemophiliac patients cannot easily form a blood clot and continue to bleed for long periods of time.

haemopoiesis (*HEE mow poy EE siss*) Formation of blood cells.

haemostasis (*HEE moh STAY siss*) The cessation of bleeding after the formation of a blood clot.

haemothorax (*HEE mow THOH rax*) Presence of blood in the thoracic cavity, usually from trauma.

hammer One of the three small bones of the ear, also known as the malleus.

helper T cell Helper T cells stimulate the production of cytotoxic T cells and B cells. Also known as a CD4 cell.

hemisphere One half of the cerebrum, either the right hemisphere or the left hemisphere. The right hemisphere deals with recognising patterns and three-dimensional structures (including faces) and the emotions of words. The left hemisphere deals with mathematical and logical reasoning, analysis, and interpreting sights, sounds and sensations. The left hemisphere is active in reading, writing and speaking.

heparin (*HEP ah rin*) Substance that inhibits coagulation of blood.

hepatic duct (*hep PAT ic*) Duct that carries bile from the liver.

hernia (*HER nee ah*) A weakness in the muscles of the abdominal wall that allows loops of intestine to balloon outward.

hilum (*HIGH lum*) 1. Indentation in the medial side of each lung where the bronchus, pulmonary artery, pulmonary vein and nerves enter the lung. 2. Indentation in the medial side of each kidney where the renal artery enters and renal vein and the ureter leave.

histamine (*HIS tar mean*) Released by basophils. Heparin dilates blood vessels and increases blood flow to damaged tissue. Allows protein molecules to leak out of blood vessels into the surrounding tissue. This produces redness and swelling.

homeostasis (*hoh mee oh STAY siss*) State of equilibrium of the internal environment of the body, including fluid balance, acid–base balance, temperature, metabolism, and so forth, to keep all the body systems functioning optimally.

horizontal plane Another name for the transverse plane.

hormone (*HOR moan*) Chemical messenger of the endocrine system that is released by a gland or organ and travels through the blood.

hyperopia (*HIGH per OH pee ah*) Farsightedness. Light rays from a far object focus correctly on the retina, creating a sharp image. However, light rays from a near object come into focus posterior to the retina, creating a blurred image.

hyperpolarised The charge across the cell membrane is more negative than resting.

hypertrophy (*high PER trow fee*) Greater than normal growth.

hypodermis (*high poh DER miss*) The fatty tissue layer below the dermis of the skin.

hypothalamus (*high poh THAL ah mus*) Endocrine gland located in the brain just below the thalamus. It produces (but does not secrete) antidiuretic hormone (ADH) and oxytocin. The hypothalamus is in the centre of the brain just below the thalamus and coordinates the activities of the pons and medulla oblongata. It also controls heart rate, blood pressure, respiratory rate, body temperature, sensations of hunger and thirst, and the circadian rhythm. It also produces hormones as part of the endocrine system. In addition, the hypothalamus helps control emotions (pleasure, excitement, fear, anger, sexual arousal) and bodily responses to emotions; regulates the sex drive; contains the feeding and satiety centres; and functions as part of the 'fight-or-flight' response of the sympathetic nervous system.

hypoxia Reduced oxygen tension in tissues, tissues poorly oxygenated.

ilium (*ILL ee um*) Most superior hip bone. Bony landmarks include the iliac crest and the anterior-superior iliac spine (ASIS). Posteriorly, each ilium joins one side of the sacrum.

incisors (*in SIGH sors*) Chisel-shaped teeth in the middle of the dental arch that cut and tear food on their incisal surface. There are eight incisors, four in the maxilla and four in the mandible.

incus (*ING kuss*) Second bone of the middle ear. It is attached to the malleus on one end and the stapes on the other end. Also known as the anvil.

infarct (*in FARKT*) Cellular death due to lack of blood flow (perfusion).

inferior Pertaining to the lower half of the body or a position below an organ or structure.

inferior vena cava (*in FEAR ee* or *VEEN ah KAH va*) The vena cava is the largest vein in the body. The inferior vena cava receives blood from the abdomen, pelvis and lower extremities and takes it to the heart.

inflammation (*in flah MAY shun*) Tissue reaction to injury that includes swelling and reddening due to increased blood flow to the area.

ingestion (*in JESS chun*) Process of taking in material (particularly food).

inguinal region (*ing WHY nal REE jun*) The groin.

innate immunity (*ih NATE im YOO nih tee*) Defence against pathogens that you are born with and that does not improve with experience or remember specific pathogens. The first line of defence.

inotropism (*EYE no TROPE izm*) An influence on the force of muscular contraction.

insula (*INS you lah*) The deep lobes of the cerebral hemispheres.

intercalated discs (*in ter CAL ate ed*) Structures that connect heart tissue cells to facilitate a smooth contraction.

interferon (*in ter FEAR on*) Substance released by macrophages that have engulfed a virus. Interferon stimulates body cells to produce an antiviral substance that keeps a virus from entering a cell and reproducing.

interleukin (*in ter LOO kin*) Released by macrophages, it stimulates B cell and T lymphocytes and activiates NK cells. It also produces the fever associated with inflammation and infection.

internal Structures deep within the body or an organ.

interneurons (*in ter NURE ons*) Neurons that facilitate communication between neurons.

iris (*EYE riss*) Coloured portion of the eye that controls the size of the opening (pupil) where light passes into the eye.

ischaemia (*iss KEE mee ah*) Tissue injury due to a decrease in blood flow.

jejunum (*ju GEE num*) Second part of the small intestine.

joint Area where two bones come together.

keratin (*CARE ah tin*) Hard protein found in the cells of the outermost part of the epidermis and in the nails.

keratinisation (*CARE ah tin eye ZAY shun*) The process of forming a horny growth such as fingernails.

kidney Organ of the urinary system that filters blood and produces urine, controlling fluid and ionic balance.

labia (*LAY bee ah*) An outer pair of vertical fleshy lips covered with pubic hair (the labia majora) and a smaller, thinner, inner pair of lips (the labia minora) that partially cover the clitoris, and urethral and vaginal openings. Part of the external female genitalia.

labyrinth (*LAB er inth*) Intricate communicating passage of the inner ear essential for maintaining equilibrium.

labyrinthitis (*LAB er inth EYE tiss*) Bacterial or viral infection of the semicircular canals of the inner ear, causing severe vertigo.

lacrimal apparatus (*LAK rim all app ah RAY tuss*) Structures involved with the secretion and production of tears.

lacteal (*LACK teal*) Pertains to milk.

large intestine Organ of absorption between the small intestine and the anal opening to the outside of the body. The large intestine includes the caecum, appendix, colon, rectum and anus. Also known as the large bowel.

laryngitis (*lah in JIGH tiss*) Hoarseness or complete loss of the voice, difficulty swallowing and cough due to swelling and inflammation of the larynx.

laryngopharynx (*lah RIN goh FAH rinks*) Relates to both the larynx and pharynx.

larynx (*LA rinks*) Triangular-shaped structure in the anterior neck (visible as the laryngeal prominence or Adam's apple) that contains the vocal cords and is a passageway for inhaled and exhaled air. Also known as the voice box.

lateral (*LAT er all*) Pertaining to the side of the body or the side of an organ or structure.

lens Clear, hard disc in the internal eye. The muscles and ligaments of the ciliary body change its shape to focus light rays on the retina.

lesion (*LEE zhun*) General category for any area of visible damage on the skin, whether it is from disease or injury.

leucocytes (*LOO koh sights*) White blood cells. There are five different types of mature leucocytes: neutrophils, eosinophils, basophils, lymphocytes and monocytes.

leucocytosis (*LOO coe sigh TOE sis*) Increase in the number of white blood cells above normal.

leucopenia (*LOO coe PEEN ee ah*) Abnormal decrease of white blood cells.

leukaemia (*loo KEEM ee yah*) Cancer of leucocytes (white blood cells), including mature lymphocytes, immature lymphoblasts, as well as myeloblasts and myelocytes that mature into neutrophils, eosinophils or basophils. The malignant leucocytes crowd out the production of other cells in the bone marrow. Leukaemia is named according to the type of leucocyte that is the most prevalent and whether the onset of symptoms is acute or chronic. Types of leukaemia include acute myelogenous leukaemia (AML), chronic myelogenous leukaemia (CML), acute lymphocytic leukaemia (ALL), and chronic lymphocytic leukaemia (CLL).

ligament (*LIG ah ment*) Fibrous bands that hold two bone ends together in a synovial joint.

limbic system (*LIM bick*) Processes memories and controls emotions, mood, motivation and behaviour. Links the conscious to the unconscious mind. The limbic system consists of the thalamus, hypothalamus, hippocampus, amygdaloid bodies and fornix.

lingual (*LING gue ahl*) Pertaining to the tongue.

lipocyte (*LYE poh sight*) Cell in the subcutaneous layer that stores fat.

lithotripsy (*LITH oh trip see*) Medical or surgical procedure that uses sound waves to break up a kidney stone.

lobes 1. Large divisions of the lung, visible on the outer surface. 2. Large area of the hemisphere of the cerebrum. Each lobe is named for the bone of the skull that is next to it: frontal lobe, parietal lobe, temporal lobe and occipital lobe.

local potential Change in the charge across a cell membrane that is proportional to the size of the stimulus.

lower oesophageal sphincter (*EH soff ah geel SFINK ter*) Ringed muscle leading into the stomach.

lumbar region (*LUM baa REE jun*) Two of the nine regions of the abdominopelvic area. The right and left lumbar regions are inferior to the right and left hypochondriac regions. Also refers to the lower back, spinal cord and spinal column.

lumen (*LOO men*) Opening in the centre of a large tube, for example the centre of the tubes in the digestive and respiratory systems and the blood vessels.

lunula (*LOON you lar*) Whitish half-moon visible under the proximal portion of the nail plate. It is the visible tip of the nail root.

lymph (*LIMF*) Fluid that flows through the lymphatic system.

lymph nodes (*LIMF nodes*) Small, encapsulated pieces of lymphoid tissue located along the lymphatic vessels. Lymph nodes filter and destroy invading micro-organisms and cancerous cells present in the lymph.

lymphatic fluid (*lim FAT ik FLOO id*) Clear and colourless fluid of the lymphatic system.

lymphatic system (*lim FAT ik*) Body system that includes lymphatic vessels, lymph nodes, lymph fluid and lymphoid tissues (tonsils and adenoids, appendix, Peyer's patches), lymphoid organs (spleen and thymus), and the blood cells, lymphocytes and macrophages.

lymphatic vessels (*lim FAT ik*) Vessels that begin as capillaries carrying lymph, continue through lymph nodes, and empty into the right lymphatic duct and the thoracic duct.

lymphocyte activation (*LIMF oh SIGHT ak tih VAY shun*) Stimulation of lymphocytes, 'waking them up' to fight a pathogen.

lymphocyte proliferation (*LIMF oh SIGHT pro liff er AY shun*) Reproduction of activated lymphocytes so there are many copies.

lysis (*LYE siss*) Destruction or breakdown.

lysosome (*LIE soh soam*) Organelle that consists of a small sac with digestive enzymes in it. These destroy pathogens that invade the cell.

macrophages (*MACK row fage ez*) Cells that take fragments of the pathogen they have eaten and present them to a B cell (lymphocyte). This stimulates the B cell to become a plasma cell and make antibodies against that specific pathogen. Macrophages also activate helper T cells in this way. Macrophages also produce special immune response chemicals: interferon, interleukin and tumor necrosis factor.

macroscopic anatomy (*MAK row scop ic ah NAH tom ee*) Study of large structures of the body.

major calyx, minor calyces (*KAY licks, KAY leh seez*) Tubes in the kidney which carry urine from the nephrons to the renal pelvis.

malignant (*mah LIG nant*) Cancerous, able to spread to distant parts of the body.

malleus (*MALL ee us*) First bone of the middle ear. It is attached to the tympanic membrane on one end and to the incus on the other end. Also known as the hammer.

mast cells Connective tissue cells that are important in cellular defence and contain heparin and histamine.

mastication (*MASS tih CAY shun*) Process of chewing, during which the teeth and tongue together tear, crush and grind food. This is part of the process of mechanical digestion.

medial (*MEE dee al*) Pertaining to the middle of the body or the middle of an organ or structure.

mediastinum (*meh dee yuh STY num*) Central area within the thoracic cavity. It contains the trachea, oesophagus, heart and other structures.

medulla oblongata (*meh DULL lar ob long GAR ta*) Most inferior part of the brainstem that joins to the spinal cord. It relays nerve impulses from the cerebrum to the cerebellum. It contains the respiratory centre. Cranial nerves IX through XII originate there.

melanin (*MELL an in*) Dark brown or black pigment that gives colour to the skin and hair.

melanocytes (*mel AN oh sights*) Cells that produced melanin.

melanoma (*mell an NO ma*) A malignant pigmented mole or tumour.

melatonin (*MELL ah TOH nin*) Hormone secreted by the pineal body. It maintains the 24-hour wake–sleep cycle known as the circadian rhythm.

membrane A thin soft pliable layer of tissue which can line a cavity or cover an organ or structure.

memory B cells Antibody-producing cells which are produced when a pathogen is encountered the first time. Memory B cells are stored until the pathogen comes again. They are responsible for secondary response.

memory T cells Memory cells responsible for cell-mediated immunity. They are produced when a pathogen is encountered for the first time. Memory T cells are stored until the pathogen comes again. They are responsible for secondary response.

Ménière's disease (*MANY erz*) Recurring and progressive disease that includes progressive deafness, ringing ears, dizziness and the feeling of fullness in the ears.

meninges (*men IN jeez*) Three separate membranes that envelope and protect the entire brain and spinal cord. The meninges include the dura mater, arachnoid and pia mater.

meningitis (*men in JYE tiss*) Inflammation of the meninges of the brain or spinal cord by a bacterial or viral infection. Initial symptoms include fever, headache, nuchal rigidity (stiff neck) lethargy, vomiting, irritability and photophobia.

mesentery (*ME sen tare ee*) Membranous sheet of peritoneum that supports the jejunum and ileum.

metabolism (*me TAH bow lizm*) Process of using oxygen and glucose to produce energy for cells.

Metabolism also produces by-products like carbon dioxide and other waste products. The ongoing cycle of anabolism and catabolism.

metastasis (*meh TASS tah siss*) Process by which cancerous cells break off from a tumour and move through the blood vessels or lymphatic vessels to other sites in the body.

metric system System of measurement based on the power of ten.

microglia (*my crow GLY ah*) Cells that move, engulf and destroy pathogens anywhere in the central nervous system.

microscopic anatomy (*MY kroh scop ic ah NAH tom ee*) Study of structures that require the aid of magnification.

midbrain An area that connects the pons and the cerebellum with the hemispheres of the cerebrum.

midsagittal plane (*mid SAJ it all*) An imaginary vertical plane that divides the entire body into right and left sides and creates a midline. The midsagittal plane is named for the sagittal suture of the skull.

mitochondria (*my toe CON dree ya*) The energy organelle of the cell.

mitral (*MY trall*) Pertains to the bicuspid or mitral valve of the heart.

mixed nerve A nerve that carries both sensory and motor information.

molars (*MOW lars*) Largest tooth, located posterior to the premolar. It crushes and grinds food on its large, flat occlusal surface.

motor neurons (*MOW ter NEW ronz*) Neurons that innervate muscle tissue.

motor system The system responsible for movement.

mucosa (*mew KOSE ah*) 1. Lining throughout the gastrointestinal system that consists of a mucous membrane that produces mucus and an underlying smooth muscle layer that contracts to move food. 2. Mucous membrane that lines the respiratory tract. It warms and humidifies incoming air. It produces mucus to trap foreign particles. 3. Mucous membrane lining the inside of the bladder. 4. Mucous membranes lining the nasal cavity that warm and moisturise the incoming air. They also produce mucus to trap foreign particles.

mucous (*MEW cuss*) Pertaining to mucus.

muscle Many muscle fascicles grouped together and surrounded by fascia.

muscular dystrophy (*MUSS CUE lar DISS troh fee*) Genetic disease due to a mutation of the gene that makes the muscle protein dystrophin. Without dystrophin, the muscles weaken and then atrophy. Symptoms appear in early childhood as weakness first in the lower extremities and then in the upper extremities. The most common and most severe form is Duchenne's muscular dystrophy; Becker's muscular dystrophy is a milder form.

muscularis externa (*muss CUE lar iss ek STERN ah*) The outside muscular layer of an organ or tubule.

myalgia (*my AL jar*) Pain in one or more muscles due to injury or muscle disease. Polymyalgia is pain in several muscle groups.

myasthenia gravis (*my as THEE nee ya GRAV iss*) Abnormal and rapid fatigue of the muscles, particularly evident in the muscles of the face; there is ptosis of the eyelids. Symptoms worsen during the day and can be relieved by rest. The body produces antibodies against its own acetylcholine receptors located on muscle fibres. The antibodies destroy many of the receptors. There are normal levels of acetylcholine, but too few receptors remain to produce sustained muscle contractions.

myelin (*MY lin*) Fatty sheath around the axon of a neuron. It acts as an insulator to keep the electrical impulse intact. Myelin around the axons of the brain and spinal cord is produced by oligodendrocytes. Myelin around axons of the cranial and spinal nerves is produced by Schwann cells. An axon with myelin is said to be myelinated.

myofibril (*my oh FYE bril*) Thin filament (actin) and thick filament (myosin) within the muscle fibre that give it its characteristic striated appearance.

myopia (*my OH pee ah*) Nearsightedness. Light rays from a near object focus correctly on the retina, creating a sharp image. However, light rays from a far object come into focus anterior to the retina, creating a blurred image.

myosin (*MY oh sin*) The thick protein filament found in muscle fibres.

nares (*NAIR eez*) The paired external openings of the nasal cavity.

nasal Pertaining to the nose.

nasal cavity (*NAY zl CAV it ee*) Hollow area inside the nose that is lined with mucosa or mucous membrane.

nasopharynx (*nay zoh FA rinks*) Uppermost portion of the throat where the posterior nares unite. The nasopharynx contains the opening for the Eustachian tubes and the adenoids.

natural active immunity Antibodies developed due to exposure to a pathogen.

natural killer (NK) cell Type of lymphocyte that matures in the red marrow and is the body's first cellular defence against invading micro-organisms. Without the help of antibodies or complement, an NK cell recognises a pathogen by the antigens on its cell wall and releases chemicals that penetrate and destroy it.

natural passive immunity Immunity due to the passage of antibodies from mother to child across the placenta or in breast milk.

neck Transitional area between the root and crown where the tooth becomes narrower. The neck of the tooth is located just above and below the gum line.

necrosis (*neh KROH siss*) Grey-to-black discoloration of the skin in areas where the tissue has died.

negative feedback loop Physiological process that works against the trend. Most often brings a variable back to set point. For example, as blood pressure rises, heart rate may decrease to bring blood pressure back to 'normal'.

negative selection The destruction of lymphocytes which react to 'self' antigens. These lymphocytes must be deleted to prevent autoimmunity.

nephron (*NEFF ron*) Microscopic functional unit of the kidney.

nephropathy (*neff ROPP ah thee*) General word for any disease process involving the kidney. Diabetic nephropathy involves progressive damage to the glomeruli because of diabetes mellitus. The tiny arteries of the glomerulus harden (glomerulosclerosis) because of accelerated arteriosclerosis throughout the body.

nerves Bundles of individual axons.

nervous system Body system that includes the brain, cranial nerves, spinal cord, spinal nerves and neurons. It receives signals from parts of the body and interprets them as pain, touch, temperature, body position, taste, sight, smell and hearing. It coordinates body movement. It maintains and interprets memory and emotion.

neuroglia (glial cells) (*new ROH gly ah*) Cells that hold neurons in place and perform specialised tasks. Includes astrocytes, ependymal cells, microglia, oligodendrocytes, satellite cells and Schwann cells.

neuromuscular (*new roh MUSS cue lar*) Pertaining to the nervous and muscular system.

neuromuscular junction (*new roh MUSS cue lar*) Area on a single muscle fibre where a nerve connects.

neuron (*NEW ron*) An individual nerve cell. The functional part of the nervous system.

neurotransmitter (*new row TRANS mit ter*) Chemical messenger that travels across the synapse between a neuron and another neuron, muscles fibre or gland.

neutrophils (*NEW trow fills*) A type of leucocyte that performs non-specific phagocytosis.

nodes of Ranvier (*ran vee AYE*) Constriction of the myelin sheath on a myelinated nerve fibre that facilitates nodal transmission of the impulse.

norepinephrine (*NOR ep ee NEFF rinn*) 1. Neurotransmitter for the sympathetic nervous system. It goes between neurons and an involuntary muscle, organ or gland. Controls the flight-or-fight response. 2. Hormone secreted by the adrenal medulla in response to stimulation by nerves of the sympathetic nervous system.

nucleolus (*new klee OH luss*) Round, central region within the nucleus. It makes ribosomes.

occipital lobe (*ok SIP it all*) Lobe of the cerebrum that receives sensory information from the eyes. Contains the visual cortex for the sense of sight.

oesophagus (*eh SOFF ah guss*) Flexible, muscular tube that moves food from the pharynx to the stomach.

oligodendrocytes (*OLLY go DEN drow sites*) Neuroglial cells which produce myelin in the CNS.

ophthalmologist (*op thal MOL oh jist*) A doctor who specialises in the eyes.

oral Pertaining to the mouth.

orbit Bony socket in the skull that surrounds all but the anterior part of the eyeball.

organ A part of the body comprised of tissues, that has a specialised function.

organelles (*ore gah NELLZ*) Small structures in the cytoplasm that have various specialised functions. Organelles include mitochondria, ribosomes, the endoplasmic reticulum, the Golgi apparatus and lysosomes.

oropharynx (*OR oh FA rinks*) Middle portion of the throat just behind the oral cavity. It begins at the level of the soft palate and ends at the epiglottis.

osmosis (*oz MOW siss*) The passage of the solvent through a semi-permeable membrane to equalise concentrations.

osmotic pressure (*oz MOTT ik*) The pressure which develops when there are two solutions of varying concentrations that are separated by a semi-permeable membrane.

osseous tissue (*OSS ee us*) Bone, a type of connective tissue.

ossicles (*OSS ickles*) Three tiny bones in the middle ear that function in the process of hearing: malleus, incus and stapes.

ossification (*OSS siff i cay shun*) Process by which cartilaginous tissue is changed into bone from infancy through puberty. Also known as osteogenesis.

osteoarthritis (*OSS tee oh arth RYE tiss*) Chronic inflammatory disease of the joints, particularly the large weight-bearing joints of the knees and hips, although it often occurs in the joints that move repeatedly like the shoulders, neck and hands.

osteoblasts (*OSS tee yoh blasts*) Osteocytes that form new bone.

osteoclasts (*OSS tee oh clasts*) Osteocytes that break down old or damaged areas of bone.

osteocytes (*OSS tee oh sites*) Bone cells. There are two types of osteocytes: osteoclasts and osteoblasts.

osteomalacia (*OSS tee oh mah LAY she ah*) Abnormal softening of the bones due to a deficiency of vitamin D. Chondromalacia is abnormal softening of the cartilage, specifically of the patella.

osteoporosis (*OSS tee oh por OH sis*) Condition of increased bone porosity that weakens the bones, usually seen in the elderly.

otitis media (*oh TYE tiss ME dee ah*) Acute or chronic bacterial infection of the middle ear.

oval window Opening in the temporal bone between the middle ear and the vestibule of the inner ear. The opening is covered by the end of the stapes.

oxytocin (*OCK see TOH sin*) Hormone secreted by the posterior pituitary gland. It stimulates the uterus to contract and begin labour. It stimulates the 'let-down reflex' to get milk flowing for breastfeeding.

pacemaker Cells or group of cells that automatically generate electrical impulses.

palate (*PAHL ate*) Roof of the mouth.

palatine tonsils (*PAL a teen TON sills*) Lymphoid tissue on either side of the throat where the soft palate arches downward in the oropharynx.

pancreas (*PAN kree ass*) A digestive and endocrine organ located in the abdominal cavity that produces digestive enzymes (amylase, lipase, protease, peptidase) and releases them into the duodenum. It also contains the islets of Langerhans (alpha, beta and delta cells) that produce and secrete the hormones glucagon, insulin and somatostatin.

pancreatitis (*PAN kree ah TYE tiss*) Inflammation or infection of the pancreas.

paralysis (*pah RALL ah siss*) Temporary or permanent loss of muscle function.

parasympathetic nervous system (*para sim pah THET ik*) Division of the autonomic nervous system that uses the neurotransmitter acetylcholine and carries nerve impulses to the heart, involuntary smooth muscles and glands while the body is at rest.

parathyroid glands (*PARA THIGH royd*) Endocrine glands, four of them, on the posterior lobes of the thyroid gland. They produce and secrete parathyroid hormone.

parietal lobe (*pa ree eh tall*) Lobe of the cerebrum that receives sensory information about temperature, touch, pressure, vibration and pain from the skin and internal organs.

parietal pleura (*pa ree tul PLOO rah*) One of the two layers of the pleura. It lines the thoracic cavity.

parotid salivary gland (*pah ROT id SAL ih vair ee*) Gland that secretes saliva that helps to lubricate food so that it is easier to chew and swallow.

passive transport The general term for the transportation of cellular material without the use of energy.

patellar (*pah TELL ar*) Pertaining to the kneecap.

pathogen (*PATH oh jenn*) Micro-organism that causes a disease. Pathogens include bacteria, viruses, protozoa and other micro-organisms, as well as plant cells like fungi or yeast.

pathology (*path ALL oh jee*) The study of disease.

pedal (*PEE dall*) Pertaining to the foot or feet.

pepsin Digestive enzyme produced by the stomach that breaks down food protein into smaller protein molecules.

perforin (*PER for in*) Chemical secreted by cytotoxic T cells which makes holes in the cell membranes of pathogens or infected cells, killing them.

perfusion (*per FEW zhun*) Blood flow to a particular region.

perilymph (*peri LIMF*) Pale lymph fluid found in the labyrinth of the inner ear.

periodontal (*perr ee oh DON til*) Literally means 'around the teeth' and may refer to the area of the gum.

periosteum (*pair ee OSS tee um*) Thick, fibrous membrane that goes around and covers the outside of a bone.

peripheral (*per RIFF er all*) Referring to 'away from centre' or the extremities.

peritubular capillaries (*peri TUBE you lar*) Capillaries surrounding the renal tubules.

pH A test of how acidic or alkaline a substance is.

phagocytosis (*FAG oh sigh TOH siss*) The process by which a phagocyte destroys a foreign cell or cellular debris; a type of endocytosis.

pharyngitis (*FA rin JIGH tiss*) Bacterial or viral infection of the throat. When the bacteria group A beta-haemolytic streptococcus causes the infection, it is known as strep throat.

pharynx (*FAR inks*) The throat; contains the passageways for food and for inhaled and exhaled air.

physiology (*fiz ee ALL oh jee*) The study of the function of the body's structures.

pia mater (*PEE ah MAY ter*) Thin, delicate innermost layer of the meninges that covers the surface of the brain and spinal cord. It contains many small blood vessels.

pineal gland (*pin EE all*) Endocrine gland in the brain that lies posterior to the pituitary gland. It secretes the hormone melatonin.

pinna (*PINN ah*) The auricle of the exterior ear that collects sound waves.

pinocytosis (*PIN oh sigh TOE siss*) Process in which a cell absorbs fluid material.

pituitary gland (*pi TURE it air ee*) Endocrine gland in the brain that is connected by a stalk of tissue to the hypothalamus. Also known as the hypophysis. It is known as the master gland of the body. It consists of the anterior and the posterior pituitary gland.

plantar Referring to sole of the foot.

plasma (*PLAZ mar*) Clear, straw-coloured fluid portion of the blood that carries blood cells and contains dissolved substances like proteins, glucose, minerals, electrolytes, clotting factors, complement proteins, hormones, bilirubin, urea and creatinine.

plasma proteins (*PLAZ mah*) Protein molecules in the plasma. The most important one is albumin.

pleural cavities (*PLERR all*) The space between the parietal and visceral layers of the pleura.

pleural effusion (*PLOO ral ef FYOO shun*) Accumulation of fluid within the pleural space due to inflammation or infection of the pleura and lungs.

plexus (*PLECK sus*) A network of nerves or vessels.

plicae circulares (*PLY kay sir cue LAR es*) One of the transverse folds in the small intestine.

pneumothorax (*NEW mow THOR aks*) Large volume of air that forms in the pleural space and progressively separates the two pleural membranes.

polycythaemia (*poll ee sigh THEE mee ah*) Increased number of erythrocytes due to uncontrolled production by the red marrow. The cause is unknown. The viscosity of the blood increases; it becomes viscous (thick), and the total blood volume is increased.

pons (*ponz*) Area of the brainstem that relays nerve impulses from the body to the cerebellum and back to the body. Area where nerve tracts cross from one side of the body to the opposite side of the cerebrum. Cranial nerves V through VIII originate there.

positive feedback Vicious cycle. During positive feedback, physiological processes send body chemistry or other attributes further and further away from equilibrium (set point). The trend will continue until something breaks the cycle.

positive selection During lymphocyte development, only those lymphocytes which can actually react to antigens will survive.

postcentral gyrus (*JIE russ*) Ridge on the surface of the cerebrum posterior to the central sulcus in each hemisphere. The postcentral gyrus contains the primary somatic sensory area for your sense of touch.

posterior The back of the body.

precentral gyrus (*JIE russ*) Ridge on the surface of the cerebrum anterior to the central sulcus in each hemisphere. The precentral gyrus contains the primary motor cortex for voluntary movements.

presbyopia (*PRESS bee OH pee ah*) Loss of flexibility of the lens with blurry near vision and loss of accommodation.

prognosis (*prog NOH siss*) The predicted outcome of a disease.

prolactin (*proh LAK tinn*) Hormone secreted by the anterior pituitary gland. It stimulates milk glands of the breasts to develop during puberty and to produce milk during pregnancy.

prone position Lying with the anterior section of the body down.

prothrombin (*pro THROM bin*) Blood clotting factor that is activated just before the thrombus is formed.

protozoa (*pro tow ZOW ah*) Unicellular organisms.

proximal (*PROK sim al*) Referring to 'near' a reference point.

proximal tubule (*PROK sim al TUBE yule*) The part of the renal tubule closest to the glomerulus. Many substances are secreted or absorbed in this part of the renal tubule.

pulmonary artery (*PULL mon air ree AH ter ree*) Artery that carries blood from the heart to the lungs. The pulmonary artery is the only artery in the body that carries blood that has low levels of oxygen.

pupil (*PYOO pill*) Round opening in the iris that allows light rays to enter the internal eye.

pustule (*PUST yule*) A small elevation of skin filled with lymph or pus.

pyloric sphincter (*pye LOR ik SFINK ter*) Muscular ring that keeps food in the stomach from entering the duodenum.

pylorus (*pye LOR uss*) Narrowing canal of the stomach just before it joins the duodenum. It contains the pyloric sphincter.

rapid eye movement (REM) The dream stage of sleep.

reabsorption (*ree ab SORP shun*) Process by which water and substances in the filtrate move out of the renal tubule and into the blood in a nearby capillary.

rectum Final part of the large intestine. It is a short, straight segment that lies between the sigmoid colon and the outside of the body.

reflex Involuntary muscle reaction that is controlled by the spinal cord. In response to pain, the spinal cord immediately sends a command to the muscles of the body to move. All of this takes place without conscious thought or processing by the brain. The entire circuit is also known as a reflex arc.

refractory period (*reh FRACT or ree*) Short period of time when the myocardium is resting and unresponsive to electrical impulses.

regulatory T cells Type of T cell that shuts down decreases immune response.

renal artery (*REE nal AHR ter ee*) Major artery that carries blood to the kidney.

renal corpuscle (*REE nal KOR puss el*) The filtration apparatus of the kidney, consists of the glomerulus and the glomerular capsule.

renal nephron (*REE nal NEFF ron*) Fundamental functional unit of the kidney, consists of the renal corpuscle and renal tubule.

renal pelvis (*REE nal PELL vis*) Funnel-shaped part of the kidney that collects urine.

renal vein (*REE nal*) Major blood vessel that carries blood away from the kidneys.

renin–angiotensin–aldosterone (*REE nen-an gee oh TEN sen-al DOSS ter ohn*) A complex hormone system that regulates blood volume and blood pressure. The system is triggered when blood flow to the kidney decreases.

repolarisation The opposite of depolarisation.

respiration (*res pir AY shun*) The process of gas exchange at the lungs or tissue sites. Oxygen and carbon dioxide are exchanged in the alveoli during external respiration. Oxygen and carbon dioxide are exchanged at the cellular level during internal respiration.

retina (*RET in ah*) Membrane lining the posterior cavity. It contains rods and cones. Landmarks include the optic disc and macula.

Rh factor A blood group discovered on the surface of erythrocytes of Rhesus monkeys and found to a variable degree in humans. Can be Rh-negative or Rh-positive.

rhinoplasty (*RYE noh plass tee*) Surgical procedure that uses plastic surgery to change the size or shape of the nose.

ribonucleic acid (RNA) (*RYE bow new KLEE ik*) Molecule contained in ribosomes and necessary for making proteins.

ribosome (*RYE bow soams*) Granular organelle located throughout the cytoplasm and on the endoplasmic reticulum. Ribosomes contain RNA and proteins and are the site of protein synthesis.

rigor mortis (*rig er MORE tiss*) A stiffness that occurs in dead bodies as a result of retained calcium and decreased ATP.

rods Light-sensitive cells in the retina. They detect black and white and function in daytime and nighttime vision.

root Part of the tooth that is hidden below the gum line. The premolars have one or more roots. The molars have multiple roots.

rotation Spin a body part on its axis.

rugae (*ROO gay*) 1. Deep folds in the gastric mucosa. 2. Folds in the mucosa of the bladder that disappear as the bladder fills with urine.

sarcomeres (*SAR koh mears*) Portion of striated muscle fibril that lies between the two adjacent dark lines.

satellite cells Neuroglia cells that enclose the cell bodies of neurons in the spinal ganglia.

Schwann cell (*SHWON*) Cell that forms the myelin sheaths around axons of the cranial and spinal nerves.

sclera (*SKLAIR ah*) White, tough, fibrous connective tissue that forms the outer layer around most of the eye. Also known as the white of the eye.

sebaceous glands (*sib AY shuss*) An exocrine gland of the skin that secretes sebum. Sebaceous glands are located in the dermis. Their ducts join with a hair and sebum coats the hair shaft as it moves toward the surface of the skin. Also known as oil glands.

sebum (*SEE bum*) Oily substance secreted by sebaceous glands.

secondary response Increased immune response mediated by memory cells when meeting a pathogen that it recognises.

secretion (*sih CREE shun*) The movement of a chemical out of a cell or gland.

segmentation Division into similar parts.

self vs. non-self recognition The ability of the immune system to distinguish between the body's cells and cells that do not belong in the body.

semi-circular canals Three canals in the inner ear that are oriented in different planes (horizontally, vertically, obliquely) that help the body keep its balance. It relays information to the brain via the vestibular branch of the vestibulocochlear nerve.

septum (*SEP tum*) 1. Wall of cartilage and bone that divides the nasal cavity into right and left sides. 2. Partitioning wall that divides the right atrium from the left atrium (interatrial septum) and the right ventricle from the left ventricle (interventricular septum).

serosa (*seh ROSE ah*) A serous membrane.

serous membrane (*SEER uss*) Double-layered membrane lining a serous cavity. The parietal layer lines the wall of the cavity and the visceral layer covers the organs in the cavity. There is a potential, fluid-filled cavity between the layers.

sinoatrial node (*sigh noh AY tree all*) Pacemaker of the heart. Small knot of tissue located in the posterior wall of the right atrium in a shallow channel near the entrance of the superior vena cava. The SA node dictates the heart rate at 70–80 beats per minute when the body is at rest. It generates the electrical impulse for the entire conduction system of the heart.

sinus (*SIGH nuss*) Hollow cavity within a bone of the cranium.

skeletal muscle (*SKELL ee tal*) One of three types of muscles in the body, but the only muscle that is under voluntary, conscious control. Under the microscope, skeletal muscle has a striated appearance.

smooth muscle One of three types of body muscles that is involuntarily controlled and found in the lining of the airways, blood vessels and uterus.

somatic nervous system (*soh MAT ick*) Division of the peripheral nervous system that uses the neurotransmitter acetylcholine and carries nerve impulses to the voluntary skeletal muscles.

spasm involuntary contraction of a muscle.

sphincter (*SFINK ter*) Muscular ring around a tube; a valve.

spinal cavity (*SPY nal*) A continuation of the cranial cavity as it travels down the midline of the back. The spinal cavity lies within and is protected by the bones (vertebrae) of the spinal column. The spinal cavity contains the spinal cord, the spinal nerves and spinal fluid.

spinal cord (*SPY nal*) Part of the central nervous system. Continuous with the medulla oblongata of the brain and extends down the back in the spinal cavity. Ends at L2 and separates into individual nerves (cauda equina).

spinal nerves (*SPY nal*) Thirty-one pairs of nerves. Each pair comes out from the spinal cord between two vertebrae. An individual spinal nerve consists of dorsal nerve roots and ventral nerve roots.

spinal roots (*SPY nal*) Axon bundles attached to each spinal cord segment. The dorsal roots are sensory and the ventral roots are motor. The dorsal and ventral roots join to form the spinal nerves.

spinocerebellar tract (*SPY no ser eh BELL ar*) Sensory pathway from the spinal cord to the cerebellum.

spinothalamic tract (*SPY no thal A mic*) Sensory pathway from the spinal cord to the thalamus and eventually the primary somatic sensory cortex. Contains pain and crude touch information.

spleen Lymphoid organ located in the abdominal cavity behind the stomach. The spleen destroys old erythrocytes, breaking their haemoglobin into haem and globins. It also acts as a storage area for whole blood. Its white pulp is lymphoid tissue that contains B and T lymphocytes.

spores A protective barrier to allow for future reproduction in a hostile environment.

sprain Overstretching or tearing of a ligament.

squamous cells (*SKWAY muss*) A flat, scaly epithelial cell.

stapes (*STAY peez*) Third bone of the middle ear. It is attached to the incus on one end and the oval window on the other end. Also known as the stirrup.

sternal Pertaining to the sternum.

sternum (*STER num*) Vertical bone of the anterior thorax to which the clavicle and ribs are attached. Also known as the breastbone.

steroids (*STARE roydz*) Ringed lipids that function as extremely powerful hormones.

strain Overstretching of a muscle, often due to physical overexertion. This causes inflammation, pain, swelling and bruising as the capillaries in the muscle tear. There can be small tears in the muscle itself. Also known as a pulled muscle.

stratified (*STRAT i fied*) Having more than one layer.

stratum basale (*STRAR tum BAY sale*) The stem cell layer of the epidermis.

stratum corneum (*STRAR tum core NEE um*) The outermost horny layer of the epidermis.

striated muscle (*stry ATE ed*) Skeletal muscle.

subarachnoid space (*sub ah RAK noyd*) Space beneath the arachnoid layer of the meninges. It is filled with cerebrospinal fluid.

subcutaneous fascia (*sub cue TAY nee us FAY shee yah*) Connective tissue layer beneath the skin.

subdural space (*sub DURE al*) The space between the arachnoid and dura matter.

sublingual salivary glands (*sub LING you all SAL ih var ee*) Smallest of salivary glands found between the tongue and the mandible, one on each side.

submandibular salivary glands (*sub MAN dib you lar SAL ih var ee*) Salivary glands beneath the mandible or jaw.

submucosa (*sub mew KOSE ah*) Layer of connective tissue under a mucous membrane.

sulcus (*SULL cuss*) One of many shallow grooves between the gyri in the cerebrum and cerebellum. Plural: sulci.

superior Pertaining to the upper half of the body or a position above an organ or structure.

superior vena cava (*soo PEER ee yor VEEN ah KAH vah*) Largest vein that drains venous blood from the upper portions of the body.

supine position (*sue PINE*) Position of lying on the posterior part of the body. Also known as the dorsal supine position.

surfactant (*sir FAC tent*) Protein-fat compound that creates surface tension and keeps the walls of the alveolus from collapsing inward with each inhalation.

sympathetic nervous system Division of the autonomic nervous system that uses the neurotransmitter norepinephrine and carries nerve impulses to the heart, involuntary muscles, and glands during times of increased activity, danger or stress.

synapse (*SIH naps*) Space between the axon of one neuron and the dendrites of the next neuron.

syndrome (*SIN drohm*) A set of symptoms and signs associated with and characteristic of one particular disease.

synergistic (*siner JISS tick*) A co-operating action of certain muscles.

synovial fluids (*sigh NO vee al*) Clear lubricating fluid that is secreted by the synovial membrane.

synovial membrane (*sigh NO vee al*) The membrane lining a capsule of the joint.

systems An organised grouping of related structures or parts that perform specific functions.

systole (*SISS toh lee*) Combined contractions of the atria and the ventricles.

T lymphocytes (*T LIMF oh sights*) White blood cells involved in adaptive immunity. There are four types of T lymphocytes: helper T cells, cytotoxic T cells, memory T cells and regulatory T cells.

tactile corpuscles (*KOR puss els*) Elongated bodies found in nerve ends that act as receptors for slight pressure or touch.

taste buds Sensory end organs that provide us with a sense of taste.

temporal lobe Lobe in the cerebrum that receives sensory information from the auditory cortex for hearing and the olfactory cortex for smelling.

tendon Cord-like white band of non-elastic fibrous connective tissue that attaches a muscle to a bone.

tendonitis (*ten dun EYE tis*) Inflammation of any tendon from injury or overuse.

testes (*TESS teez*) Small, egg-shaped glands in the scrotum. Also known as the testicles. They contain interstitial cells that secrete testosterone. They also contain the seminiferous tubules that produce spermatozoa.

tetanus (*TET an nuss*) An acute infectious disease caused by a bacterium that can lead to severe spasms of voluntary muscles.

thalamus (*THAL a muss*) Relay station in the brain that receives sensory nerve impulses from the optic nerves and sends them to the visual centres in the occipital lobes of the brain.

thoracic cage (*thor AS sick*) The portion of the skeleton to include the ribs, sternum and thoracic vertebrae that house and protect the lungs, heart and great vessels.

thoracic cavity (*thor A sick CAV ih tee*) Hollow space within the thorax that is filled with the lungs and structures in the mediastinum.

thoracic duct (*thor A sick*) The main lymph duct of the body.

thrombin (*THROM bin*) An enzyme that reacts with fibrinogen, converting it to fibrin which forms a clot.

thrombocyte (*THROM boh sight*) Cell fragment that is flat and does not have a nucleus. It is active in the blood clotting process. Thrombocytes are also known as blood platelets.

thrombocytopenia (*THROM boh sigh TOH PEE nee ah*) Deficiency in the number of thrombocytes. This can be due to exposure to radiation or toxic chemicals or drugs that damage the stem cells in the bone marrow.

thrombus (*THROM buss*) A blood clot.

thymus (*THIGH muss*) Lymphoid organ in the thoracic cavity. As an endocrine gland, it releases hormones known as thymosins. The thymosins cause lymphoblasts in the thymus to mature into T lymphocytes.

thyroid gland (*THIGH royd*) Endocrine gland in the neck that produces and secretes the hormones T3, T4 and calcitonin. Its two lobes and narrow connecting bridge (isthmus) give it a shield-like shape.

tinnitus (*tin IT uss*) Sounds (buzzing, ringing, hissing or roaring) that are heard constantly or intermittently in one or both ears, even in a quiet environment.

tissues Collection of similar cells that perform a particular function.

tongue Large muscle that fills the oral cavity and assists with eating and talking. It contains taste buds and receptors for the sense of taste.

tonus (*TONE us*) A partial steady contraction of a muscle; firmness.

trabecula (*tra BECK yoo la*) A fibrous cord of connective tissues that acts as a supporting fibre.

trachea (*TRACK ee uh*) Rigid tubular air pipe between the larynx and the bronchi that is a passageway for inhaled and exhaled air.

transitional (*tranz ISH on al*) Moving from one state to another.

transverse plane (*tranz VERS*) Plane that divides the body into top and bottom sections, superior and inferior.

tricuspid (*try CUSS pid*) Pertains to having three cusps or points as in the tricuspid valve of the right heart.

tuberculosis (TB) (*TUE ber cue low siss*) Lung infection caused by the bacterium *Mycobacterium tuberculosis* and spread by airborne droplets expelled by coughing.

tubular reabsorption (*tue bue yoo lar ree ab SORP shun*) Movement of substances out of the renal tubule and back into the blood. The substances will be retained in the body.

tubular secretion (*TUBE yoo lar see CREE shun*) Movement of substances from the blood into the renal tubule. These substances will leave the body in the urine.

tumour necrosis factor (TNF) (*neck ROW siss*) Released by macrophages, it destroys endotoxins produced by certain bacteria. It also destroys cancer cells.

tunica externa (*TUE nik ah ex TERN ah*) The outer layer of an artery.

tunica interna (*TUE nik ah in TERN ah*) The inner lining of an artery.

tunica media (*TUE nik ah mee DEE ah*) The middle muscular layer of an artery.

turbinates (*TER bin ates*) Three long projections (superior, middle, inferior) of the ethmoid bone that jut into the nasal cavity: superior, middle and inferior. They break up the stream of air as it enters the nose. Also known as the nasal conchae.

tympanic membrane (*tim PAN ik*) Membrane that divides the external ear from the middle ear. Also known as the eardrum.

universal donor A person who has the type O blood that can be transfused to any individual of the ABO blood groups.

universal recipient A person who has type AB blood who can receive blood from the ABO blood groups.

upper respiratory infection (URI) Bacterial or viral infection of the nose that can spread to the throat and ears. The nose is a part of the respiratory system as well as the ENT system. Also known as a common cold or head cold.

ureter (*you REE ter*) Tube that connects the pelvis of the kidney to the bladder.

urethra (*you REE thrah*) Tube that connects the bladder to the outside of the body.

urinary bladder (*You rin air ree blah der*) Holding receptacle for urine before it is expelled (voided) from the body.

vagus nerve (*VAY gus*) Cranial nerve X. Sensation and movement of the throat. Sensory and motor for thoracic and abdominal organs.

vasculature (*VAZ kew lah chor*) Network of blood vessels in a particular organ.

vasoconstriction (*vaz oh kon STRIK shun*) Constriction of the smooth muscle in the artery wall causes the artery to become smaller in diameter.

vasodilation (*vaz oh DYE lay shun*) Relaxation of the smooth muscle in the artery wall causes the artery to become larger in diameter.

vein Blood vessel that carries oxygen-poor blood as well as carbon dioxide and waste products of cellular metabolism away from the cells and back to the heart. Veins have one-way valves that keep blood from flowing backwards, away from the heart. The exception is the pulmonary vein, which carries oxygenated blood from the lungs to the heart.

ventilation (*vent ill AY shun*) The bulk movement of gas into and out of the lungs.

ventral Pertaining to the anterior of the body particularly the abdomen.

ventricles (*VEN trik lz*) 1. Two lower chambers of the heart. Intraventricular structures are located in the ventricles. 2. Four hollow chambers within the brain that contain cerebrospinal fluid. The two lateral ventricles are within the right and left hemispheres of the cerebrum. The third ventricle is small and connects the lateral ventricles to the fourth ventricle, which is at the level of the pons and medulla oblongata.

venules (*VEN yules*) Smallest branch of a vein.

vertebrae (*VERT eb ray*) One of 33 irregularly shaped bony segments of the spinal column.

vertigo Sensation of being off balance when the body is not moving. Caused by upper respiratory infection, middle or inner ear infection, head trauma, or degenerative changes of the semicircular canals.

vesicle (*VEE sickle*) A small bladder or blister; a membrane-bound storage sac inside a cell.

vestibule chamber (*VESS ti bule*) A small cavity or space at the beginning of a canal.

vestibulocochlear nerve (*VESS tib you low COCK lee are*) The eighth cranial nerve responsible for hearing and balance. Sometimes called the acoustic nerve.

villi (*VILL eye*) Microscopic projections of the mucosa within the lumen of the small intestine.

virus A parasitic micro-organism that depends on other cells for its metabolic and reproductive needs.

visceral (*VISS er all*) Pertaining to organs.

visceral pleura (*VISS er all PLOO rah*) One of the two membranes of the pleura. It covers the surface of the lung.

vital signs Medical procedure during a physical examination in which the temperature, pulse and respirations (TPR), as well as the blood pressure, are measured.

vitreous humour (*VITT ree uss HUW mer*) Clear, gel-like substance that fills the posterior cavity of the eye.

vocal cords Connective tissue bands in the larynx that vibrate and produce sounds for speaking and singing.

white blood cells (WBCs) Leucocytes; responsible for the immune response.

z lines Lines visible on the surface of skeletal muscle that mark the ends of each sarcomere.

Index

abbreviations used in medicine 8–10
  see also Appendix C
abdominal cavity 30
abdominal region 35
  anatomical divisions 32
  quadrants 33–4
abdominopelvic cavity 30
abnormal function *see*
  pathophysiology
acardia 8
acetylcholine 151, 196, 198, 199, 231
acid/base balance in the body 46–7, 440
acids 46–7
actin 151–3
action potentials of neurons 191–3
active transport in and out of cells 59, 61–4
active transport pumps 62, 63–4
Addison's disease 166, 276, 345
adenine 49, 50
adenosine triphosphate (ATP) 49, 68, 152–3
adrenal glands 264, 267
adrenaline (epinephrine) 198, 231, 264, 274, 292
adrenocorticosteroids 274
aetiology of a disease 12
afferent pathways 14
ageing process, changes in the body 484–7
AIDS (acquired immune deficiency syndrome) 74, 337, 338, 345
alcohol, and antidiuretic hormone (ADH) 269
alcoholism, effects of 490
alkalis (bases) 46–7
allergies 345–6
amazing body facts 497–9
amino acids 49
anabolic steroids 275, 493
anabolism 13
anaemia 311
anaesthesia 200, 201, 232, 250, 251
anaphylaxis 333, 346
anatomical position 24
anatomy
  definition 2
  gross (macroscopic) 2
  links with physiology 3–4
  microscopic 2
angiotensin-converting enzyme (ACE) 49, 440
anions 46
anorexia nervosa 13
antebrachial region 35–6
antecubital region 35–6
anterior, directional term 27
antibiotics 74, 494–5
antibodies (immunoglobulins, Igs) 326, 327

antidiuretic hormone (ADH, vasopressin) 267, 268–9, 438, 440
antigens 326
apocrine sweat glands 167–8
appendicitis, psoas test 33
arteriosclerosis 309
arthritis 119, 131–2
asexual cellular reproduction 69–72
asthma 154, 346, 381–2, 491, 495
astrocytes 189, 190
ataxia 157
atelectasis 372, 380–1
atherosclerosis 309–11
atmospheric pressure, and the respiratory system 14
atoms 42–4, 56
  chemical bonds 44–5
ATP (adenosine triphosphate) 49, 68, 152–3
autoimmune disorders 337, 345
autonomic nervous system 229–32, 292
axillary region 35–6
axons 90, 91, 189–91, 195–6

B lymphocytes 329, 330, 334–44
bacteria 69, 73–4, 75
balance, sense of 247, 248
bases (alkalis) 46–7
benign tumours 72
bile salts 49
blood, as connective tissue 86 *see also* cardiovascular system
blood glucose regulation 272–4
blood pressure 11, 154
blood vessels *see* cardiovascular system
body cavities 30
body composition, elements 42, 43
body planes 25–7
body positions 24–5
body regions 32–6
body temperature
  a vital sign 11
  core body temperature 156
  effects on enzymes 13–14
  regulation 13, 15–17, 175–6, 262–3
bone cells (osteocytes) 113–14
bone marrow 113, 114, 115
bone tissue 113–14, 115
bones
  anatomy 113, 114
  classification by shape 112–13
  disorders 130–2
  epiphyseal growth plate 118
  fractures 130–1
  growth and repair 117–18
  surface structures 114, 116
  see also skeletal system
botulinus toxin 157

botulism 207
brachial region 35–6
bradycardia 8
brain anatomy and functions, 214–32
  brainstem 215, 216–17, 218, 219
  cerebellum 215, 216, 218, 219, 221
  cerebrospinal fluid (CSF) 220
  cerebrum 214–16, 218–19
  diencephalon 218, 219, 220
  effects of ageing 484
  external anatomy 214–17
  internal anatomy 218–21
  lobes of the brain 215–16
  meninges 90, 200–2, 203, 217
brain injuries 207–8
buccal cavity 30
buccal region 35–6
burns to the skin 170–2
bursa 119

caffeine, and antidiuretic hormone (ADH) 269
calcitrol 85
calcium
  in the diet 132
  role in muscle movement 152–3
cancellous (spongy) bone 113–14, 115
cancer 72
  of the large bowel 415
  prevention and treatment 495–7
  skin cancer 93, 489–90, 496–7
  stages 324–5
carbohydrates 48
carbon dioxide, and acid/base balance 47
cardiac muscle 89, 90, 289
cardiologist 8
cardiology 6, 8
cardiomegaly 8
cardiopathy 8
cardiorespiratory system 354
cardiovascular system 92, 100–1, 284–313
  arteries 284–5, 286, 287–8, 300–3
  blood clotting 297, 306–7
  blood components and functions 294–7
  blood flow through the heart 287–90
  blood groups and transfusions 298–300
  blood pressure 301, 302–3
  blood problems 311–12
  blood vessel disorders 309–11
  blood vessel types 284–5
  blood vessels (vascular system) 300–5
  capillaries 284–5, 301–5
  cardiac cycle 287–90
  cardiac pacemaker cells 291–2

components of the system 284–5
coronary circulation 289–90
disorders and diseases 308–12
heart contraction and conduction system 291–4
heart disorders 308–9, 310, 312, 405
heart structure and function 285–94
links with the lymphatic system 312–13
veins 284–5, 286, 287–8, 300–3
wellness and personal choices 490
carditis 8
carotene in the skin 166
carpal region 35–6
carpal tunnel syndrome 206, 208, 488, 489
cartilage 119, 120, 130–1
cartilaginous joints 120
catabolism 13
cations 46
cauda equina 200–1
caudal, directional term 26
caudal region 35
cavities of the body 30
cell cycle 70–2
cell cytoplasm 64, 65
cell division 69–72
cell membrane 57–64
cell motility structures 68–9
cell organelles 64–9
cell structure 57–69
cells
  building blocks 56
  energy supply 68
  features of cells 56
  junctions 62–3
  types in epithelial tissues 84–5
  types within the human body 56, 58
cellular level of the body 4, 5
cellular reproduction 69–72
cellular respiration 68
central, directional term 29
centrioles 66
centrosomes 65, 66
cephalic, directional term 26
cephalic region 35
cerebellum 229
cerebral palsy 232–3
cerebrospinal fluid (CSF) 200, 220
cerebrovascular accident (CVA, stroke) 232, 405
cervical region 35–6
chemistry 42–5
chicken pox 74–5
childbirth *see* reproductive system
cholesterol 48–9
chondrosarcoma 132
chromatin in the cell nucleus 65

chromosomes 65, 69–72
chronic bronchitis 381, 382
chronic fatigue syndrome 156
chronic obstructive pulmonary disease (COPD) 381
cilia 69, 361–3
circulatory system *see* cardiovascular system
colostomy 411
columnar epithelial cells 84–5
coma 232
compact bone 113–14, 115
compound (chemical), definition 42, 44–5
congenital deformity of the spine 127–8
connective tissue 85–8
control centres for homeostatic responses 14
conus medullaris 200–1
core body temperature *see* body temperature
coronal plane 27
covalent bonds (electron sharing) 45
cramp in muscles 157
cranial, directional term 26
cranial cavity 30
cranial nerves and functions 221–3
cranial region 35
cross-sectional view of the body 26
CT (computed tomography) scans 28
cuboidal epithelial cells 84–5
Cushing's syndrome 276
cutaneous membranes 86
cytokinesis 70
cytology 2
cytoplasm of a cell 64, 65
cytosine 49, 50
cytoskeleton 68

da Vinci, Leonardo 2
deep, directional term 28–9
dendrites 89, 91, 189–91, 195
desmosomes 62–3
diabetes insipidus 444
diabetes mellitus 173, 206, 274, 275, 310–11, 345, 433, 434, 498, 499
diabetic nephropathy 434
diagnosis of disease 12
diaphragm 30, 148
diaphysis (of a bone) 113, 114, 115
diarrhoea 415
diffusion 59, 63
digestion *see* gastrointestinal system
digital region 35–6
directional terms 26–9
disaccharides 48
disease
    definition 11
    signs and symptoms 15
    see also pathology
disease-related terms 11–12
distal, directional term 28–9
DNA (deoxyribonucleic acid) 49, 50, 65
dorsal, directional term 27
dorsal cavities 30
Down's syndrome 452
dysentery 77

E. coli 73
ears
    disorders 252–3, 493–4
    structures and functions 245–8

eccrine sweat glands 167–8
ECG (electrocardiogram) 293
effectors, role in homeostasis 14
efferent pathways 14
electrocardiogram 8
electrolytes 45–6
electrons 42–5
elements 42–4
    composition of the body 42, 43
emotions and the limbic system 232
emphysema 381, 382
endocrine system 92, 98–9, 258–76
    adrenal glands 264, 267, 274–5
    anabolic steroids 275
    blood glucose regulation 272–4
    body temperature regulation 262–3
    control of endocrine activity 261–5
    control of hormone levels 263–6
    disorders and diseases 275–6
    functions of endocrine organs 258–9, 260
    gonads 275
    homeostasis 261–5
    hormonal control hierarchy 264–5
    hormones produced by endocrine glands 259–61, 275
    humoral control of hormone levels 264–6
    hypothalamus 262–3, 264–5, 266, 267–9
    negative feedback 261–5
    neural control of hormone levels 264
    pancreas 272–4
    parathyroid glands 272
    pineal gland 272
    pituitary gland 266–70
    positive feedback 263
    sex hormones 275
    steroid abuse in sport 275
    thymus gland 272
    thyroid gland 270–2
    wellness and personal choices 493
endocytosis 62, 63–4
endoplasmic reticulum 67
energy supply
    for cells 68
    for muscles 156
enzymes 13–14, 49, 390, 394, 398, 400, 402, 403–7, 413–15
ependymal cells 189, 190
epidural anaesthesia 201
epigastric region 32
epinephrine (adrenaline) 198, 231, 264, 274, 292
epiphyseal growth plate 118
epiphysis (of a bone) 113, 114, 115
epithelial membranes 85–6
epithelial tissues 84–6
Epstein–Barr virus 206
erythropoiesis (red blood cell production) 113, 114, 115
eukaryotic cells 69–70
exocytosis 62, 63–4
external, directional term 28–9
extracellular fluid (ECF) 45
eyes
    disorders 252–3, 493–4
    structures and functions 241–4

facial skeletal muscles 144
facial twitching 157
facilitated diffusion 59–60, 63
fat cells (lipocytes) 168

femoral region 35–6
fever 31, 263
fibromyalgia 156
fibrous joints 120
fight-or-flight response 188, 230–2, 436
filtration (passive transport) 60–1, 63
flagella 68
forensics and hair 175
Fowler's position 24–5
frontal (coronal) plane 27
fugu (puffer fish) poison 194
function, focus of physiology 3–4
fungal infections 75–6
fungi 75–6

gallbladder 413, 414
gap junctions between cells 62–3
gastrointestinal system 92, 103, 390–418
    accessory organs 412–15
    bacteria in the large intestine 411–12
    chemical breakdown of food 390, 394, 398, 400, 402, 403–7, 413–15
    digestion (definition) 390
    disorders and diseases 407, 409, 411, 414, 415–18
    enzymes 390, 394, 398, 400, 402, 403–7, 413–15
    gallbladder 413, 414
    gastric secretions 399–403
    hormones 401–2, 405
    ingestion 390
    large intestine 409–12
    liver 412–13
    mechanical breakdown of food 390, 392–7
    mouth and oral cavity 391–7
    nutrition 407–9
    oesophagus 397
    pancreas 413–15
    pharynx 397
    salivary glands 394
    small intestine 403–7
    stages in the digestive process 390
    stomach 399–403
    structure 390, 391
    taste 392–4
    teeth 395–7
    tissue types in the alimentary canal walls 398–9
    tongue 392–4, 395
    wellness and personal choices 491–3
genes 65
giardiasis 77
glial cells (neuroglia) 89, 91, 189, 190
glucagon 272–3
glucose 48, 49, 68, 156, 272–4
glycogen 48, 156
Golgi apparatus 67–8
gonads, production of sex hormones 275
Graves' disease 271, 276, 345
groin regions 32
gross anatomy 2
guanine 49, 50
Guillain–Barré syndrome 157, 206, 207, 232
gustatory sense (taste) 240, 249, 250, 361, 392–4

haemophilia 312
hair colour 175
hair growth 173–5
Hashimoto's disease 275–6
Haversian systems (osteons) 113–14, 115
head injuries 232, 233
hearing
    care of 493–4
    sense of 240, 245–8
heart *see* cardiovascular system
heart rate (pulse), a vital sign 11
heat produced by muscles 156
Helicobacter pylori 416
hernias 33, 157
herpes viruses 74, 206
histology 2
HIV (human immunodeficiency virus) 74, 206, 337, 338
homeostasis 13–17, 232, 261–5, 484
horizontal plane 26
hormones 268, 269
    comparison with neurotransmitters 260
    control of hormone levels 263–6
    female reproduction 275, 459, 462–5, 471
    male reproduction 275, 468–70, 471
    produced by endocrine glands 259–61, 275
    produced by the kidneys 438–40
human body
    map 24–36
    structural hierarchy 4, 5
Huntington's disease 233
hydrocephalus 220
hydrogen atom 42–3
hydrogen bonds 45, 50
hydrolysis 48
hypersensitivity reactions 345–6
hypocalcaemia 85
hypochondriac regions 32
hypogastric region 32
hypophysis *see* pituitary gland
hypothalamus 262–3, 264–5, 266, 267–9

iliac regions 32
immune system 92, 102, 325–46
    acquisition of immunity to pathogens 342
    adaptive immunity (specific) 326–8, 334–41
    antibodies (immunoglobulins, Igs) 326, 327
    antigens 326
    barriers 328
    body's defence system 343–4
    cells involved 328–30, 332, 334–44
    chemical components 330–1
    components of the immune system 328–33
    disorders and diseases 337, 344–6
    fever 332
    immune response 333–41
    inflammatory response 331–3
    innate immunity (non-specific) 326–8, 334
    leucocytes (white blood cells) 328–30, 332, 334–41
    links with the lymphatic system 343–4
    wellness and personal choices 494–5

immunisation 342, 494
immunodeficiency disorders 337, 345
  see also AIDS; HIV
inferior (caudal), directional term 26
inguinal hernia 33
inguinal regions 32
injection method of drug delivery 170
insulin 48, 272–4
integumentary system 92, 94–5, 97,
  164–79
  area and weight of skin 164
  body temperature regulation 175–6
  burns to the skin 170–2
  carotene in the skin 166
  components 164
  dermis 164, 166–8
  disorders and diseases 176–9
  drug delivery via 170
  epidermis 164–6
  factors affecting skin colour 165–6
  functions 164
  hair colour 175
  hair growth 173–5
  layers of the skin 164–8
  lipocytes (fat cells) 168
  melanin in hair 175
  melanin in skin 165–6
  nail growth 172
  peripheral perfusion 173
  sebaceous (oil) glands 166, 167–8,
    173–5
  skin disorders and diseases 93,
    176–9, 489–90, 496–7
  skin grafting 170
  skin healing processes 168–70
  subcutaneous fascia (hypodermis)
    164, 168
  sudoriferous (sweat) glands 167–8
  vitamin D production 85
  wellness and personal choices
    488–90
  wound repair 168–70
internal, directional term 28–9
interphase of the cell cycle 70
interstitial fluid 45
intracellular fluid (ICF) 45
involuntary muscles 148, 153–5
ionic bonds (electron transfer) 44–5
isotope, definition 44

joints 119–21
  disorders 130–2
  movement classification 122–3

kidneys *see* urinary system
kinesiology 149
kyphosis 127–8

lactose 48
lactose intolerance 407
language of anatomy and physiology
  see medical terms
language of disease 11–12
lateral, directional term 26–7
left and right directional terms 28
leucopenia 312
leukaemia 311, 345
leukocytosis 311–12
ligaments 120
limbic system 232
lipids 48–9
liver, structure and functions 412–13
lordosis 127–8
lower extremities, bones of 128–30

lumbar region 32, 35–6
lupus erythematosus 345
Lyme disease 206
lymph, as connective tissue 86
lymph nodes 102
lymphatic system 92, 102, 320–5
  functions 320
  links with the cardiovascular
    system 312–13, 320–2
  links with the immune system
    343–4
  role in the body's defence system
    343–4
  structures 320–4
  T lymphocytes 324
lysosomes 68

macroscopic anatomy 2
magnetotaxis in bacteria 75
malaria 76–7
malignant tumours 72
maltose 48
medial, directional term 26–7
median plane 26–7
medical terms 4–10
  abbreviations 8–10
  combining forms 6–8
  prefixes 5–8
  suffixes 5–9
  word roots 5–8
  see also Appendix B
medicine delivery via the skin 170
medullary cavity (of a bone) 113, 114,
  115
meiosis, distinction from mitosis 72
melanin
  in hair 175
  in skin 165–6
melanoma 93
melatonin 272
membranes
  epithelial types 85–6
  meninges 90, 200–2, 203, 217
  synovial 87–8
meninges 90, 200–2, 203, 217
meningitis 208, 217
mental health 488
metabolic syndrome 15
metabolism 12–13
metastasis of cancer cells 72
metric system, use in healthcare 9–10
microglia 189, 190
microorganisms 72–7
microscopic anatomy 2
midsagittal plane 26–7
mitochondria 66–7, 68
mitosis 69–72
molecules 43–4, 56
monosaccharides 48
motor system 226–9
movement of joints, classification
  122–3
movement of skeletal muscles 147–53
MRI (magnetic resonance imaging)
  28
MRSA (methicillin-resistant
  Staphylococcus aureus) 73, 495
mucous membranes 86–7
mucus/mucous distinction 89
multiple sclerosis 195, 345
mumps 74
muscular dystrophy 157
muscular system 92, 94, 96, 140–57
  cardiac muscle 141, 154–5

connections with the nervous
  system 151, 157
diaphragm muscle 148
disorders and diseases 156–7
energy supply for muscles 156
fibromyalgia 156
functions of muscles 156
heat production 156
involuntary muscles 148, 153–5
ligaments 142
maintaining core body temperature
  156
movement of skeletal muscles
  147–53
muscle movement at cellular level
  151–3
muscle tissue 89, 90
muscle tone (tonus) 155
myalgia 156
neuromuscular diseases 157
  role of actin and myosin in
    movement 151–3
  role of ATP in movement 152–3
  role of calcium in movement 152–3
  sarcomeres (functional units within
    muscle) 151–2
  skeletal (striated) muscles 140–1,
    142–53
  skeletal muscle movement 147–53
  smooth (visceral) muscle 140–1,
    153–4
  sphincters 154
  tendons 142
  types of muscle 140–1
  visceral (smooth) muscle 140–1,
    153–4
  voluntary muscles 142 *see also*
    skeletal (striated) muscles
  wellness and personal choices 488,
    489
myalgia 156
myasthenia gravis 157, 207, 345
myelin 189, 190, 195–6
myocardial infarction 310, 312, 405
myofibrils 151–3
myosin 151–3

nail growth 172
nasal cavity 30
nasal region 35–6
natural killer (NK) cells 330, 334
necrosis 33
negative feedback mechanisms 14–17,
  261–5
nervous system 93, 95–8, 186–208
  action potentials of neurons 191–3
  anaesthesia 200
  autonomic nervous system 187–8
  axons 189–91, 195
  basic operation 186–8
  brain anatomy and functions
    214–32
  central nervous system (CNS)
    186–7, 188
  cerebrospinal fluid (CSF) 200
  components 186–8
  cranial nerves and functions 221–3
  dendrites 189–91, 195
  disorders and diseases 206–8, 232–3
  enteric nervous system 188
  fight-or-flight response 188, 230–2,
    436
  input from the sensory system
    186–8

meninges 200–2, 203
motor (efferent) system 187–8
myelin 189, 190, 195–6
nerve cell types 89–90, 91
nervous tissue 188–93
neuroglia (glial cells) 89, 91, 189,
  190
neuromuscular junction 189, 198–9
neurons 89–90, 91, 189–99
neurotransmitters 196–200, 231–2,
  260
parasympathetic nervous system
  187–8
peripheral nervous system (PNS)
  186–8
reflexes 204–6
somatic nervous system 187–8
speed of response 188, 196
spinal cord anatomy 200–6
spinal cord injuries 206–7, 232,
  233
spinal nerves 200–1, 203, 204, 205
sympathetic nervous system 187–8,
  264
synapses 189, 196–9
wellness and personal choices 488
nervous system integration 224–32
  autonomic nervous system 229–32
  limbic system 232
  motor system 226–9
  reticular system 232
  role of the cerebellum 229
  somatic sensory system 224–6
neuroglia 89, 91, 189, 190
neuromuscular diseases 157
neuromuscular junction 189, 198–9
neuromuscular system 151
neurons 89–90, 91, 189–99
neurotransmitters 196–200, 231–2
  comparison with hormones 260
neutrons 42–4
noradrenaline (norepinephrine) 197,
  198, 231, 232, 264, 274
nucleic acids 49, 50
nucleolus of a cell 65
nucleotides 49, 50
nucleus of a cell 64–5
nutrients 13
nutrition 132, 407–9, 485, 491–3

occipital region 35
oestrogen 48
older people, ageing processes 484–7
oligodendrocytes 189, 190, 195
oral cavity 30
oral region 35–6
orbital cavity 30
orbital region 35–6
organ level of the body 4, 5
organ system level of the body 4, 5
organic compounds 48–50
organismal level of the body 4, 5
organs of the human body 90, 92–3
osmosis 60, 63
ossification process (osteogenesis)
  117–18
osteoarthritis 119
osteoblasts 117–18
osteoclasts 117–18
osteocytes 113–14, 117–18
osteomalacia 132
osteons 113–14, 115
osteoporosis 131–2
osteoprogenitor cells 117

oxygen requirement of the body 13
oxytocin 17, 263, 268, 269

pain
    as a symptom 12
    in older patients 485–6
    sensing 250, 251
    somatic sensory system 224–6
    spinal cord pathways 227–9
pancreas 267–8, 269, 413–15
paralysis 157, 232
parasympathetic nervous system 230,
    231
parathyroid glands 272
parietal membranes 86
passive transport in and out of cells
    58–61, 63
patellar region 35–6
pathogens 73
pathology 4
pathophysiology 4
pedal region 35–6
pelvic cavity 30
pelvic girdle, bones of 128–30
peptic ulcer disease 415–16, 418
pericardial cavity 30
periosteum 113, 114, 115
peripheral, directional term 29
peripheral neuropathy 206–7
peripheral perfusion 173
peripheral vascular disease 173
persistent vegetative state 217
pH scale 46–7
phaeochromocytoma 276
phagocytosis 62, 63–4
phospholipids 48
physiology
    definition 2–3
    links with anatomy 3–4
    sub-specialties 3
pineal gland 272
pinocytosis 62, 63–4
pituitary gland 266–70
plantar region 35–6
plasma 45
pleural effusion 382
pneumonia 371, 381
pneumothorax 382
polio 206, 232
polypharmacy 486–7
polysaccharides 48
popliteal region 35
positive feedback in the body 17, 263
posterior, directional term 27
prednisone (hydrocortisone) 275
progesterone 48
prognosis of a disease 12
prone position 24–5
proteins in the body 49
protons 42–4
protozoa 76–7
proximal, directional term 28–9
psoas test 33
pubic region 35–6
pulse, a vital sign 11

quadrants of the abdominal region
    33–4

radiology 28
receptors, role in homeostasis 14
red blood cells 3, 113, 114, 115
reflexes 204–6
regions of the body 32–6

renal system see urinary system
reproductive system 93, 104–5,
    450–77
    contraception 473
    early stages of the human life cycle
        453–4
    erection and ejaculation 470–1
    female disorders 456, 473–5, 476
    female reproductive anatomy
        453–9
    female reproductive physiology
        459–65
    hormonal control (female) 459,
        462–5, 471
    hormonal control (male) 468–70,
        471
    human chromosomes 451–2
    male disorders 470, 475, 477
    male reproductive anatomy 465–7
    male reproductive physiology
        468–71
    meiosis for sexual reproduction
        451–2
    menstrual cycle 459–64
    mitosis for growth and repair 450–1
    pregnancy and labour 17, 464–5,
        471–2, 495
    production of gametes 451–2
    sex chromosomes 452
    spermatogenesis 468
    sterilisation 473
    wellness and personal choices 495
respiratory rate, a vital sign 11
respiratory system 92, 101, 354–83
    alveolar gas exchange 357, 358,
        369–72
    bony thorax 374–5
    breathing process 375–80
    components of the system 354–5
    disorders and diseases 380–3
    functions of the system 354
    gases in the atmosphere 355–6
    larynx 365–6
    lower respiratory tract 367–80
    lungs 357–74
    mucociliary escalator 361–3
    nose and nasal cavity 359–63
    oxygen carrying 369–70
    pharynx 364–5
    role in acid/base balance 47
    sinuses 363–4
    structures of the thoracic cavity
        372–4
    upper airway structures and
        functions 358–66
    ventilation versus respiration
        355–7
    wellness and personal choices
        490–1
reticular system 232
rheumatoid arthritis 345
ribosomes 65–6, 67
ribs 126
right and left directional terms 28
rigor mortis 153
RNA (ribonucleic acid) 49, 65
rubella (German measles) 74

sagittal plane 26–7
sarcomeres 151–2
Sarin nerve gas 199
satellite cells (PNS glial cells) 189,
    190
Schwann cells 189, 190, 195

SCID (severe combined immune
    deficiency) 345
scoliosis 127–8
sebaceous (oil) glands 166, 167–8,
    173–5
selective serotonin re-uptake
    inhibitors (SSRIs) 197
sensation 96
senses 93, 97
    balance 247, 248
    cutaneous senses 240
    disorders of the eye and ear 252–3,
        493–4
    effects of ageing 485
    general senses 240
    hearing 240, 245–8, 252–3, 493–4
    range of senses 240
    sight 240, 241–4, 252–3, 493–4
    smell 240, 249–50, 360–1
    special senses 240
    taste 240, 249, 250, 361, 392–4
    temperature 250–1
    touch 250–1
    visceral senses 240
    wellness and personal choices
        493–4
    see also pain
serotonin 197, 198
serous membranes 86–7
set point, control of body features
    261–3
sexually transmitted diseases 495, 496
shin splints 157
shingles 206
SI system of measurement 10
sickle cell anaemia 3
sight
    care of 493–4
    sense of 240, 241–4
signs of disease 11–12, 15
simple epithelial tissue 84–5
skeletal (striated) muscle 89, 90,
    140–1, 142–53
skeletal system 92, 94, 95, 112–32
    appendicular skeleton 123, 128–30
    axial skeleton 123–8
    bone anatomy 113, 114
    bone classification by shape
        112–13
    bone fractures 130–1
    bone growth and repair 117–18
    bone marrow 113, 114, 115
    bone tissue 113–14, 115
    bones of the bony thorax 124, 126
    bones of the lower extremities
        128–30
    bones of the skull 123–4, 125
    bones of the spinal column 126–8
    bones of the upper extremities
        128–9
    cartilage 119, 120
    cartilaginous joints 120
    disorders and diseases 130–2
    effects of ageing 130–2
    fibrous joints 120
    functions 112
    joints 119–21
    ligaments 120
    movement classification 122–3
    ossification process (osteogenesis)
        117–18
    pelvic girdle 128–30
    red blood cell production 113, 114,
        115

spinal disfigurements 127–8
surface structures of bones 114, 116
synovial joints 119, 120–1
tendons 120
wellness and personal choices 488,
    489
skin see integumentary system
skull, bones of 123–4, 125
sleeping sickness 77
smell, sense of 240, 249–50, 360–1
smoking, harmful effects 383, 488,
    490, 495, 496
smooth (visceral) muscle 89, 90,
    140–1, 153–4
solute 45
solvent 45
somatic sensory system 224–6
spasms in muscles 157
special senses 93, 97, 240 see also
    hearing; sight; smell; taste
sphincters 154
spinal cavity 30
spinal column 31, 126–8
    disfigurements 127–8
spinal cord
    anatomy 200–6
    injuries 206–7, 232, 233
spinal nerves 200–1, 203, 204, 205
spleen 323, 324
spongy (cancellous) bone 113–14,
    115
sprains 157
squamous epithelial cells 84–5
starches 48
statins 405
stature disorders 270
sternal region 35–6
sternum 112, 124, 126
steroid hormones 261, 274–5
    abuse in sport 275, 493
steroids 48–9
stimuli 14
strains 157
stratified epithelial tissue 84–5
stress management 488
striated muscle tissue 89, 90
stroke (cerebrovascular accident,
    CVA) 232, 405
structure, focus of anatomy 2, 3–4
structural hierarchy of the body 4, 5
subdural haematoma 233
sucrose 48
sudoriferous (sweat) glands 167–8
sugars 48
superficial, directional term 28–9
superior (cranial or cephalic),
    directional term 26
supine position 24–5
sweating 16, 17
sympathetic nervous system 230–1,
    264
symptoms of disease 11–12, 15
synapses 189, 196–9
syndrome X 15
syndromes of disease 12
synovial fluid 119
synovial joints 87–8, 119, 120–1
synovial membranes 87–8
systems of the human body 92–105
    see also specific systems

T lymphocytes and helper cells 324,
    329, 330, 334–44
tachycardia 6, 8

taste, sense of 240, 249, 250, 361, 392–4
teeth, structure and function 395–7
temperature sensors in the skin 250–1
tendonitis 157
tendons 120
testosterone 48
tetanus 157
thermoregulation 13, 15–17
thoracic cage, bones of 124, 126
thoracic cavity 30
thoracic region 36
thrombocytopenia 312
thrombus formation 307
thymine 49, 50
thymus gland 272, 323, 324
thyroid gland 270–2
tight junctions between cells 62–3
tissue level of the body 4, 5
tissue types 84–90
tonus (muscle tone) 155

touch, sense of 250–1
trabeculae 114, 115
transdermal patches 170
transitional epithelial cells 84–5
transverse plane 26
triglycerides 48
tuberculosis (TB) 382–3
tumours 72

umbilical hernia 33
umbilical region 32, 35
upper extremities, bones of 128–9
uracil 49
urinary system 93, 104, 424–44
    bladder structure and function 441–3
    components 424
    disorders and diseases 430, 434, 435, 442, 443–4
    functions 47, 425
    hormones secreted by the kidneys 438–40

kidney structure and function 425–31
nephron structure and function 428–31
renal blood flow 427–8
tubular reabsorption and secretion 436–40
ureters 424
urethra 424
urination reflex 441–3
urine formation 425, 431–40, 444
urinary tract infection (UTI) 442, 443

vaccination 342
vasoconstriction 153–4, 175–6
vasodilation 153–4, 175–6
vasopressin (antidiuretic hormone, ADH) 267, 268–9, 438, 440
ventral, directional term 27
ventral cavities 30
vertebrae 31, 112 see also spinal column

vesicles formed by cells 62, 63–4, 68
viruses 74–5
visceral (smooth) muscle 140–1, 153–4
visceral membranes 86
vital organs 90
vital signs 11
vitamin D 48, 85, 132, 425, 489, 491–3
vitamins 491–3
voluntary muscles 142 see also skeletal (striated) muscles

water requirement of the body 13
wellness
    and personal choices 487–95
    for each body system 487–95
wound repair processes 168–70

X-rays 28

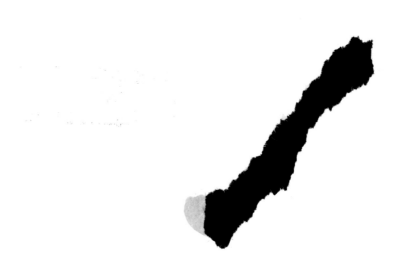